HANDBOOK OF APPLICABLE MATHEMATICS

Volume VI: Statistics

PART A

HANDBOOK OF APPLICABLE MATHEMATICS

Chief Editor: Walter Ledermann

Editorial Board: Robert F. Churchhouse
Harvey Cohn
Peter Hilton
Emlyn Lloyd
Steven Vajda

Assistant Editor: Carol van der Ploeg

Volume I: ALGEBRA
Edited by Walter Ledermann, *University of Sussex*
and Steven Vajda, *University of Sussex*

Volume II: PROBABILITY
Emlyn Lloyd, *University of Lancaster*

Volume III: NUMERICAL METHODS
Edited by Robert F. Churchhouse, *University College Cardiff*

Volume IV: ANALYSIS
Edited by Walter Ledermann, *University of Sussex*
and Steven Vajda, *University of Sussex*

Volume V: GEOMETRY AND COMBINATORICS
Edited by Walter Ledermann, *University of Sussex*
and Steven Vajda, *University of Sussex*

Volume VI: STATISTICS—Parts A and B
Edited by Emlyn Lloyd, *University of Lancaster*

HANDBOOK OF

APPLICABLE MATHEMATICS

Chief Editor: Walter Ledermann

Volume VI: Statistics

PART A

Edited by

Emlyn Lloyd

University of Lancaster

A Wiley–Interscience Publication

JOHN WILEY & SONS

Chichester – New York – Brisbane – Toronto – Singapore

Library of Congress Cataloging in Publication Data:
(Revised for vol. 6)
Main entry under title:
Handbook of applicable mathematics.
    'A Wiley–Interscience publication.'
    Includes bibliographies and indexes.
    Contents:—v. 2. Probability—v. 3. Numerical
methods—v. 4. Analysis—[etc.]—v. 6 Statistics.
1. 1. Mathematics—1961–   . I. Ledermann, Walter,
1911
QA36.H36   510   79-42724
ISBN 0 471 27821 1 (v. 2)
ISBN 0 471 90274 8 (PART A)
ISBN 0 471 90272 1 (PART B)
ISBN 0 471 90024 9 (SET)

British Library Cataloguing in Publication Data:

Handbook of applicable mathematics.
    Vol. 6: Statistics
    1. Mathematics
    I. Ledermann, Walter   II. Lloyd, Emlyn
    510   QA36
ISBN 0 471 90274 8 (PART A)
ISBN 0 471 90272 1 (PART B)
ISBN 0 471 90024 9 (SET)

Typeset and printed by J. W. Arrowsmith Ltd, Bristol BS3 2NT

Contributing Authors

David Cooper, Institute of Hydrology, Wallingford, U.K.

Ian Dunsmore, The University, Sheffield, U.K.

John Gower, Rothamsted Experimental Station, Harpenden, Hertfordshire, U.K.

Peter Jones, University of Keele, Staffordshire, U.K.

Emlyn Lloyd, University of Lancaster, Lancaster, U.K.

Enda O'Connell, Institute of Hydrology, Wallingford, U.K.

Adrian Smith, University of Nottingham, Nottingham, U.K.

Granville Tunnicliffe-Wilson, University of Lancaster, Lancaster, U.K.

David Warren, University of Lancaster, Lancaster, U.K.

Joe Whittaker, University of Lancaster, Lancaster, U.K.

Contents

PART A

Part B contains Chapters 11–20.

Introduction
to the
Handbook of Applicable Mathematics

Today, more than ever before, mathematics enters the lives of every one of us. Whereas, thirty years ago, it was supposed that mathematics was only needed by somebody planning to work in one of the 'hard' sciences (physics, chemistry), or to become an engineer, a professional statistician, an actuary or an accountant, it is recognized today that there are very few professions in which an understanding of mathematics is irrelevant. In the biological sciences, in the social sciences (especially economics, town planning, psychology), in medicine, mathematical methods of some sophistication are increasingly being used and practitioners in these fields are handicapped if their mathematical background does not include the requisite ideas and skills.

Yet it is a fact that there are many working in these professions who do find themselves at a disadvantage in trying to understand technical articles employing mathematical formulations, and who cannot perhaps fulfil their own potential as professionals, and advance in their professions at the rate that their talent would merit, for want of this basic understanding. Such people are rarely in a position to resume their formal education, and the study of some of the available textbooks may, at best, serve to give them some acquaintance with mathematical techniques, of a more or less formal nature, appropriate to current technology. Among such people, academic workers in disciplines which are coming increasingly to depend on mathematics constitute a very significant and important group.

Some years ago, the Editors of the present Handbook, all of them actively concerned with the teaching of mathematics with a view to its usefulness for today's and tomorrow's citizens, got together to discuss the problems faced by mature people already embarked on careers in professions which were taking on an increasingly mathematical aspect. To be sure, the discussion ranged more widely than that—the problem of 'mathematics avoidance' or 'mathematics anxiety', as it is often called today, is one of the most serious problems of modern civilization and affects, in principle, the entire community—but it was decided to concentrate on the problem as it affected professional effectiveness. There emerged from those discussions a novel format for presenting mathematics to this very specific audience. The intervening years have been spent in putting this novel conception into practice, and the result is the Handbook of Applicable Mathematics.

THE PLAN OF THE HANDBOOK

The 'Handbook' consists of two sets of books. On the one hand, there are (or will be!) a number of *guide books*, written by experts in various fields in which mathematics is used (e.g. medicine, sociology, management, economics). These guide books are by no means comprehensive treatises; each is intended to treat a small number of particular topics within the field, employing, where appropriate, mathematical formulations and mathematical reasoning. In fact, a typical guide book will consist of a discussion of a particular problem, or related set of problems, and will show how the use of mathematical models serves to solve the problem. Wherever any mathematics is used in a guide book, it is cross-referenced to an article (or articles) in the *core volumes*.

There are 6 core volumes devoted respectively to Algebra, Probability, Numerical Methods, Analysis, Geometry and Combinatorics, and Statistics. These volumes are texts of mathematics—but they are no ordinary mathematical texts. They have been designed specifically for the needs of the professional adult (though we believe they should be suitable for any intelligent adult!) and they stand or fall by their success in explaining the nature and importance of key mathematical ideas to those who need to grasp and to use those ideas. Either through their reading of a guide book or through their own work or outside reading, professional adults will find themselves needing to understand a particular mathematical idea (e.g. linear programming, statistical robustness, vector product, probability density, round-off error); and they will then be able to turn to the appropriate article in the core volume in question and *find out just what they want to know*—this, at any rate, is our hope and our intention.

How then do the content and style of the core volumes differ from a standard mathematical text? First, the articles are designed to be read by somebody who has been referred to a particular mathematical topic and would prefer not to have to do a great deal of preparatory reading; thus each article is, to the greatest extent possible, self-contained (though, of course, there is considerable cross-referencing within the set of core volumes). Second, the articles are designed to be read by somebody who wants to get hold of the mathematical ideas and who does not want to be submerged in difficult details of mathematical proof. Each article is followed by a bibliography indicating where the unusually assiduous reader can acquire that sort of 'study in depth'. Third, the topics in the core volumes have been chosen for their relevance to a number of different fields of application, so that the treatment of those topics is not biased in favour of a particular application. Our thought is that the reader—unlike the typical college student—will already be motivated, through some particular problem or the study of some particular new technique, to acquire the necessary mathematical knowledge. Fourth, this is a handbook, not an encyclopedia—if we do not think that a particular aspect of a mathematical topic is likely to be useful or interesting to the kind of reader we have in mind, we have omitted it. We have not set out to include everything known on a particular topic, and we are *not* catering for the professional mathematician! The Handbook has been written as a contribution to the practice of mathematics, not to the theory.

The reader will readily appreciate that such a novel departure from standard textbook writing—this is neither 'pure' mathematics nor 'applied' mathematics as traditionally interpreted—was not easily achieved. Even after the basic concept of the Handbook had been formulated by the Editors, and the complicated system of cross-referencing had been developed, there was a very serious problem of finding authors who would write the sort of material we wanted. This is by no means the way in which mathematicians and experts in mathematical applications are used to writing. Thus we do not apologize for the fact that the Handbook has lain so long in the womb; we were trying to do something new and we had to try, to the best of our ability, to get it right. We are sure we have not been uniformly successful; but we can at least comfort ourselves that the result would have been much worse, and far less suitable for those whose needs we are trying to meet, had we been more hasty and less conscientious.

It is, however, not only our task which has not been easy. Mathematics itself is not easy! The reader is not to suppose that, even with his or her strong motivation and the best endeavours of the editors and authors, the mathematical material contained in the core volumes can be grasped without considerable effort. Were mathematics an elementary affair, it would not provide the key to so many problems of science, technology and human affairs. It is universal, in the sense that significant mathematical ideas and mathematical results are relevant to very different 'concrete' applications—a single algorithm serves to enable the travelling salesman to design his itinerary, and the refrigerator manufacturing company to plan a sequence of modifications of a given model; and could conceivably enable an intelligence unit to improve its techniques for decoding the secret messages of a foreign power. Given this universality, mathematics cannot be trivial! And, if it is not trivial, then some parts of mathematics are bound to be substantially more difficult than others.

This difference in level of difficulty has been faced squarely in the Handbook. The reader should not be surprised that certain articles require a great deal of effort for their comprehension and may well involve much study of related material provided in other referenced articles in the core volumes—while other articles can be digested almost effortlessly. In any case, different readers will approach the Handbook from different levels of mathematical competence and we have been very much concerned to cater for all levels.

THE REFERENCING AND CROSS-REFERENCING SYSTEM

To use the Handbook effectively, the reader will need a clear understanding of our numbering and referencing system, so we will explain it here. Important items in the core volumes or the guidebooks—such as definitions of mathematical terms or statements of key results—are assigned sets of numbers according to the following scheme. There are six categories of such mathematical items, namely:

    (i) Definitions
   (ii) Theorems, Propositions, Lemmas and Corollaries
  (iii) Equations and other Displayed Formulae
  (iv) Examples
   (v) Figures
  (vi) Tables

Items in any one of these six categories carry a triple designation a.b.c. of arabic numerals, where 'a' gives the *chapter* number, 'b' the *section* number, and 'c' the number of the individual *item*. Thus items belonging to a given category, for example, definitions are numbered in sequence within a section, but the numbering is independent as between categories. For example, in Section 5 of Chapter 3 (of a given volume), we may find a displayed formula labelled (5.3.7) and also Lemma 5.3.7. followed by Theorem 5.3.8. Even where sections are further divided into *subsections*, our numbering system is as described above, and takes no account of the particular subsection in which the item occurs.

As we have already indicated, a crucial feature of the Handbook is the comprehensive cross-referencing system which enables the reader of any part of any core volume or guide book to find his or her way quickly and easily to the place or places where a particular idea is introduced or discussed in detail. If, for example, reading the core volume on Statistics, the reader finds that the notion of a *matrix* is playing a vital role, and if the reader wishes to refresh his or her understanding of this concept, then it is important that an immediate reference be available to the place in the core volume on Algebra where the notion is first introduced and its basic properties and uses discussed.

Such ready access is achieved by the adoption of the following system. There are six core volumes, enumerated by the Roman numerals as follows:

    I Algebra
   II Probability
  III Numerical Methods
  IV Analysis
   V Geometry and Combinatorics
  VI Statistics: Parts A and B

A reference to an item will appear in square brackets and will *typically* consist of a pair of entries [see A, B] where A is the volume number and B is the triple designating the item in that volume to which reference is being made. Thus '[see II, (3.4.5)]' refers to equation (3.4.5) of Volume II (Probability). There are, however, two exceptions to this rule. The first is simply a matter of economy!—if the reference is to an item in the same volume, the volume number designation (A, above) is suppressed; thus '[see Theorem 2.4.6]', appearing in Volume III, refers to Theorem 2.4.6. of Volume III.

The second exception is more fundamental and, we contend, wholly natural. It may be that we feel the need to refer to a substantial discussion rather than to a single mathematical item (this could well have been the case in the reference to 'matrix', given as an example above). If we judge that such a comprehensive reference is appropriate, then the second entry B of the reference may carry only two numerals—or even, in an extreme case, only one. Thus the reference '[see I, 2.3]' refers to Section 3 of Chapter 2 of Volume I and recommends the reader to study that entire section to get a complete picture of the idea being presented.

Bibliographies are to be found at the end of each chapter of the core volumes and at the end of each guide book. References to these bibliographies appear in the text as '(Smith (1979))'.

It should perhaps be explained that, while the referencing *within* a chapter of a core volume or *within* a guide book is substantially the responsibility of the author of that part of the text, the cross-referencing has been the responsibility of the editors as a whole. Indeed, it is fair to say that it has been one of their heaviest and most exacting responsibilities. Any defects in putting the referencing principles into practice must be borne by the editors. The successes of the system must be attributed to the excellent and wholehearted work of our invaluable colleague, Carol van der Ploeg.

CHAPTER 1

Introduction to Statistics

1.1. THE MEANING OF THE WORD 'STATISTICS'

The most common everyday meaning of the word 'statistics' is: 'Numerical facts or data collected and classified.' (The quotation is from the *Oxford English Dictionary*.) Thus for example one speaks of educational statistics, medical statistics, tax statistics, the statistics of the glass industry, and so on.

An alternative definition is the following, also from the *O.E.D.* 'In early use, that branch of political science dealing with the collection, classification, and discussion of facts bearing on the condition of a state or community. In recent use, the department of study that has for its object the collection and arrangement of numerical facts or data, whether relating to human affairs or to natural phenomena.'

The 'in early use' definition, purged of its restriction to political science, would not be too far from modern usage, but the 'in recent use' meaning is curiously old-fashioned, completely omitting what for our purposes is the key aspect: the *interpretation* of data.

A (partial) definition that would be reasonably acceptable to most contemporary practitioners is (to paraphrase the *O.E.D.*): 'In current use, the department of study that has for its object the collection and interpretation of numerical data', the interpretative aspect being the one that is now regarded as the essence of the subject. It is not to be expected of such a broad discipline that it should be capable of concise definition or summary description, but as a first approximation one might say that the main object of Statistics is to draw sensible conclusions from discrepant data.

It is the fact that, except in trivial cases, real data are always discrepant that underlies the whole of statistical methodology. The discrepancies, or variations between individual observations, might (for example) be due to *experimental error*, as in reading the position of a pointer when it lies between two calibrated scale positions on the dial of an instrument; or it might be the effect of fluctuations imposed by external circumstances, as in the changes imposed by atmospheric variation on the recorded intensity of light from a star, or by the non-uniformity of electronic equipment on a message transmitted by radio or

1

by wire, in which case it is called *noise*: yet again, it might be that the entities under observation form part of a population whose members exhibit *inherent variability* in the character being measured, as in the heights of 20-year old male students.

In most cases the situation which produces the data is too complex to allow of an analysis based on anything like a complete description that reflects all its details. It is therefore replaced by a mathematical model which reproduces what are thought to be its salient features, and excludes what is considered to be irrelevant. This model will exploit such scientific laws as are thought to apply, and will usually involve deterministic and 'random' elements. The latter, in turn, are represented by a probabilistic model. It is to the elucidation of this and to the testing of its validity that statistical inference, properly speaking, applies.

EXAMPLE 1.1.1.   *Milk and children's weight.*   Consider as an example the question whether the regular drinking of milk improves the physique of school children. Before an answer can be attempted we must decide what quantities of milk (a half-pint a day?) are to be taken, over what period (a year?), by what age of children ($9\frac{1}{2}$ to $10\frac{1}{2}$ years?), and which aspect or aspects of their physique are to be measured (their weight?). The naive technique of merely weighing a child before and after the period of regular milk-drinking will not do, since one cannot then disentangle the increase in weight attributable to milk from that which would have occurred without the milk. To sort this out, a group of children subjected to a milk-enriched diet must be compared with a control group on a standard diet (which must be defined). A typical method would be to select the children allocated to the milk-fed and control group by a randomized procedure which would allow one to regard the individual weight changes as realizations of independent random variables [see II, Definition 4.4.1]. Imprecisely, but with sufficient accuracy, these differences in weight might be taken to be Normally distributed [see II, § 11.4] with standard deviation σ, and with expectation μ_1 for those on the ordinary diet, and μ_2 for those on the milk-rich diet. Here μ_1, μ_2 and σ are 'parameters' of unknown magnitude. Suitable approximations to their values, called 'estimates', would have to be inferred from the data, and the original question: 'Does additional milk lead to an increase in weight?' translated to the following: 'Is the difference between the estimate of μ_1 and that of μ_2 *significant*—that is, is it large enough to allow us to discount chance effects and conclude that μ_2 really is larger than μ_1, and, if so, by how much; and how accurate is the estimated difference?'

Simple as this example is in principle, it illustrates some of the main features of statistical inference. In the first place, the collection of the data must be organized in such a way as to validate the subsequent probabilistic arguments: the enquiry must be properly designed and the sampling technique must fulfil the desired purpose. Secondly, features of the children's life considered to be irrelevant to the enquiry are to be excluded, and the model founded on the simplifying assumption that, subject to proper design, all the variability may be explained in terms of the chosen family of probability distributions [see II, Chapters 4 and 11]. The choice of this family (the Normal, in our example) will have been the result of a compromise between the complexity required by realism and the simplicity needed to draw useful quantitative inferences unencumbered by irrelevant mathematical difficulties. Further investigations may be needed to guarantee the validity of this choice. In this example the model is a simple one: the observations are the subjects' weights, and the 'effect' being investigated is taken to be the difference between the final and initial weights. In a more elaborate experiment the observations might be expressed as a more complicated function of the variables and/or parameters suggested or demanded by the relevant field of study, and some or all of these variables would be regarded as suffering random variation, calling for their representation as random variables. Appropriate families of distributions would have to be invoked and their parameters estimated, and this procedure followed, as above, by some validating exercise to verify the suitability of the model as a whole.

For example, the observations might be the total rainfall measured at each of 20 neighbouring sites during each of 100 consecutive quarter-hour intervals. The model, based on radiometeorology, might be one that specified the rainfall intensity at given locations as functions of time, in terms of the initiation, growth and decay of moving cloud-cells. The variables would then include the rate of cloud-cell initiation (possibly as a two-dimensional Poisson process [see II, § 20.1.7]) and the parameters would control the shape of the clouds and their rates of growth and decay.

EXAMPLE 1.1.2.   *Dosage-mortality experiments.*   Another example, described in detail in section 6.6, is the assessment of the mortality of insects as a function of the dose of insecticide applied. In this example, the potency of the insecticide at various dosage levels is measured by the number of insects killed after exposure to the prescribed dosage. At very low dosage levels no insects are killed and, at very high levels all are killed; while, at intermediate dosages the percentage kill, which is subject to a good deal of experimental variation, increases on average with the dose. It is necessary (a) to postulate a plausible parametric model that is capable of describing the 'growth curve' of proportion killed versus dosage level, (b) to estimate the parameters of this curve, and to verify that the resulting curve does indeed provide an acceptable model, and (c) to provide a value for the dosage level required to effect a 50% kill (this being the conventional measure of toxicity), together with an assessment of the reliability of this estimated value.

The examples show that we need techniques for obtaining good approximations ('estimates') to the values of the parameters that specify which particular member of the chosen family of probability distributions we are concerned with, and for assessing the accuracy of these estimates in a way that would enable us to tell, for example, whether numerically different estimates are significant of a real difference between the actual (unknown) parameter values; and, further, for testing whether or not a proposed family of distributions can provide an acceptable model for the observed data. These are perhaps the most important aspects of statistical inference and in the chapters that follow they and other techniques based upon them are described in some detail. A fuller account of the contents of these chapters is given in section 1.3.

Further reading: An illuminating introduction to statistical inference will be found in Barnett (1982), Chapter 1.

1.2. SAMPLING DISTRIBUTION; STATISTIC; ESTIMATE

It was explained in section 1.1 that the statistical approach to a question about the real world replaces that question by an approximately equivalent one about aspects of probability distributions that occur in an appropriate model of that world. Thus the effect of milk on the weight of children who drink it is discussed in terms of the value of a parameter θ (or set of parameters) of a probability distribution representing the weight increases of individual children. A finite number of children will have been involved, and they—or more properly the increases in their weights—form the *sample*. This sample, of size n, say, will provide n pieces of data, namely the individual weight increases $x_1, x_2, \ldots, x_n$. We shall suppose that this forms the whole of the data available to the statistician. His *estimate* of the value of the unknown parameter θ—assuming for simplicity that there in only one—must, therefore, necessarily be a number calculated by some rule from the sample values $x_1, \ldots, x_n$, say

$$\theta^* = t_n(x_1, x_2, \ldots, x_n). \tag{1.2.1}$$

For example, there might be good reasons to take as estimate θ^* the *sample mean*

$$(x_1 + x_2 + \ldots + x_n)/n.$$

Any such combination of observed values is called a *statistic* (thus generalizing the meaning given to this word in non-technical usage, where it is taken to mean a single one of a collection of data or a single numerical fact used in an argument). A statistic is a number computed from a sample. If it is to be used as an estimate of a parameter θ, it must in some sense be an approximation to θ. The question is, in what sense?

In attempting to answer this question we must remember that the particular value that we happened to get for the statistic from our sample of children's weights would have been different had we chosen a different group of children, as we might well have done. Indeed, if the inference we hope to draw is to apply to a wider group of children than those directly involved in our sample, it is essential that the sample should have been selected from that wider group by a method involving some element of random choice. It is plausible to argue that the question whether our particular estimate is a good one or not can be answered in terms of another question involving the entire class of estimate values (1.2.1) capable of being generated by the sampling procedure.

EXAMPLE 1.2.1.   *Industrial sampling.*   In a 'load' consisting of about 10,000 nominally identical manufactured articles it is known that some of the articles are defective in that a critical dimension lies outside the required tolerances. It is required to estimate the proportion (θ, say) of these defective articles in the load, from precise measurements carried out on a sample of 20 articles taken from the load.

Consider first the sampling procedure. Suppose this to be the following: 20 articles are to be chosen 'at random', i.e. in such a way that at each choice every article in the load has an equal chance of being chosen. (This is not always easy to achieve, and practical techniques will vary according to the size and nature of the load.) Since in our case the load is very large in comparison with the sample, the proportion of defective articles in the remainder after removing the sample will not differ sensibly from the original proportion θ. In these circumstances the statistical properties of the sample are not discernably different from those resulting from 'random sampling with replacement' (see II, § 3.6.3). It follows that, to an acceptable degree of approximation, the probability that the sample will contain r defective articles, for $r = 0, 1, 2, \ldots, 19$ or 20, is the Binomial probability [see II; § 5.2.2]

$$\binom{20}{r} \theta^r (1-\theta)^{20-r}. \tag{1.2.2}$$

Now consider the estimate. Criteria for formulating rules that produce 'good' estimates will be discussed later (see Chapter 3); for the present we rely on the argument that the proportion of defective items in the sample would seem, intuitively, to be a reasonable approximation to the proportion of defective articles in the load. Suppose therefore that we adopt, as our estimate of θ, the number θ^* defined by

$$\theta^* = r/20, \tag{1.2.3}$$

where r represents the number of defective articles in the sample (of size 20), and that the observed value of r is 8, so that

$$\theta^* = 0{\cdot}40.$$

The expression (1.2.3) is a special case of (1.2.1); it is a rule for the construction of the estimate from sample data. Now, according to (1.2.2), the value of r in (1.2.3) is a realization [see II, Chapter 4] of a random variable R whose distribution is Binomial $(20, \theta)$ [see II, § 5.2.2]. It follows that θ^* is a realization of a random variable, T, say, where

$$T = R/20, \tag{1.2.4}$$

with possible values

$$0, 1/20, 2/20, \ldots, 19/20, 1,$$

and whose probability distribution is given by

$$P(T = r/20) = P(R/20 = r/20)$$
$$= P(R = r)$$
$$= \binom{20}{r} \theta^r (1 - \theta)^{20-r}, \qquad r = 0, 1, \ldots, 20 \qquad (1.2.5)$$

by (1.2.2).

In the example, the question 'What is the proportion of defective items in the load?' has been replaced by the question 'What is the value of the parameter θ in the probability distribution (1.2.2)?' The estimate $\theta^* = 0.40$ is regarded as a realization of a random variable T, whose distribution has been obtained in (1.2.5). It is convenient to have a name for the random variable of which the estimate is a realization. It is called the *estimator,* and its probability distribution is called the *sampling distribution* of the estimate (or of the estimator). (A similar meaning attaches to the sampling distribution of any statistic, whether or not it is used directly as an estimate.)

The above discussion suggests that this question: 'Is θ^* a good estimate of θ?' may be regarded as a shorthand form of the following: 'In the sampling distribution of θ^*, is there a high probability that the observed value will be nearly equal to θ?' In the industrial sampling example discussed above we can at least make a subjective judgment on this by examining the sampling

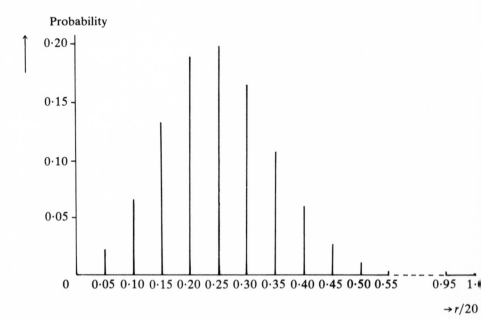

Figure 1.2.1: Graph of the probabilities shown in Table 1.2.1.

distribution in question. For the special values $n = 20$, $\theta = 0.25$, this is tabulated below and illustrated graphically. It will be seen that, for this value of θ at least, the probability distribution is indeed concentrated in the neighbourhood of θ. Realizations of the estimator are then rather likely to be near to θ, and, in this sense, θ^* appears to be a not unreasonable estimate of θ.

For further information on industrial sampling inspection, see: Hald (1981), Wetherill (1969)—Bibliography H. (The bibliography referencing system is explained in section 1.3.6.)

Value	Prob	Value	Prob
0	0·003	0·35	0·112
0·05	0·021	0·40	0·061
0·10	0·067	0·45	0·027
0·15	0·134	0·50	0·010
0·20	0·190	0·55	0·003
0·25	0·202	0·60	0·000
0·30	0·169	0·65	0·000
		⋮	⋮
		1·00	0·000

Table 1.2.1: Sampling distribution of the proportion of defective items in random samples of size 20, when the proportion θ of defective items in the population is 0.25. Equivalently, the probability distribution of T:

$$P(T = r/20) = \binom{20}{r}(0.25)^r(0.75)^{20-r},$$

$$r = 0, 1, \ldots, 20.$$

1.3.   THE SUBJECT-MATTER OF THIS BOOK

Although 'Statistics' means a good deal more than 'Applied Probability Theory', the concepts and methods of Statistics do depend heavily on those of Probability. An ideal exposition would perhaps develop Probability-and-Statistics as a single integrated discipline. In the present Handbook however it has been thought convenient to devote one volume (Volume II) to Probability and one to Statistics. This does not mean that the whole of Volume II must be mastered before attempting the present volume! On the contrary, the methods expounded here are in many cases self-standing, and in others require only a small background of Probability. In all cases where some such background is required, whether of an elementary or more advanced nature, references are supplied to the relevant section(s) of Volume II. Similarly, of course, for references to other volumes of the Handbook.

It will be clear from what has been said that, whilst there are topics that clearly belong to 'Probability' and others that, equally clearly, belong to 'Statistics', there remain topics which can properly be classified only as belonging to both. In our case, for example, questions relating to the independence of quadratic forms in Normally distributed random variables, which might well have been discussed in the Probability volume (Volume II), have in fact been treated as having mainly statistical interest and allocated to the present volume. The same applies to 'non central' sampling distributions [see § 2.8].

The topics covered in this volume are briefly described in the sections 1.3.1 to 1.3.5 which follow, while section 1.3.6 mentions some of the topics not covered. Before giving this outline, however, it is perhaps necessary to say a word or two about the order in which topics are treated.

One of the primary purposes of this Handbook [see the Introduction to the present volume] is to provide a convenient collection of the mathematical procedures and results that are utilized in the guidebooks and on which the reader of a guidebook might be expected to require further information. This being the case, it might be thought that an alphabetical arrangement of topics would best serve the purpose, as in an encyclopedia. Such an arrangement would necessarily have led to a large number of rather short and heavily interdependent entries. In view of the partially sequential structure of mathematics it seemed to the editors that their purpose would be better served by grouping topics in coherent chapters, thus providing much of the continuity and sense of development found in a good traditional textbook, but retaining the virtue of an encyclopedia by extensive internal cross-referencing. Since, however, the object has not been to write a textbook of the traditional kind, some liberties have been taken with the order of presentation, and the editors have felt free to refer the reader forward to later chapters as well as back to earlier ones. In particular, we have gathered together in an early chapter (Chapter 2) most of the material relating to sampling distributions and associated topics that is required later throughout the book. This makes Chapter 2 a somewhat miscellaneous one. For the most part, each of the other chapters presents a unified treatment of a single broadly defined topic; in one or two cases it has been thought convenient to spread the main topic over two chapters.

We now proceed with the promised outline of the statistical topics covered in this volume.

1.3.1. Sampling Distributions; Distribution-free Methods

The statistician has to make his inferences from a sample of observations. Each observation is a realization of a random variable; there is a known set of possible values which an observation on this random variable may take, some having a greater probability of occurring than others. The one that happens to manifest itself on a particular occasion is the *realization* in question. The probabilities of the possible realizations are specified by a function called the *probability distribution* of the random variable. In exceptional cases the probability distribution might specify the probability of a given realization in direct numerical terms, but more commonly the function involves one or more parameters, of unknown magnitude. This raises the problem of finding which combinations of the sample values serve best as approximations to these parameters. Each such combination is a *statistic*, and, like each individual observation, the statistic is a realization of a random variable. If x_1, x_2 and x_3 are independent observations on a Normal (μ, σ) distribution, that is one with expectation μ and standard deviation σ (these being the *parameters* of the Normal family of distributions) we may regard x_1 as a realization of a random variable X_1, x_2 as a realization of X_2, and x_3 as a realization of X_3, where X_1, X_2, and X_3 are independent Normal (μ, σ) variables. We may call X_1 the random variable *induced* by x_1, X_2 that induced by x_2, and X_3 that induced by x_3. The *statistic* $\bar{x} = (x_1 + x_2 + x_3)/3$, the 'sample mean', is a realization of the random variable $\bar{X} = (X_1 + X_2 + X_3)/3$, which may similarly be described as the random variable induced by $\bar{x}$. From the known properties of the Normal distribution [see II, § 11.4.5] it follows that the probability distribution of this induced random variable $\bar{X}$ is Normal $(\mu, \sigma/\sqrt{3})$. This is the *sampling distribution* of the statistic $\bar{x}$. In terms of μ and σ it tells us the probabilities of the various possible realizations of $\bar{X}$ (one of which, of course, is the sample mean $\bar{x}$ in *our* sample): in particular, the probability density takes its largest value at μ, whence $\bar{x}$ is a not unreasonable estimate of μ; and the probability that our value of $\bar{x}$ lies beyond a specified distance from μ is also given (in terms of σ) by the sampling distribution.

Thus the sampling distribution of a statistic can enable us to judge whether or not a suggested statistic might be used as an estimate of a parameter of interest; and it can also give information about the variability of the statistic and hence of the reliability of the estimate.

[Here, and throughout the book, we have used the convention of denoting a random variable by a capital Latin letter (X, for example), and a realization of that random variable by a small Latin letter (x or x_l or x_j, for example).]

Sampling distributions are therefore of high technical importance and we have devoted a chapter (Chapter 2) to a collection of information about the sampling distributions of various statistics of practical importance.

Procedures that depend heavily on sampling distributions are however open to the criticism that the sampling distribution of a statistic depends on the underlying distributional assumptions in the probability model itself, so that if the latter are not firmly based the whole edifice is unsafe. In practice most of the widely used procedures are *robust*, that is they are rather insensitive to the kind of variations that ought reasonably to be allowed for in the probability model.

Clearly the most robust procedures of all are those—if such exist—which make effectively no distributional assumptions at all. Such procedures do exist, and are called '*distribution-free*' (or, '*non-parametric*'). Chapter 14 provides an account of these methods.

1.3.2. Estimates, Tests, Decisions

The deceptively brief heading of this section describes what is in fact the largest part of the book.

The problem of estimation has been briefly described in section 1.3.1. Chapter 3 expands this description and provides an introduction to systematic approaches to the problem of finding a good estimate, including graphical methods of displaying the information in a sample, and formal criteria such as, for example, the requirements that the estimate of a parameter should have the same physical dimensions as the parameter in question, that the estimate should in some sense relate to the parameter of interest and not some other parameters, and that the estimate should have as small a variability (as measured by its standard deviation) as possible.

It turns out that in certain cases it is possible to concentrate into a single ('sufficient') statistic the whole of the information about a parameter that is available in the sample. This concept is also discussed in Chapter 3, which ends with a short section on practical methods of actually finding or constructing estimates that have desirable properties.

A meaningful estimation statement should, clearly, not be restricted to giving a single numerical approximation to the unknown value of the parameter; it should also provide information about the reliability of that approximation. Whilst these two aspects of what is a single problem are intimately interconnected, it is nevertheless sometimes convenient to discuss them separately, in which case they are referred to as '*point estimation*' and '*interval estimation*' respectively. Chapter 4 is concerned mainly with interval estimation, concentrating firstly on (a) 'confidence intervals', which are related to the behaviour of a statistic in repeated sampling, the theory of which therefore depends heavily on sampling distributions; secondly (b) on 'likelihood intervals', aspects of the *likelihood function*, which seeks to assess the relative plausibilities of various suggested parameter values in the light of the unique sample actually obtained; and thirdly (c) on Bayesian intervals, constructed in terms of a method of inference in which the sample under discussion is regarded as updating and improving the precision of prior information already available before the sample was obtained. (This last approach is considered in greater depth in Chapter 15.)

Since the whole *raison d'être* of Statistics as an intellectual discipline is the existence of variability, every estimate is subject to error; if two differing estimates of a parameter are obtained, one under one set of circumstances and one under another, it is not immediately obvious whether or not the difference between them corresponds to a real difference between the parameters in question. For example, the parameter might be the probability of being infected with a given disease after taking drug A (one circumstance) or drug B (the other circumstance). Such questions are assessed by procedures called statistical *tests*, or *significance tests*, described in Chapter 5.

One approach to statistical tests, an approach associated particularly with the name of R. A. Fisher [see Box (1978)—Bibliography D], regards them as sometimes tentative steps in the progress of scientific research, providing the scientist with objective criteria by which to judge the claims of a hypothesis that is being investigated. Another, associated primarily with the names of

J. Neyman and E. S. Pearson, regards the testing procedure as one in which a *decision* is to be made as between one course of action and another, or between accepting the validity of one hypothesis rather than another. In the field of ordinary statistical practice the actual procedures produced by these opposing viewpoints do not differ much, but in more recent developments decision theory has become a subject in its own right in which the penalties for making wrong decisions and the rewards for avoiding them are stringently analysed, leading to a corpus of knowledge that has unifying consequences in the theory of estimation and testing as well as in broader fields. The subject is described in Chapter 19.

One particular category of tests is concerned with the task of assessing the appropriateness of the probabilistic model assumed to underly one's data. One has assumed, on apparently good grounds, that a sequence of irregularly occurring events (e.g. responses of a Geiger counter to radioactive impulses) is a Poisson process [see II: § 20.1]. The relevant parameters having been estimated from the data on this assumption, one has to ask whether the fitted model is in fact consistent with the sample. Are the sample values actually obtained near enough to what would be expected of the fitted model? The most widely used procedure to deal with this kind of question is to compute a certain statistic invented by Karl Pearson, and to base the test on the sampling distribution of that statistic. This is Pearson's 'chi-squared (χ^2) test of goodness of fit', described in Chapter 7.

Of the various conceivable systematic methods of constructing 'point' esti-mates and finding their reliability, the most useful is the method of *maximum likelihood*, which is described and exemplified in Chapter 6. Another widely used method which can be regarded either as a special case of this, or as an independently justified procedure, is the method of *least squares*. This, together with a more or less systematized set of rules for carrying out the associated tests (called the *analysis of variance*, 'ANOVA' for short) is described in Chapter 8.

The estimation and test procedures referred to above utilize the data in a 'completed' sample: that is, the sampling procedure has been completed before the data is utilized. There are circumstances in which individual pieces of data or sub-collections of data become available seriatim; tests have been devised which use sampling procedures of this sort, in which evidence for or against a hypothesis of interest accumulates as the sampling proceeds, until the evidence becomes convincing, when the sampling stops. Such testing pro-cedures are called *sequential*; they are described in Chapter 13.

In few areas has the variability of the world been more forcibly demonstrated than in scientific agriculture, which in its early days faced great difficulties from this cause in its efforts to compare different varieties of seed, or the same variety under differing manurial treatments. The importance of sound agricul-ture to the human condition, together with the cost and duration of field trials, played a powerful role in concentrating attention on the need to design the trials and experiments in the most efficient way. This need called into existence the science (or art) of 'the design of comparative experiments', nowadays of course not restricted to agriculture. Chapter 9 provides an introduction to this vast discipline, and Chapter 10 is devoted to methods of analysing data from such experiments.

These techniques are founded on the 'linear model', in which the response of the system (e.g. the yield of wheat) to given stimuli (e.g. the application of various quantities of specified fertilizers) is taken to be a linear function of the variables. The concept of linearity can however be profitably extended to models that are not so obviously linear, as in most analyses of variance. For example the toxicity of certain drugs appears to be zero at and below a threshold value; then to increase with increasing dosage, at first slowly, then faster, then more slowly again, levelling off at a saturation dosage beyond which the drug effects a 100% kill (cf. Example 1.1.2). The response curve, as measured by the number of experimental animals killed by the stated dose, is a sigmoid (S-shaped). A transformation can be found which sends this into a straight line, so that, unexpectedly perhaps, we obtain a linear model, to which least squares methods may be applied (complicated however by the unequal variabilities of the responses.)

Such generalizations of the linear model are discussed in Chapters 11 and 12.

1.3.3.   Bayesian Inference

Reference has already been briefly made to *Bayesian* statistics, so-called after the Rev. Thomas Bayes, an eighteenth-century English mathematician. (For biographical information see Pearson and Kendall (1970)—Bibliography D.) To oversimplify, the Bayesian approach to estimation regards the parameter to be estimated as a random variable in the sense that the most useful way of describing one's information about it is in terms of a probability distribution.

In the industrial sampling procedure discussed in Example 1.2.1 the proportion of defectives in a load was estimated as the value of a certain statistic based on a sample from that load, and on nothing else. Now suppose the load in question had been one of a sequence of loads in which it was known from experience that the values of the proportion of defectives (θ) varied, independently from one load to another, in a known way: for example, that in 3% of the loads the value of θ was 0·01, in 5% of them it was 0·025, etc. The value of θ in a given load could then be regarded as a realization of a random variable with known ('prior') probability distribution. The sample values from this load could then be combined with this prior distribution, using Bayes's theorem [see II, § 16.10], to provide an improved probabilistic specification (a 'posterior distribution') of the uncertainties concerning the value of θ in the sampled load. The 'modern Bayesian' approach to statistical inference takes the view that there is always *some* prior information about an unknown parameter, less precise perhaps than in the case quoted above, but still capable of being expressed as a prior probability distribution from which to construct a posterior distribution. This is the subject matter of Chapter 15.

1.3.4.   Multivariate Analysis

It is only in the simplest of situations that the statistician is concerned with only one random variable. More commonly, each entity that is sampled yields a number of separate measurements, as for example the height, waist-girth, and weight of a member of a human population. The statistician is then concerned with such questions as: do the separate components of an observation vector behave independently of each other; if not, how may one describe their joint behaviour: is it the case that some of the components are more informative than others in discriminating between different categories of the entities studied; and so forth. The classical approach to these questions is discussed in Chapter 17, while Chapter 18 provides a survey of contemporary thinking on the subject.

1.3.5.   Time Series

The final category in this description of statistical topics is concerned with the analysis of a sequence of observations (each subject to random variation) arising from a source that is itself developing, changing, or fluctuating. The observations might be for example a sequence of daily readings of the water level in the Thames at Marlow, the total weekly rainfall values in San Francisco, hourly readings of the concentration of a specified chemical in a pressure vessel in a chemical engineering process, monthly accident statistics on a nominated stretch of motorway, ... The variations in the data are a compound, in unknown proportions, of an unknown deterministic scheme (such as for example a purely sinusoidal seasonal effect) with fluctuations following some (unknown and possibly changing) probability distributions in which moreover the behaviour of the system at time t is not independent of its behaviour at earlier times $t-1, t-2, \ldots$. The purposes for which the system is being studied would typically include prediction and forecasting.

The very broad and rapidly developing subject of Time Series is treated in Chapter 18. The important estimation technique known as Kalman filtering is described in Chapter 20.

1.3.6.   Bibliography and References

Related topics in the text are interconnected by a system of cross-references, and extensive references are also made to underlying or otherwise relevant mathematical material in other volumes of the Handbook, as explained in the Introduction to the Handbook. References to sources outside the Handbook are organized in two distinct ways, one by chapters and one for the volume as a whole.

Chapter references are listed at the end of each chapter under the heading 'Further reading'. These give further information on specific topics that arise in that particular chapter, and are indicated at the appropriate point in the text as (e.g.) 'see Barnett (1982), Chapter 1'.

Volume references are collected in a General Bibliography at the end of the volume. This list is sub-divided into sections: A for bibliographies, B for dictionaries, encyclopedias and handbooks, C for general texts covering a wide field, D for historical and biographical material, E for guides to statistical tables, F for tables of random deviates having specified distributions, G for tables of statistical functions, and H for special topics not dealt with (or dealt with very briefly) in the text. References to these lists are indicated in the text by, for example, 'For further information, see Kendall and Buckland (1971)—Bibliography B'.

1.3.7.   Appendix: Statistical Tables

Serious statistical work requires the use of extensive tables (see Bibliography G). For many purposes however, the reader will find that the small collection of tables in the Appendix will suffice. The collection includes tables of the Binomial, Poisson, Normal, Student, χ^2 and F distributions, 5000 random digits, 500 Standard Normal deviates, and charts for determining confidence intervals for the Binomial and Poisson parameters.

1.3.8.   Some Topics that are not Covered

The ideal book on statistics would be a balanced exposition of theory and practice, covering all aspects of the subject, comprehensible to all readers, and contained in a single volume of moderate size. The editors are aware that this ideal has not been attained: in particular, some topics may well have been given too much space, others too little, and yet others none at all. The main thrust of the book is towards the interpretation of data. For reasons of space, the practical details of the *gathering* of data have been given less emphasis: an abbreviated introduction to the design of comparative experiments is given in Chapter 9, while the important practical techniques of sample survey design have regretfully had to be relegated to the Bibliography. [See Arkin (1963), Barnett (1974), Cochran (1963), Deming (1950), Hanson, Hurwitz and Madow (1953), Stuart (1976) and Yates (1960)—all in Bibliography H.)

Other topics have suffered the same fate, on the grounds of being too specialized for the present volume, too near the edges of the field, and/or being the subject matter of proposed future publications in the series. These include the foundations and general principles of uncertain inference, the application of mathematical programming and optimization techniques in statistics, the analysis of special kinds of data sets such as directional data or extreme values, the use and availability of statistical computer packages, statistical simulation and Monte Carlo methods, and industrial sampling inspection and quality control. References to texts in some of these topics will be found in the General Bibliography, Section H. ('Bibliography H').

1.4. CONVENTIONS AND NOTATIONS

We conclude this chapter with some miscellaneous remarks on notational and other conventions employed in the book. Some of these are standard but others are idiosyncratic and require explanation.

1.4.1. Mathematical Conventions

logarithms: unless otherwise specified, 'log x' always means 'ln x', that is 'log$_e$ x', the natural logarithm, or log to base e.

set membership: $\in$: '$x \in \mathcal{A}$' means 'x is a member of the set (or class) $\mathcal{A}$'. [see I, § 1.1.]

O notation: we frequently deal with a statistic [see Definition 2.1.1], t_n say, evaluated from a sample of size n, for which some property might be expressed as being equal to $h_n + e_n$, where h_n is a specified function and e_n is an error which diminishes with increasing n. To say that the error is $O(n^{-1})$, for example, means that e_n has 'the same order of magnitude' as n^{-1}; this means that, for large values of n, e_n behaves approximately like an^{-1}, for some constant a. Similarly $O(n^{-1/2})$, etc. [see IV, Definition 2.3.3].

1.4.2. Statistical and Probabilistic Conventions and Abbreviations

(i) *Abbreviations*

c.d.f.: *cumulative distribution function* [see II, §§ 4.3.2, 10.3].

distr ( ): *the distribution of* ( ), as in 'distr (X) = distr (Y)', meaning that X and Y have a common distribution; c.f. 'twiddle', below.

d.f.: *degrees of freedom* [see, e.g., § 2.5.4].

i.i.d.: *independent and identically distributed*, as in 'the i.i.d. variables X_1, $X_2, \ldots, X_n$'.

p.d.f.: *probability density function*, also called 'frequency function'. We use 'p.d.f.' in this book for discrete as well as for continuous distribution. Purists who object to the use of 'density' in a discrete context may prefer to translate 'p.d.f.' as 'point distribution function' [see II, § 4.3.1, 10.1].

r.v.: *random variable* [see II, Chapter 4].

'twiddle' ($\sim$): *distributed as*. Thus '$X \sim$ Normal (μ, σ)' means 'the distribution of X is Normal (μ, σ).' Some writers resent this usage because '$\sim$' is already overworked in other branches of mathematics, for example it is used to denote an equivalence relation [see I, § 1.3.3] and also asymptotic equality [see IV, Definition 2.3.2]. For others its convenience outweighs this objection.

(ii) *Nomenclature of standard distributions*

Bernoulli (θ): The Bernoulli distribution with success-parameter θ, that is the distribution of a r.v. R with p.d.f.

$$P(R = r) = \theta^r(1-\theta)^{1-r}, \qquad r = 0, 1.$$

[see II, § 5.2.1.]
Binomial (n, θ): The distribution of a r.v. R for which

$$P(R = r) = \binom{n}{r}\theta^r(1-\theta)^{n-r}, \qquad r = 0, 1, \ldots, n$$

[see II, § 5.2.2].
Gamma (α, β): The distribution of a r.v. X with p.d.f. at x given by

$$\{x^{\alpha-1} e^{-x/\beta}\}/\beta^\alpha\Gamma(\alpha), \quad x > 0.$$

Here α is called the scale parameter and β the shape parameter. [see II, § 11.3.1.]
MVN: multivariate Normal [see II, § 13.4].
Normal (μ, σ): The Normal distribution with expected value μ and standard deviation σ. Sometimes abbreviated to '$N(\mu, \sigma)$'. (The variance is σ^2. Some authors therefore use the alternative designation 'Normal (μ, σ^2)', or '$N(\mu, \sigma^2)$'.)
[see II, § 11.4.]
Poisson (θ): The distribution of the r.v. R for which

$$P(R = r) = e^{-\theta}\theta^r/r!, \qquad r = 0, 1, \ldots.$$

[see II, § 5.4].
Uniform (a, b): The distribution of the r.v. X with p.d.f. at x given by

$$f(x) = \begin{cases} 1/(b-a), & a \le x \le b \\ 0, & \text{otherwise.} \end{cases}$$

[see II, § 10.7.1.]

(iii) *Capital letter convention for random variables*

We employ the general convention of reserving upper case Latin letters to denote random variables, using the corresponding lower case letters for realizations (= observed values). Thus we speak of a sample ($x_1, x_2, \ldots, x_n$) of observations on a r.v. X. Established usage dictates occasional departures from this convention, as for example in the use of 'F' as the name of a particular distribution. Strict adherence to the convention is somewhat pedantic, and professional statisticians do not always bother. For learners and non-experts, however, its consistent use in strongly recommended.

(iv) *Notation for moments and related quantities*

We use the symbol $E(X)$ to denote the expected value or *expectation* of a r.v. [see II, Chapter 8]. Variants of this usage are $\mathbf{E}$, $\mathbb{E}$, $\mathscr{E}$.

Our abbreviation for the *variance* of X is var (X); the symbol $V(x)$ is also widely used. For the *standard deviation* of X we use 's.d. (X)'; for the covariance of X and Y, 'cov (X, Y)'; for the correlation coefficient of X and Y, 'corr (X, Y)'; and for the skewness of X, 'skew (X)'. [see II, Chapter 9.]

(v) *Non-standard usages: induced r.v., statistical copy*

A sample of mutually independent observations $x_1, x_2, \ldots, x_n$ on a r.v. X may alternatively be regarded as a set consisting of an observation x_1 on a r.v. X_1, an observation x_2 on a r.v. X_2, etc., where $X_1, X_2, \ldots, X_n$ are *statistical copies* of X: that is, they are i.i.d., and their common distribution is the distribution of X:

$$\text{distr } (X_j) = \text{distr } (X), \qquad j = 1, 2, \ldots, n.$$

The statement 'x is a realization (=observed value) of X' may be turned round. Thus: 'X is the r.v. *induced* by x', meaning that the sampling distribution [see Chapter 2] of x is distr (X). The statistic $\bar{x} = \sum_1^n x_j/n$ (the sample mean), which has a particular known numerical value in a given sample, has a sampling distribution which may be derived by standard procedures from the common sampling distribution of the x_j, and the random variable that has this as its probability distribution is the r.v. *induced* by $\bar{x}$. It will naturally be denoted by the symbol $\bar{X} = \sum_1^n X_j/n$, where $X_1, X_2, \ldots, X_n$ are statistical copies of X.

To speak of the *sampling* distribution of a statistic, meaning the probability distribution of the corresponding induced random variable, is pedantic to about the same extent as is the notational distinction between a r.v. X and a realization x of that r.v.; as in that situation, it is an aid to clarity of thought.

(vi) *An ambiguity in the meaning of $P(A|K)$: 'the probability of A on K'*

One meaning of '$P(A|B)$' is 'the conditional probability of the proposition A, given that proposition B is true' [see II, § 6.5]. Then $P(A|B) = P(A \cap B)/P(B)$; both '$P(A)$' and '$P(B)$' are meaningful.

However, we often use '$P(A|H)$' to mean 'the probability of the proposition A, calculated on the assumptions H', usually abbreviated to 'the probability of A on H,' H being a 'hypothesis'. Here is an example: A is the proposition '$X > x_0$', X being a Normal $(\mu, 1)$ variable where the value of μ is unknown, and H is the hypothesis that $\mu = 0$.

Yet another shade of meaning occurs in the usage '$P(N = n|\theta)$' or '$\phi(n|\nu)$'; here N is a r.v. whose distribution involves a parameter θ of unknown magnitude, and $P(N = n|\theta)$ means 'the probability of obtaining the observed value n, expressed in terms of the parameter θ'. Similarly $E(N|\theta)$, etc. In practice it is usually clear from the context which meaning is intended.

(vii) *Nomenclature for tabular values: percentage points*

Statistical work requires frequent use of tables of the c.d.f. of probability distributions. For some of the distributions in common use, what might be described as ordinary tables are available. For example, Appendix T3 gives an 'ordinary' table of the function $\Phi(u)$, the Standard Normal integral, and Appendix T5 a similar table for Student's distribution. In order to economise on space, however, many tables are readily available only in an abbreviated 'inverse' format. In the case of the Standard Normal integral $\Phi(u)$ mentioned above, for example, instead of tabulating $\Phi(u)$ against u, the inverse format tabulates u against Φ; that is, it gives values u_α such that $1 - \Phi(u_\alpha) = \alpha$, as in Appendix T4.

In a random variable Z, values z_α such that

$$P(Z \geq z_\alpha) = \alpha$$

are called upper 100α *percentage points* of the distribution of Z; values ζ_β such that

$$P(Z \leq \zeta_\beta) = \beta$$

are called lower percentage points, and

$$\zeta_\beta = z_{1-\beta}.$$

The designation 'percentage points', without the prefix 'upper' or 'lower', usually means 'upper percentage points'.

The percentage point method is used, for example, in most commonly available tables of Student's distribution (but not in ours), and in tables of the χ^2 and F distribution (as in Appendix T6, T7).

Lower percentage points are also called *fractiles* or *quantiles*. Special cases are the lower and upper quartiles, which are the 25% and the 75% fractiles: the median of course is the 50% point.

E.H.L.

1.5. FURTHER READING

Dictionaries and encyclopedias of statistics are listed in Bibliography B; of the 'general texts' listed in Bibliography C, one that is of particular interest in the present context is Barnett (1982); and historical works are listed in Bibliography D.

CHAPTER 2

Sampling Distributions

2.1. MOMENTS AND OTHER STATISTICS

2.1.1. Statistic

As explained in Chapter 1, we adopt the viewpoint that, in attempting to describe aspects of the world that are subject to variability and uncertainty, it is sensible to work in terms of random variables and their probability distributions [see II, Chapter 4], these usually being postulated as belonging to specified families suggested or implied by the context. One of the objects of statistical inference is then to identify the particular member of the specified family of distributions in question; or, less ambitiously, to eliminate (tentatively, at least) some of the possibilities in the family; or to eliminate or confirm the postulated family as a whole. This is to be done by carrying out an appropriate analysis of the available data. It then turns out that particular combinations of the data values, each called a *statistic*, play a major rôle in the analysis. The particular combinations that deserve consideration depend on the nature of the probability distributions involved and the type of inference that is attempted.

EXAMPLE 2.1.1. *Sampling inspection.* Consider a collection (called a 'load' or 'batch') of more or less similar items consisting of individual entities that differ one from another in respect of a certain measurable or observable quality: they might for example be machined rods of nominal length 50 mm, whose individual lengths vary somewhat due to manufacturing fluctuations. It is desired to estimate the proportion of rods for which the lengths lie in a specified range, say between 49 and 51 mm. Such rods will be called *satisfactory*, while the others will be called *defective*. It is usually impracticable to examine all the rods in the load. Instead, a sample of rods may be examined, of predetermined number, say 100. The information potentially available would be an allocation of one of the labels 'satisfactory', 'defective', to each of the 100 rods examined. If, however, the sample were *random* (and precautions would normally be taken to ensure that this was so)—that is, drawn in such a way that each of the distinguishable (unordered) subsets of 100 rods had the same chance of being selected—one would not require the whole of this set of 100 pieces of information: only the total number of defectives in the sample (four, say) would be used in the subsequent analysis.

In this example, the statistic is simply the total number of defective items in the sample.

For a sample of size s taken from a batch of size b containing d defectives (where d is unknown), the number of defectives in the sample is a random variable (R, say) and, the probability of getting a specified number (r, say) of defectives in a particular sample is

$$P(R=r) = \binom{d}{r}\binom{b-d}{s-r}\bigg/\binom{b}{s}, \qquad r = 0, 1, \ldots, \min(s, d). \quad (2.1.1)$$

This is a member of the Hypergeometric family of distributions [see II, § 5.3]. The unknown parameter that identifies the particular member of the family that refers to our batch is d. Inferences about the value of d will be made on the basis of the statistic r (= four in our example), that is the total number of defectives in the sample [see Example 1.2.1].

EXAMPLE 2.1.2 (*continuation*).   Use of simplified approximating families of distributions:

If in Example 2.1.1 the batch size b were very much larger than the sample size s (e.g. $b = 10,000$, $s = 100$), the Hypergeometric distribution (2.1.1) could be replaced, with negligible error, by the Binomial [see II, § 5.2.2]:

$$P(R = r) = \binom{s}{r}p^r(1-p)^{s-r}, \qquad r = 0, 1, \ldots, s, \qquad (2.1.2)$$

$$p = d/b.$$

The relevant statistic would still be the total number of defectives in the sample.

EXAMPLE 2.1.3 (*continuation*).   In Example 2.1.1 the family of distributions involved was prescribed by the sampling procedure. Now suppose that, instead of seeking to estimate the proportion—π (49, 51), say—of rods whose lengths x lay in the specified interval $\{x : 49 \leqslant x \leqslant 51\}$, it was required to estimate the proportions $\pi(u, v)$ of rods whose lengths x belonged to the interval $\{x : u \leqslant x \leqslant v\}$ for *all* pairs of values of u and v ($u < v$). This is equivalent to the following: regard the measured length x of a particular rod as a realization of a continuous random variable X [see II, § 10.1], and estimate the probability distribution of X. This in turn might be interpreted as follows: postulate a Normal distribution with expected value μ and standard deviation σ for X [see II, § 11.4] and estimate the values of the parameters μ and σ of that distribution. [This might be thought to be an untenable postulate, since in principle arbitrarily large observed values of $|X|$ might be obtained if X

were Normally distributed, whereas the length of our rods cannot be less than zero and will not in practice be greater than, say, 60 mm. In fact however the Normal postulate might well be a very reasonable one if the standard deviation were small [see II, § 9.2 and § 11.4.3], since then the probability of very large deviations from the mean would be negligible.] In this case, given as data the lengths $x_1, x_2, \ldots, x_n$ of the rods in a sample of size n, where n is substantially smaller than the size of the batch, the appropriate statistics would be $\sum_1^n x_r$ and $\sum_1^n x_r^2$ [see § 6.4.1].

EXAMPLE 2.1.4. *A non-static situation.* Examples 2.1.1, 2.1.2 and 2.1.3 dealt with samples taken, in each case, from a fixed distribution. Such cases may be called static ones. As an example of a non-static case consider the following. A spring, fixed at one end, has a known weight x_i attached to its free end. The length y_i of the spring is then measured, and an observation obtained which is affected by errors of measurement. The procedure is repeated for $i = 1, 2, \ldots, k$. The weights $x_1, x_2, \ldots, x_k$ are accurately known constants. Because they correspond to different ('varying') weights they are often called *non-random variables*. The corresponding spring lengths y_i are subject to error. A convenient model is then, for each i, to regard y_i as a realization of a Normally distributed random variable Y_i, with expectation [see II, § 8.1] $E(Y_i) = \alpha + \beta x_i$ (Hooke's Law), and variance [see II, § 9.1] given by var $(Y_i) = \sigma^2$ (the same for all i). The object of the experiment is to estimate the elastic modulus β. Relevant statistics for this case turn out to be $\sum_1^k y_i$ and $\sum_1^k x_i y_i$ [see Example 4.5.3]. These are combinations of the observed values y_i of random variables and of certain associated non-random variables x_i.

In the light of these examples we now summarize in the form of the following definition.

DEFINITION 2.1.1. *Statistic.* Let $y_1, y_2, \ldots, y_k$ denote a set of observed values of random variables and $x_1, x_2, \ldots, x_m$ a set of (known) values of associated non-random variables. Any function of these, $h(y_1, \ldots, y_k; x_1, \ldots, x_m)$ say, whose numerical value can be computed once the sample values y_r and the associated variables x_s are given, is called a *statistic*.

Every inferential procedure must necessarily make use of a statistic, or of several statistics. In the theory of estimation, for example, one is concerned with identifying which member of a given family of probability distributions has given rise to the sample. This identification requires the allocation of a numerical value ('estimate') to each of the parameters that appears in the mathematical formula specifying the family [see Chapter 3]. Each such numerical value is the value of a statistic. Practical estimation procedures are concerned with the best ways of finding suitable statistics for this purpose.

The statistics that are thrown up by the theory of estimation and the theory of tests are often combinations of a simple system of statistics known as *sample moments*, which are sample analogues of the population moments.

2.1.2. Moments

(a) *Population moments*

An important set of constants associated with a random variable and its probability distribution is the set of (population) moments [see II, § 9.11]. The *r*th moment ($r = 1, 2, \ldots$) of the random variable X is the quantity

$$\mu'_r = E(X^r). \tag{2.1.3}$$

The first moment, μ'_1, is simply the expectation of X, often denoted by the symbol μ:

$$\mu = \mu'_1 = E(X). \tag{2.1.4}$$

Related to the moments are the *central* moments

$$\mu_r = E(X - \mu)^r_1 \qquad r = 1, 2, \ldots \tag{2.1.5}$$

The first central moment is identically zero. The second is the variance (a measure of variability). The third is related to skewness, (a measure of asymmetry), the coefficient of skewness of X being defined as

$$\text{skew}(X) = \mu_3 / \mu_2^{3/2}. \tag{2.1.6}$$

The fourth central moment μ_4 is related to the curvature of the p.d.f. near its maximum. Higher central moments do not have any direct interpretation.

Obvious generalizations are possible in multivariate distributions. For example, for a population each of whose members possess *two* attributes of interest, such as height and weight, we invoke a *bivariate* pair of random variables, (X, Y) say, realizations $(x_1, y_1), (x_2, y_2), \ldots$ of which represent the pair (height, weight) for individual members of the population. The probabilistic behaviour of X and Y is described by their joint probability distribution.

The (bivariate) moments (or product moments) of this distribution are the quantities

$$\mu'_{r,s} = E(X^r Y^s), \qquad r_1, s = 1, 2 \ldots;$$

and the central moments are defined as

$$\mu_{r,s} = E\{(x - \xi)^r (Y - \mu)^s\},$$
$$\xi = E(X), \qquad \mu = E(Y). \tag{2.1.7}$$

The most important of these 'mixed' or bivariate moments in the *covariance*, defined as

$$\text{cov}(x, y) = \mu_{1,1} = E\{(X - \xi)(y - \mu)\}.$$

The scaled version of this,

$$\rho(X, Y) = \mu_{1,1} / \sigma_X \sigma_Y. \tag{2.1.8}$$

is the *correlation coefficient* corr (X, Y), a quantity which in appropriate circumstances is a measure of the degree of association between X and Y. Here $\sigma_X^2 = \text{var}(X)$, $\sigma_Y^2 = \text{var}(Y)$.

(b) *Sample moments*

The sample analogues of population moments are the *sample moments*. For a sample $(x_1, x_2, \ldots, x_n)$ the rth sample moment is

$$m_r' = \sum_{j=1}^{n} x_j^r / n, \qquad r = 1, 2, \ldots. \tag{2.1.9}$$

Correspondingly, if the sample is given in frequency table format, that is, if $x_1, x_2, \ldots, x_k$ is a list of possible distinct observed values of X, and $f_1, f_2, \ldots, f_k$ the frequencies with which they occur in the sample, we have

$$m_r' = \sum_{j=1}^{k} f_j x_j^r / n,$$

where

$$n = \sum_{j=1}^{k} f_r$$

is the sample size.

Similarly we have *central* sample moments, also known as sample moments *about the mean*, given by

$$m_r = \sum_{j=1}^{n} (x_j - \bar{x})^r / n, \qquad r = 1, 2, \ldots \tag{2.1.10}$$

where

$$\bar{x} = m_1'$$

is the sample mean; with the corresponding version

$$m_r = \sum_{j=1}^{k} f_j (x_j - \bar{x})^r / n, \qquad r = 1, 2, \ldots \tag{2.1.11}$$

for the frequency table format.

Relation between sample moments about the mean and those about the origin: The sample moments m_r about the mean are related to the corresponding quantities m_r' about the origin by the following formulae:

$$\left. \begin{aligned} m_2 &= m_1' - \bar{x}^2, \\ m_3 &= m_3' - 3m_2'\bar{x} + 2\bar{x}^3, \\ m_4 &= m_4' - 4m_3'\bar{x} + 6m_2'\bar{x}^2 - 3\bar{x}^4. \end{aligned} \right\} \tag{2.1.12}$$

and so on.

The *r*th sample moments m_r, m'_r are possible estimates, though not necessarily very good ones, of the corresponding population moments μ_r, μ'_r.

The special case of the *second* sample moment is discussed further in the next section, (c).

(c) *Sample variance and sample standard deviation*

The second sample moment about the mean is one version of the *sample variance*. More usually however the latter is defined as

$$s^2 = nm_2/(n-1)$$

$$= \sum_{j=1}^{n} (x_j - \bar{x})^2/(n-1) \qquad (2.1.13)$$

or, equivalently, in the frequency table format, as

$$s^2 = \sum_{j=1}^{k} f_j(x_j - \bar{x})^2/(n-1), \qquad n = \sum_1^k f_j.$$

The positive square root of this, s, is the *sample standard deviation* of the sampled variable.

The idea of dividing by $n-1$ instead of n is motivated by one or several of the following arguments.

(i) Bias: s^2 is an *unbiased estimate* of the population variance σ^2; that is, the average of a large number n of sample values approaches σ^2 as n becomes arbitrarily large [see § 3.3.2]. Against this it must be recorded that s is then not an unbiased estimate of σ [see § 2.3.5].

(ii) Makes sense when $n = 1$: When $n = 1$, s^2 is undefined, which is just what is required of a sample estimator of σ^2, since with a sample of size 1 there is no information available about variability. The value of m_2, however, is zero—not a good estimate of σ^2.

(iii) Don't rock the boat: Standard estimation and test procedures, and the associated tables, use the divisor $n-1$ [see, e.g., § 2.5.5].

(iv) Degrees of freedom: The sum of squares $\sum_1^n (x_j - \bar{x})^2$ may be expressed as a sum of squares of $n-1$ algebraically independent variables: in other words the quadratic form $\sum_1^n (x_r - \bar{x})^2$ has $n-1$ degrees of freedom ($=$ rank). This makes it algebraically convenient as well as logically appealing to divide by $(n-1)$.

(d) *Bivariate samples*

In a sample (x_1, y_1), (x_2, y_2), ... (x_n, y_n) from a bivariate population, where x_r denotes the height (for example) and y_r the weight of the *r*th individual in the sample, the *sample covariance* is

$$m_{1,1} = \sum_1^n (x_r - \bar{x})(y_r - \bar{y})/n.$$

$$= \left\{ \sum_1^n x_r y_r - n\bar{x}\bar{y} \right\} \Big/ n \qquad (2.1.14)$$

where $\bar{x} = \sum_1^n x_r/n$ and $\bar{y} = \sum_1^n y_r/n$. In the frequency table format this becomes

$$m_{1,1} = \sum_1^k f_r(x_r - \bar{x})(y_r - \bar{y})/n.$$

For reasons similar to those listed in relation to sample variance, in § 2.1.2 (i), the usually adopted *estimate of population covariance* $\mu_{1,1}$ is not $m_{1,1}$ but

$$c_{1,1} = nm_{1,1}/(n-1)$$

$$= \sum_1^n (x_i - \bar{x})(y_i - \bar{y})/(n-1). \qquad (2.1.15)$$

More generally the mixed 'product-moment' of order r, s in the bivariate sample is

$$m'_{r,s} = \sum_1^n x_i^r y_i^s/n, \qquad r, s = 1, 2, \ldots \qquad (2.1.16)$$

and the corresponding central moment

$$m_{r,s} = \sum_1^n (x_i - \bar{x})^r (y_i - \bar{y})^s/n. \qquad (2.1.17)$$

The special cases $s = 0$, $r = 0$ give

$$m_{r,0} = \sum_1^n (x_i - \bar{x})^r/n, \qquad r = 1, 2, \ldots,$$

$$\qquad (2.1.18)$$

$$m_{0,s} = \sum_1^n (y_i - \bar{y})^s/n. \qquad s = 1, 2, \ldots,$$

These are, respectively, the marginal rth central moment of the x-values and the marginal sth central moment of the y-values. Obvious modifications are required for data in the frequency table format (compare (2.1.10), (2.1.11).)

Correlation coefficient

The scaled version

$$r(x, y) = m_{1,1}/\sqrt{(m_{1,0}m_{0,1})} \qquad (2.1.19)$$

of the sample covariance is called the *sample (product-moment) correlation coefficient*, or sometimes Pearson's correlation coefficient. It is an estimate of the population correlation coefficient $\rho(x, y)$ of (2.1.8) Notice that the following is equivalent to (2.1.19):

$$r(x, y) = c_{1,1}/s(x)s(y)$$

where $s^2(x)$ is the sample variance (2.1.13) of the x values and $s^2(y)$ the sample variance of the y values.

2.2. SAMPLING DISTRIBUTIONS: INTRODUCTORY DEFINITIONS AND EXAMPLES

Having chosen a particular statistic such as the sample mean (the average value of the observations) and noted its numerical value, we have to recognize that in a repetition of the sampling procedure the numerical value of the same statistic in the second sample would probably differ from that in the first. A sequence of such repetitions would generate a sequence of numerical values of the statistic, some values occurring rarely and others more frequently. We may thus conceive of a population of values, with an associated probability distribution. This is the *sampling distribution* of the statistic.

A particularly simple case occurs in Examples 2.1.1 and 2.1.2 where we consider a statistic r, the number of defective items in a sample. In Example 2.1.2 this was treated as a realization of (i.e. an observation on) a random variable R [see II, Chapter 4] whose distribution was Binomial (s, p) [see II, § 5.2.2]. Here s, the number of rods inspected, is also the number of 'trials' in the sense of the Binomial distribution, and p is the unknown proportion of defectives in the batch.

The statistic r is a realization of the random variable R. The *sampling distribution* of the statistic r is the probability distribution (2.1.2) [see II, § 4.3] of the corresponding random variable R.

In Example 2.1.3 the statistics considered were $\sum_1^n x_r$ and $\sum_1^n x_r^2$, the x_r being realizations of the random variable X which is Normal (μ, σ) [see II, § 11.4].

It is notationally convenient here to regard x_1 as a realization of a random variable X_1, x_2 as a realization of another random variable $X_2, \ldots$, and x_n as a realization of X_n, where the random variables $X_1, X_2, \ldots, X_n$ are i.i.d. (that is, mutually independent [see II, § 4.4] and identically distributed), their common distribution being that of the original random variable X. The mutual independence [see II, Definition 4.4.1] of the observed events $X = x_1$, $X = x_2, \ldots$ implied by the sampling procedure is reflected in the imputation of mutual independence to the 'carrier' random variables X_r, and the fact that all the x_r are observed values from the same distribution is reflected in our assigning to the X_r a common distribution, namely that of X. We may speak of the X_r as 'statistical copies' of X. Formally, we have the following definition:

DEFINITION 2.2.1.   *Statistical copies*; *induced random variables*; *random sample*.   Random variables $X_1, X_2, \ldots$ are said to be *statistical copies* of a given random variable X if the X_r are mutually independent and identically distributed, their common distribution being that of X. A set of independent observations $(x_1, x_2, \ldots, x_k)$ on X is called a *random sample*. We may conveniently regard x_1 as an observation on X_1, x_2 as an observation on X_2 and so on. These random variables $X_1, X_2, \ldots$ are *induced* by the observations $x_1, x_2, \ldots$. Similarly the statistic $y = h(x_1, x_2, \ldots, x_k)$ *induces* the random variable $Y = h(X_1, X_2, \ldots, X_k)$.

[For the definition of a random sample from a finite population, see, e.g. II, § 5.3.]

Thus in Example 2.1.3 the statistics $\sum_1^n x_r, \sum_1^n x_r^2$ may be regarded as realizations of the induced random variables $\sum_1^n X_r, \sum_1^n X_r^2$ respectively, where $x_1, x_2, \ldots, x_n$ are statistical copies of x. Now the random variable $\sum_1^n X_r$ is the sum of n mutually independent Normal (μ, σ) variables, and is therefore itself Normal with expectation μ and standard deviation $\sigma/\sqrt{n}$ [see § 2.5.3 (a)]. This—the Normal $(\mu, \sigma/\sqrt{n})$—is therefore the sampling distribution of the statistic $\sum_1^n x_r$ for the sample in question. Similarly the sampling distribution of the statistic $\sum_1^n x_r^2$ is the distribution of the induced random variable $\sum_1^n X_r^2$.

In Example 2.1.4 we had k mutually independent random variables $Y_1, Y_2, \ldots, Y_k$, but these no longer necessarily had a common distribution; associated with the observed values y_r of the Y_r, we had non-random variables x_r whose magnitudes could be regarded as known exactly. The statistic $\sum_1^k x_r y_r$ is regarded as a realization of the random variable $\sum_1^k x_r Y_r$, a weighted sum of the independent random variables $Y_1, Y_2, \ldots, Y_k$. The sampling distribution of the statistic $\sum x_r y_r$ is the probability distribution of the induced random variable $\sum_1^k x_r Y_r$. As it happens, in this example the Y_r are independent Normals, with $E(Y_r) = \alpha + \beta x_r$, and var $(Y_r) = \sigma^2, r = 1, 2, \ldots, k$. It follows [see § 2.5.3] that the sampling distribution of $\sum x_r y_r$ is Normal, with expectation $\alpha \sum x_r + \beta \sum x_r^2$ and variance $\sigma^2 \sum x_r^2$.

In the light of these examples we now give a formal definition of a sampling distribution.

DEFINITION 2.2.2. *Sampling distribution of a statistic.* Let $y_1, y_2, \ldots, y_n$ represent a collection of data in which, for each j, Y_j may be regarded as a realization of a random variable Y_j. Let $x_1, x_2, \ldots, x_m$ be a set of associated non-random variables whose values are known. (This may include, for instance, the sample size.) Let the statistic under consideration be

$$h(y_1, y_2, \ldots, y_n; x_1, x_2, \ldots, x_m).$$

The *sampling distribution* of the statistic is the probability distribution of the induced random variable

$$h(Y_1, Y_2, \ldots, Y_n; x_1, x_2, \ldots, x_n).$$

[Here the y_r may be scalars or vectors [see I, § 5.1, 5.2]. In the latter case the Y_r will be vector random variables [see II, § 13.3.1]. The x_r, likewise, may be scalars or vectors. The statistic h may be a scalar-valued function of vectors or it may itself be a vector. In the latter case its sampling distribution will be a multivariate probability distribution [see II, § 13.1].

Further examples follow.

EXAMPLE 2.2.1.   *The sample mean.*   Suppose $x_1, x_2, \ldots, x_n$ to be a random sample [see Definition 2.2.1] of observations from a Poisson distribution with parameter θ [see II: § 5.4]. Consider the statistic $\bar{x} = (x_1 + x_2 + \ldots + x_n)/n$. This, the sample mean, induces the random variable

$$\bar{X} = (X_1 + \ldots + X_n)/n,$$

where the X_r are i.i.d. [see § 1.4] Poisson (θ) variables. The distribution of the sum S_n of n independent Poisson (θ) variables is Poisson $(n\theta)$ [see II, § 7.2], so that

$$P(S_n = r) = e^{-n\theta}(n\theta)^r/r!, \qquad r = 0, 1, \ldots$$

whence

$$P(\bar{X} = v) = P(S_n = nv)$$
$$= e^{-n\theta}(n\theta)^{nv}/(nv)!, \qquad v = 0, 1/n, 2/n, \ldots.$$

This is the sampling distribution of the statistic $\bar{x}$.

EXAMPLE 2.2.2.   *Random sample from a bivariate distribution.* (*Sampling distribution of a vector-valued statistic.*)   Suppose $\boldsymbol{X}$ is the two-dimensional vector random variable defined by $\boldsymbol{X} = (Y, Z)$, where Y and Z are one-dimensional variables whose joint distribution is bivariate Normal [see II, § 13.4.6]   with $E(X) = \lambda$, $E(Y) = \mu$,   corr $(X, Y) = \rho$, var $(X) = \sigma^2$, var $(Y) = \omega^2$. A random sample of size n from this distribution will consist of n ordered pairs $(y_1, z_1), (y_2, z_2), \ldots, (y_n, z_n)$, these being independent realizations of the pair (Y, Z). The statistic $\bar{x} = (\bar{y}, \bar{z})$, where $\bar{y} = \sum_1^n y_1/n$ and $\bar{z} = \sum_1^n z_1/n$, is of interest in connection with inferences concerning λ and μ. To discuss the sampling distribution of the vector-valued statistic $\bar{x} = (\bar{y}, \bar{z})$ we introduce the induced 2-dimensional random variables $\boldsymbol{X}_1 = (Y_1, Z_1)$, $\boldsymbol{X}_2 = (Y_2, Z_2), \ldots, \boldsymbol{X}_n = (Y_n, Z_n)$, which are statistical copies of $\boldsymbol{X}$ in the sense that the $\boldsymbol{X}_r$ are mutually independent, and, for each r, $\boldsymbol{X}_r$ has the same distribution as $\boldsymbol{X}$. Thus, the pairs (Y_j, Z_j) and (Y_k, Z_k) are mutually independent whenever $j \neq k$, and, for each j, Y_j and Z_j have the same bivariate Normal distribution as Y and Z. It follows that the induced random variable $\bar{\boldsymbol{X}} = (\bar{Y}, \bar{Z})$ has the bivariate Normal distribution with $E(\bar{Y}) = \lambda$, $E(\bar{Z}) = \mu$,   corr $(\bar{Y}, \bar{Z}) = \rho$, var $(\bar{Y}) = \sigma^2/n$, var $(\bar{Z}) = \omega^2/n$. This is the sampling distribution of $\bar{x} = (\bar{y}, \bar{z})$.

'Artificial' samples: simulation

Published collections of independent realizations of random variables with specified distribution provide a means for producing small 'simulated' random samples. When samples of substantial size are required this procedure is inefficient, and computer-generation of realizations ('compute simulation') is preferable.

Published collections of 'random numbers' belonging to various distributions are exemplified in Bibliography F: For the uniform distribution, see (e.g.) The RAND Corporation (1955); for the Normal distribution, see Wold (1954); for Bivariate Normal pairs, see Fieller, Lewis and Pearson (1957); and for the Exponential distribution, see Clark and Holtz (1960) or Barnett (1965). Since sums of exponentials have a gamma distribution, simulated samples of gamma (and therefore of χ^2) variables may be obtained from the exponential 'random numbers'.

For information on generating random realizations see Newman and Odell (1971), (also in Bibliography F) or section 26.8 of Abramowitz and Stegun (Bibliography G).

2.3. SAMPLING MOMENTS OF A STATISTIC

The moments [see § 2.1.2] of the sampling distribution [see Definition 2.2.2] of a statistic are called the sampling moments of that statistic; similarly for the central sampling moments. (It should particularly be noted that sampling moments are not the same as sample moments: see section 2.1.2 (b).)

DEFINITION 2.3.1. *Sampling moments.* The rth sampling moment of a statistic t is the rth moment of the sampling distribution of t ($r = 1, 2, \ldots$). Equivalently, the rth sampling moment of t is

$$E(T^r), \qquad r = 1, 2, \ldots$$

where T is the random variable induced [see § 1.4.2 (v)] by t.

The rth *central* sampling moment of t is

$$E(T - \tau)^r, \quad r = 1, 2, \ldots,$$

where $\tau = E(T)$.

The first sampling moment is called the sampling expectation, the second central sampling moment is called the sampling variance, and so on, in conformity with the general usage for population moments. Thus we may speak of the sampling expectation of the sample mean: this is the expectation of the sampling distribution of $\bar{x}$.

The standard deviation is not a moment but it is a very important related statistical quantity.

DEFINITION 2.3.2. *Sampling standard deviation. Standard error.* The sampling standard deviation of a statistic t is the standard deviation (i.e. the positive square root of the variance) of the sampling distribution of t.

An appropriate *estimate* of the sampling standard deviation of t is called the *standard error* of t. [see § 4.1.2.]

For example, if t is the sample mean $\bar{x}$ of a sample of size n from a distribution having variance σ^2, the sampling variance of $\bar{x}$ is σ^2/n [see (2.3.1)], and the sampling standard deviation of $\bar{x}$ therefore $\sigma/\sqrt{n}$.

The sample standard deviation [see (2.5.23)] is a statistic, which has its own sampling distribution, and has therefore a sampling variance and a sampling standard deviation. This sampling distribution, for the case of a Normal sample, in discussed in section 2.5.4(e).

It is sampling moments that form the subject matter of the present section. Of particular interest are the sampling expectation, the sampling variance, and the sampling skewness; these are, respectively, the expectation [see II: § 8.1], the variance [see II, § 9.2.1] and the skewness [see II, § 9.10.1] of the sampling distribution of the statistic in question.

2.3.1. Lower Sampling Moments of the Sample Mean

Let $x_1, x_2, \ldots, x_n$ be a sample of n independent observations on a random variable X. The sample mean $\bar{x}$ is defined as $(x_1 + x_2 + \ldots + x_n)/n$, and its sampling distribution is the distribution of the induced random variable $\bar{X} = (X_1 + X_2 + \ldots + X_n)/n$, where the X_r are statistical copies of X [see Definition 2.2.1]. Straightforward calculations then show that

$$E(\bar{X}) = E(X),$$

$$\text{var}\,(\bar{X}) = n^{-1}\,\text{var}\,(X),$$

$$\text{skew}\,(\bar{X}) = n^{-1/2}\,\text{skew}\,(X).$$

(2.3.1)

The statistic $\bar{x}$ has strong intuitive claims to serve as an estimate [see § 1.3.2] of the parameter $E(X)$: its sampling expectation (that is, its long-term average in repetitions of the sampling procedure) is identically equal to the parameter in equation, a property called *unbiasedness* [see § 3.3.2] and its sampling variance diminishes with increasing sample size, whence, by the Chebychev inequality [see II, § 9.5], in a sufficiently large sample it is highly probable that $\bar{x}$ will be very close to $E(X)$.

2.3.2. Lower Sampling Moments of the Sample Variance

With the notation of section 2.3.1, the sample variance v is sometimes defined as $\sum_1^n (x_i - \bar{x})^2/n$, sometimes as $\sum_1^n (x_i - \bar{x})^2/(n-1)$ [see § 2.1.2(c), 2.5.4(d)]. In the present section we initially use the former definition. The first sampling moment, and the second and third *central* sampling moments of v, are, respectively,

$$E(V),\qquad \text{var}\,(V),\qquad E\{V - E(V)\}^3,$$

where V is the random variable induced by v, (so that v is a realization of V):

$$V = \sum_1^n (X_i - \bar{X})^2/n,$$

$\bar{X}$ and the X_i being as defined in section 2.3.1. Similarly the sampling skewness of the sample variance v is

$$\text{skew}(V) = E\{(V - E(V))\}^3/\{\text{var}(V)\}^{3/2}.$$

The evaluation of $E(V)$ is fairly straightforward, but the second and third central moments are more demanding. The results are given below, expressed in terms of the central moments [see II, § 9.11] μ_r of X; that is, in terms of

$$
\begin{aligned}
\mu_2 &= \text{var}(X) \\
&= E(X - \mu)^2 \quad &&\text{(where } \mu = E(X)) \\
\mu_3 &= E(X - \mu)^3, \quad &&(= \mu_2^{3/2}\,\text{skew}(X)) \\
\mu_4 &= E(X - \mu)^4, \quad &&\text{etc.}
\end{aligned}
\tag{2.3.2}
$$

It is found that

$$
\begin{aligned}
E(V) &= \frac{n-1}{n}\mu_2, \\[2mm]
\text{var}(V) &= \frac{\mu_4 - \mu_2^2}{n} - \frac{2(\mu_4 - 2\mu_2^2)}{n^2} + \frac{\mu_4 - 3\mu_2^2}{n^3} \\[2mm]
&= \frac{\mu_4 - \mu_2^2}{n} + O\!\left(\frac{1}{n^2}\right), \\[2mm]
E\{V - E(V)\}^3 &= \frac{\mu_6 - 3\mu_2\mu_4 - 6\mu_3^2 + 2\mu_2^3}{n^2} + O\!\left(\frac{1}{n^3}\right), \\[2mm]
\text{skew}(V) &= \frac{\mu_6 - 3\mu_2\mu_4 - 6\mu_3^2 + 2\mu_2^3}{(\mu_4 - \mu_2^2)^{3/2}n^{1/2}} + O\!\left(\frac{1}{n^{3/2}}\right).
\end{aligned}
\tag{2.3.3}
$$

(For the 'O' notation, see § 1.4.1.)

If follows that $v' = nv/(n-1) = \sum_1^n (x_i - \bar{x})^2/(n-1)$ is an unbiased [see § 3.3.2] estimate of the variance μ_2 of the sampled population, and that the sampling variance of v', like that of v, is $n^{-1}(\mu_4 - \mu_2^2) + O(n^{-2})$, a quantity that diminishes with increasing n.

Further details of these results, and of those given in §§ 2.3.3–2.3.6 will be found in Cramér (1946)—Bibliography C.

2.3.3. Sampling covariance between the Sample Mean x̄ and the Sample Variance v

The sampling covariance [see II, § 9.6.1] between $\bar{x}$ and v is

$$\text{cov}(\bar{X}, V) = \frac{n-1}{n}\mu_3. \tag{2.3.4}$$

In particular, $\bar{X}$ and V are uncorrelated in the case where the distribution of X is symmetrical, since then $\mu_3 = 0$.

2.3.4   Sampling Moments of Higher Sample Moments

Detailed calculations become very elaborate for the sample moments $m_k = \sum_1^n (x_i - \bar{x})^k / n$, for large values of k. For $k = 3$ the sampling expectation is given by

$$E\left\{\sum_1^n (X_i - \bar{X})^3 / n\right\} = (n-1)(n-2)\mu_3 / n^2. \tag{2.3.5}$$

Generally, for $k = 2, 3, \ldots,$

$$E\left\{\sum_1^n (X_i - \bar{X})^k / n\right\} = \mu_k + O(n^{-1}) \tag{2.3.6}$$

and

$$\mathrm{var}\left\{\sum_1^n (X_i - \bar{X})^k / n\right\} = c(k, n)/n + O(n^{-2}), \tag{2.3.7}$$

where

$$c(k, n) = \mu_{2k} - 2k\mu_{k-1}\mu_{k+1} - \mu_k^2 + k^2\mu_2\mu_{k-1}^2.$$

2.3.5.   Sampling Moments of the Sample Standard Deviation

Defining the sample standard deviation as $s = \sqrt{\{\sum_1^n (x_i - \bar{x})^2 / n\}}$ we have

$$E(\sqrt{V}) = \sqrt{\mu_2} + O(n^{-1}) \tag{2.3.8}$$

and

$$\mathrm{var}(\sqrt{V}) = \frac{\mu_4 - \mu_2^2}{4\mu_2 n} + O(n^{-2}). \tag{2.3.9}$$

2.3.6.   Sampling Moments of the Sample Skewness

In conformity with the definition

$$\mathrm{skew}(X) = E(X - \mu)^3 / \{\mathrm{var}(X)\}^{3/2} \qquad (\text{where } \mu = E(X))$$

for the skewness of the random variable X, we define the sample skewness of a sample $(x_1, x_2, \ldots, x_n)$ of observations on X as

$$g = m_3 / m_2^{3/2} \qquad (= \sqrt{\{m_3^2 / m_2^3\}}) \tag{2.3.10}$$

where

$$m_3 = \sum_1^n (x_1 - \bar{x})^3 / n, \qquad m_2 = \sum_1^n (x_1 - \bar{x})^2 / n.$$

It then turns out that the sampling expectation of g is

$$\text{skew}(X) + O(n^{-1}),\qquad\qquad (2.3.11)$$

and the sampling variance of g is

$$d(n)/n + O(n^{-3/2}),\qquad\qquad (2.3.12)$$

where $d(n)$ is defined by

$$4\mu_2^5 d(n) = 4\mu_2^2\mu_6 - 12\mu_2\mu_3\mu_5 - 24\mu_2^3\mu_4$$
$$+ 9\mu_3^2\mu_4 + 35\mu_2^2\mu_3^2 + 36\mu_2^5.\qquad (2.3.13)$$

When the distribution of X is symmetrical this last expression reduces to

$$4\mu_2^2\mu_6 - 24\mu_2^3\mu_4 + 36\mu_2^5.\qquad\qquad (2.3.14)$$

2.4. DISTRIBUTIONS OF SUMS OF INDEPENDENT IDENTICALLY DISTRIBUTED VARIABLES

The statistic $\sum_1^n x_r$ is of widespread occurrence. Its sampling distribution is, of course, the distribution of the sum $S_n = \sum_1^n X_r$ of the i.i.d. induced random variables $X_1, \ldots, X_n$, which we define as statistical copies of the sampled random variable X [see Definition 2.2.1]. We list the distributions of $\sum X_r$ for various random variables X in Tables 2.4.1 and 2.4.2 (see pages 34 and 35).

2.5. SAMPLING DISTRIBUTIONS OF FUNCTIONS OF NORMAL VARIABLES

A large body of statistical theory is based on the behaviour of samples from Normal distributions. In this section we summarize some of the basic properties of Normal variables and of related statistics. Further details may be found in (for example) Hogg and Craig (1965) Chapter 4, 13; Kendall and Stuart (1969) Volume 1; Mood, Graybill and Boes (1974), Chapter V, VI; and Wilks (1961) Chapter 8; all in Bibliography C.

2.5.1. The Normal Distribution [see II, § 11.4]

A random variable X is said to have the Normal (μ, σ) distribution, or the Normal distribution with parameters (μ, σ)—that is with expectation μ and standard deviation σ—if its probability density function (p.d.f.) at x [see II, § 10.1] is

$$f(x) = (2\pi)^{-1/2}\sigma^{-1}\exp\{-(x-\mu)^2/2\sigma^2\}.\qquad (2.5.1)$$

This is illustrated in Figure 3.5.2.

Distribution of X	Distribution of S_n
Description $f(x) = P(X = x)$	Description $f_n(x) = P(S_n = x)$
Bernoulli	Binomial (n, p)
$\left.\begin{array}{l} f(0) = 1 - p \\ f(1) = p \end{array}\right\} 0 < p < 1,$	$\binom{n}{x} p^x (1-p)^{n-x}, \quad x = 0, 1, \ldots, n$
Binomial (k, p)	Binomial (nk, p)
$\binom{k}{x} p^x (1-p)^{k-x}, \quad 0 < p < 1,$ $x = 0, 1, \ldots, k$	$\binom{nk}{x} p^x (1-p)^{nk-x}, \quad x = 0, 1, \ldots, nk$
Poisson (θ)	Poisson $(n\theta)$
$e^{-\theta} \theta^x / x!, \quad \theta > 0, \quad x = 0, 1, 2, \ldots$	$e^{-n\theta} (n\theta)^x / x!, \quad x = 0, 1, 2, \ldots$
Geometric	Negative Binomial (Pascal)
$p(1-p)^x, \quad 0 < p < 1, \quad x = 0, 1, 2, \ldots$	$\binom{-n}{x} p^n (1-p)^x, \quad x = 0, 1, 2, \ldots$
Negative binomial	Negative binomial
$\binom{-\alpha}{x} p^\alpha (1-p)^x, \quad \alpha > 0, \quad 0 < p < 1,$ $x = 0, 1, 2, \ldots$	$\binom{-n\alpha}{x} p^{n\alpha} (1-p)^x, \quad x = 0, 1, 2, \ldots$

Note: for $\alpha > 0$ and $x = 0, 1, \ldots,$

$$\binom{-\alpha}{x} = \underbrace{(-\alpha)(-\alpha - 1) \ldots (-\alpha - x + 1)/x!}_{x \text{ factors}}$$

$$= \text{coefficient of } t^x \text{ in the power series expansion of } (1 + t)^{-\alpha}.$$

Table 2.4.1: Distribution of sums of statistical copies of X (*X discrete*). [For details of these distributions see II, Chapter 5.]

2.5.2.   Effect of a Linear Transformation; Standardization

If

$$Y = aX + b$$

where X is Normal (μ, σ), and a $(a \neq 0)$ and b are given constants, Y is also Normally distributed, but with parameters $(a\mu + b, \sigma|a|)$. This result follows from an application of Theorem 10.7.1 of Volume II.

Distribution of X	Distribution of S_n
Description $f(x) =$ p.d.f. of X at x	Description $f_n(x) =$ p.d.f. of S_n at x

Exponential (a) $a^{-1} e^{-x}, \quad x > 0$	Erlang with n stages $=$ Gamma (n, a) $a^{-n} x^{n-1} e^{-x/a}/(n-1)!, \quad x > 0$ (Equivalently, if $a = 1, 2S_n$ has the χ^2 distribution on $2n$ d.f.) [see § 2.5.41a).]
Gamma (α, β) $\dfrac{x^{\alpha-1} e^{-x/\beta}}{\beta^{\alpha} \Gamma(\alpha)}, \quad x > 0$	Gamma $(n\alpha, \beta)$ $\dfrac{x^{n\alpha-1} e^{-x/\beta}}{\beta^{n\alpha} \Gamma(n\alpha)}, \quad x > 0$
Normal (μ, σ) $(2\pi)^{-1/2} \sigma^{-1} \exp\{-(x-\mu)^2/2\sigma^2\}$	Normal $(n\mu, \sigma\sqrt{n})$ $(2\pi n)^{-1/2} \sigma^{-1} \exp\{-(x-n\mu)^2/2n\sigma^2\}$

Uniform $(0, 1)$

$$f(x) = \begin{cases} 1, & 0 < x < 1 \\ 0, & \text{otherwise} \end{cases}$$

$$(n-1)! f_n(x) = \begin{cases} g_{n,1}(x), & 0 < x < 1 \\ g_{n,1}(x) + g_{n,2}(x), & 1 < x < 2 \\ \quad\vdots \\ g_{n,1}(x) + \ldots + g_{n,r}(x), & \\ \qquad\qquad r-1 < x < r \\ \quad\vdots \\ g_{n,1}(x) + \ldots + g_{n,n}(x), & \\ \qquad\qquad n-1 < x < n, \end{cases}$$

where

$$g_{n,r+1}(x) = (-1)^{r+2} \binom{n}{r} (x-r)^{n-1},$$

$$r = 0, 1, \ldots, n-1.$$

Table 2.4.2: Distribution of sums of statistical copies of X, (X continuous). [For details of these distributions see II, Chapter 11.]

If X is Normal (μ, σ), the particular linear transformation

$$U = (X - \mu)/\sigma \tag{2.5.2}$$

defines a r.v. U that is Normal $(0, 1)$. This is the *standard* Normal variable. Its density function at u, usually denoted by $\phi(u)$, is

$$\phi(u) = (2\pi)^{-1/2} \exp\{-\tfrac{1}{2}u^2\}, \qquad -\infty < u < \infty. \tag{2.5.3}$$

Values of $\phi(u)$ are widely tabulated, as are values of the '*standard Normal integral*', that is the cumulative distribution function (c.d.f.) $\Phi(u)$, given by

$$\Phi(u) = P(U \le u)$$

$$= \int_{-\infty}^{u} \phi(z)\, dz. \tag{2.5.4}$$

A table of this function is given in the Appendix (T3).

With the aid of section 2.5.1, tables of $\Phi(u)$ can be utilized to provide values of the c.d.f. of any Normal variable. If X is Normal (μ, σ) we write

$$X = \sigma U + \mu$$

whence

$$P(X \leq x) = P(\sigma U + \mu \leq x)$$
$$= P\{U \leq (x - \mu)/\sigma\} \quad \text{(since } \sigma > 0)$$
$$= \Phi\{(x - \mu)/\sigma\}. \quad (2.5.5)$$

Thus, in terms of Φ, the probability that X lies in a stated interval (x_1, x_2) is given by

$$P(x_1 \leq X \leq x_2) = P(X \leq x_2) - P(X \leq x_1)$$
$$= \Phi\{(x_2 - \mu)/\sigma\} - \Phi\{(x_1 - \mu)/\sigma\}. \quad (2.5.6)$$

2.5.3. Linear Functions of Normal Variables

(a) *Linear functions of independent Normals*

Let

$$Y = a_1 X_1 + a_2 X_2 + \ldots + a_n X_n + b$$

where the X_r are independent Normals, the parameters of X_r being (μ_r, σ_r), $r = 1, 2, \ldots, n$. Then Y is Normal (λ, ω), where

$$\left.
\begin{aligned}
\lambda &= a_1 \mu_1 + a_2 \mu_2 + \ldots + a_n \mu_n + b \\
\text{and} & \\
\omega^2 &= a_1^2 \sigma_1^2 + a_2^2 \sigma_2^2 + \ldots + a_n^2 \sigma_n^2.
\end{aligned}
\right\} \quad (2.5.7)$$

[see II, § 11.4.5].

It is a consequence of the Central Limit Theorem [see II, 11.4.2] that Y is *approximately* Normal (λ, ω) even when the X_r are not themselves Normally distributed.

The most important application of the result (2.5.7) occurs when the X_r are all Normal (μ, σ). Then

$$\left.
\begin{aligned}
\lambda &= (a_1 + \ldots + a_n)\mu + b \\
\text{and} & \\
\omega^2 &= (a_1^2 + \ldots + a_n^2)\sigma^2.
\end{aligned}
\right\} \quad (2.5.8)$$

(b) *Sampling distribution of the sample mean*

A particular case of this last result gives the distribution of the arithmetic mean $\bar{X}$ of the X_r. If

$$\bar{X} = (X_1 + X_2 + \ldots + X_n)/n,$$

the distribution of $\bar{X}$ is Normal $(\mu, \sigma/\sqrt{n})$. This important result gives the sampling distribution of the sample mean $\bar{x}$ of a sample of n observations on a random variable X that is Normal (μ, σ).

In view of the importance of the sample mean in estimation theory the remark that follows (2.5.7), relating to approximate Normality, acquires particular relevance here: that is, the distribution of $\bar{X}$ is approximately Normal $(\mu, \sigma/\sqrt{n})$ regardless (within wide limits) of the actual distribution of the X_r [see II, § 17.3].

(c) *Linear function of correlated Normals*

Suppose that $X_1, X_2, \ldots, X_n$ are jointly distributed in the multivariate Normal distribution [see II, § 13.4] in which $E(X_r) = \mu_r$ and var $(X_r) = \sigma_r^2$, $r = 1, 2, \ldots, n$, and corr $(X_r, X_s) = \rho_{rs}$, $r, s = 1, 2, \ldots, n$; then

$$Y = a_1 X_1 + a_2 X_2 + \ldots + a_n X_n + b$$

is Normally distributed with parameters (λ, ω), where

$$\lambda = a_1 \mu_1 + a_2 \mu_2 + \ldots + a_n \mu_n + b,$$

$$\omega^2 = \sum_i \sum_j a_i a_j \rho_{ij} \sigma_i \sigma_j \qquad (2.5.9)$$

$$= \sum_i a_i^2 \sigma_i^2 + 2 \sum_i \sum_{\substack{j \\ i<j}} a_i a_j \rho_{ij} \sigma_i \sigma_j$$

$$= \mathbf{a'Va}$$

where $\mathbf{a'} = (a_1, a_2, \ldots, a_n)$, and $\mathbf{V} = (\rho_{rs}\sigma_r\sigma_s)$ is the dispersion matrix of the X's. [For information about vectors and matrices, see I, Chapters 5 and 6.]

(d) *Several linear functions of correlated Normals*

Suppose that $X_1, X_2, \ldots, X_n$ are jointly multivariate Normal as in section 2.5.3(c). The basic result on linear functions of the X_r is as follows: let

$$Y_r = a_{r1} X_1 + a_{r2} X_2 + \ldots + a_{rn} X_n + b_r, \qquad r = 1, 2, \ldots, n \quad (2.5.10)$$

where the matrix of coefficients $\mathbf{A} = (a_{rs})$ is non-singular [see I, Definition 6.4.2]. Then the Y_r are themselves multivariate Normal, with

$$E(Y_r) = a_{r1} \mu_1 + a_{r2} \mu_2 + \ldots + a_{rn} \mu_n + b_r, \qquad r = 1, 2, \ldots, n \quad (2.5.11)$$

and with dispersion matrix [see II, Definition 9.6.3] given by

$$\mathbf{AVA'}, \tag{2.5.12}$$

where $\mathbf{A'}$ is the transpose [see I, § 6.5] of $\mathbf{A}$, and $\mathbf{V}$ is the dispersion matrix of the X_r; that is $\mathbf{V}$ is the matrix (v_{rs}) where

$$v_{rs} = \rho_{rs}\sigma_r\sigma_s, \qquad r, s = 1, 2, \ldots, n,$$

with $\rho_{rs} = \rho_{sr}$ for all r and s, and $\rho_{rr} = 1$ for each r.

If, instead of having a full set $Y_1, Y_2, \ldots, Y_n$ of n linearly independent linear functions of $X_1, X_2, \ldots, X_n$, we are concerned only with a subset $Y_1, Y_2, \ldots, Y_k$ $(k < n)$ of linearly independent linear functions, the members of this subset will still have the multivariate (k-variate) Normal distribution, with expectations as in (2.5.11). Their dispersion matrix will be the leading $(k \times k)$ submatrix [see I, § 6.13] of the matrix $\mathbf{AVA'}$ of (2.5.12).

(e) *Independent linear functions of correlated Normals*

Let $X_1, X_2, \ldots, X_n$ be multivariate Normal as in section 2.5.3(c) and let their dispersion matrix $\mathbf{V}$ be factorized [see I, § 6.1] as

$$\mathbf{V} = \mathbf{SS'},$$

where $\mathbf{S}$ is a non-singular matrix [see I, Definition 6.4.2].

Let the matrix $\mathbf{A}$ of section 2.5.3(d) be

$$\mathbf{A} = \mathbf{S}^{-1}$$

[see I, § 6.4] then the Y_r, defined by the linear transformation (2.5.11), are multivariate Normal with dispersion matrix

$$\mathbf{AVA'} = \mathbf{S}^{-1}(\mathbf{SS'})(\mathbf{S}^{-1})' = \mathbf{I},$$

where $\mathbf{I}$ is the unit matrix [see I, § 6.2 (vii)]. It follows that in this case the Y_r are mutually independent standard Normal variables.

(f) *Independent linear functions of i.i.d. Normals*

The case where the X_r are mutually independent can be derived from sections 2.5.3 (c), (d) and (e) by taking $\mathbf{V}$ to be a diagonal matrix [see I, § 6.7(iv)]. The most important special case of this kind occurs when the X_r are identically distributed independent Normals, with common expectation μ, say, and variance σ^2, so that the dispersion matrix is $\mathbf{V} = \sigma^2\mathbf{I}$. Then, if the linear transforms $Y_1, Y_2, \ldots, Y_n$ are defined by (2.5.10) where the matrix $\mathbf{A}$ is orthogonal, the dispersion matrix of the Y_r is $\mathbf{AA'}\sigma^2 = \mathbf{I}\sigma^2$, so that the Y_r are *mutually independent standard Normals*.

More generally, if the rows of the matrix $\mathbf{A}$ are mutually orthogonal [see I, § 10.2], without necessarily being orthonormal, so that the product $\mathbf{AA'}$ is diagonal [see I, § 6.7(iv)]

$$\mathbf{AA}' = \mathrm{diag}\,(b_1^2, b_2^2, \ldots, b_n^2),$$

the Y_r will be multivariate Normal with dispersion matrix

$$\sigma^2 \,\mathrm{diag}\,(b_1^2, b_2^2, \ldots, b_n^2),$$

that is the Y_r will be mutually independent Normals with expectations given by (2.5.11) and with variances given by

$$\mathrm{var}\,(Y_r) = b_r^2 \sigma^2, \qquad r = 1, 2, \ldots n.$$

In particular, any k linear functions ($k \leq n$)

$$Y_j = a_{j1}X_1 + \ldots + a_{jn}X_n + b_j, \qquad j = 1, 2, \ldots, k, \qquad (2.5.13)$$

of the i.i.d. variables $X_1, X_2, \ldots, X_n$, whose common distribution is Normal (μ, σ), will be mutually independent Normals provided

$$\sum_r a_{ir}a_{jr} = 0, \qquad i \neq j.$$

Then $E(Y_1)$, $E(Y_2)$ etc. are as given in (2.5.11), and the variances are

$$\mathrm{var}\,(Y_1) = \sigma^2 \sum a_{1r}^2,$$
$$\mathrm{var}\,(Y_2) = \sigma^2 \sum a_{2r}^2, \quad \text{etc.} \qquad (2.5.14)$$

2.5.4.   Quadratic Functions of Normal Variables

a) *The chi-squared distribution. Sums of squares of independent standard Normals. Quadratic forms in Normal variables*

One of the most important classes of quadratic functions in sampling theory is the class of functions reducible to sums of squares of independent standard Normals. Let $U_1, U_2, \ldots, U_\nu$ be independent standard Normal variables [see I, § 11.4.1], and let

$$K_\nu = U_1^2 + U_2^2 + \ldots + U_\nu^2. \qquad (2.5.15)$$

This is called a chi-squared (χ^2) variable [see II, § 11.4.11] with ν degrees of freedom ('d.f.'), a statement usually abbreviated to 'a $\chi^2(\nu)$ variable', or 'a χ_ν^2 variable'. The p.d.f. of K_ν at z is

$$f_\nu(z) = z^{(\nu/2)-1}\, e^{-z/2}/2^{\nu/2}\Gamma(\tfrac{1}{2}\nu), \qquad z > 0, \qquad (2.5.16)$$

unimodal distribution [see II, § 10.1.3] with its maximum at $z = \nu - 2$ [see Figure 2.5.1], with expectation, variance and skewness given by

$$E(K_\nu) = \nu, \qquad \mathrm{var}\,(K_\nu) = 2\nu, \qquad \mathrm{skew}\,(K_\nu) = 2\sqrt{2}/\sqrt{\nu}.$$

Note on notation: Let it be freely admitted that the notation K_ν used above is a highly pedantic insistence on using a capital Latin letter to denote a random

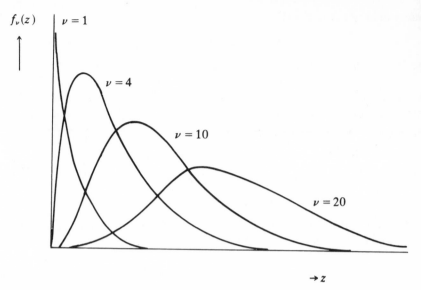

Figure 2.5.1: The χ_ν^2 p.d.f. $f_\nu(z)$ of (2.5.16) for various values of the degrees-of-freedom parameter ν_1. (Reproduced by permission of John Wiley & Sons Inc. from *Statistical Theory with Engineering Applications* by A. Hald. © 1952 John Wiley & Sons Inc.)

variable. In practice the usage is always χ^2 or $\chi^2(\nu)$, or χ_ν^2. The same symbol, χ^2, is also commonly used to denote a realization or a particular value of K_ν. The context has to be relied upon to avoid ambiguity.

Thus if Z is a gamma variable with unit scale parameter, and with shape parameter α, $2Z$ is a χ^2 variable with d.f. $\nu = 2\alpha$, or, equivalently, 'a gamma (α) variable is a $\frac{1}{2}\chi^2(\nu)$ variable, with $\nu = 2\alpha$'.

The special case $\nu = 2$ is worth noting. The p.d.f. is

$$f_2(z) = \tfrac{1}{2}e^{-z/2}, \qquad z > 0.$$

Thus the χ_2^2 distribution is the exponential distribution with expectation equal to 2.

Additive property of χ^2 variables:

A sum of independent χ^2 variables is a χ^2 variable.

One important and useful property of the χ^2 family is the fact that it is closed under addition: as is obvious from (2.5.15), we have the following addition rule:

If C_1 and C_2 are independent χ^2 variables with m, n degrees of freedom (d.f.) respectively, $C_1 + C_2$ is a χ^2 variable with $m + n$ d.f.

The rule extends to sums of more than two variables.

A convenient usage: The $k\chi^2_\nu$ variable

It frequently happens that we are concerned with a random variable Z such that Z/k has a χ^2_ν distribution. In this case we say that Z is a '$k\chi^2_\nu$' variable.

Quadratic forms having the chi-squared distribution.

The definition given in (2.5.15) may be reformulated as follows. Let $\mathbf{u}' = (U_1, U_2, \ldots, U_\nu)$. Then [see I, § 9.1] $\mathbf{u}'\mathbf{u} = \sum_1^\nu U_r^2$ has the χ^2 distribution on ν degrees of freedom (d.f.).

It is well-known that quadratic forms which are not immediately expressible as sums of squares may, by suitable transformations, be reduced to sums of squares of transformed variables [see I, § 9.1]. It is natural, therefore, to ask whether such forms may have a χ^2 distribution. The main result of such enquiries, which has many applications, is contained in the following theorem:

THEOREM 2.5.1. *Necessary and sufficient conditions for a quadratic form in independent standard Normals to have a chi-squared distribution. Let* $\mathbf{u}' = (U_1, U_2, \ldots, U_k)$, *where the* U_r *are independent standard Normal variables. Let* $\mathbf{A} = \mathbf{A}'$ *denote a symmetric matrix* [*see* I, § 6.7(v)] *with real constants as entries. Then the non-negative quadratic form* $\mathbf{u}'\mathbf{Au}$ *has a chi-squared distribution if, and only if,* $\mathbf{A}^2 = \mathbf{A}$. *In this case the number of degrees of freedom equals* rank $(\mathbf{A})$ = trace $(\mathbf{A})$ [*for rank see* I, § 5.6; *for trace see* I, § 6.2(vii)].

EXAMPLE 2.5.1. *Sampling distribution of 'sample sum of squares'.* The simple sum of squares of standard Normals rarely occurs as a statistic, but a related statistic is of common occurrence and is very important. This is the sum of squared deviations of observations $x_1, x_2, \ldots, x_n$ from their sample mean $\bar{x}$. This quantity

$$d^2 = \sum_1^n (x_i - \bar{x})^2$$

is frequently called, simply, the *sample sum of squares*. When the observations x_r form a sample from a Normal (μ, σ) population, the distribution of d^2/σ^2 is that of chi-squared on $n-1$ degrees of freedom.

To see how this comes about, we regard $x_1, x_2, \ldots, x_n$ as realizations of induced random variables $X_1, X_2, \ldots, X_n$ respectively, where the X_r are statistical copies of X, that is, they are mutually independent Normal (μ, σ) variables; and $\bar{x}$ is similarly regarded as a realization of the induced r.v. $\bar{X} = \sum_1^n X_r/n$.

The r.v. induced [see Definition 2.2.1.] by d^2 is then

$$D^2 = \sum_1^n (X_i - \bar{X})^2.$$

Then $U_i = (X_i - \mu)/\sigma$ is a standard Normal variable, $i = 1, 2, \ldots, n$; and $\bar{U} = \sum_1^n U_i/n = (\bar{X} - \mu)/\sigma$, whence

$$D^2/\sigma^2 = \sum_1^n (U_r - \bar{U})^2.$$

Although the variables $U_1, U_2, \ldots, U_n$ are mutually independent, the variables $U_1 - \bar{U}, U_2 - \bar{U}, \ldots, U_n - \bar{U}$ are not, since they all involve $\bar{U} = \sum_1^n U_r/n$. In terms of the vector $\mathbf{u}' = (U_1, U_2, \ldots, U_n)$ we have

$$D^2/\sigma^2 = \sum_1^n U_r^2 - n\bar{U}^2$$

$$= \mathbf{u}'\mathbf{u} - n(\mathbf{u}'\mathbf{1}/n)^2 \quad \text{where } \mathbf{1} = (1, 1, \ldots, 1)'$$

$$= \mathbf{u}'\mathbf{u} - n(\mathbf{u}'\mathbf{1})(\mathbf{1}'\mathbf{u})/n^2$$

$$= \mathbf{u}'A\mathbf{u},$$

where

$$\mathbf{A} = \mathbf{I} - \mathbf{11}'/n.$$

On squaring, it is seen that $\mathbf{A} = \mathbf{A}^2$; and, by inspection of the diagonal entries of $\mathbf{A}$, namely $(1 - 1/n, 1 - 1/n, \ldots, 1 - 1/n)$, we see that trace $(\mathbf{A}) = n - 1$, whence it follows from Theorem 2.5.1 that D^2/σ has the χ^2 distribution, with $n - 1$ degrees of freedom.

(b) *Independence of the sample sum of squares and the sample mean in Normal samples*

The result discussed in Example 2.5.1 is part of the following theorem.

THEOREM 2.5.2.   *Orthogonal resolution of $\sum (X_r - \mu)^2$. Let $X_1, X_2, \ldots, X_n$ be independent Normal (μ, σ) variables and let $\bar{X} = \sum_1^n X_r/n$. Then*

$$\sum (X_r - \mu)^2/\sigma^2 = \sum (X_r - \bar{X})^2/\sigma^2 + n(\bar{X} - \mu)^2/\sigma^2,$$

and the two terms on the right are mutually independent χ^2 variables, with degrees of freedom $n - 1$ and 1 respectively.

This is a special case of a more general result given in Theorem 2.5.5 in section 2.5.8, and is essential for the understanding of Student's 't' (see § 2.5.5) and for the Analysis of Variance (see Chapter 8).

(c) *Tables of the χ^2 distribution*

To apply the above result, and others like it, one needs a table of the cumulative distribution function of the χ^2 distribution for each value of the number of degrees of freedom. Tables of this c.d.f. exist (see Bibliography) but are widely available only in terms of 'percentage points' [see § 1.4.2 (vii)]. A version of this is given in the Appendix (T6). It gives values of the quantity $\chi^2(\alpha; \nu)$ such that

$$P\{K_\nu \geq \chi^2(\alpha, \nu)\} = \alpha, \tag{2.5.17}$$

for various values of α.

In the Biometrika Tables, for instance [Pearson and Hartley (1966)— Bibliography G], $\chi^2(\alpha, \nu)$ is tabulated for

$$100\alpha = 0\cdot1, \, 0\cdot5, \, 1, \, 2\cdot5, \, 5, \, 10, \, 25, \, 50, \, 75, \, 90, \, 95, \, 97\cdot5, \, 99, \, 99\cdot5,$$

and for

$$\nu = 1(1)30(10)100.$$

The same volume also has a table (Table 7) of the probability integral (i.e. the c.d.f. in direct form) of the χ^2 distribution, giving $P(K_\nu \leq \chi^2)$ for $\nu = 1(1)30(2)70$ and

$$\chi^2 = 0.001(0\cdot001)0\cdot010(0\cdot01)0\cdot1(0\cdot1)2(0\cdot2)10(0\cdot5)20(1)40(2)134.$$

χ^2 tables and the incomplete gamma function

The incomplete gamma function [see Abramowitz & Stegun (1970)— Bibliography G] is the function

$$G(x, a) = \int_0^x e^{-t} t^{a-1} \, dt / \Gamma(a), \qquad a > 0.$$

It follows from (2.5.15) that

$$G(x, a) = P(K_\nu \leq k), \qquad \nu = 2a, \qquad k = 2x.$$

χ^2 tables and the Poisson distribution

If the random variable R has the Poisson distribution [see Appendix (T2)] with parameter θ, we have

$$P(R \geq c) = \sum_c^\infty e^{-\theta} \theta^j / j!$$

$$= P(K_\nu \leq k), \qquad \nu = 2c, \qquad k = 2\theta. \tag{2.5.18}$$

This follows on performing repeatedly the operation of integration by parts [see IV, § 4.3] on the integral expressing $P(K_\nu \geq k)$. This integral is

$$\int_{2\theta}^{\infty} z^{c-1} e^{z/2} \, dz/2^c \Gamma(c) = \int_{\theta}^{\infty} u^{c-1} e^{-u} \, du/\Gamma(c)$$

$$= A(c, \theta), \text{say}$$

$$= \frac{1}{(c-1)!} \left\{ [-u^{c-1} e^{-u}]_{\theta}^{\infty} + (c-1) \int_{\theta}^{\infty} u^{c-2} e^{-u} \, du \right\}$$

$$= \theta^{c-1} e^{-\theta}/(c-1)! + A(c-1, \theta)$$

$$= \theta^{c-1} e^{-\theta}/(c-1)! + \theta^{c-2} e^{-\theta}/(c-2)! + A(c-2, \theta)$$

etc. whence the result quoted.

This property is utilized in Table 7 of the Biometrika Tables mentioned above, which serves as cumulative distribution table both for the χ^2 distribution and for the Poisson distribution.

(d) Sampling distribution of the sample variance

In a sample $x_1, x_2, \ldots, x_n$ on a Normal (μ, σ) variable, the sample variance is variously defined as

$$\sum_1^n (x_i - \bar{x})^2/n \qquad (= v_0, \text{say})$$

and as

$$\sum_1^n (x_i - \bar{x})^2/(n-1) \qquad (= v_1, \text{say}) \tag{2.5.19}$$

The second definition is the more usual, having the property of being an unbiased estimate [see II, § 3.3.2] of σ^2.

More generally, we consider the statistic

$$v = \sum_1^n (x_i - \bar{x})^2/a(n), \tag{2.5.20}$$

where the divisor $a(n)$ is an arbitrary function of the sample size n. This is a realization of the r.v.:

$$V = \sum_1^n (X_i - \bar{X})^2/a(n),$$

where the X_r are independent Normal (μ, σ) variables. Its distribution may be obtained from Theorem 2.5.1, according to which $a(n) V/\sigma^2$ is a chi-squared variable on $n-1$ d.f. Thus the sampling p.d.f. of $\sum_1^n (x_i - \bar{x})^2/a(n)$ at z is

$$g_n(z) = \frac{\{a(n)\}^{(n-1)/2} z^{(n-3)/2} \exp\{-a(n)z/2\sigma^2\}}{2^{(n-1)/2} \sigma^{n-1} \Gamma\{\tfrac{1}{2}(n-1)\}}, \qquad (z > 0), \qquad n = 2, 3, \ldots.$$

$$\tag{2.5.21}$$

with $a(n) = n$ for the definition v_0 and $a(n) = n - 1$ for the unbiased estimate v_1 as in (2.5.19). It follows that

$$E(V) = (n-1)\sigma^2/a(n) = \begin{cases} (n-1)\sigma^2/n, & a(n) = n \\ \sigma^2, & a(n) = n - 1 \end{cases}$$

and that

$$\mathrm{var}\,(V) = 2(n-1)\sigma^4/\{a(n)\}^2 = \begin{cases} 2(n-1)\sigma^4/n^2, & a(n) = n \\ 2\sigma^4/(n-1), & a(n) = n - 1. \end{cases}$$

Thus the sampling variance of the unbiased estimate v_1 of σ^2, based on a sample of size n, is

$$2\sigma^4/(n-1). \tag{2.5.22}$$

(e) *Sampling distribution of the sample standard deviation*

The sample standard deviation may be defined as

$$w_n = \sqrt{\left\{ \sum_1^n (x_r - \bar{x})^2/a(n) \right\}} \tag{2.5.23}$$

where $a(n)$ is an appropriate divisor. The maximum likelihood method [see § 6.4.1] gives $a(n) = n$, while the value of $a(n)$ that makes w_n^2 an unbiased estimate of σ^2 is $(n-1)$ [see Example 3.3.5]. Both values make w_n a biased estimate of σ, w_n underestimating σ in both cases. It is shown in the sequel that, for $n \geq 2$, the choice of $a(n)$ as

$$a(n) = n - \tfrac{3}{2}$$

produces an estimate of σ that is very nearly unbiased.

Let V be defined as above, and let $W_n(>0)$ be defined as

$$W_n = V^{1/2}.$$

Thus

$$W_n = \sqrt{\left\{ \sum_1^n (X_r - \bar{X})^2/a(n) \right\}}.$$

The p.d.f. of this induced r.v. at w is

$$h_n(w) = 2wg_n(w^2), \qquad w > 0,$$

where $g_n(z)$ is given by (2.5.21); hence [see II, § 4.7]

$$h_n(w) = \frac{\{a(n)\}^{(n-1)/2}}{2^{(n-3)/2}\sigma^{n-1}\Gamma\{\tfrac{1}{2}(n-1)\}}\, w^{n-2}\exp\{-a(n)w^2/2\sigma^2\}, \qquad (w > 0),$$

$$\tag{2.5.25}$$

whence the rth moment of W_n is

$$E(W_n^r) = \left\{\frac{2\sigma^2}{a(n)}\right\}^{r/2} \frac{\Gamma\{\frac{1}{2}(r+n-1)\}}{\Gamma\{\frac{1}{2}(n-1)\}}, \qquad r = 1, 2, \ldots \qquad (2.5.26)$$

Bias

In particular

$$E(W_n) = c_n\sigma$$

where

$$c_n = \sqrt{\frac{2}{a(n)}} \frac{\Gamma(\frac{1}{2}n)}{\Gamma\{\frac{1}{2}(n-1)\}}. \qquad (2.5.27)$$

When $a(n) = n-1$, corresponding to the case where the sample variance, taken as $\sum (x_i - \bar{x})^2/(n-1)$, is an unbiased estimate of σ^2, the sample standard deviation defined as

$$\sqrt{\{\sum (x_i - \bar{x})^2/(n-1)\}} \qquad (2.5.28)$$

has sampling expectation $c_n'\sigma$, where

$$c_n' = \frac{\Gamma(\frac{1}{2})}{\Gamma\{\frac{1}{2}(n-1)\}} \sqrt{\frac{2}{n-1}}. \qquad (2.5.29)$$

This is always less than unity. The estimate (2.5.28) is therefore a biased estimate of σ, the magnitude of the bias being illustrated by the following table (Table 2.5.1). In the same table we give the values of the divisor $a(n)$ that makes (2.5.23) an unbiased estimate of σ: this is

$$a_0(n) = 2\Gamma^2(\tfrac{1}{2}n)/\Gamma^2\{\tfrac{1}{2}(n-1)\}. \qquad (2.5.30)$$

Sample size n	Sampling expectation of $\sigma^{-1}\sqrt{\left\{\sum\limits_{1}^{n} (x_i - \bar{x})^2/(n-1)\right\}}$ c_n'	Values of $a_0(n)$ such that $\sqrt{\left\{\sum\limits_{1}^{n} (x_i - \bar{x})^2/a_0(n)\right\}}$ is an unbiased estimate of σ	$n - \tfrac{3}{2}$
5	0·9400	3·534	3·5
10	0·9727	8·515	8·5
25	0·9896	23·502	23·5
50	0·9949	48·502	48·5
100	0·9975	98·501	98·5
200	0·9987	198·501	198·5

Table 2.5.1: The bias in estimates of σ.

The second column (c_n') of the Table 2.5.1 shows that, with $n = 10$ for example, the divisor $n-1$ used in (2.5.24) produces an estimate of σ whose sampling expectation is 0.9727σ. The third column $(a_0(n))$ shows the divisor required to produce an unbiased estimate of σ.

It will be seen from the table that the 'unbiased' divisor is very nearly $n - \frac{3}{2}$ (compare with the final column). It follows that the estimate $\sqrt{\{\sum_1^n (x_i - \bar{x})^2/(n - 3/2)\}}$ is an excellent approximation to an unbiased estimate of σ.

Sampling Variance of the Estimate (2.5.23) *of* σ

It follows from (2.5.26) that the sampling variance of the estimate w_n of σ defined in (2.5.23) is

$$\left(\frac{n-1}{a(n)} - c_n^2\right)\sigma^2, \tag{2.5.31}$$

where c_n is defined in (2.5.27). With $a(n) = n$, $n-1$ or $n - \frac{3}{2}$ this is approximately equal to $\sigma^2/2n$, the error being a term of order n^{-2}. Numerical values are given, for representative values of n, for $a(n) = n-1$ and $n - \frac{3}{2}$, together with the approximate values $\sigma^2/2n$, in Table 2.5.2.

n	Approximate value $\sigma^2/2n$	Accurate values $a(n) = n-1$	$a(n) = n - \frac{3}{2}$
10	0·05	0·0539	0·0588
25	0·02	0·0207	0·0213
50	0·01	0·0101	0·0103
100	0·005	0·0050	0·0051
200	0·0025	0·0025	0·0025

Table 2.5.2: Sampling variance of $w_n = \sqrt{\{\sum_1^n (x_i - \bar{x})^2/a(n)\}}$ as given in (2.5.31).

It will be seen that the biased estimate has a slightly smaller variance than the unbiased one, but that the approximation $\sigma^2/2n$ is sufficiently accurate for most purposes.

Probability that estimate lies in a specified interval

Probability computations involving the random variable W_n defined in (2.5.24) may be carried out with the aid of tables of the chi-squared distribution [see Appendix T6], since $K_{n-1} = a(n) W_n^2/\sigma^2$ is a chi-squared variable on $n-1$

d.f. For example, to find the value of $P(0\cdot98\sigma \leq W_n \leq 1\cdot02\sigma)$, with $n = 25$, we have (taking $a(n) = n - 1 = 24$):

$$K_{n-1} = (n-1) W_n^2 / \sigma^2 = 24 W_{25}^2 / \sigma^2,$$

whence

$$P(0\cdot98 \leq W_{25}/\sigma \leq 1\cdot02) = P\{24(0\cdot98)^2 \leq K_{24} \leq 24(1\cdot02)^2\}$$

$$= P\{23\cdot05 \leq K_{24} \leq 24\cdot96\}.$$

(For this kind of application the percentage point format for tables as used in Appendix T6 is particularly unsatisfactory. Using the (more extensive) ordinary cumulative distribution tables of χ^2 in the Biometrika collection (see Pearson and Hartley (1966) Bibliography G) we find the probability is $0\cdot115$.)

2.5.5.   Student's Distribution (the '*t* Distribution')

Suppose X is a Normal (μ, σ) variable [see II, § 11.4.3] and that $x_1, x_2, \ldots, x_n$ is a sample of observations on X, that the sample mean $\bar{x} = \sum_1^n x_r / n$ is taken as an estimate of μ and that

$$s^2 = \sum_1^n (x_r - \bar{x})^2 / (n-1)$$

is taken as the estimate of σ^2 [see § 2.5.4(d)]. As usual we introduce the variables $X_1, X_2, \ldots, X_n$ as statistical copies of X, x_r being regarded as a realization of X_r for $r = 1, 2, \ldots, n$. Then $\bar{x}$ and s^2 are realizations of, respectively,

$$\bar{X} = \sum_1^n X_r / n,$$

$$S^2 = \sum_1^n (X_r - \bar{X})^2 / (n-1).$$

It follows from section 2.5.4(c) that $n(\bar{X} - \mu)^2 / \sigma^2$ and $(n-1)S^2 / \sigma^2$ are mutually independent chi-squared variables having 1 d.f. and $n - 1$ d.f. respectively.

'Student' (W. S. Gossett) introduced the statistical quantity

$$t = \frac{\bar{x} - \mu}{s / \sqrt{n}},$$

now always called Student's t or Student's ratio. [For biographical information, see Pearson and Kendall (1970) Chapter 24—Bibliography D.] Its sampling distribution, of fundamental importance in statistical inference, is called Student's distribution on $n - 1$ degrees of freedom (d.f.). It will be seen that

t, as defined above, is a realization of the r.v.

$$T = n^{1/2}(\bar{X} - \mu)/S$$

$$= \frac{n^{1/2}(\bar{X} - \mu)/\sigma}{S/\sigma}$$

so that

$$T^2 = \frac{(\bar{X} - \mu)^2/(\sigma^2/n)}{S^2/\sigma^2} = \frac{K_1}{K_{n-1}/(n-1)}$$

where

$$K_1 = n(\bar{X} - \mu)^2/\sigma^2$$

and

$$K_{n-1} = (n-1)S^2/\sigma^2$$

$$= \sum_1^n (X_r - \bar{X})^2/\sigma^2.$$

Thus K_1 and K_{n-1} are mutually independent χ^2 variables with 1 d.f. and $n-1$ d.f. respectively, and Student's T as defined above is

$$T = \sqrt{\left\{ \frac{K_1}{K_{n-1}/(n-1)} \right\}}.$$

More generally we define Student's ratio and its distribution as follows:

DEFINITION 2.5.1.   *Student's ratio.*   A random variable T_ν which can be expressed in the form $T_\nu = \sqrt{(\nu K_1/K_\nu)}$, where K_1 and K_ν are mutually independent chi-squared variables with 1 d.f. and ν d.f. respectively, is called 'Student's ratio on ν d.f.', and its distribution is called 'Student's distribution on ν d.f.'

Since the numerator and denominator of T_ν^2 are mutually independent and are proportional to chi-squared variables it is a straightforward matter to derive the distribution of T_ν^2, and hence that of T_ν. [see § 2.5.6.] It turns out that the p.d.f. at w of Student's ratio T_ν on ν degrees of freedom is

$$s_\nu(w) = b_\nu(1 + w^2/\nu)^{-(\nu+1)/2}, \qquad \nu = 1, 2, \ldots$$

where

$$b_\nu = \Gamma\{(\nu+1)/2\}/\Gamma(\nu/2)\sqrt{(\nu\pi)}.$$

The cumulative distribution of T_ν is widely tabulated (A version is given in the Appendix T5).

The p.d.f. is symmetrical about the origin. It qualitatively resembles the standard Normal p.d.f., but has 'larger' (i.e. less slowly converging) tails, the effect being more pronounced for smaller than for larger values of ν [see Figure 4.5.1]. In the special case $\nu = 1$ it coincides with the Cauchy distribution [see II, § 11.7], while for values of ν exceeding (about) 40 it fairly closely approximates the standard Normal.

The lower moments of T_ν are given by

$$E(T_\nu) = 0$$

$$\text{var}\,(T_\nu) = 1 + 2/(\nu - 2), \qquad \nu > 2,$$

$$\text{skew}\,(T_\nu) = 0.$$

2.5.6.   The Variance-ratio (F) Distribution

(a) *F as a weighted ratio of χ^2 variables*

The most widely applied of all statistical techniques, the analysis of variance, depends heavily on the comparison of mutually independent 'sums of squares' which are proportional to chi-squared variables. The basic statistic is a realization of a random variable $F_{m,n}$ of the form

$$F_{m,n} = \frac{K_m/m}{K_n/n}.$$

where K_m and K_n are mutually independent chi-squared variables having m d.f. and n d.f. respectively. Its distribution is called the variance-ratio or F-distribution with m and n degrees of freedom. The symbol F stands for Fisher—a tribute to Sir Ronald Aylmer Fisher. Fisher himself, however, preferred to use $z = \frac{1}{2}\log_e F_\nu$.

It will be seen from Definition 2.5.1 that T_n^2, the square of Student's ratio on n d.f., has the F-distribution with 1 and n d.f.

[Why, it might be asked, do statisticians not work in terms of the apparently simpler random variable K_m/K_n? The answer is that the factor n/m in the definition of $F_{m,n}$ serves as a convenient scaling factor: the expectation of $F_{m,n}$ is nearly unity—it is $n/(n-2)$, to be precise—for all values of m and n $(n > 2)$.]

The distribution of $F_{m,n}$ may be derived by straightforward manipulations (following II, § 4.7). The p.d.f. of K_m at z is $f_m(z)$ as defined in (2.5.15) and that of K_m/m at x is therefore

$$h_m(x) = mf_m(mx).$$

Similarly the p.d.f. of n/K_n at y is

$$g_n(y) = ny^{-2}f_n(n/y).$$

Finally, the p.d.f. of $F_{m,n} = (K_m/m)(n/K_n)$ is the convolution [see II, Chapter 7] of $h_m(\cdot)$ and $g_n(\cdot)$, whence the p.d.f. of $F_{m,n}$ at z is

$$a(m, n)z^{(m-2)/2} \bigg/ \left\{ 1 + \frac{m}{n} z \right\}^{(m+n)/2}, \qquad z > 0$$

where

$$a(m, n) = \frac{\Gamma\{(m+n)/2\}}{\Gamma(m/2)\Gamma(n/2)} \left(\frac{m}{n}\right)^{m/2}.$$

The p.d.f. is illustrated in Figure 2.5.2.

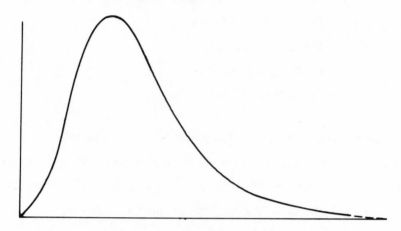

Figure 2.5.2: The $F_{m,n}$ p.d.f. for typical values of m and n.

The expected value and the variance are given by

$$E(F_{m,n}) = n/(n-2), \qquad n > 2$$

and

$$\text{var}\,(F_{m,n}) = 2n^2(m+n-2)/m(n-2)^2(n-4), \qquad n > 4.$$

Note that the expectation depends only on n.

The c.d.f. of $F_{m,n}$ is widely tabulated, but only in the 'percentage points' format. Our own table (see Appendix T7) gives upper percentage points $x_p(m, n)$ such that

$$P\{F_{m,n} \geq x_p(m, n)\} = p$$

for $p = 0\cdot05$, $0\cdot01$ and $0\cdot001$ and for $m = 1(1)$, 10, 12, 15, 20, 24, 30, 40, 60, 120, ∞ and $n = 1(1)30$, 40, 60, 120, ∞.

The Biometrika tables, Table 7 (see Pearson and Hartley (1966) Appendix T7) extend these to $p = 0\cdot25$, $0\cdot10$, $0\cdot05$, $0\cdot025$, $0\cdot01$, $0\cdot005$, $0\cdot001$.

Published tables give values of x_p such that $x_p \geq 1$ only. For values of x_p such that $x_p < 1$ one uses the result that

$$x_{1-p}(n, m) = \frac{1}{x_p(m, n)}.$$

This enables one to obtain the lower percentage points. For example the lower

5% point of $F(20, 10)$ is $x_{0 \cdot 95}(20, 10) = 1/x_{0 \cdot 05}(10, 20)$. Since the (upper) 5% point of $F(10, 20)$ is given in the table as $2 \cdot 35$, the required lower 5% point of $F(20, 10)$ is $1/2 \cdot 35 = 0 \cdot 426$.

These results follow from the fact that

$$p = P\{F_{m,n} \geq x_p(m, n)\} = P\{nK_m/mK_n \geq x_p(m, n)\}$$

$$= P\{mK_n/nK_m \leq 1/x_p(m, n)\}$$

$$= P\{F_{n,m} \leq 1/x_p(m, n)\}$$

$$= 1 - P\{F_{n,m} \geq 1/x_p(m, n)\}$$

$$= P\{F_{n,m} \geq x_{1-p}(n, m)\}.$$

(b) *Relation between the F distribution and the Beta distribution*

A r.v. Y is said to have the Beta (k, m) distribution if its p.d.f. at y is

$$f(y; k, m) = y^{k-1}(1-y)^{m-1}/B(k, m), \qquad 0 \leq y \leq 1, \qquad (k > 0, m > 0),$$

where $B(k, m)$ is the Beta function with parameters k and m:

$$B(k, m) = \int_0^1 u^{k-1}(1-u)^{m-1}$$

$$= \Gamma(k)\Gamma(m)/\Gamma(k+m)$$

[see II, § 11.6].

If U and V are independent χ^2 variables with $2k$ and $2m$ degrees of freedom, respectively, and $Y = U/(U+V)$, then Y has a Beta (k, m) distribution [see II, § 11.6.3]. It follows that

$$\frac{m}{k} \cdot \frac{Y}{1-Y} = \frac{U/2k}{V/2m}$$

and therefore that $mY/k(1-Y)$ has the $F_{2k,2m}$ distribution.

(c) *Approximation to the F distribution when one degree of freedom is very much larger than the other*

If $F_{m,n}$ has the F distribution with degrees of freedom m, n, it is not difficult to establish that

$$\lim_{n \to \infty} P\{F_{m,n} \geq f\} = P\{\chi_m^2 \geq mf\},$$

where, as usual, χ_m^2 has the χ^2 distribution on m d.f. The practical import of this is that, if $n \gg m$,

$$P\{F_{m,n} \geq f\} \doteq P(\chi_m^2 \geq mf).$$

As an example, we are concerned in Example 5.10.1 with the problem of evaluating

$$P\{F_{2,13061} \geq 0 \cdot 66).$$

This is described as a 'problem' because the d.f. values are well beyond what is available in any published table. Using the above approximation we have

$$P(F_{2,13061} \geq 0 \cdot 66) \doteqdot P(\chi_2^2 \geq 1 \cdot 32)$$

$$= 0 \cdot 6$$

approximately, (interpolating in the χ_2^2 table).

Similarly if in $F_{m,n}$ we have $m \gg n$,

$$P(F_{m,n} \geq f) \doteqdot P(\chi_n^2 \leq n/f).$$

2.5.7. The Sample Correlation Coefficient

Let $(x_1, y_1), (x_2, y_2), \ldots, (x_n, y_n)$ be a sample of n pairs of observations from the pair (X, Y) of random variables which have, jointly, the bivariate Normal distribution with correlation coefficient ρ [see II, § 13.4.6(i)]. The sample correlation coefficient r is defined as

$$r = A/\sqrt{(BC)}$$

where

$$A = \sum_1^n \{(x_i - \bar{x})(y_i - \bar{y})\} = \sum_1^n x_i y_i - n\bar{x}\bar{y},$$

$$B = \sum_1^n (x_i - \bar{x})^2 = \sum_1^n x_i^2 - n\bar{x}^2,$$

$$C = \sum_1^n (y_1 - \bar{y})^2 = \sum_1^n y_i^2 - n\bar{y}^2.$$

The sampling distribution of the sample correlation coefficient has p.d.f. at r given by

$$f_n(r, \rho) = \frac{2^{n-3}}{\pi(n-3)!} (1-\rho^2)^{(n-1)/2}(1-r^2)^{(n-4)/2} a(n, r, \rho), \qquad -1 < r < 1,$$

where

$$a(n; r, \rho) = \sum_{s=0}^{\infty} \Gamma^2\{n+s-1)/2\}(2\rho r)^s/s!$$

The sampling expectation of r is [see § 1.4]

$$\rho + O(n^{-1})$$

[for the 'O' notation, see § 1.4], and the sampling variance

$$(1-\rho^2)^2/n + O(n^{-3/2}).$$

The p.d.f. is unimodal [see II, § 10.1.3] (for $n > 4$). Some representative cases are illustrated in Figure 2.5.3.

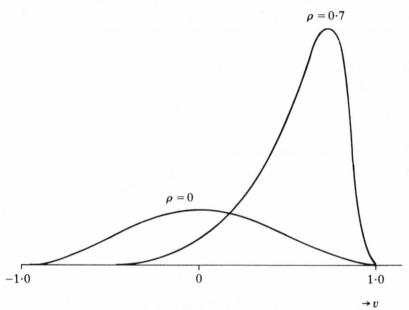

Figure 2.5.3: Density function of the sample correlation coefficient in Normal distributions with (a) $\rho = 0$; (b) $\rho = 0.7$. Reprinted with permission of Macmillan Publishing Company from *Statistical Methods for Research Workers*, 14th edition by Sir Ronald A. Fisher. Copyright © 1970 University of Adelaide.

The inverse hyperbolic tangent transformation.

For practical purposes these distributions are regarded as being too complicated to work with (except when $\rho = 0$) and instead use is made of the following remarkable transformation discovered by Fisher. Let

$$z = \operatorname{arctanh} r = \tfrac{1}{2} \ln \frac{1+r}{1-r},$$

$$\zeta = \operatorname{arctanh} \rho = \tfrac{1}{2} \ln \frac{1+\rho}{1-\rho}.$$

Then the sampling distribution of z is nearly Normal with expectation $\zeta + \rho/[2(n-1)]$ and variance $1/(n-3)$. [see Example 5.2.2.]

When ρ is zero this (approximating) sampling distribution becomes Normal with expectation zero and variance $1/(n-3)$. However, when $\rho = 0$, the exact sampling distribution of r reduces to

$$f_n(r; 0) = b_n(1 - r^2)^{(n-4)/2}, \qquad -1 < r < 1.$$

where

$$b_n = \Gamma\{(n-1)/2\}/[\sqrt{\pi}\,\Gamma\{(n-2)/2\}],$$

whence the sampling distribution of

$$(n-2)^{1/2} r / \sqrt{(1 - r^2)}$$

is Student's distribution on $n - 2$ d.f. [see also § 2.7.5 and Example 5.2.1.] In the foregoing, $\Gamma(\cdot)$ denotes the gamma function [see IV, § 10.2].

For further information, see Fisher, Chapter VI—see Bibliography C.

2.5.8. The Independence of Quadratic Forms: the Fisher–Cochran Theorem; Craig's Theorem

In the statistical technique called the analysis of variance, mentioned in section 2.5.6 (and applied in Chapters 8 and 10) it is frequently necessary to establish whether or not certain 'sums of squares' are or are not mutually independent. The principal criterion is supplied by the following theorem, known as Cochran's theorem or as the Fisher-Cochran theorem.

THEOREM 2.5.3 (*Fisher–Cochran*). *Let* $U_1, U_2, \ldots, U_n$ *be independent standard Normals. Let* $Q_1, Q_2, \ldots, Q_k$ *be non-negative quadratic forms in the variables* $U_1, U_2, \ldots, U_n$, *with ranks* $n_1, n_2, \ldots, n_k$ *respectively* [*see I, § 5.6*], *such that*

$$\sum_1^n U_1^2 = Q_1 + Q_2 + \ldots + Q_k.$$

Then the Q_r *are mutually independent chi-squared variables if, and only if,*

$$n_1 + n_2 + \ldots + n_k = n.$$

In this case Q_r *has* n_r *d.f.,* $r = 1, 2, \ldots, k$.

Two related theorems that are also useful are the following:

THEOREM 2.5.4. *With the notation of Theorem 2.5.3, suppose*

$$\sum_1^n U_r^2 = Q_1 + Q_2$$

where Q_1 *is a chi-squared variable on* m *d.f. and* Q_2 *is non-negative,. Then* Q_2 *is a chi-squared variable on* n–m *d.f., and is independent of* Q_1.

THEOREM 2.5.5.　*With the notation of Theorem 2.5.3, suppose that Q, Q_1 and Q_2 are non-negative quadratic forms in the variables U_1, U_2, ..., U_n, such that*

$$Q = Q_1 + Q_2,$$

and that Q and Q_1 are chi-squared variables with n d.f. and m d.f. respectively. Then Q_2 is a chi-squared variable with $n-m$ d.f., and is independent of Q_1.

EXAMPLE 2.5.2.　*Orthogonal resolution of $\sum (U_r - \mu)^2$.*　Consider the following algebraic identity in the independent standard Normal variables U_1, U_2, ..., U_n (with $\bar{U} = \sum_1^n U_r/n$):

$$\sum_1^n U_r^2 = n\bar{U}^2 + \sum_1^n (U_r - \bar{U})^2$$

$$= Q_1 + Q_2.$$

The quadratic forms Q_1 and Q_2 are clearly non-negative. The rank of Q_1 is obviously 1, and that of Q_2 is $n-1$. It follows by Theorem 2.5.3 that Q_1 and Q_2 are independent chi-squared variables with 1 d.f. and $n-1$ d.f. respectively. (This result was established by another argument in Example 2.5.1.)

Alternatively and more simply we may argue that, as $\bar{U}$ is Normal $(O, n^{-1/2})$, the variable $\bar{U}n^{1/2}$ is a standard Normal, whence $Q_1 = n\bar{U}^2$ is a chi-squared variable on 1 d.f. It follows on applying Theorem 2.5.4 to the identity that $Q_2 = \sum_1^n (U_r - \bar{U})^2$ is a chi-squared variable on $n-1$ d.f. The corresponding result for independent Normal (μ, σ) variables $X_1, X_2, ..., X_n$ is based on the identity

$$\sum_1^n (X_r - \mu)^2 = n(\bar{X} - \mu)^2 + \sum_1^n (X_r - \bar{X})^2$$

or

$$Q = Q_1 + Q_2,$$

say. Here Q/σ^2 is a χ^2 variable with n d.f., and Q_1/σ^2 is a χ^2 variable with 1 d.f., whence, by the theorem, Q_2/σ^2 is a χ^2 variable on $n-1$ d.f., independent of Q_1. This result is important in the theory of Student's t-test for the significance of a sample mean [see § 5.8.2].

EXAMPLE 2.5.3.　*Analysis for comparison of two means.*　In Student's test for the comparison of the means $\bar{y}$ and $\bar{z}$ of two independent samples [see § 5.8.4], say $(y_1, y_2, ..., y_p)$ and $(z_1, z_2, ..., z_q)$, from Normal populations with a common variance σ^2 but with possibly unequal expected values, we need the following identity:

$$\sum_1^{p+q} (x_r - \bar{x})^2 = \left\{ \sum_1^p (y_r - \bar{y})^2 + \sum_1^q (z_r - \bar{z})^2 \right\} + \frac{pq}{p+q} (\bar{y} - \bar{z})^2.$$

Here

$$x_r = \begin{cases} y_r, & r = 1, 2, \ldots, p, \\ z_{r-p}, & r = p+1, p+2, \ldots, p+q, \end{cases}$$

and $\bar{x} = \sum_1^{p+q} x_r/(p+q) = (p\bar{y}+q\bar{z})/(p+q)$ is the overall mean.

If we write the identity in the form

$$Q = Q_1 + Q_2,$$

it is easy to see that Q/σ^2 is a χ^2 variable with $p+q-1$ d.f., Q_1/σ^2 is a χ^2 variable with $(p-1)+(q-1)$ d.f., and Q_2/σ^2 is a χ^2 variable with 1 d.f. It follows that Q_1 and Q_2 are independent of each other.

The above theorems are useful in cases where a quadratic form has been split into two or more quadratic forms. Where this resolution does not arise, as for example when we are simply given two quadratic forms and the question is, are they independent or not, the main theorem is the following:

THEOREM 2.5.6. *Craig's Theorem.* *Let Q_1, Q_2 be quadratic forms in the independent Normal (μ, σ) variables $X_1, X_2, \ldots, X_n$, the matrix of Q_1 being* **A** *and that of Q_2 being* **B**. *Then Q_1 and Q_2 are mutually independent if, and only if,* **AB** $= 0$.

EXAMPLE 2.5.4. *Independence of $\bar{x}$ and s^2.* Let $x_1, x_2, \ldots, x_n$ be a random sample from a Normal (μ, σ) distribution. The sampling independence of $\bar{x} = \sum_1^n x_r/n$ and $s^2 = \sum_1^n (x_r - \bar{x})^2/(n-1)$, effectively established in Example 2.5.2, can be demonstrated with the aid of Theorem 2.5.6 as. follows: we consider the quadratic forms $Q_1 = (\sum_1^n X_r)^2$ and $Q_2 = \sum_1^n (X_r - \bar{X})^2 = \sum_1^n X_r^2 - n\bar{X}^2$, where, as usual, X_r is the r.v. induced by $x_r (r = 1, 2, \ldots, n)$, and $\bar{X}$ by $\bar{x}$. The matrix (**A**) of Q_1 is $\mathbf{11}'/n$ and the matrix (**B**) of Q_2 is $\mathbf{I} - \mathbf{11}'/n$. (**I** is the unit matrix, and $\mathbf{1}' = (1, 1, \ldots, 1)$.) Then $\mathbf{AB} = \mathbf{11}'(\mathbf{I} - \mathbf{11}'/n) = \mathbf{11}' - \mathbf{1}(\mathbf{1}'\mathbf{1})\mathbf{1}'/n = \mathbf{11}' - \mathbf{11}'$ (since $\mathbf{1}'\mathbf{1} = n) = \mathbf{0}$, whence Q_1 and Q_2 are mutually independent. Suppose however that we are interested in the independence or otherwise of a linear form $Z = \sum_1^n a_r X_r = \mathbf{a}'\mathbf{X}$, say, and $Q_2 = \sum_1^n (X_r - \bar{X})^2$. We construct the quadratic form $Z^2 = (\mathbf{a}'\mathbf{X})^2 = \mathbf{X}'\mathbf{a}\mathbf{a}'\mathbf{X} = \mathbf{X}'\mathbf{A}\mathbf{X}$ where the matrix **A** of the quadratic form is $\mathbf{aa}'$. As before, the matrix **B** of Q_2 is $\mathbf{I} - \mathbf{11}'/n$. Then $\mathbf{AB} = \mathbf{aa}'(\mathbf{I} - \mathbf{11}'/n) = \mathbf{aa}' - \mathbf{a}(\mathbf{a}'\mathbf{1})\mathbf{1}'/n = \mathbf{aa}' - \alpha\mathbf{a}\mathbf{1}'$ (where $\alpha = \mathbf{a}'\mathbf{1}/n = \sum_1^n a_r/n, = \bar{a}$ say). The (r, s) entry in this matrix is $a_r(a_s - \alpha)$, and this is non-zero for all (r, s) unless $a_s = \alpha$ for all s, that is, unless Z is proportional to the sample mean $\bar{X}$. Thus it is not merely the case that $\bar{X}$ is independent of S^2: it is also the case that $\bar{X}$ (or a form proportional to $\bar{X}$) is the *only* linear form in the X_r that is independent of S^2. For example, in a sample of size 3, the sum $X_1 + X_2 + X_3$ is independent of the sample sum of squares $\sum (X_r - \bar{X})^2$, but the linear combination $X_1 - 2X_2 + X_3$ is *not* independent of the sample sum of squares.

2.5.9.   The Range and the Studentized Range

The *range r_n* of a sample of observations $x_1, x_2, \ldots, x_n$ on a random variable X is the difference $x_{(n)} - x_{(1)}$ between the largest observation $x_{(n)}$ and the smallest $x_{(1)}$ [c.f. § 14.3]. If $X_{(n)}$ and $X_{(1)}$ denote the random variables induced by $x_{(n)}$ and $x_{(1)}$, the sampling distribution of the range is the distribution of the r.v.

$$R_n = X_{(n)} - X_{(1)}.$$

If, for example, X is Uniformly distributed on $(0, a)$, the p.d.f. at r of R_n is

$$h(r) = n(n-1)r^{n-2}(a-r)/a^n, \qquad 0 \le r \le a,$$

with expected value

$$E(R) = (n-1)a/(n+1)$$

[see II, § 15.5].

The range is used in applied statistics as an estimate of dispersion in small samples. In a sample of size 2 the range is exactly equivalent, in information content, to the sample standard deviation: when $n = 2$,

$$\sum (x_r - \bar{x})^2/(n-1) = \tfrac{1}{2}(x_1 - x_2)^2,$$

so that the sample standard deviation in this case is just $r_2/\sqrt{2}$; thus r_2 is an estimate of $\sigma\sqrt{2}$. For samples of size greater than 2, but not exceeding, say, 10 or 12, the range is an inefficient [see Definition 3.3.5 and Example 3.3.10] but reasonably acceptable estimate of an appropriate multiple of the population standard deviation σ [see, for example, Hald, (1957) chapter 12—Bibliography C.] With increasing sample size the relative efficiency of the estimate decreases, and it is not recommended for use when the sample size exceeds 12. This point is discussed in some detail of section 3.3.5 of O. L. Davies, [see Davies (1957)—Bibliography C]. Tables of the sampling distribution of the so called standardized range r_n/σ are available for Normal samples [see, e.g. Hald (1952) or Owen (1962)—both in Bibliography C].

Studentized range

The formation of the standardized range r_n/σ from the range r_n of a sample requires that the population standard deviation σ be known. When (as is nearly universally the case) σ is unknown, its replacement by a suitable estimate is suggested. If a statistic v is available that is independent of r_n and such that v/σ^2 is a χ^2 valuable on, say, m d.f., the quantity $\sqrt{v/m}$ is a 'suitable estimate' of σ, and the statistic

$$r_n / \sqrt{\left(\frac{v}{m}\right)}$$

is called the Studentized range [c.f. Definition 2.5.1]. Its *raison d'être* is that it forms the basis of a method, due to Tukey, of obtaining simultaneous confidence intervals for several parameters. [see Table T7 of Graybill (1976)].

2.6. ASYMPTOTIC SAMPLING DISTRIBUTION OF $\bar{x}$ AND OF NON-LINEAR FUNCTIONS OF $\bar{x}$

The richest results in sampling theory (as exemplified in § 2.5) are obtained when the underlying distribution is Normal. Although exact Normality is never encountered in practice, Normal sampling theory may still apply, to an acceptable degree of approximation, because of the following results (and of multivariate generalizations of them). These are in many cases stated in terms of 'asymptotic Normality', which we now define.

DEFINITION 2.6.1. *Asymptotic Normality for large values of n.* A statistic s_n based on a sample of size n is said to be asymptotically Normal, with expectation μ and variance v_n, if

$$\lim_{n\to\infty} P\left(\frac{S_n - \mu}{\sqrt{v_n}} \le y\right) = (2\pi)^{-1/2} \int_{-\infty}^{y} e^{-(u^2)/2}\, du$$

$$= \Phi(y)$$

where S_n is the random variable induced by s_n [see Definition 2.2.1], and Φ is the c.d.f. of the standard Normal distribution. [For 'lim' see IV, Definition 1.2.1; for 'standard Normal' see II, § 11.4.1.]

The practical interpretation is that, for large n, the distribution of s_n is reasonably well approximated by the Normal $(\mu, \sqrt{v_n})$. For example, Student's statistic on n degrees of freedom is asymptotically Normal $(0, 1)$, and indeed, for most practical purposes, may be treated as being Normal $(0, 1)$, to a good approximation, for $n \ge 40$.

THEOREM 2.6.1 (*Khintchine's Theorem*). *If* $(x_1, x_2, \ldots, x_n)$ *is a sample from a distribution which has a finite expected value* μ, *the sample mean* $\bar{x}$ *converges in probability to* μ [*see Definition 3.3.1*].

This means that, for large values of n, the value of $\bar{x}$ is unlikely to differ substantially from μ.

THEOREM 2.6.2 (*Lindeberg's Theorem*). *If* $(x_1, x_2, \ldots, x_n)$ *is a sample from a distribution which has a finite expected value* μ *and finite variance* σ^2, *the sampling distribution of the sample mean* $\bar{x}$ *is asymptotically Normal with expectation* μ *and variance* σ^2/n, *for large n.*

This result is a special case of the Central Limit Theorem [see II, § 17.3]. The practical interpretation is that, for large n, the sampling distribution of the sample mean $\bar{x}$ is reasonably well approximated by the Normal $(\mu, \sigma/\sqrt{n})$ distribution.

THEOREM 2.6.3 (*The de Moivre–Laplace Theorem*). *If the random variable R has the Binomial (n, θ) distribution [see II, § 5.2.1], R is asymptotically Normal with expected value $n\theta$ and variance $n\theta(1-\theta)$. [see also II, § 11.4.7].*

Of the various applications of these and other similar theorems, one which has outstanding utility is the following (here stated in a form due to Wilks (1961)—Chapter 9—see Bibliography C.].)

THEOREM 2.6.4. *Asymptotic sampling distribution of $g(\bar{x})$. If $(x_1, x_2, \ldots, x_n)$ is a sample from a distribution which has a finite expected value μ and finite variance σ^2, and $g(x)$ is a prescribed function of x, then, subject to conditions stated below, the sampling distribution of $g(\bar{x})$ is asymptotically Normal with expectation $g(\mu)$ and variance $\{g'(\mu)\}^2\sigma^2/n$ [see Definition 2.6.1].*
The conditions on the function g are that $g'(x)$ should exist in some neighbourhood of $x = \mu$, and that $g'(\mu) \neq 0$.

With the aid of this result it may be shown, for example, that, in samples of size n from various distributions, the corresponding asymptotic Normal sampling distributions of the sample mean $\bar{x}$, or of a specified function $g(\bar{x})$, is as shown in Table 2.6.1.

Distribution of X	$g(\bar{x})$	Asymptotic Normal sampling distribution of $g(\bar{x})$:	
		Expected value	Variance
Poisson (θ) [see II, § 5.4]	$2\sqrt{\bar{x}}$	θ	$1/n$
Gamma (θ) (i.e. with shape parameter θ: [see II, § 11.3.1])	$2\sqrt{\bar{x}}$	$2\sqrt{\theta}$	$1/n$
Binomial (n, θ) [see II: § 5.2.2]	$\sin^{-1}(2\bar{x}-1)$	$\sin^{-1}(2\theta-1)$	$1/n$
Geometric (θ) (i.e. with p.d.f. $P(X=x) = (1-\theta)\theta^{x-1}$, $x = 1, 2, \ldots$ [see II, § 5.2.3]	$\log\{\bar{x}(1+\sqrt{1-1/\bar{x}}\}$	$\log\left(\dfrac{1+\sqrt{\theta}}{1-\theta} - \dfrac{1}{2}\right)$,	$1/n$
Uniform $(-\frac{1}{2}\theta, \frac{1}{2}\theta)$ [see II, § 11.1]	$\sqrt{12}\log(2\bar{x})$	$\sqrt{12}\log\theta$,	$4/n$

Table 2.6.1

2.7. APPROXIMATION TO SAMPLING EXPECTATION AND VARIANCE OF NON-LINEAR STATISTICS; VARIANCE-STABILIZING TRANSFORMATIONS; NORMALIZING TRANSFORMATIONS

2.7.1. Approximations

(a) *Functions of a single* r.v.

It is inherent in the concept of a limit in mathematics that, in a sequence $\{z_n\}$, $n = 1, 2, \ldots$, if the sequence converges to a limit a as $n \to \infty$, z_n must be nearly equal to a for all sufficiently large values of n. Asymptotic results quoted in section 2.6, which are strictly valid only in the limit as $n \to \infty$, will be nearly true for all sufficiently large *finite* values of n. Unfortunately it is not often easy to tell how large 'sufficiently large' is. In practice one may be forced to use the asymptotic result (or something based on it) as an approximation when n is only moderately large, or even not large at all. Further, one will need approximations to the sampling distribution (or at least to the expected value and variance of that distribution) of non-linear functions of statistics other than the sample mean discussed in section 2.6.4. In these cases it is usual to rely hopefully on the somewhat crude approximations to the expectation and variance of a suitably smooth function $h(X)$ of a random variable X obtainable from the first few terms (or even the first term itself) of the Taylor expansion [see IV, § 3.6] of the function $h(\cdot)$ about the point $\mu = E(X)$. These approximations are

$$\left. \begin{aligned} E\{h(X)\} &\doteq h(\mu) + \tfrac{1}{2}h''(\mu)\sigma^2 \\ \\ \text{var}\,\{h(X)\} &\doteq \{h'(\mu)\}^2\sigma^2 \end{aligned} \right\} \tag{2.7.1}$$

and

where $\mu = E(X)$ and $\sigma^2 = \text{var}\,(X)$.

These approximations are frequently taken in the simpler versions

$$E\{h(X)\} \doteq h(\mu),$$
$$\text{s.d.}\,\{h(X)\} \doteq |h'(\mu)|\sigma \tag{2.7.2}$$

Thus, for example, with $h(x) = 1/X$, we have

$$E(1/X) \doteq 1/\mu, \qquad \text{s.d.}\,(1/X) \doteq \sigma/\mu^2.$$

Coefficient of variation

When X is a positive variable, its variability may conveniently be specified in terms of the coefficient of variation, defined as

$$\text{c.v.}\,(X) = \text{s.d.}\,(X)/E(X) = \sigma/\mu$$

[see II, § 9.2.6].

For the function $h(x) = X^\alpha$ the approximations (2.7.2) are most neatly expressed in terms of the coefficient of variation. We have

$$E(X^\alpha) \doteq \mu^\alpha,$$

$$\text{s.d. } (X^\alpha) \doteq |\alpha| \mu^{\alpha-1} \sigma$$

and so

$$\text{c.v. } (X^\alpha) \doteq |\alpha| \text{ c.v. } (X).$$

In particular

$$\text{c.v. } (1/X) \doteq \text{c.v. } (X).$$

EXAMPLE 2.7.1.   *Sample standard deviation.*   Suppose Y is Normal (μ, σ), and consider the sample variance $v = \sum_1^n (y_r - \bar{y})^2 / (n-1)$ of a sample $(y_1, y_2, \ldots, y_n)$. It was shown in sections 2.5.4(d) and (e) that the induced random variable V has expected value σ^2 and variance $2\sigma^4 / (n-1)$. For the random variable $S = V^{1/2}$ induced by the sample standard deviation $s = v^{1/2}$, (2.7.1) leads to the following approximations for $E(S)$ and var (S) when we replace the X of (2.7.1) by V, μ by σ^2, and σ^2 by $2\sigma^4 / (n-1)$, and take $h(\mu)$ to be $\mu^{1/2}$:

$$E(S) \doteq \sigma + \frac{1}{2} \left(\frac{2\sigma^4}{n-1} \right) (-\tfrac{1}{4} \sigma^{-3})$$

$$= \{1 - 1/4(n-1)\} \sigma,$$

$$\text{var } (S) \doteq \left(\frac{2\sigma^4}{n-1} \right) (\tfrac{1}{4} \sigma^{-2})$$

$$= \sigma^2 / 2(n-1).$$

The exact values [see § 2.5.4(e)] are

$$E(S) = c(n)\sigma, \qquad c(n) = \left(\frac{2}{n-1} \right)^{1/2} \frac{\Gamma(n/2)}{\Gamma[(n-1)/2]}$$

$$(\doteq \sqrt{\{2n-3)/(2n-2)\}} \text{ for large } n);$$

$$\text{var } (S) = d(n)\sigma^2, \qquad d(n) = 1 - \frac{2}{n-1} \frac{\Gamma^2(n/2)}{\Gamma^2[(n-1)/2]}$$

$$(\doteq 1/(2n-4) \text{ for large } n).$$

The approximations (2.7.1) are reasonably accurate in this case, as may be seen from the selection of numerical values given in Table 2.7.1.

Sample size	Expected value		Variance	
n	Exact	Approximate	Exact	Approximate
5	$0 \cdot 940\sigma$	$0 \cdot 938\sigma$	$0 \cdot 116\sigma^2$	$0 \cdot 125\sigma^2$
10	$0 \cdot 973\sigma$	$0 \cdot 972\sigma$	$0 \cdot 054\sigma^2$	$0 \cdot 056\sigma^2$
20	$0 \cdot 987\sigma$	$0 \cdot 986\sigma$	$0 \cdot 026\sigma^2$	$0 \cdot 028\sigma^2$
50	$0 \cdot 995\sigma$	$0 \cdot 995\sigma$	$0 \cdot 010\sigma^2$	$0 \cdot 0102\sigma^2$

Table 2.7.1: Sampling expectation and variance of sample standard deviation from a Normal population. Exact and approximate values.

It must not be thought that the approximations (2.7.1) are always as accurate as this. [See II, § 9.9 for some less favourable cases.]

(b) *Functions of two random variables*

The formulae in (2.7.2) generalise to the case of two variables as follows. Let $h(X_1, X_2)$ be a specified differentiable function of the random variables X_1, and X_2, where

$$E(X_1) = \mu_1, \qquad E(X_2) = \mu_2,$$

$$\text{var}(X_1) = \sigma_1^2, \quad \text{var}(X_2) = \sigma_2^2, \qquad \text{corr}(X_1, X_2) = \rho.$$

Then

$$E\{h(X_1, X_2)\} \doteqdot h(\mu_1, \mu_2) \tag{2.7.2}$$

and

$$\text{var}\{h(X_1, X_2)\} \doteqdot h_1^2 \sigma_1^2 + 2h_1 h_2 \rho \sigma_1 \sigma_2 + h_2^2 \sigma_2^2 \tag{2.7.3}$$

where

$$h_j = \partial\{h(\mu_1, \mu_2)\}/\partial\mu_j, \qquad j = 1, 2.$$

When X_1 and X_2 are uncorrelated (and, a fortiori, when they are *independent*), the variance approximation reduces to

$$\text{var}\{h(X_1, X_2)\} \doteqdot h_1^2 \sigma_1^2 + h_2^2 \sigma_2^2. \tag{2.7.4}$$

EXAMPLE 2.7.2.   *Variance of a product and of a quotient.*   For the product $X_1 X_2$ of independent random variables the approximation (2.7.2) is exact, while the variance formula (2.7.4) gives the approximation

$$\text{var}(X_1 X_2) \doteqdot \mu_2^2 \sigma_1^2 + \mu_1^2 \sigma_2^2;$$

The (neater) corresponding formula for the coefficient of variation [see § 2.7.1] is

$$\{\text{c.v.}(X_1 X_2)\}^2 = \{\text{c.v.}(X_1)\}^2 + \{\text{c.v.}(X_2)\}^2.$$

In this example it is easy to see that the exact formulae are

$$E(X_1, X_2) = \mu_1 \mu_2,$$
$$\text{var}(X_1 X_2) = E(X_1 X_2)^2 - (\mu_1 \mu_2)^2$$
$$= E(X_1^2) E(X_2^2) - \mu_1^2 \mu_2^2$$
$$= (\sigma_1^2 + \mu_1^2)(\sigma_2^2 + \mu_2^2) - \mu_1^2 \mu_2^2$$
$$= \sigma_1^2 \sigma_2^2 + \mu_1^2 \sigma_2^2 + \mu_2^2 \sigma_1^2,$$

whence

$$\{\text{c.v.}(X_1 X_2)\}^2 = \{c.v.(X_1)\}^2 + \{\text{c.v.}(X_2)\}^2.$$
$$+ \{\text{c.v.}(X_1)\}^2 \{\text{c.v.}(X_2)\}^2.$$

For a quotient X_1/X_2 the approximations are

$$E(X_1/X_2) \fallingdotseq \mu_1/\mu_2,$$
$$\text{var}(X_1/X_2) \fallingdotseq (\mu_1^2/\mu_2^2)\{\sigma_1^2/\mu_1^2 + \sigma_2^2/\mu_2^2\}$$

and

$$\{\text{c.v.}(X_1/X_2)\}^2 \fallingdotseq \{\text{c.v.}(X_1)\}^2 + \{\text{c.v.}(X_2)\}^2.$$

Thus the approximations coincide for the coefficients of variation of $X_1 X_2$ and of X_1/X_2.

2.7.2. Variance-stabilizing Transformations

(a) *A general formula*

When data consists of *counts* ('how many sterile specimens?') the relevant sampling distribution is often Binomial or Poisson. Difficulties are then introduced into the analysis not only by the fact that the observations are integers, but also by the fact that the sampling variance depends on the unknown parameter. For example if the observed proportions of successes in two populations are r_1/n_1 and r_2/n_2 respectively, inferences about the success probabilities θ_1 and θ_2 in the separate distributions are complicated by the fact that the sampling variances of r_1 and r_2 are quadratic functions of the parameters, namely $n_1 \theta_1 (1 - \theta_1)$ and $n_2 \theta_2 (1 - \theta_2)$ respectively.

Life would be easier if a transformation could be found which changed all Binomials into new variables which all had the same variance; and similarly for Poissons. Transformations which go some way towards this ideal may be found with the aid of (2.7.1). If X has expected value θ and variance $\sigma^2(\theta)$, and $Y = h(X)$ is a transform of X we have

$$\text{var}(Y) \fallingdotseq \sigma^2(\theta)\{h'(\theta)\}^2.$$

We can arrange that var (Y) is a constant (approximately!) by choosing $h(\theta)$ so that

$$\sigma(\theta)h'(\theta) = k \quad \text{(a constant)}.$$

This will be the case if

$$h(\theta) = k \int \frac{1}{\sigma(\theta)} \, d\theta, \tag{2.7.4}$$

the transformed variable $Y = h(X)$ having variance k^2 (approximately).

(b) *Poisson data*: *the square root transformation*

Suppose X has a Poisson (θ) distribution, so that $\sigma^2(\theta) = \theta$. Then (2.7.2) becomes

$$h(\theta) = k \int \theta^{-1/2} \, d\theta$$

$$= \sqrt{\theta} \quad \text{(on taking } k = \tfrac{1}{2}).$$

We are thus led to the transformation from x to

$$\sqrt{x}. \tag{2.7.5}$$

The observed counts $x_1, x_2, \ldots$ are transformed to $\sqrt{x_1}, \sqrt{x_2}, \ldots$, the transformed data having sampling variance approximately equal to $1/4$.

The accuracy of the approximation may be assessed by a direct numerical evaluation of var ($\sqrt{X}$), using:

$$\text{var}(\sqrt{X}) = \theta - \{E(\sqrt{X})\}^2,$$

$$E(\sqrt{X}) = \sum_0^\infty r^{1/2} f_r, \quad (f_r = e^{-\theta}\theta^r/r!, r = 0, 1, \ldots).$$

Some representative values are given in Table 2.7.2

θ	Variance of the untransformed variable X	Variance of the transformed variable $\sqrt{X}$
0·2	0·2	0·164
0·4	0·4	0·272
0·6	0·6	0·334
0·8	0·8	0·381
1	1	0·402
2	2	0·390
5	5	0·287
10	10	0·259
20	20	0·255

Table 2.7.2: The variance-stabilizing effect of the square root transformations on Poisson data with parameter θ.

Whilst the transformation may not at first strike one as being outstandingly successful in achieving a constant variance of 0·25, it does in fact do a remarkable job in reducing the oscillation in the variance for values of θ less than about 10, and maintains a practically constant value thereafter.

Anscombe has shown that the transformation to

$$\sqrt{(X+3/8)} \qquad\qquad (2.7.6)$$

is somewhat better. For this, with $\theta = 2$ for example, the variance is 0·2315, already quite close to the target value of 0·25. [see Wetherill (1981), Chapter 8—Bibliography C.]

(c) *Binomial data: the arc sine (or angular) transformation*

If X is Bernoulli (n, θ), observations $x_1, x_2, \ldots, x_n$ will often be combined to form the statistic $\bar{x} = r/n$ $(r = x_1 + \ldots + x_n)$, this being the usual estimate of the parameter θ. The sampling distribution of r is Binomial (n, θ), with expected value $\mu = n\theta$ and variance $\sigma^2 = n\theta(1-\theta)$; whence the corresponding values for r/n are θ and $\theta(1-\theta)/n$. We now seek a transform $z = h(r/n)$ with constant sampling variance. By (2.7.1) we have

$$\mathrm{var}\,(z) = \{h'(\theta)\}^2 \theta(1-\theta)/n$$

of which the right-hand member is constant $(= k^2)$ if

$$h'(\theta) = kn^{1/2}/\sqrt{\{\theta(1-\theta)\}}.$$

This is satisfied by taking

$$h(\theta) = 2kn^{1/2} \sin^{-1}\sqrt{\theta},$$

i.e. by taking

$$z = 2kn^{1/2} \sin^{-1}\sqrt{(r/n)}.$$

Equivalently and more conveniently we may transform from r to $z/2k\sqrt{n}$, i.e. to

$$\sin^{-1}\sqrt{(r/n)}, \qquad\qquad (2.7.7)$$

which has sampling variance approximately equal to $1/4n$. The modification

$$\sin^{-1}\sqrt{\frac{r+3/8}{n+1/4}}, \qquad\qquad (2.7.8)$$

due to Anscombe, is better. The approximate sampling variance for this is $1/(4n+2)$. [See Wetherill (1981). Chapter 8—Bibliography C.]

(d) *Stabilizing the variance of a regressed variable by weighting*

If in a proposed linear regression analysis [see § 6.5] the scatter of the values of the regressed variable y appears on a data plot to vary systematically with the regressor variable x, a simple weighting procedure will stabilize the variance. For example if the sampling standard deviation of $y(x)$—that is of values of y corresponding to the value x of the regressor variable—increases in proportion to the value of x, a transformation from $y(x)$ to the weighted data

$$z(x) = y(x)/x$$

will stabilize the variance.

2.7.3. Normalizing Transformations

Because of the relative tractability of data from a Normal universe, it is often helpful to make a transformation which will bring non-Normal data to a condition of approximate Normality.

(a) *The log transform*: *positively skewed positive variables*

Essentially positive unbounded data frequently come from a distribution with a positively skewed p.d.f. [see II, § 9.10.1] resembling that of a log-Normal [see II, § 11.5] or a Gamma [see II, § 11.3] or Chi-squared [see II, § 11.4.11] distribution. If a random variable X is log-Normally distributed then its logarithm is (by definition) Normal. The log transformation will however achieve approximate Normality when applied to most random variables having a distribution qualitatively resembling the log-Normal, simply as a consequence of its sending the zero value to minus infinity. The effect is illustrated in Figure 2.7.1, where the c.d.f.'s of chi-squared distributions with various degrees of freedom values are plotted on graph paper on which the c.d.f. of the log-Normal is a straight line. (This graph paper is called 'log probability paper'. c.f. § 3.2.2.(d).)

(b) *The log transform for variables bounded above and below; the inverse hyperbolic tangent transformation for the correlation coefficient*

If a random variable X is constrained to lie in the interval (a, b), the transform $Y = \log\{(X-a)/(X-b)\}$ will have values lying in the doubly infinite interval $(-\infty, +\infty)$, a fact which suggests that Y might be approximately Normal.

This transformation is in fact outstandingly successful in the case of the correlation coefficient. The sample correlation coefficient r computed from a sample of n pairs (x_j, y_j) from a bivariate Normal distribution [see II: § 13.4.6], viz:

$$r = \frac{\sum_1^n x_j y_j - \left(\sum_1^n x_j\right)\left(\sum_1^n y_j\right)\Big/ n}{\sqrt{[\{\sum x_j^2 - (\sum x_j)^2/n\}\{\sum y_j^2 - (\sum y_j)^2/n\}}}$$

[see § 2.5.7] has values lying in the interval $(-1, +1)$. The sampling distribution

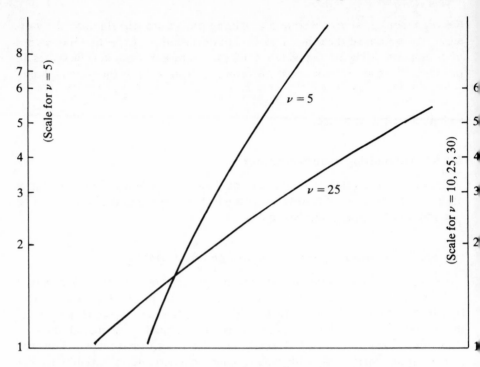

Figure 2.7.1: Graph of c.d.f. of (a) χ^2_5; (b) χ^2_{25} on log probability paper.

of r is highly skewed, the exact form depending on the population value ρ of the correlation coefficient in the parent distribution. The transformed statistic [see § 2.5.7]

$$z = \tfrac{1}{2}\log \frac{1+r}{1-r} \, (= \tanh^{-1} r) \qquad\qquad (2.7.9)$$

has a nearly Normal sampling distribution, with expected value

$$\tfrac{1}{2}\log \frac{1+\rho}{1-\rho} + \frac{\rho}{2(n-1)}$$

and variance

$$1/(n-3),$$

approximately.

This transformation greatly simplifies the problems of specifying the accuracy of a computed value of r, taken as an estimate of σ [see Example 5.2.2].

(c) *Normalizing transforms of χ^2*

Although the χ^2 distribution [see § 2.5.4(a) and Chapter 7] is well tabulated, it is sometimes convenient to work in terms of an approximately Normal function of the χ^2 variable. The following two are commonly used for χ^2 with ν degrees of freedom:

(i) for $\nu > 100$:

$$X = \sqrt{(2\chi^2)} - \sqrt{(2\nu - 1)} \qquad\qquad (2.7.10)$$

is approximately standard Normal; this is a reasonably good approximation also for $30 < \nu \leq 100$, but a more accurate version is the following:

(ii) for $\nu > 30$:

$$X = \{(\chi^2/\nu)^{1/3} - (1 - 2/9\nu)\}/\sqrt{(2/9\nu)} \qquad\qquad (2.7.11)$$

is approximately standard Normal.

For example, with $\nu = 40$ the probability that χ^2 should exceed $51\cdot805$ is $0\cdot100$; the approximation (2.7.11) gives $1 - \Phi(x)$ where $x = (1\cdot0900 - 0\cdot9944)/0\cdot0745 = 1\cdot283$, whence $1 - \Phi(x) = 0\cdot998$, which has an error of only $0\cdot2\%$. The approximation (2.7.10) for which $\nu = 40$ is allegedly too small yet gives the approximate value $0\cdot98$, the error here being only 2%.

(d)   *The probability integral transform*: *probits*

In principle any continuous random variable is (exactly) Normalizable, with the aid of the probability integral transform: if the c.d.f. of X at x is $F(x)$, the transformed variable $U = F(X)$ is Uniformly distributed on $(0, 1)$ [see II, § 1.4 and § 10.7]. If we denote by $\Phi(y)$ the c.d.f. at y of the standard Normal variable [see II, § 11.4.1], it follows that the random variable $\Phi(Y)$ is Uniform $(0, 1)$. Thus the transformation from X to Y given by

$$\Phi(Y) = F(X),$$

or

$$Y = \Phi^{-1}\{F(X)\}, \qquad\qquad (2.7.12)$$

will transform X into a standard Normal variable.

The transform

$$y = \Phi^{-1}(z)$$

or, more commonly (to avoid negative values)

$$y = 5 + \Phi^{-1}(z),$$

is called the *probit* of z.

An application of this idea is given in § 2.7.4 and in § 6.6. A detailed study of its application in dosage-mortality investigations will be found in Finney (1971). For an account of another transformation of the same general nature, the 'logit' transformation, see Ashton (1972). In this a probability p is replaced by $z = \log p/(1-p)$.

2.7.4.   Curve-straightening Transformations

In situations where, according to a given hypothesis, observations $y(x)$ should lie on a curve $q(x)$, with superimposed random errors, it is usually worthwhile making a visual judgement of the adherence of the plotted points to this hypothesis before embarking on a more elaborate assessment. It is easier for the eye to detect departures from a straight line than from a curve, and it is therefore useful to transform the data so that the curve $q(x)$ becomes a line.

A common example of this idea occurs in testing whether a sample comes from a Normal distribution. Here, for a set of values of x, $y(x)$ denotes the proportion of observations which are less than (or equal to) x in value, and $q(x)$ is the Normal c.d.f. Special graph paper is available, called '(arithmetic) probability paper' on which the scales are such that $q(x)$ becomes a straight line [see § 3.2.2(d); also II, § 11.4.8].

Another example occurs in the analysis of dosage-mortality data, in which (for example) it is required to assess the toxicity of a reagent which, when administered to specified organisms, has no effect when sufficiently dilute, is a moderately effective poison at low concentrations, and more so at higher concentrations, until it kills the entire sample when the concentration is high enough. A conventional measure of the degree of toxicity is the dosage which effects a 50% kill. Observations on the mortality $y(x)$ of organisms subjected to increasing dosages (or possibly logarithms of dosages) x would be expected to lie on a sigmoid curve such as that shown in Figure 2.7.2. A convenient model for such a curve is a Normal (μ, σ) cumulative distribution function. The problem of estimating the toxicity reduces then to the problem of estimating the values of μ and σ. The direct approach to this involves successive approximations which may be simplified by using a curve-straightening 'probit' transformation [see § 6.6].

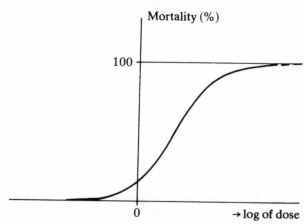

Figure 2.7.2: A sigmoid curve.

2.7.5. A Student-distributed Transform of the Sample Correlation Coefficient when $\rho = 0$

For completeness, we give here a transformation of the sample correlation coefficient in Normal samples that has already been referred to in section 2.5.7. This is as follows:

If r denotes the sample correlation coefficient calculated as in section 2.5.7 from a sample of n independent pairs $(x_1, y_1), (x_2, y_2), \ldots (x_n, y_n)$, where the x_r are observations on a Normally distributed random variable X, and the y_r are observations on a Normally distributed random variable Y, *where X and Y are independent*, the sampling distribution of the transform

$$(n-2)^{1/2} r / \sqrt{(1-r^2)}$$

is Student's distribution in $n-2$ d.f. [see Example 5.2.1].

2.7.6. A χ^2-distributed Transform of a Uniformly Distributed Variable

Straightforward application of standard transformation theory [see II, § 10.7] shows that, if a r.v. x has c.d.f. $F(x)$, the transform

$$Y = F(X)$$

is Uniformly distributed on $(0, 1)$ [see II, § 11.1]. This is the famous probability integral transform [c.f. § 2.7.3(d)].

Take the case where X has the exponential distribution with expected value 2. The p.d.f. at x is

$$f(x) = \tfrac{1}{2} e^{-x/2}, \qquad x \geq 0,$$

and the c.d.f. is

$$F(x) = 1 - e^{-x/2}, \qquad x \geq 0.$$

It follows from the Probability Integral Transform property quoted above that the transform

$$Y = 1 - e^{-X/2}$$

is a Uniform $(0, 1)$ variable.

Hence, of course, the transform

$$Z = e^{-X/2} (= 1 - Y)$$

also has that Uniform distribution. Conversely, if Z is Uniform $(0, 1)$, and

$$X = -2 \log Z,$$

then X is exponentially distributed with expected value 2; that is [see § 2.5.4(a)], X is a χ^2 variable on 2 degrees of freedom.

Fisher has exploited this result to produce a method of combining the significance levels of a set of significance tests [see § 5.9].

2.8. NON-CENTRAL SAMPLING DISTRIBUTIONS

2.8.1. Non-central Chi-squared

Let
$$W_m = V_1^2 + V_2^2 + \ldots + V_m^2$$

where V_r is Normal $(\mu_r, 1)$, and $V_1, V_2, \ldots, V_m$ are mutually independent. The distribution of W_m is called the *non-central chi-squared distribution* with m degrees of freedom and with non-centrality parameter

$$\lambda = \tfrac{1}{2} \sum_1^m \mu_r^2$$

If $\lambda = 0$ the distribution reduces to the ordinary 'central' chi-squared [see § 2.5.4(a)]. An alternative representation of W_m is

$$W_m = \sum_1^m (U_r + \mu_r)^2$$

where the U_r are independent standard Normals. Thus the special case $m = 1$ has the representation
$$W_1 = (U + \mu)^2$$

where U is a standard Normal. The non-centrality parameter is $\lambda = \tfrac{1}{2}\mu^2$.

The fundamental properties of this distribution are the following:

(i) $E(W_m) = m + 2\lambda$.

(ii) $\mathrm{var}\,(W_m) = 2(m + 4\lambda)$.

(iii) The moment generating function of W_m is

$$E(\exp W_m \theta) = (1 - 2\theta)^{-m/2} \exp\{2\lambda\theta/(1 - 2\theta)\}, \qquad 0 < \theta < \tfrac{1}{2}.$$

(iv) The p.d.f. of W_m at w is

$$f(w; \lambda) = \sum_{r=0}^{\infty} g_r(\lambda) h_r(w; \lambda)$$

where $g_r(\lambda) = e^{-\lambda} \lambda^r / r! = $ p.d.f. at r of a Poisson (λ) distribution, [see II, § 5.4], and

$$h_r(w; \lambda) = \{w^{\alpha(r)-1} e^{-w/2}\}/2^{\alpha(r)}\Gamma\{\alpha(r)\}, \qquad w > 0,$$

where
$$\alpha(r) = \tfrac{1}{2}(m + 2r)$$

so that h_r is the p.d.f. at w of a central chi-squared distribution with $m + 2r$ degrees of freedom [see II, § 11.4.11];

(v) If $W_{m'}$ and $W_{m''}$ are independent noncentral chi-squared variables with m', m'' degrees of freedom and with non-centrality parameters λ', λ'' respectively, $W_{m'} + W_{m''}$ is also non-central chi-squared, with $m = m' + m''$ degrees of freedom and with parameter $\lambda = \lambda' + \lambda''$.

The most important application of the distribution occurs in the discussion of the power function or sensitivity function of analysis of variance tests [see § 5.3.1]. For further information see Graybill (1976) Chapter 4. Tables of the non-central χ^2 distribution may be found in Harter and Owen (1970) Volume 1—see Bibliography G.

2.8.2.   Non-central F

The noncentral F distribution with degrees of freedom m, n, and with noncentrality parameter λ is the distribution of the ratio

$$nW/mZ.$$

where W has the non-central χ^2 distribution with m d.f. and parameter λ [see § 2.8.1], while Z, which is independent of W, has the ordinary χ^2 distribution on n d.f.

Applications are similar to those for non-central χ^2. For further information and tables see Graybill (1976) Chapter 4 and Table T11. Tables are also·given in Harter and Owen (1974) Volume 2—see Bibliography G.

2.8.3.   The Non-central Student Distribution

The distribution of the ratio

$$\frac{U+\lambda}{\sqrt{(V/m)}},$$

where U is a standard Normal and V is a (central) chi-squared variable with m degrees of freedom, (U, V being mutually independent) is called the *non-central Student distribution* with m d.f. and parameter λ.

Thus, for example, in a sample $(x_1, x_2, \ldots, x_n)$ from a Normal (μ, σ) distribution, with sample mean $\bar{x}$ and sample variance $s^2 = \sum_1^n (x_j - \bar{x})^2/(n-1)$, the sampling distribution of the statistic

$$t' = (\bar{x} - \mu + \delta)n^{1/2}/s$$

is non-central Student with $n-1$ d.f. and parameter

$$\lambda = n^{1/2}\delta/\sigma.$$

This may be verified by noting that

$$t' = \frac{(\bar{x} - \mu + \delta)/(\sigma/\sqrt{n})}{s/\sigma}$$

$$= \frac{\{(\bar{x} - \mu)/(\sigma/\sqrt{n})\} + \lambda}{\sqrt{\{\sum (x_i - \bar{x})^2/\sigma^2\}}/\sqrt{(n-1)}}.$$

This is a realization of $\{(U+\lambda)\sqrt{\nu}\}/\sqrt{(\chi_\nu^2)}$, where $\lambda = \delta n^{1/2}/\sigma$ and $\nu = n-1$, since the sampling distribution of $\bar{x}$ is Normal $(\mu, \sigma/\sqrt{n})$ [see § 2.5.3(b)], and that of $\sum_1^n (x_i - \bar{x})^2/\sigma^2$ is (central) χ^2 on $\nu = n-1$ d.f. [see Example 2.5.1], and the numerator and denominator are mutually independent [see § 2.5.4(c)].

An application to the sensitivity (or power) of a significance test is given in section 5.3.2. For further information see Owen (1976). For tables see Resnikoff and Liebermann (1957)—Bibliography G.

2.9. THE MULTINOMIAL DISTRIBUTION IN SAMPLING DISTRIBUTION THEORY

2.9.1. Binomial, Trinomial and Multinomial (m) distributions

(i) *Binomial.* A simple Bernoulli trial is a statistical experiment in which there are two possible outcomes, A_1 and A_2 (often called 'success' and 'failure'). Let

$$P(A_1) = p_1, \qquad P(A_2) = p_2$$

where

$$p_1 + p_2 = 1.$$

An important random variable is the total number R_1 of occurrences of A_1 in n independent trials of this kind; alternatively we might be interested in the total number R_2 of occurrences of A_2. We need discuss only one of them, however, since

$$R_1 + R_2 = n.$$

The distribution of R_1 is Binomial $(n; p_1)$:

$$P(R_1 = r_1) = \binom{n}{r} p_1^{r_1} (1 - p_1)^{n - r_1}, \qquad r_1 = 0, 1, \ldots, n \qquad (0 \le p_1 \le 1).$$

We might prefer to write this in the more symmetrical form

$$p(r_1, r_2) = \frac{n!}{r_1! r_2!} p_1^{r_1} p_2^{r_2},$$

$$r_1 = 0, 1, \ldots, n; \qquad r_2 = 0, 1, \ldots, n; \qquad r_1 + r_2 = n; \qquad (2.9.1)$$

$$(0 \le p_1 \le 1; \qquad 0 \le p_2 \le 1; \qquad p_1 + p_2 = 1).$$

Note that the occurrence of two symbols, r_1 and r_2, in (2.9.1) must not delude us into regarding the formula as being a bivariate probability function. It gives us the *uni*variate probability that $R_1 = r_1$, in which case $r_2 = n - r_1$; or the *uni*variate probability that $R_2 = r_2$, in which case $r_1 = n - r_1$.

(ii) *Trinomial.* In a trial which has three possible outcomes A_1, A_2, A_3, with

$$P(A_s) = p_s, \qquad s = 1, 2, 3$$

and

$$p_1 + p_2 + p_3 = 1,$$

the joint distribution of R_1 and R_2, the total numbers of occurrences of A_1 and of A_2 respectively in n independent trials, is the obvious generalization of (2.9.1), namely

$$p(r_1, r_2, r_3) = \frac{n!}{r_1! r_2! r_3!} p_1^{r_1} p_2^{r_2} p_3^{r_3},$$

$$r_1, r_2, r_3 = 0, 1, \ldots, n; \qquad r_1 + r_2 + r_3 = n; \qquad (2.9.2)$$

$$(0 \le p_s \le 1, \qquad s = 1, 2, 3; \qquad p_1 + p_2 + p_3 = 1.)$$

This gives the value of the *bivariate* probability $P(R_1 = r_1, R_2 = r_2)$, in which case $r_3 = n - r_1 - r_2$, or of $P(R_1 = r_1, R_3 = r_3)$, in which case $r_2 = n - r_1 - r_3$, or, similarly, of $P(R_2 = r_2, R_3 = r_3)$.

(iii) *Multinomial* (m). Now suppose that there are m possible outcomes to the trial; we call them $A_1, A_2, \ldots, A_m$, and let

$$P(A_s) = p_s, \qquad s = 1, 2, \ldots, m,$$

with

$$p_1 + p_2 + \ldots + p_s = 1.$$

Let R_j denote the total number of occurrences of the outcome A_j in n independent trials, $j = 1, 2, \ldots, m$. The joint distribution of $R_1, R_2, \ldots, R_{m-1}$ is then

$$p(r_1, r_2, \ldots, r_{m-1}, r_m) = \frac{n!}{r_1! r_2! \ldots r_m} p_1^{r_1} p_2^{r_2} \ldots p_m^{r_m},$$

$$r_s = 0, 1, \ldots, n; \qquad s = 1, 2, \ldots, m;$$

$$r_1 + r_2 + \ldots + r_m = n;$$

$$(0 \le p_s \le 1, \qquad s = 1, 2, \ldots, m; \qquad p_1 + p_2 + \ldots + p_m = 1).$$

$$(2.9.3)$$

This gives the value of

$$P(R_1 = r_1, R_2 = r_2, \ldots, R_{m-1} = r_{m-1}) \quad (\text{with } r_m = 1 - r_1 - r_2 - \ldots - r_{m-1}),$$

or, equivalently, of

$$P(R_1 = r_1, R_2 = r_2, \ldots, R_{m-2} = r_{m-2}, R_m = r_m),$$

with

$$r_{m-1} = 1 - r_1 - r_2 - \ldots - r_{m-2} - r_m, \text{ etc.}$$

Thus (2.9.3) gives the probability distribution of any $m - 1$ of the m random variables $R_1, R_2, \ldots, R_m$, with $R_1 + R_2 + \ldots + R_m = n$. It is called the Multinomial (m) distribution, with index n and probability parameters $p_1, p_2, \ldots, p_m$. The Multinomial (2) is the Binomial, the Multinomial (3) is the Trinomial, etc.

2.9.2. Properties of the Multinomial (m)

(a) *Low-order moments*
The expectations are

$$E(R_j) = np_j, \qquad j = 1, 2, \ldots, m.$$

The variances are

$$\text{var}\,(R_j) = np_j(1 - p_j), \qquad j = 1, 2, \ldots, m$$

and the covariances are

$$\text{cov}\,(R_j, R_k) = -np_j p_k, \qquad j, k = 1, 2, \ldots, m; \qquad j \neq k.$$

(b) *Marginal distributions*

All the marginal distributions are themselves Multinomial, as follows: the marginal distribution of R_j is

$$\text{Binomial}\,(n_j, p_j), \qquad j = 1, 2, \ldots, m;$$

the joint marginal distribution of R_j and R_k is

$$\text{Trinomial}\,(n; p_j, p_k), \qquad j, k = 1, 2, \ldots, m \;(j \neq k),$$

etc.

2.9.3. The Multinomial as a Conditional of the Joint Distribution of Independent Poisson Variables

A result which is sometimes useful in sampling theory is the following:
Suppose that $X_1, X_2, \ldots, X_k$ are independent Poisson variables, with parameters $\mu_1, \mu_2, \ldots, \mu_k$ respectively. Then the distribution of $X_1 + \ldots + X_k$ is Poisson (μ), $\mu = \mu_1 + \ldots + \mu_k$, and the conditional distribution of $X_1, X_2, \ldots, X_k$, given $X_1 + X_2 + \ldots + X_k = x$, is

$$P(X_1 = x_1, \ldots, X_k = x_k | X_1 + \ldots + X_k = x)$$

$$= \frac{\Pi(e^{-\mu_r}\mu_r^{x_r}/x_r!)}{e^{-\mu}\mu^n/x!} = \frac{x!}{\Pi(x_r!)}\Pi\left(\frac{\mu_r}{\mu}\right)^{x_r}.$$

In the trivial case where $\sum x_r \neq x$ this is of course zero, but when $\sum x_r = x$ it is equivalent to (2.9.3), and shows that the conditional distribution is Multinomial (k) with index x and probability parameters $p_1, p_2, \ldots, p_k$, where $p_s = (\mu_s/\mu)$, $s = 1, 2, \ldots, k$.

2.9.4. Frequency Tables

The joint sampling distribution of the frequencies is Multinomial.
Suppose that a sample of n observation on a continuous random variable

yields a frequency table [see § 3.2.2(b)] as follows:

Cell label	1	2	...	k
Frequency	f_1	f_2	...	f_k

The cell boundaries need not be equally spaced: if the smallest and largest observations are d', d'' respectively, the 'cells' might be the intervals (a_j, a_{j+1}), $j = 0, 1, \ldots, k-1$, of x-values for any partition

$$d' = a_0 < a_1 < a_2 < \ldots < a_k = d''.$$

An observed value x is placed in the cell labelled j if

$$a_{j-1} \leqslant x < a_j, \qquad j = 1, 2, \ldots, k.$$

It follows that the joint sampling distribution of the frequencies $f_1, f_2, \ldots, f_2$ is Multinomial (k) with index n and probability parameters $p_1, p_2, \ldots, p_k$:

$$P(f_1, f_2, \ldots, f_k) = \frac{n!}{f_1! f_2! \ldots f_k!} p_1^{f_1} p_2^{f_2} \ldots p_k^{f_k},$$

where

$$p_j = P(a_{j-1} \leq X < a_j), \qquad j = 1, 2, \ldots, k.$$

(The same applies, with obvious modifications, when the random variable X is discrete.)

E.H.L.

2.10. FURTHER READING AND REFERENCES

References have been given in the chapter to works listed in the general Bibliography. Further references are given below.

Ashton, W. D. (1972). *The Logit Transformation with Special Reference to its Uses in Bioassay*, Griffin.

Finney, D. J. (1977). *Probit Analysis*, Third edition, Cambridge University Press.

Graybill, F. A. (1976). *Theory and Application of the Linear Model*, Duxbury Press, Mass.

Owen, D. B. (1968). A Survey of Properties and Applications of the Non-central *t*-distribution, *Technometrics* **10**, 445–478.

CHAPTER 3

Estimation: Introductory Survey

3.1. THE ESTIMATION PROBLEM

When statisticians speak of the estimation problem they usually intend the restricted meaning indicated in section 1.2: the available data are assumed to be observations from one or more of a specified family of probability distributions in which the individual members of the family are distinguished from one another by the values of one or more 'parameters', and the estimation problem is to find from the data the best statistical approximation to the unknown value(s) of the parameter(s) appropriate to the data, together with an objective measure of the accuracy of the approximation.

The choice of the family of distributions appropriate to the problem in hand may in certain cases be determined more or less uniquely by circumstances, but in many situations this choice is far from being unique. When one is concerned with the estimation of the unknown proportion of defective items in a load, based on a random sample of predetermined size [see Example 2.1.1], it is pretty clear that the effective datum is the number of defective items in the sample and it follows from the sampling procedure that this is a realization of a Hypergeometric distribution [see II, § 5.3] which may under known conditions be satisfactorily approximated by a Binomial distribution [see II, § 5.2.2]. If however, one is concerned with, say, the average length of the items in the load, it is necessary to work in terms of a family of length distributions. Which family? For the example quoted it may well be reasonable to assume that the manufacturing fluctuations which generate the distribution of lengths have an additive cumulative effect, and to argue from this, by invoking the Central Limit Theorem [see II, § 11.4.2 and § 17.3], that the distribution will be (approximately, at least) Normal [see II, § 11.4.3].

In other situations there may be no physical or scientific reason for adopting one family rather than another, the only basis for choice being the evidence supplied by the sample itself. In such cases there may well be more than one apparently suitable family. This phenomenon makes it clear that the theory of estimation requires for its completion a method of assessing whether or

79

not the 'fitted' distribution is one from which the observed sample might reasonably have been drawn. This question, 'Goodness of Fit', is discussed in Chapter 7.

Obviously, any estimate of an unknown parameter must be based on the sample. The particular combination of observations (called a *statistic* [see Definition 2.1.1]) that is to serve as our estimate will be chosen with various criteria in mind. There are two main approaches. In one, the observed value of the statistic is regarded as an observation on the *sampling distribution* of the statistic: one takes into account not only the observations actually made but also potential observations that might have been made. In the other approach the properties of the estimate are developed solely in terms of the values actually observed; this is the *likelihood* approach.

It is the sampling distribution approach that is the more fully developed and widely used, in terms of unbiased estimates and minimal sampling variance or of Bayesian criteria or of decision theory, or of a combination of these; and this is the approach to which the greater part of the present book is devoted. Likelihood methods are described in less detail [see §§ 3.5.4, 4.13.1, 6.2.1].

As an introductory example of statistical inference suppose we have ten equally reliable but discrepant observations $x_1, x_2, \ldots, x_{10}$ on the weight of a given specimen. What is the magnitude (μ, say) of its actual weight? The statistical approach is to postulate that the discrepancies among the observations are due to randomly fluctuating experimental conditions, and to regard the observations as realizations [see II, § 4.1] of a set of random variables $X_1, X_2, \ldots, X_{10}$. A particular family of probability distributions to which these random variables belong is then proposed. In the light of a knowledge of the experimental procedure it might, for example, seem appropriate to assume this to be the Normal family, that is that the X_r are independent random variables, with a common Normal distribution, and that no information other than the ten observations should be taken into account. (For an alternative viewpoint see Chapter 15, on Bayesian methods.) The particular member of the Normal family that applies to our data has parameters (μ, σ) with unknown values which we have to estimate. In the present example, we may identify the parameter μ with the unknown 'actual value' of the weight; and the parameter σ is a measure of the variability in the observations induced by the observational technique [see II, § 11.4.3].

The next stage is to find a combination $h_1(x_1, x_2, \ldots, x_{10})$ of the observations to use as the *statistic* [see § 2.1] whose numerical value supplies an appropriate approximation to the value of μ, and a second statistic to serve a similar purpose for σ. These numerical values provide the estimates of μ and σ.

For reasons that will become clear the statistics

$$\hat{\mu} = h_1(x_1, x_2, \ldots, x_{10}) = \tfrac{1}{10} \sum_{1}^{10} x_r = \bar{x} \qquad (3.1.1)$$

(the sample mean) and

$$\hat{\sigma} = h_2(x_1, x_2, \ldots, x_{10}) = \left\{ \sum_1^{10} (x_r - \bar{x})^2/9 \right\}^{1/2} = s \qquad (3.1.2)$$

(the sample standard deviation) have good claims to be used. (Here we have adopted the widely used convention of denoting an estimate of a parameter θ by $\hat{\theta}$; we might equally have used a notation such as θ^* or $\tilde{\theta}$.)

Estimate and estimator

The numerical value of the estimate $\bar{x}$ given above may be regarded as a realization of the induced [see Definition 2.2.1] random variable

$$\bar{X} = (X_1 + X_2 + \ldots + X_{10})/10,$$

which is called the *estimator* corresponding to the estimate $\bar{x}$. Similarly the estimator corresponding to the sample standard deviation s is

$$S = \left\{ \sum_0^{10} (X_r - \bar{X})^2/9 \right\}^{1/2}.$$

This idea may be summed up in the following definition.

DEFINITION 3.1.1.   *Estimate and estimator; standard error.*   An *estimate* $\hat{\theta}$ of a parameter θ based on a sample $x_1, x_2, \ldots, x_n$ is a statistic [see Definition 2.1.1], say

$$\hat{\theta} = h(x_1, x_2, \ldots, x_n)$$

whose numerical value may be used as an approximation to the unknown value of θ. The sampling distribution of $\hat{\theta}$ is the distribution of the random variable

$$T = h(X_1, X_2, \ldots, X_n),$$

where the X_r are the random variables induced [see Definition 2.2.1] by the x_r. Thus T is the random variable induced by $\hat{\theta}$. This random variable T is the *estimator* corresponding to the estimate $\hat{\theta}$. An appropriate estimate of the standard deviation of T is called the *standard error* of $\hat{\theta}$ [c.f. § 4.1.2].

In the case exemplified in (3.1.1) and (3.1.2) there are two parameters, μ and σ, estimated by the statistics $\hat{\mu} = \bar{x}$ and $\hat{\sigma} = s$ respectively. The induced random variables are

$$\bar{X} = (x_1 + \ldots + x_n)/n$$

(taking the sample now to be of size n), and

$$S = \left\{ \sum_1^n (X_r - \bar{X})^2/(n-1) \right\}.$$

$\bar{X}$ and S are mutually independent [see § 2.5.4(c)]; $\bar{X}$ is Normal with expected value μ and variance σ^2/n; the distribution of S is most easily described by the statement that $9S^2/\sigma^2$ is a chi-squared variable [see § 2.5.4(a)] on 9 degrees of freedom.

In what sense can it be said that $\bar{x}$ is a good estimate of μ? The most direct answer is provided by the fact that a realization of $\bar{X}$ is unlikely to be very different from μ: for any fixed positive Δ and any fixed λ, the probability that $\bar{X}$ will lie in the interval $\lambda \pm \Delta$ is greatest when $\lambda = \mu$.

This shows that the estimate $\bar{x}$ has some claim to be regarded as a good estimate of μ. That is not to say that better estimates might not exist: if, for example, $u = h(x_1, x_2, \ldots, x_n)$ is a statistic such that the induced variable U has a higher probability of lying in the interval $\mu + \Delta$ than has $\bar{X}$ we should necessarily regard U as being better than $\bar{X}$ as an estimator of μ. If the inequality

$$P(U \in \mu \pm \Delta) > P(\bar{X} \in \mu \pm \Delta)$$

held for all values of μ and Δ, U would be *uniformly* better than $\bar{X}$ as far as this criterion is concerned.

This line of argument leads to the idea that maximal *concentration* ought to be a required property of a good estimator. The estimator T would be said to have uniformly maximal concentration with respect to the parameter θ if, for every other estimator T',

$$P(\theta - \lambda_1 < T < \theta + \lambda_2) \geq P(\theta - \lambda_1 < T' < \theta + \lambda_2) \qquad (3.1.3)$$

for all positive λ_1 and λ_2. Unfortunately estimates having this property do not in general exist, and a less ambitious approach has to be made: see section 3.3.

Returning now to our example involving the mean of a Normal (μ, σ) sample of size n, we use the well-known fact that the probability is 0·95 that the realized value $\bar{x}$ of $\bar{X}$ will be in the interval $\mu \pm 1.96\sigma/\sqrt{n}$. [see Appendix (T3), (T4).]

If we replace the unknown $\sigma/\sqrt{n}$ by its estimated value $s/\sqrt{n}$ (the *standard error* of $\hat{\mu}$), we may paraphrase the last statement as follows: 'There is a high probability that the realized value $\bar{x}$ will lie at a distance not exceeding $2s/\sqrt{n}$ from μ, and hence that the unknown μ will lie at a distance not exceeding $2s/\sqrt{n}$ from the observed $\bar{x}$.'

This is a rather rough and ready statement, capable of considerable refinement—for example in terms of confidence intervals [see Example 4.5.2]. Nevertheless it is intuitively appealing. In practical statistics the standard error of an estimate is widely used as a measure of the accuracy of that estimate, on the one hand in the above intuitive sense and, on the other, as permitting the evaluation of a more sophisticated measure.

As to what that measure should be there are various schools of thought. Some statisticians regard the problem as being satisfactorily solved by the

Bayesian approach (see Chapter 15) in which, instead of regarding the sample as the only relevant information, one also takes into account additional 'prior' information, that is, information that existed before the detailed observations became available—e.g. the knowledge that μ is certainly not less than 100 g and not more than 400 g: in other cases the prior information might be a good deal more precise. The accretion of this prior information is carried out by regarding the values of μ that were conceivable a priori as realizations of a random variable having a certain distribution—the 'prior' distribution—and regarding the corresponding values that are to be considered in the light of the data as realizations of a second distribution, called the posterior distribution, this being a conditional distribution [see II, § 6.5] obtained by conditioning the prior distribution on the data. Using Bayes' Theorem [see Chapter 15], one therefore ends up with a probability distribution of possible values of μ, from which, in terms of the fixed observed value of $\bar{x}$, it is possible to construct probability intervals of μ-values of given 'size' (e.g. 0·95). This kind of result is precisely what one would like to get. There is of course a certain degree of arbitrariness in the choice of prior distribution. The more serious difficulty however is that to many statisticians the concept of a prior distribution in circumstances of this kind is philosophically unacceptable.

Some of these will claim that it is not only the idea of a non-objective prior distribution that sticks in their gullet: they also argue that the unknown value of μ is an unknown *constant*, and so cannot be treated as a random variable. This objection is however not necessarily fatal. In terms of our example, if σ were, say, equal to 10 gm, and $\bar{x}$ equal to 125 g, the postulated value 200 for μ could be regarded as unlikely in the sense that one would be very unlikely to obtain a realized value equal to 125 from a Normal (μ, σ) distribution with $\mu = 200$ and $\sigma = 10$. R. A. Fisher himself, an arch-opponent of Bayesian inference, accepted this viewpoint and advocated a theory of 'fiducial' inference in which probabilities were assigned in this way to possible values of μ, without invoking a prior distribution. In spite of Fisher's penetrating analytical power and profound statistical intuition, however, the theory of fiducial inference has never been presented to the statistical world in a completely convincing fashion, and it has not become part of the accepted canon. [See, e.g. Kendall and Stuart (1973), vol. 2, Chapter 21; and Barnett (1982); both in Bibliography C.]

If one is unable to accept the view that probabilities can be assigned to intervals of possible μ-values, one is reduced to the standard techniques of making probability statements about $\bar{x}$ in ways that involve an indirect statement about μ, namely those associated with (i) confidence intervals and/or significance tests, (ii) making decisions with known risks of error, or (iii) allocating to postulated values of the unknown μ degrees of relative plausibility that are proportional to their relative likelihoods.

These concepts are discussed in greater detail in Chapter 4. A selection of references is given in section 3.6.

Point estimation and interval estimation

A practical estimation procedure must have as its aim not only the selection of a particular statistic whose numerical value will provide the required approximation ('estimate') of the parameter in question, but also the construction of an appropriate measure of the *accuracy* of the estimate. These are of course two aspects of one and the same problem. Nevertheless it is often convenient to discuss them separately, in which case the first aspect, namely the selection of a statistic, is called 'point estimation', and the second, namely the specification of its accuracy, is called 'interval estimation'.

3.2. INTUITIVE CONCEPTS AND GRAPHICAL METHODS

3.2.1. Introduction

Perhaps the most fundamental intuitive concept of estimation theory is the idea that a sample resembles the population from which it is taken. To exploit this somewhat vague idea one needs to specify methods of describing a sample that are in some sense analogous to methods of describing a population. The principal such methods for populations are the following:

(i) Direct or indirect specification of the complete distribution—direct specification being in terms of the p.d.f. or c.d.f. [see § 1.4.2(i)], and indirect in terms of one of the standard generating functions [see II, Chapter 12].

(ii) Specification of particular aspects of the complete distribution such as the lower moments [see II, § 9.11], selected percentage points [see § 2.5.4(c)] and so forth.

Sample analogues of all of these have been devised. Sample versions of the generating function of a probability distribution (p.g.f., m.g.f., etc.) have not been widely used and will not be discussed further in this volume. Sample analogues of the moments of a distribution have been dealt with in some detail in Chapter 2. What remains as the subject matter of the present section is the sample version of the p.d.f. and the c.d.f. We proceed to discuss these in section 3.2.2.

3.2.2. Frequency Tables, Histograms and the Empirical c.d.f.

(a) *Discrete data*

The basic method of organizing the information in a sample is to form a frequency table. For a *discrete* univariate random variable [see II, Chapter 5] R, defined on, say, the non-negative integers, this is simply a table giving the number of times the value r occurs in the sample, for $r = 0, 1, 2, \ldots$, or, equivalently, that number expressed as a proportion of the sample size, n. These are called respectively the *frequency* f_r of the observation r, and its *relative frequency* f_r/n. The *cumulative frequencies* c_r are the numbers of observations x for which $x \le r$; these numbers, when expressed as a proportion of the sample size n, are the *relative cumulative frequencies* c_r/n.

EXAMPLE 3.2.1. Rutherford and Geiger's data on the number of α-particles emitted from a radio-active source in $7 \cdot 5$ seconds consist of the information tabulated in columns 1 and 3 of Table 3.2.1.

Column number					
1	2	3	4	5	6
Cell no. r	Number of particles emitted	Frequency f_r	Relative frequency (%) $100f_r/n$	Cumulative frequency $c_r = \sum_0^r f_j$	Relative cumulative frequency (%) $100c_r/n$
0	0	57	2·19	57	2·19
1	1	203	7·78	260	9·97
2	2	383	14·69	643	24·65
3	3	525	20·13	1168	44·79
4	4	532	20·40	1700	65·18
5	5	408	15·64	2108	80·83
6	6	273	10·46	2381	91·30
7	7	139	5·33	2520	97·01
8	8	45	1·73	2565	98·35
9	9	27	1·04	2592	99·39
10	10	10	0·38	2602	99·77
11	11	4	0·15	2606	99·92
12	12–14	2* $(f_{12}+f_{13}+f_{14}=2)$	0·08*	2608 $(=c_{14})$	100·00
	Total n	2608	100·00		

Table 3.2.1: Frequency table showing Rutherford and Geiger's data.

The basic items of information, namely the frequencies, are given in column 3. The sum of the entries in this column equals the sample size n ($=2608$). One possibly disconcerting feature of the table is that the frequencies f_{12}, f_{13} and f_{14} are not separately provided. Instead we are given the 'grouped frequency' $f_{12}+f_{13}+f_{14}=2$, emphasized as 2* in the frequency column (column 3). It is a common practice in frequency tables to group low frequencies in this way. In this particular table there is only one grouped entry in the frequency column, but in general there may be several.

Sample analogue of p.d.f.

Column 4 gives the values of the *relative frequencies* expressed as percentages of the total, n ($=2608$). (The starred entry 0·08* is the grouped frequency

$f_{12}+f_{13}+f_{14}$ expressed as a percentage of the total.) The table of relative frequencies is the sample analogue of a tabulation of the probability density function [see II, § 4.3.1] of the underlying random variable R.

The cumulative frequencies in column 5 are partial sums of the frequency column. The effect on the cumulative frequencies of the grouping '12–14' is to deprive us of the values of c_{12} and c_{13}, but the value of c_{14} is of course correct. Finally the last column gives the values of the c_r as a percentage of the sample size. Naturally the final entry in this column is 100, since 100% of the observations satisfy the condition $x \leq 14$. This column provides the sample analogue of the c.d.f. (cumulative distribution function) [see II, § 4.3.2] of the underlying random variable.

(The initial column, 'column 1', is there solely to provide a means of identifying particular rows.)

In the example considered above there are 13 entries in the frequency column, say $z_0, z_1, \ldots, z_{12}$, where

$$z_r = f_r, \qquad r = 0, 1, \ldots, 11$$

and

$$z_{12} = f_{12} + f_{13} + f_{14}.$$

The sampling distribution [see § 2.2] of this vector of 13 entries is Multinomial [see II, § 6.4.2]. It follows that the sampling expectation [see § 2.3.1] of the frequency z_r is $n\pi_r, r = 0, 1, \ldots, 12$, where $n = 2608$ is the sample size, and π_r is the probability that an observation falls into cell 'r'. Similarly the sampling expectation of the relative frequency z_r/n is π_r. In our example if R is the number of particles emitted in a randomly chosen interval of length 7·5 seconds, we have

$$\pi_0 = P(R = 0),$$
$$\pi_1 = P(R = 1), \ldots, \pi_{11} = P(R = 11),$$
$$\pi_{12} = P(12 \leq R \leq 14).$$

In the present case it may plausibly be assumed that R has a Poisson distribution [see II, § 5.4]. If the frequency table had not involved any grouped cells the appropriate estimate of the Poisson parameter λ would have been the sample mean. The amalgamation of f_{12}, f_{13} and f_{14} makes the estimation problem more complicated in principle, but the frequencies involved are so small in proportion to the sample size (2 out of 2608) that it is intuitively obvious that the effect of this grouping on the value of the estimate must be small. The detailed computation, using the method of maximum likelihood, is carried out in Example 6.7.1.

For our present purposes we take the estimate, with sufficient accuracy, to be the sample mean, calculated as if the 'grouped' observations were in the middle of the grouped cell, i.e. at $r = 13$. The estimate then turns out to be 3·871.

It is then illuminating to juxtapose the tabulated frequencies and the expected values attributed to them by the assumed (Poisson) distribution. This is done in the following table, in which the expected frequencies have been rounded to the nearest integer, taking

$$\pi_r = \begin{cases} P(R=r) = e^{-\lambda}\lambda^r/r!, & r = 0, 1, \ldots, 11 \\ P(12 \le R \le 14) = e^{-\lambda}\left(\dfrac{\lambda^{12}}{12!} + \dfrac{\lambda^{13}}{13!} + \dfrac{\lambda^{14}}{14!}\right), \end{cases}$$

with

$$\lambda = 3.871.$$

Number of particles emitted	Frequency f_r	Expected frequency (rounded) $n\pi_r$
0	57	54
1	203	211
2	383	407
3	525	525
4	532	508
5	408	394
6	273	254
7	139	140
. 8	45	68
9	27	29
10	10	11
11	4	4
12–14	2	1

Total $n = 2608$

Table 3.2.2: Frequencies compared with their expected values. [see Example 3.2.1].

The apparent concordance between the frequencies and their expected values supplies intuitive evidence in favour of the assumed Poisson distribution. A similar agreement would of course be evident between the relative frequencies and their expected relative frequencies π_r, justifying the claim that the relative frequencies are the natural sample analogue of the p.d.f.

[For details of an objective test of the closeness of the apparent agreement see Chapter 7.]

(b) *Bar chart and histogram of discrete data*

Consider the following frequency table, derived from Table 3.2.1 by deleting the row corresponding to cell No. 12. (This is the frequency table that would have been obtained if no observations exceeding 11 in magnitude had been recorded.) In such a frequency table, in which no frequencies have been amalgamated ('grouped'), a completely satisfactory graphical representation would be given by a 'bar chart', that is a sequence of ordinates ('bars') of height f_r, plotted against r, for $r = 0, 1, \ldots, 11$. This is illustrated in Figure 3.2.1.

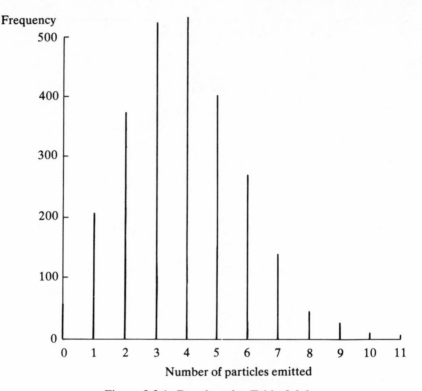

Figure 3.2.1: Bar chart for Table 3.2.3.

As an aid to the eye the bars may be thickened until they touch each other. The bar representing the frequency f_r has now become a rectangle of height f_r with its centre at the abscissa r and its left and right-hand boundaries at $r - \frac{1}{2}$, $r + \frac{1}{2}$ respectively. (See Figure 3.2.2.)

The height of the bar is numerically equal to the area of the rectangle: the height scale has become an area scale, in the sense that the total frequency of the events $r = 6$, $r = 7$, $r = 8$ (for example) is represented by the sum of the areas of the rectangles centred on $r = 6$, $r = 7$ and $r = 8$. The graph is an example of a *histogram* for discrete data 'with equal groupings'.

Now suppose that, instead of having each frequency individually specified as in Table 3.2.3, some are *grouped*, as in Table 3.2.4. Here the cells corresponding to $r = 0$ and $r = 1$ have been grouped, as have $r = 6$ and $r = 7$, and $r = 8$, 9, 10 and 11. In addition the data for $r = 12$, 13 or 14, which had already

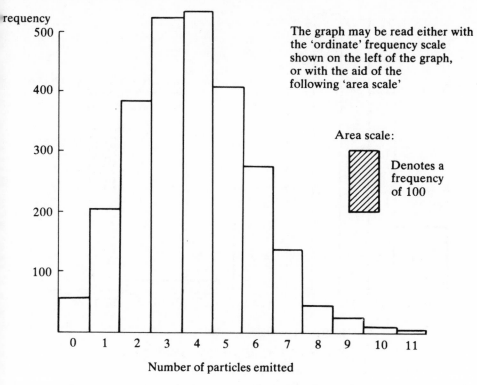

The graph may be read either with the 'ordinate' frequency scale shown on the left of the graph, or with the aid of the following 'area scale'

Area scale:

Denotes a frequency of 100

Number of particles emitted

Figure 3.2.2: Bar chart for Table 3.2.3, with thickened bars.

Number of particles emitted r	0	1	2	3	4	5	6	7	8	9	10	11	Total
Frequency f_r	57	203	383	525	532	408	273	139	45	27	10	4	2606

Table 3.2.3

been grouped in the original frequency table (Table 3.2.1) have been included. In this frequency table, with unequal groupings, a sensible graphical representation would, as far as possible, retain the main features of Figure 3.2.2. The graphical representation of the grouped frequencies 57 and 203, corresponding to $r = 0$ and $r = 1$, ought to present a visual suggestion of merging the separate rectangles for $r = 0$ and $r = 1$ into a combined rectangle whose height is the *average* of the two separate heights. In addition to its obvious advantages for visual interpretation, this device preserves the desirable *area scale* convention of Figure 3.2.3. This exemplifies the convention adopted in drawing the histogram of grouped discrete data with unequal groupings. The histogram of the frequency table of Table 3.2.4 is shown in Figure 3.2.3.

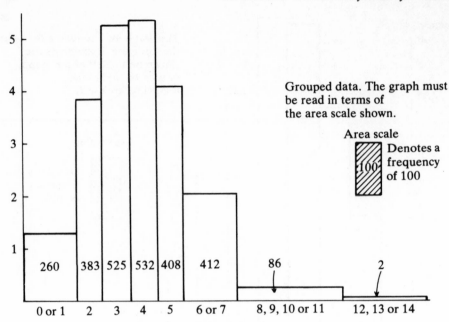

Figure 3.2.3: Histogram for Table 3.2.4.

Number of particles emitted	Frequency
0 or 1	$57 + 203 = 260$
2	383
3	525
4	532
5	408
6 or 7	$273 + 139 = 412$
8, 9, 10 or 11	$45 + 27 + 10 + 4 = 86$
12, 13 or 14	2

Table 3.2.4

(c) *Continuous data*

Similar but slightly more elaborate considerations apply to 'continuous data', that is to observations on a *continuous* random variable X [see II, § 10.1]. To form a frequency table for a sample of observations on X, one partitions the range (a, b) of sample values into k 'cells' or classes (h_{r-1}, h_r) separated by boundaries h_r, for $r = 1, 2, \ldots, k$:

$$a = h_0 < h_1 < h_2 < \ldots < h_k = b,$$

the first, second, ... cells being the intervals (h_0, h_1), (h_1, h_2), etc. One then

tabulates the *frequencies,* that is the numbers of observations falling into the
various cells:

f_r = number of observations x such that $h_{r-1} < x \le h_r$, $r = 1, 2, \ldots, k$.

Published tables sometimes adopt an alternative convention in which the
frequency f_r is taken as the number of observations such that $h_{r-1} < x < h_r$,
augmented by half a unit for each observation that (to the accuracy of the
measurements) coincides with h_{r-1} or h_r. An example is shown in Table 3.2.5.

Cell boundaries (inches)	Central height (inches)	Frequency
(59, 60)	59·5	1
(60, 61)	60·5	2·5
(61, 62)	61·5	1·5
(62, 63)	62·6	9·5
(63, 64)	63·5	31
(64, 65)	64·5	56
(65, 66)	65·5	78·5
(66, 67)	66·5	127
(67, 68)	67·5	178·5
(68, 69)	68·5	189
(69, 70)	69·5	137
(70, 71)	70·5	137
(71, 72)	71.5	93
(72, 73)	72·5	52·5
(73, 74)	73·5	39
(74, 75)	74·5	17
(75, 76)	75·5	6·5
(76, 77)	76·5	3·5
(77, 78)	77·5	1
(78, 79)	78·5	2
(79, 80)	79·5	1
		1164

Table 3.2.5: Frequency table showing heights of men: equal
cell-widths.

Half-unit frequencies: when an individual falls on a cell
boundary, the convention is to allot half a unit of frequency
to each of the adjoining cells. (Reprinted with permission
of Macmillan Publishing Company from *Statistical Methods
for Research Workers* by Sir Ronald A. Fisher. Copyright
© 1970 University of Adelaide.)

The *cumulative frequencies* are defined as

c_r = number of observations x such that $x \le h_r$

$= f_1 + f_2 + \ldots + f_r$, $r = 1, 2, \ldots, k$.

The number k of cells and the positions h_r of the cell boundaries are to some extent arbitrary. In published tables these may result from a compromise between economy and detail. Cell widths are often all (or nearly all) equal, as in a table of heights of persons, measured to the nearest inch. Non-uniform widths are however often required, as for example in a table of deaths from whooping cough, tabulated against age at death, where the low, not very age-sensitive death rate that applies for ages exceeding, say, 15, call for broad cells each spanning perhaps 5 or 10 years, while the larger and more highly age-sensitive death rates that apply to children call for narrow cells of width perhaps 6 months or 1 year. An example of a frequency table with non-uniform cell widths is shown in Table 3.2.6, obtained by amalgamating cells in Table 3.2.4.

(For the joint sampling distribution of the frequencies, see section 2.9.4.)

(i)	Cell boundaries (inches)	Frequency
	(59, 64)	45·5
	(64, 66)	134·5
	(66, 70)	631·5
	(70, 72)	230
	(72, 80)	122·5
		1164

(ii)		
	(59, 62)	5
	(62, 65)	96·5
	(65, 67)	205·5
	(67, 69)	367·5
	(69, 72)	367
	(72, 75)	108·5
	(75, 80)	14
		1164

Table 3.2.6: Grouped frequency tables with non-uniform cell widths. Data from Table 3.2.5 retabulated with two different selections of cell boundaries.

The corresponding histograms are shown in Figure 3.2.4.

(d) *Histogram of continuous data*

The most informative graphical version of the frequency table is a special kind of graph called a histogram. This was introduced, in Figures 3.2.2 and 3.2.3, for discrete data. The concept is however more usually applied to

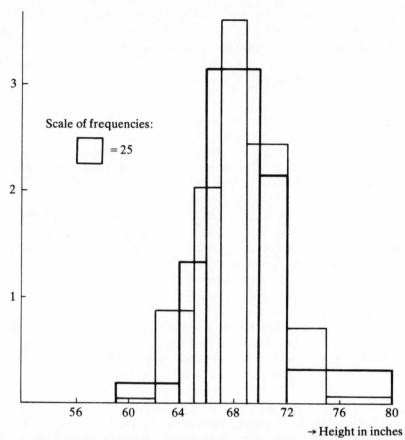

Figure 3.2.4: Histograms of frequency table of continuous data, grouped in cells of unequal widths. The two histograms shown are both drawn from the same data, using different cell boundaries in the two cases. (Data from Table 3.2.6.)

continuous data. In a histogram, for each cell (h_{r-1}, h_r), a rectangle of height proportional to $f_r/(h_r - h_{r-1})$ is erected on the interval (h_{r-1}, h_r) as base. Thus the area of the cell is proportional to the frequency f_r, and the area of that part of the histogram that lies between abscissae h_j and h_m is proportional to the number of observations x such that $h_j < x \le h_m$. If the unit of area is chosen so that the total area enclosed by the histogram is unity, we may interpret the area between h_j and h_m as a rough estimate of $P(h_j < X \le h_m)$. Thus the histogram is a sample analogue of the graph of the p.d.f.

Examples are shown in Figure 3.2.4.

(e) *Sample analogue of c.d.f.; probability graph papers*

As in Example 3.2.1, the relative frequencies in a frequency table of continuous data form a natural sample analogue of the p.d.f., and, again as in that Example, the relative cumulative frequencies form a natural sample analogue of the c.d.f. [see § 1.4.2(i) for these abbreviations]. The *relative cumulative frequency* ('r.c.f.') *function*, defined on the h_r, is

$$\text{r.c.f. } (h_r) = \sum_1^r f_j / n, \qquad r = 1, 2, \ldots, k.$$

This function is also known as the *empirical c.d.f.* [see § 14.2]. Not often invoked in the case of discrete data, it is frequently used as the basis for subjective tests and comparisons in the case of continuous data. Were it not for sampling fluctuations, this function would coincide (where it is defined) with the c.d.f. $F(x) = P(X \leq x)$ of the underlying random variable X [see II, § 10.1.1].

A useful graphical method of assessing the extent of this coincidence is based on the following idea. Since $F(x)$ is necessarily an increasing function, it is always possible to rescale the ordinates nonuniformly in such a way that the graph of $F(x)$ against x becomes a straight line. Starting with ordinary uniformly-ruled graph paper one imposes a new non-uniform ordinate scale as follows: for each of a set of suitable values of y (e.g. $y = 0$, $0 \cdot 01$, $0 \cdot 02, \ldots, 0 \cdot 99$) one attaches the label $F(y)$ to the point on the ordinate axis whose distance from the origin is y, and one plots points with reference to the newly labelled scale. When $F(x)$ is a function for which, in its explicit form, the argument at x has the form $(x - \lambda)/\omega$ for constants λ and ω, this rescaling will have the effect that the graph is a straight line whatever the values of λ and ω.

This is very convenient, since a plot of the relative cumulative frequency function on this specially prepared graph paper would then consist of a set of points lying near a straight line, thus making possible a quick subjective judgement of whether or not the plot in question approximates that of the c.d.f. of the putative distribution. For an illustration, see Example 3.5.1.

Graph paper of this kind is commercially available for the Normal, ('arithmetic probability paper' [see II, § 11.4.8]) log-Normal, ('logarithmic probability paper' [see Figure 2.7.1]), and certain other distributions.

Numerous examples of the use of probability papers will be found in Hald (1952)—see Bibliography C.

3.3. SOME GENERAL CONCEPTS AND CRITERIA FOR ESTIMATES

3.3.1. Introduction: Dimension, Exchangeability, Consistency, Concentration

Section 3.2 opened with a reference to the fundamental intuitive idea that samples resemble their parent populations. We open this section with a further set of intuitively attractive principles, this time ones that govern our attitude to a statistic that might be a candidate for the rôle of estimate for a specified parameter of a given probability distribution.

(a) *Dimensions*

The first of these, which might be called the principle of right dimensions, states that, when θ is not a non-dimensional quantity but possesses physical dimensions such as time or length, the estimate $\hat{\theta}$ should have the same physical dimensions as θ. Suppose for example we assume that the successive instants of emission of particles from a radioactive source form a Poisson process with intensity θ. [see II, § 5.4, § 11.2, § 20.1.] Then the inter-emission intervals are exponentially distributed [see II, § 11.2] with p.d.f. at x given by $\theta e^{-\theta x}$, $x > 0$. Given a sample $x_1, x_2, \ldots, x_n$ of such intervals, a statistic $t(x_1, x_2, \ldots, x_n)$ intended to estimate θ must have the same physical dimension as θ, namely [time]$^{-1}$. A possible such estimate would be $n/\sum_1^n x_r$, the reciprocal of the sample mean. The sample mean itself is excluded by this criterion for consideration as a possible estimate of θ.

When the dimensions of an estimator are not obvious from the definition it is often useful to examine the expected value.

EXAMPLE 3.3.0.   *The Negative Binomial distribution.*   If N is the number of Bernoulli (θ) trials required to achieve a fixed number x of sucesses, we have

$$P(N = n) = \binom{n-1}{x-1} \theta^x (1-\theta)^{n-x},$$

$n = x, x+1, x+2, \ldots$ (where x is a positive integer, and $0 < \theta < 1$)

[see II, § 5.2.4]. We may ask whether $1/n$ is a possible estimate of θ.

The required dimension is the same as that of $E(1/N)$. It is easier to evaluate $E\{1/(N-1)\}$ than $E(1/N)$ and we find

$$E\{1/(N-1)\} = \sum_{n=x}^{\infty} \frac{1}{n-1} \binom{n-1}{x-1} \theta^x (1-\theta)^{n-x}$$

$$= \theta/(x-1) \qquad (x \geq 2).$$

This shows that $1/(n-1)$ does have the correct dimensions, and, indeed, that $(x-1)/(n-1)$ is an unbiased estimate [see § 3.3.2] of θ.

(b) *Exchangeability*

A second principle, the principle of exchangeability, states that, when an estimate $t(x_1, x_2, \ldots, x_n)$ is based on a random sample $(x_1, x_2, \ldots, x_n)$ of equally accurate observations on a given random variable X, such that the order in which the observations occur is judged to be irrelevant, the estimate should be a symmetric function [see I: § 14.16] of the observations. This is exemplified in such widely used statistics as $\bar{x} = \sum_1^n x_r/n$ and $s^2 = \sum_1^n (x_r - \bar{x})^2/(n-1)$.

(c) *Consistency*

A further, and more profound principle is that of *consistency*. This is an attempt to formalize the idea that an estimate $\hat{\theta}$ of a parameter θ ought in some objective sense to be recognizably an approximation to θ itself rather than to, say 2θ, $1/\theta$, $\exp(\theta)$, etc., or to some quite other parameter ϕ. The idea is easier to state than to formalize. Fisher, who first made the idea explicit, produced a formulation along the following lines. The sample data are supposed to be in the form of a frequency table. It is conceivable that, by chance, one might obtain a sample that is a perfect replica of the population, in the sense that the frequencies $f_1, f_2, \ldots, f_k$ in the sample are exactly proportional to the corresponding probabilities $\pi_1, \pi_2, \ldots, \pi_k$ in the population. In such a sample, the value of the estimate ought to coincide exactly with the estimated parameter. Thus, if we denote the estimate $\hat{\theta}$ by $t(f_1, f_2, \ldots, f_k)$, the principle of consistency requires that

$$t(n\pi_1, n\pi_2, \ldots, n\pi_k) = \theta. \tag{3.3.1}$$

Here $n = \sum_1^k f_r$ denotes the sample size.

EXAMPLE 3.3.1.   *Consistency of an estimate of the parameter of a Geometric distribution.*   Consider the case where θ is the parameter of a Geometric variable S, with p.d.f.

$$P(S = s) = \pi_s = \theta(1-\theta)^{s-1}, \qquad s = 1, 2, \ldots$$

[see II, § 5.2.3]. Based on a sample in which the observed value s occurred with frequency f_s, $s = 1, 2, \ldots, k$, the maximum-likelihood estimate [see § 3.5.4] of θ would be

$$\hat{\theta} = 1/\bar{s} = n \bigg/ \sum_1^k s f_s, \qquad \left(n = \sum_1^k f_s \right).$$

In our case $f_s = 0$ for $s > k$, whence we can write this as

$$\hat{\theta} = n \bigg/ \sum_1^\infty s f_s,$$

The consistency requirement in this case is, then, that

$$\theta = 1 \bigg/ \sum_1^\infty s \pi_s$$

$$= 1/E(S).$$

Since in fact $E(S) = 1/\theta$, the condition is satisfied. $\hat{\theta}$ is a *consistent estimate* (in Fisher's sense) of θ.

This attractive concept unfortunately loses a good deal of its direct simplicity when we attempt to apply it to a continuous distribution, and, probably for this reason, it has not become part of the accepted canon. Instead, standard statistical wisdom uses the following related (but different) criterion, also (somewhat confusingly) called consistency.

DEFINITION 3.3.1.   *Consistency of an estimate. Convergence in probability.*   An estimate $\hat{\theta}_n = t_n(x_1, x_2, \ldots, x_n)$, based on a sample of n observations $(n = 1, 2, \ldots)$ on a random variable X, is regarded as belonging to a sequence $t_1, t_2, \ldots$ in which the explicit form of t_n as a function of $x_1, x_2, \ldots, x_n$ is specified for each value of n. Correspondingly, we envisage a sequence of random variables $\hat{\Theta}_n = t_n(X_1, X_2, \ldots, X_n)$, $n = 1, 2, \ldots$, where the X_r are statistical copies of X [see Definition 2.2.1]. Then $\hat{\theta}_n$ is said to be a consistent estimate of θ if, as $n \to \infty$, $\hat{\Theta}_n$ *converges in probability* to θ, that is, if, for every $h > 0$, however small,

$$P(\theta - h \le \hat{\Theta}_n \le \theta + h) \to 1 \qquad (3.3.2)$$

as $n \to \infty$ [see IV, § 1.2]. A convenient sufficient condition for this convergence is that

$$E(\hat{\Theta}_n) \to \theta$$

and

$$\left.\begin{array}{l} \\ \\ \\ \end{array}\right\} \qquad (3.3.3)$$

$$\mathrm{var}\,(\hat{\Theta}_n) \to 0$$

as $n \to \infty$.

On this definition, therefore, a consistent estimate has a high probability of being nearly equal to the parameter that is being estimated, provided the sample is sufficiently large.

EXAMPLE 3.3.2.   *Consistency of an estimate of the variance.*   The estimate of the variance σ^2 of a Normal distribution, given by (2.5.20) with $a(n) = n - 1$, usually denoted by the symbol s^2, has sampling expectation identically equal to σ^2, and it has sampling variance equal to $2\sigma^4/(n-1)$, which converges to zero. It follows that s^2 is a consistent estimate of σ^2 (in the sense of convergence in probability.)

Underlying the concept of consistency is the idea that, for sufficiently large values of the sample size x, the sampling distribution of the estimate should have a unimodal p.d.f. [see II, § 10.1.3], with as high and narrow a peak as possible, the maximum occurring near θ, (see Figure 3.3.1), such that, as n increases, the peak becomes ever higher and narrower, with its maximum approximating ever more closely to θ.

The weakness of the consistency concept of section 3.3.1 lies in the fact that in practice one is usually interested in the sampling distribution of an estimate based on a sample of small or moderate size, and there is no guarantee that an estimate that is consistent in the sense described will have a p.d.f. resembling Figure 3.3.1 when n is not large. For such samples the principle of consistency needs to be supplemented by another intuitive concept which we might call the principle of high local probability.

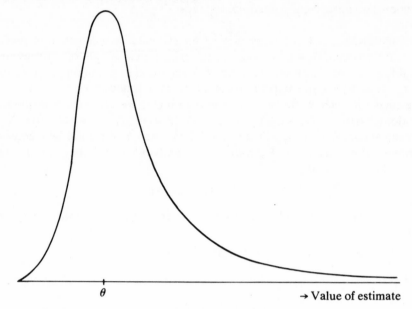

θ

→ Value of estimate

Figure 3.3.1: Sampling p.d.f. of an acceptable estimate of θ.

(d) *Concentration (high local probability)*

This states that an estimate $\hat{\theta}$ of θ should have a high probability of being nearly equal to θ, and a low probability of being very different from θ.

The following example illustrates the concept.

EXAMPLE 3.3.3.   *Sampling distributions of $\bar{x}$ and s^2 in a Normal sample.*   In the case of the Normal (μ, σ) distribution [see II, § 11.4.3], consider the statistics

$$t_1 = \bar{x} = \sum_1^n x_r/n,$$

$$t_2 = s^2 = \sum_1^n (x_r - \bar{x})^2/(n-1),$$

as estimates of μ and σ^2 respectively. They have the following sampling distributions [see § 2.2]. For t_1, the sampling distribution is Normal $(\mu, \sigma/\sqrt{n})$. This is unimodal, with mode at μ. The mode locates the maximum of a peak whose width (as measured by the sampling standard deviation) is proportional to $n^{-1/2}$. Thus the larger the value of n, the narrower is the peak.

In the case of t_2, the sampling distribution of $(n-1)s^2/\sigma^2$ is a chi-squared distribution with $n-1$ degrees of freedom [see § 2.5.4(a)]. The p.d.f. of s^2 at z is given by

$$\frac{(n-1)^{(n-1)/2}z^{(n-3)/2}\exp\{-(n-1)z/2\sigma^2\}}{2^{(n-1)/2}\sigma^{n-1}\Gamma\{(n-1)/2\}},\qquad z>0 \qquad (3.3.4)$$

(c.f. (2.5.21)).

This is a unimodal distribution with mode at the point

$$\left(1-\frac{2}{n-1}\right)\sigma^2,$$

near the desired value σ^2. The width of the peak, as measured by the sampling standard deviation, is $\sigma^2\sqrt{2/(n-1)}$—narrow when n is large (c.f. (2.5.22)).

Both estimates satisfy the requirements of the intuitive principle of high local probability enunciated above. They are in fact generally accepted as the 'best' estimates of μ and σ^2.

The requirement that an estimator of θ should be 'concentrated' in the neighbourhood of θ, in the sense implied in the opening paragraph of section 3.3.1(d), is a watering down of the (usually unrealistic) concept of 'maximal concentration' mentioned in (3.1.1). Some progress towards maximal concentration might be expected by adopting a criterion of *minimal mean square error*, a property enjoyed by an estimator T of θ when, for every other estimator T', the mean square errors $E(T-\theta)^2$, $E(T'-\theta)^2$ satisfy the inequality

$$E(T-\theta)^2\le E(T'-\theta)^2$$

for all θ. Unfortunately, estimators having minimal mean square error do not in general exist. (There are, nonetheless, strong intuitive grounds for regarding one estimator as better than another if it has smaller mean square error.)

If we restrict attention to *unbiased* estimators, (an estimator T being unbiased for θ if $E(T)=\theta$ for all θ) minimal mean square error becomes *variance*. Minimal variance unbiased estimators often do exist, and some of their properties are discussed in section 3.3.2. The related concept of efficiency is treated in section 3.3.3.

The one remaining important concept not so far explicitly mentioned is the idea that some statistics might be capable of extracting more information from the data than others. This leads to the concept of sufficiency, a *sufficient* estimate of θ being a statistic which in a defined sense extracts *all* the information in the sample that is relevant to θ. Some consequences of this idea are discussed in section 3.4.

3.3.2.   Unbiased Estimates and Minimum-variance Unbiased Estimates

The principles discussed in section 3.3.1 provide necessary conditions that a good estimate must satisfy but they do not tell us how to find such an estimate. What we need is a tighter specification of required properties in the form of a constructive definition; or, alternatively, a general method that leads to estimates with desirable properties.

Such methods are discussed in section 3.5 and in Chapter 6. In the present section we confine ourselves to the 'constructive definition' approach.

The requirements of the principle discussed in section 3.3.1 can be achieved by requiring that the 'centre' of the sampling distribution should be close to θ, and that the 'spread' of the sampling distribution should be as small as possible. In this formulation, the 'centre' is undefined: it could in principle be the mode [see II, § 10.1.3] the median [see II, § 10.3.3] or the expected value [see II, § 10.4.1]. For reasons of tractability it is the expected value that is usually chosen. Again, the requirement that the sampling expectation should be 'close' to θ leaves 'close' undefined. The criterion finally adopted, is that the sampling expectation of the estimate $\hat{\theta}$ should *coincide* with θ. The estimate is then said to be *unbiased*. The spread or dispersion of the sampling distribution is then conveniently measured by its variance. We are thus led to the concept of a '*minimum-variance unbiased estimate*'.

DEFINITION 3.3.2. *Unbiased estimate.* Let $\hat{\theta} = t(x_1, x_2, \ldots, x_n)$ be an estimate of θ, based on data in which x_r is a realization of a random variable X_r, $r = 1, 2, \ldots, n$. This is an *unbiased* estimate of θ if

$$E\{t(X_1, X_2, \ldots, X_n)\} \equiv \theta,$$

identically in θ. An estimate that is not unbiased is said to be *biased*, with *bias* $b_n(\theta)$ defined by

$$b_n(\theta) = E\{t(X_1, X_2, \ldots, X_n)\} - \theta.$$

A *minimum-variance unbiased* estimate is one which is unbiased and which, subject to this condition and, possibly, to additional conditions defining its functional form as a function of the x_r (such as, for example, linearity), has minimal variance.

It will be seen from this introduction that the property of being unbiased is not self-evidently one of the highest importance. Perhaps the most damaging criticism to which it is open is that the property is not invariant under a functional transformation (unless the transformation is linear). Suppose, for example, that we are interested, in a certain technological process, in estimating the probability that our r.v. (X, say) does not exceed a specified value x_0, so that we wish to estimate the value of $F(x_0, \theta)$ where F denotes the c.d.f. of X. If $\hat{\theta}$ is unbiased for θ, $F(x_0, \hat{\theta})$ will *not* (in general) be unbiased for $F(x_0, \theta)$; for such an application as this, therefore, unbiasedness is not particularly helpful.

As another example of the same failing, consider this: if n Bernoulli (θ) trials yield r successes, the statistic r/n is unbiased for θ but n/r is not unbiased for $1/\theta$.

Again, unbiased estimates are not necessarily more concentrated than biased ones. A familiar example is the unbiased estimate $s^2 = \sum_1^n (x_j - \bar{x})^2/(n-1)$ of a population variance. The mean square error of the biased estimate $(n-1)s^2/(n+1)$ is smaller than that of s^2.

Why, then, has unbiasedness acquired the central importance that it bears in statistical procedures? The answer lies in its tractability, linearity, and consistency: the expectation operator $E(\cdot)$ is a familiar one with many pleasant properties which make for tractability; a great number of important statistics are linear functions of the observations, and unbiasedness *is* invariant under *linear* transformations; and independent unbiased estimates may be combined to produce a more accurate unbiased estimate.

The construction of minimal variance unbiased estimates (MVUEs) in general is related to sufficiency. For further information on this see section 3.4.

The examples which now follow relate to the minimization of variance in restricted classes of estimates (linear, quadratic, etc.).

EXAMPLE 3.3.4. *Minimum-variance linear unbiased (MVLU) estimate.* Let $(x_1, x_2, \ldots, x_n)$ be a random sample of observations on a variable X with $E(X) = \theta$, and var $(X) = \sigma^2$. To find the MVLU estimate of θ we first construct the most general linear combination of the observations, say

$$\hat{\theta} = \sum_1^n a_r x_r + b.$$

The sampling expectation [see § 2.3] of this is

$$\theta \sum_1^n a_r + b.$$

For this to be identically equal to θ, that is to $1\theta + 0$, we must have

$$\sum_1^n a_r = 1, \qquad b = 0.$$

The final step is to choose the a_r so as to minimize the sampling variance $\sigma^2 \sum a_r^2$. Now it is easy to see that, subject to the condition $\sum a_r = 1$, $\sum a_r^2$ attains its minimal value when $a_1 = a_2 = \ldots = a_n = 1/n$. [For a discussion of constrained minimization, see IV, § 15.1.3.] Finally, then,

$$\hat{\theta} = \sum x_r/n = \bar{x}$$

is the MVLU estimate of θ. The general theory of minimum-variance linear unbiased estimation is explored in some detail under the heading 'Least Squares Estimation' in Chapters 8, 10.

EXAMPLE 3.3.5. *Minimum-variance quadratic unbiased (MVQU) estimate.* The variance σ^2 of X has the dimensions of X^2 and, invoking the principle of the right dimension [see § 3.3.1(a)] we require that our estimate of σ^2

should also have the dimensions of X^2. The simplest such function is a quadratic form [see I, Chapter 9], in the sample values $x_1, x_2, \ldots, x_n$, say

$$v = \mathbf{x}'\mathbf{Q}\mathbf{x},$$

where $\mathbf{x}' = (x_1, x_2, \ldots, x_n)$, $\mathbf{x}$ is its transpose, and $\mathbf{Q} = (q_{rs})$ is a symmetric $(n \times n)$ matrix [see I, § 6.2]. By the principle of exchangeability (§ 5.3.1(b)), v must be of the form

$$\alpha \sum_i x_i^2 + \beta \sum\sum_{i \neq j} x_i x_j.$$

The sampling expectation is therefore [see II, § 9.2.1]

$$\alpha n(\sigma^2 + \theta^2) + \beta n(n-1)\theta^2 = \alpha n\sigma^2 + \{\alpha n + \beta n(n-1)\}\theta^2.$$

For unbiasedness, this must be identically equal to σ^2, whence (unless θ is known to be zero)

$$\alpha n = 1, \qquad \alpha n + \beta n(n-1) = 0$$

i.e.

$$\alpha = 1/n, \qquad \beta = -1/n(n-1).$$

Thus α and β have been identified uniquely, solely by invoking exchangeability and unbiasedness. In this particular case, given the functional form of the estimate, the imposition of exchangeability is in fact equivalent to minimizing the sampling variance [see Theorem 3.3.1]. It follows that the UVQU estimate of σ^2 is

$$\frac{1}{n}\left\{\sum_1^n x_i^2 - \frac{1}{n-1}\sum_i\sum_j{}_{i \neq j} x_i x_j\right\}.$$

After a little reduction it can be shown that this is simply the familiar estimate

$$s^2 = \sum (x_i - \bar{x})^2/(n-1).$$

EXAMPLE 3.3.6   (*continuation*).   In the apparently simpler special case of the above where θ is known to be zero the principle of exchangeability operates in the same way but the condition for unbiasedness becomes simply

$$\alpha n = 1.$$

We are thus left with the estimate

$$\sum_1^n x_i^2/n + \beta \sum_i\sum_j{}_{i \neq j} x_i x_j,$$

with β so far undetermined. Routine calculations show that the sampling variance of this statistic is

$$\left\{\frac{2}{n} + n(n-1)\beta^2\right\}\sigma^4.$$

This is minimized by taking $\beta = 0$, so that, in this case, the MVUB estimate is $\sum_1^n x_i^2/n$, with sampling variance $2\sigma^4/n$.

There is a theorem (due to Halmos) that relates the concept of minimum-variance unbiased estimates to the principle of exchangeability (§ 3.2.1(b)). This is as follows:

THEOREM 3.3.1. *Symmetric estimates. Let* $x_1, x_2, \ldots, x_n$ *be a sample of exchangeable observations on X and let* $g(x_1, x_2, \ldots, x_n)$ *be an estimate of* θ, *a parameter of the distribution of X. If g is not a symmetric function of the* x_r *define the symmetrized version of g as* $\bar{g} = \sum_1^{n!} g_i(x_1, x_2, \ldots, x_n)/n!$, *where*

$$g_i(x_1, x_2, \ldots, x_n) = g\{P_i(x_1, x_2, \ldots, x_n)\}, \qquad i = 1, 2, \ldots, n!,$$

and, for each i, $P_i(x_1, x_2, \ldots, x_n)$ *is the ith (in an arbitrary order) of the* $n!$ *permutations of natural order [see I, § 8.1] applied to the ordered set* $(x_1, x_2, \ldots, x_n)$. *We take* P_1 *as the identity permutation so that* $g_1 = g$. *Then* $\bar{g}$ *is an unbiased estimate of* θ, *and the sampling variance of the statistic* $\bar{g}$ *is less than that of g.*
If g is a symmetric function, $\bar{g}$ *coincides with g.*

This means, for example, that of all linear unbiased estimates of the expectation θ of a distribution, the sample mean is the best in the sense of having smallest variance. (For the most general linear estimate is $\sum_1^n a_r x_r + b$, with arbitrary coefficients $a_1, a_2, \ldots, a_n, b$. The sampling expectation of this is $\theta \sum a_r + b$, and for this to be identically equal to θ, as required for unbiasedness, we must have $\sum_1^n a_r = 1$, and $b = 0$. Thus the most general unbiased linear estimate of θ is $\sum_1^n a_r x_r$, with $\sum_1^n a_r = 1$. The symmetrized version of this is clearly $\bar{x} = \sum_1^n x_r/n$ and, according to Halmos' Theorem, this has smaller variance than every other unbiased linear estimate $\sum a_r x_r + b$ (c.f. Example 3.3.4).

3.3.3. Efficiency: the Cramér–Rao ('C–R') bound

(a) *The C–R inequality: random sample from a univariate one-parameter distribution*

Examples 3.3.4, 3.3.5 and 3.3.6 show how to achieve minimal sampling variance in an estimate of specified functional form but leave unanswered the question whether an estimate having some other functional form might not have a still smaller sampling variance. This question is answered by the following theorem, which, under fairly general conditions, provides a lower bound to the sampling variance of an unbiased estimate.

THEOREM 3.3.2. *Lower bound to the sampling variance of an unbiased estimate. Let* $(x_1, x_2, \ldots, x_n)$ *be a sample of independent observations on a* *random variable X whose p.d.f. at x is f(x, θ), where θ is a parameter of unknown magnitude. Let $\hat{\theta}_n = t_n(x_1, x_2, \ldots, x_n)$ be an unbiased estimate of θ. Then, subject to certain regularity conditions on f, the sampling variance v_n of $\hat{\theta}_n$ satisfies the inequality*

$$v_n \geq 1/I_n(\theta) = 1/nI(\theta) \tag{3.3.5}$$

where

$$I_n(\theta) = nI(\theta),$$

and

$$I(\theta) = E\{\partial \log f(X, \theta)/\partial\theta\}^2$$
$$= -E\{\partial^2 \log f(X, \theta)/\partial\theta^2\}. \tag{3.3.6}$$

$I_n(\theta)$ has been called by Fisher the amount of information in the sample, and $I(\theta)$ the amount of information in a single observation. *The equality sign in* (3.3.5) *holds if, and only if,*

$$\sum_1^n \partial \log f(x_i, \theta)/\partial\theta = nI(\theta)(\hat{\theta}_n - \theta). \tag{3.3.7}$$

It will be seen from (3.3.5) that, subject to the regularity conditions referred to, the lower bound to the sampling variance of an unbiased estimate based on *n* observations is proportional to $1/n$.

The earliest known version of this theorem was produced by Fisher. Successive improvements and generalizations are associated with the names of Frechet, Dugué, Cramér, Rao. To attach Fisher's name to every statistical concept or theorem originated by him would be as confusing as it would be to use Gauss's name similarly in analysis. For convenience, therefore, the present theorem is usually referred to as 'the *Cramér–Rao inequality*'.

[It may be noted that the condition (3.3.7) holds if, in particular,

$$\partial \log f(x, \theta)/\partial\theta = I(\theta)\{A(x) - \theta\}.$$

In this case

$$f(x, \theta) = A(x) \int I(\theta)\, d\theta - \int \theta I(\theta)\, d\theta + D(x)$$
$$= A(x)B(\theta) + C(\theta) + D(x), \tag{3.3.8}$$

say; then

$$\sum_1^n \partial \log f(x_r, \theta)/\partial\theta = I(\theta) \sum A(x_r) - n\theta I(\theta)$$

and the condition (3.3.7) reduces to

$$\hat{\theta}_n = \sum A(x_r)/n,$$

which is required to be an unbiased estimate of θ. A p.d.f. of the form (3.3.8) is said to belong to the 'exponential family': see section 3.4.2.]

EXAMPLE 3.3.7.   *Attainability of the C–R bound in the case of the Binomial distribution.*   Suppose X has the Bernoulli distribution [see II, § 5.2.1] with parameter θ, so that the p.d.f. of X is

$$f(x, \theta) = \begin{cases} 1-\theta, & x=0 \\ \theta, & x=1 \end{cases} = (1-\theta)^{1-x}\theta^x, \qquad x=0, 1.$$

Then

$$\log f(X, \theta) = (1-X)\log(1-\theta) + X\log\theta,$$

$$\partial \log f(X, \theta)/\partial\theta = -(1-X)/(1-\theta) + X/\theta$$
$$= (X-\theta)/\theta(1-\theta),$$

and

$$\partial^2 \log f(X, \theta)/\partial\theta^2 = -(1-X)/(1-\theta)^2 - X/\theta^2.$$

Since $E(X) = \theta$, we have

$$I_n(\theta) = n\{1/(1-\theta) + 1/\theta\} = n/\theta(1-\theta),$$

and the lower bound is therefore $\theta(1-\theta)/n$.
    We see that

$$\sum \partial \log f(x_i, \theta)/\partial\theta = \sum (x_i - \theta)/\theta(1-\theta)$$
$$= \{\sum x_i/n - \theta\}/\{\theta(1-\theta)/n\}.$$

Now this is equal to $I_n(\theta)\{\hat\theta_n - \theta\}$ provided we take $\hat\theta_n = \sum x_i/n$ $(=r/n$, say where $r = \sum x_i$ is the total number of "successes" in the sample). Thus the Cramér–Rao bound is attainable in this example, the attaining estimate being r/n.

EXAMPLE 3.3.8.   *The C–R bound for σ^2 in a Normal $(0, \sigma)$ distribution.*   Suppose X is Normally distributed, with $E(X) = 0$ and var $(X) = \theta$. The p.d.f. is

$$f(x, \theta) = a\theta^{-1/2} \exp(-x^2/2\theta), \quad a = (2\pi)^{-1/2},$$

whence

$$\partial \log f(X, \theta)/\partial\theta = -1/2\theta + X^2/2\theta^2,$$
$$\partial^2 \log f(X, \theta)/\partial\theta^2 = 1/2\theta^2 - X^2/\theta^3.$$

As $E(X) = 0$, it follows that $E(X^2) = $ var $(X) = \theta$, whence

$$I_n(\theta) = -n(1/2\theta^2 - 1/\theta^2) = n/2\theta^2.$$

The lower bound to the sampling variance is thus $2\theta^2/n$. This is in fact attainable, since

$$\partial \sum \log f(x_i, \theta)/\partial\theta = \sum (x_i^2 - \theta)/2\theta^2$$
$$= I_n(\theta)\{\hat{\theta}_n - \theta\}$$

provided that we take $\hat{\theta}_n = \sum x_i^2/n$. (c.f. Example 3.3.6).

EXAMPLE 3.3.9.   *The C–R bound for σ in a Normal $(0, \sigma)$ distribution.*   Suppose, as in Example 3.3.8, that X is Normally distributed, with $E(X) = 0$, but that this time the parameter θ to be estimated is the *standard deviation*, so that $\mathrm{var}\,(X) = E(X^2) = \theta^2$. In this case we have

$$f(x, \theta) = (2\pi)^{-1/2}\theta^{-1} \exp\,(-x^2/2\theta^2),$$

whence

$$\partial \log f(X, \theta)/\partial\theta = -1/\theta + X^2/\theta^3$$

and

$$\partial^2 \log f(X, \theta)/\partial\theta^2 = 1/\theta^2 - 3X^2/\theta^4$$

so that

$$I_n(\theta) = 2n/\theta^2.$$

Thus the Cramér–Rao lower bound is $\theta^2/2n$. This is however *not* attainable under the conditions of the theorem since

$$\partial \sum \log f(x_i, \theta)/\partial\theta = (\sum x_i^2 - n\theta^2)/\theta^3$$
$$= I_n(\theta)\{\sum x_i^2/2n\theta - 2\theta\},$$

which is not of the required form (3.3.7).

There is a modified version of Theorem 3.3.2 that applies to a biased estimate. This is as follows.

THEOREM 3.3.3.   *The C–R bound for a biased estimate.*   *Suppose, with the same notation in Theorem 3.3.2, that $\theta_n^* = \theta_n^*(x_1, x_2, \ldots, x_n)$ is an estimate of θ with bias $b_n(\theta)$ [see Definition 3.3.2], that is, with sampling expectation $\theta + b_n(\theta)$. Then the sampling variance of θ_n^* is not less than*

$$\{1 + db_n(\theta)/d\theta\}^2/I_n(\theta).$$

(b)   *Attainment of the C–R bound. Efficiency of an estimate*

DEFINITION 3.3.3.   *Efficient estimates.*   An unbiased estimate of a parameter θ is said to be *efficient* if its sampling variance equals the Cramér–Rao bound.

It is somewhat disconcerting to find, as in Examples 3.3.8 and 3.3.9, that there may be an efficient estimate of a parameter σ^2 but not of its square root σ. This is part of the price we pay for the general convenience of working with expectations.

It follows from (3.3.7) that only in exceptional cases does a family of distributions admit of efficient estimation of its parameters. Where an efficient estimate $\hat{\theta}_n$ exists, it is obviously worth using. In terms of the variance criterion, no better estimate can be found (within the conditions of the theorem). An alternative unbiased estimate that is not efficient, say θ_n^*, may be said to exploit the sample less efficiently, since its precision (as measured by the reciprocal of its sampling variance) is less than that of the efficient estimate. Its 'efficiency' may be defined as the ratio $\mathrm{var}\,(\hat{\theta}_m)/\mathrm{var}\,(\theta_n^*)$. This concept of efficiency has however been extended to situations not admitting an efficient estimate, and, in general, we have the following generally accepted definition:

DEFINITION 3.3.4. *Efficiency of an estimate.* The *efficiency* of an unbiased estimate $\tilde{\theta}_n$ of a parameter θ, based on a sample of size n, is

$$\mathrm{eff}\,(\tilde{\theta}_n) = \frac{v_n(\min)}{\mathrm{var}\,(\tilde{\theta}_n)} \quad (\leq 1)$$

where

$$v_n(\min) = 1/I_n(\theta)$$

is the Cramér–Rao lower bound (Theorem 3.3.2).

This use of the words 'efficient', 'efficiency', and so on is somewhat dubious since it is not necessarily always the case that small sampling variance implies high precision. Even if we ignore this objection, the use of the use of the word 'efficiency' in cases where the Cramér–Rao bound is not attainable is somewhat ambiguous since the best estimate that can be found could well have an efficiency less than 100% according to this definition. A more generally useful concept is that of *relative* efficiency, defined as follows:

DEFINITION 3.3.5. *Relative efficiency.* The relative efficiency of two unbiased estimates θ_n^*, θ_n^{**} of a parameter θ, both based on the same sample, is defined as

$$\mathrm{var}\,(\theta_n^{**})/\mathrm{var}\,(\theta_n^*),$$

the efficiency of θ_n^* relative to θ_n^{**} being greater than unity if $\mathrm{var}\,\theta_n^* < \mathrm{var}\,\theta_n^{**}$.

EXAMPLE 3.3.10. *Relative efficiency of mean deviation and sample standard deviation as estimates of σ.* In the case of a Normal distribution with zero expectation, and standard deviation σ, the Cramér–Rao bound for the sampling variance of an unbiased estimate of σ based on a sample of size n is $\sigma^2/2n$.

This bound is not attainable, as shown in Example 3.3.9. The mean deviation

$$\sigma_n^* = (\pi/2)^{1/2} \sum_1^n |x_i|/n$$

is unbiased and has sampling variance $(\pi-2)\sigma^2/2n$, so that its efficiency is $1/(\pi-2)=0\cdot88$.

The standard estimate of σ^2 is

$$s_n^2 = \sum_1^n (x_i - \bar{x})^2/(n-1),$$

and from this we may construct the unbiased estimate of σ given (see (2.5.29)) by

$$\tilde{\sigma}_n = c_n s_n, \qquad c_n = \frac{2}{n-1}\frac{\Gamma(n/2)}{\Gamma\{(n-1)/2\}}.$$

This has sampling variance $(1-c_n^2)\sigma^2$ (see (2.5.31)) and its efficiency is therefore

$$(1-c_n^2)/2n.$$

Unlike the case of the mean deviation estimate σ_n^*, this depends on the sample size n.

Some representative values of the relative efficiency of σ_n^* and $\tilde{\sigma}_n$ (taken from Table 2.5.2) are as follows:

n	5	10	25	50
Efficiency of $\tilde{\sigma}_n$	0·86	0·93	0·97	0·99
Efficiency of σ_n^* relative to that of $\tilde{\sigma}_n$	1·02	0·94	0·90	0·88

(c) Regularity conditions

The regularity conditions under which Theorem 3.3.2 holds, which are said to define a 'regular estimation case', are related to the conditions for the validity of the identity

$$\frac{\partial}{\partial\theta} \int_{-\infty}^{\infty} \cdots \int_{-\infty}^{\infty} t(x_1, x_2, \ldots, x_n) g_n(x_1, x_2, \ldots, x_n; \theta)\, dx_1 \ldots dx_n$$

$$= \int_{-\infty}^{\infty} \cdots \int_{-\infty}^{\infty} t(x_1, x_2, \ldots, x_n) \frac{\partial}{\partial\theta} g_n(x_1, x_2, \ldots, x_n; \theta)\, dx_1 \ldots dx_n$$

$$(3.3.9)$$

where

$$g_n(x_1, x_2, \ldots, x_n; \theta) = \prod_1^n f(x_i, \theta),$$

since in the proof of the theorem one has to perform this differentiation under the integral sign. Difficulties arise if the p.d.f. has a corner or a discontinuity [see IV, § 2.3] at a point which is itself a function of θ. For example, if

$$f(x, \theta) = \begin{cases} g(x, \theta), & 0 \le x \le \theta \\ 0, & x > \theta \end{cases}$$

we have

$$\int_{-\infty}^{\infty} t(x)f(x, \theta) \, dx = \int_{0}^{\theta} t(x)g(x, \theta) \, dx$$

and

$$\frac{\partial}{\partial \theta} \int_{-\infty}^{\infty} t(x)f(x, \theta) \, dx = \frac{\partial}{\partial \theta} \int_{0}^{\theta} t(x)g(x, \theta) \, dx$$

$$= \int_{0}^{\theta} t(x) \frac{\partial g}{\partial \theta} \, dx + t(\theta)g(\theta, \theta)$$

$$\neq \int_{-\infty}^{\infty} t(x) \frac{\partial}{\partial \theta} f(x, \theta) \, dx, \qquad \text{[see IV, § 4.7]}$$

with analogous effects in the multiple integral (3.3.8). This occurrence of a discontinuity in $f(x, \theta)$ at an x-value which is a function of θ is in practice the most important example of failure of the regularity conditions required by the theorem.

EXAMPLE 3.3.11.   *Extremity of a uniform distribution.*   Suppose X has the continuous uniform distribution on $(0, \theta)$, so that the p.d.f. is

$$f(x, \theta) = \begin{cases} 0, & x \le 0 \\ 1/\theta, & 0 < x \le \theta \\ 0, & x > \theta. \end{cases}$$

The p.d.f. at u $(0 < u \le \theta)$ of the largest observation $x_{(n)}$ in a sample of size n is nu^{n-1}/θ^n, [see II, Example 15.2.1] and this has expected value $n\theta/(n+1)$. Consequently $\theta_n^* = (n+1)x_{(n)}/n$ is an unbiased estimate of θ. However one cannot apply the concept of efficiency to this estimate, since the distribution under discussion has a discontinuity at θ and is therefore not covered by the theorem. In fact the sampling variance of θ_n^* is $n\theta^2/(n+2)(n+1)^2$; this diminishes far more rapidly with increasing n than does the Cramér–Rao lower bound in a regular estimation case—the former like n^{-2}, the latter like n^{-1}.

(d) *The C–R inequality for independent vector observations on a 1-parameter multivariate distribution*

Theorem 3.3.2 as stated applied to a sample of independent scalar observations $(x_1, x_2, \dots, x_n)$ on a univariate distribution. It remains valid however if each observation x_r in the statement is interpreted as a vector observation on a multivariate distribution, as, for example, in the case of the bivariate distribution of the pair (Y, Z), with n mutually independent pairs of observations $(y_1, z_1), (y_2, z_2), \ldots, (y_n, z_n)$. The random variable '$X$' of the theorem is replaced by the pair (Y, Z), with joint p.d.f. $f(y, z; \theta)$ at (y, z), and the observation x_r by the pair (y_r, z_r), $r = 1, 2, \ldots, n$.

Subject to regularity conditions, the inequality (3.3.5) still applies, with (3.3.6) replaced by

$$I_n(\theta) = nE\{\partial \log f(Y, Z; \theta)/\partial \theta\}^2$$
$$= -nE\{\partial^2 \log f(Y, Z; \theta)/\partial \theta^2\}. \tag{3.3.9}$$

The bound is attained if, and only if,

$$\sum_1^n \partial \log f(x_i, y_i; \theta)/\partial \theta = I_n(\theta)(\hat{\theta}_n - \theta), \tag{3.3.10}$$

this being the analogue of (3.3.7). Here, as there, $\hat{\theta}_n$ must be a *statistic* which provides an unbiased estimate of θ.

EXAMPLE 3.3.12.   *A 1-parameter trinomial distribution.*   Suppose Y and Z are trinomial $(k; \theta, \theta)$ [see II, § 6.4.1], so that

$$f(y, z; \theta) = P(Y = y, Z = z) = \frac{k!}{y! \, z! \, (k - y - z)!} \, \theta^{y+z}(1 - 2\theta)^{k-y-z}$$

whence

$$\partial \log f(Y, Z; \theta)/\partial \theta = \frac{Y + Z}{\theta} - \frac{2(k - Y - Z)}{1 - 2\theta}$$
$$= \{(Y + Z)/2k - \theta\}/\{\theta(1 - 2\theta)/k\}, \tag{3.3.11}$$

and

$$\partial^2 \log f(Y, Z; \theta)/\partial \theta^2 = -(Y + Z)/\theta^2 - 4(k - Y - Z)/(1 - 2\theta)^2.$$

Since $E(Y) = E(Z) = k\theta$, we have

$$E\{\partial^2 \log f(Y, Z; \theta)/\partial \theta^2\} = 2k/\theta + 4k/(1 - 2\theta)$$
$$= 2k/\theta(1 - 2\theta).$$

On the basis of a sample of n independent pairs of observations $(y_1, z_1), \ldots, (y_n, z_n)$, in (Y, Z), the Cramér–Rao lower bound to the sampling variance of an unbiased estimate of θ is

$$I_n(\theta) = \theta(1 - 2\theta)/2nk.$$

This lower bound is in fact achieved by the estimate

$$\hat{\theta}_n = \sum_1^n (y_i + z_i)/2kn$$

$$= \tfrac{1}{2}(\bar{y} + \bar{z}),$$

since, using (3.3.11), we have

$$\sum \partial \log f(y_i, z_i; \theta)/\partial\theta = \{\sum (y_i + z_i)/2k - n\theta\}/\{\theta(1 - 2\theta)/k\}$$

$$= I_n(\theta)\{\sum (y_i + z_i)/2kn - \theta\}$$

$$= I_n(\theta)(\hat{\theta}_n - \theta).$$

(e) *The C–R inequality for observations that are not independent and/or not identically distributed*

With appropriate modification, the Cramér–Rao inequality applies also in the case where the observations are not independent and/or are not identically distributed, as for example where, for $r = 1, 2, \ldots, n$, x_r is a realization of a random variable X_r which is Normal with $E(X_r) = r\theta$ (a simplified version of linear regression). Let $\hat{\theta}_n$ be defined as in Theorem 3.3.2. The appropriate version of the Cramér–Rao inequality for its sampling variance v_n is then

$$v_n \geq 1/I_n(\theta)$$

where

$$I_n(\theta) = E\{\partial \log g_n(X_1, X_2, \ldots, X_n; \theta)/\partial\theta\}^2$$

$$= -E\{\partial^2 \log g_n(X_1, X_2, \ldots, X_n; \theta)/\partial\theta^2\},$$

$g_n(u_1, u_2, \ldots, u_n)$ being the joint p.d.f. at $(u_1, u_2, \ldots, u_n)$ of the random variables $(X_1, X_2, \ldots, X_n)$.

(f) *A generalization of the C–R inequality to the case of several parameters. The information matrix*

Let X have p.d.f. at x given by $f(x; \theta_1, \theta_2, \ldots, \theta_k)$, the θ_r being unknown parameters, and let $x_1, x_2, \ldots, x_n$ be a sample of n observations on X. (Here, X, and similarly the x_r, may be scalar or vector quantities. We shall treat the scalar case. The modifications required for the vector case are as exemplified above in section 3.3.3(d).) The likelihood function [see § 4.13.1] is

$$l = l(\theta_1, \ldots, \theta_k; x_1, \ldots, x_n) = \prod_{r=1}^n f(x_r; \theta_1, \ldots, \theta_k).$$

Let $\theta_r^* = \theta_r^*(x_1, \ldots, x_n)$, be an unbiased estimate of θ_r, $r = 1, 2, \ldots, k$. The analogue for this multiparameter case of the quantity of information $I_n(\theta)$ of the one-parameter theory is the *information matrix* $\mathbf{I}(\boldsymbol{\theta})$, a symmetric $k \times k$ matrix [see I, § 6.7] whose (r, s) entry is

$$-E\left\{\frac{\partial^2 \log l(\theta_1, \ldots, \theta_k; X_1, \ldots, X_n)}{\partial \theta_r \, \partial \theta_s}\right\}$$

$$= -nE\left\{\frac{\partial^2 \log f(X; \theta_1, \theta_2, \ldots, \theta_k)}{\partial \theta_r \, \partial \theta_s}\right\}, \qquad r, s = 1, 2, \ldots, k. \quad (3.3.12)$$

The multivariate analogue of the sample variance of a single estimate θ^* is the sampling dispersion matrix $\mathbf{V}$ of the estimates $\theta_1^*, \ldots, \theta_k^*$, that is, a symmetric $k \times k$ matrix whose (r, s) entry is the sampling covariance of θ_r^* with θ_s^*, $(r, s = 1, 2, \ldots, k$: when $r = s$ the entry is the sampling variance of θ_r^*).

The analogue for this situation of the Cramér–Rao inequality (3.3.5) is the following:

the matrix

$$\mathbf{V} - \{\mathbf{I}(\boldsymbol{\theta})\}^{-1}$$

is positive semi-definite: that is, for every non-zero $(k \times 1)$ vector $\boldsymbol{\lambda}$ we have

$$\boldsymbol{\lambda}'\mathbf{V}\boldsymbol{\lambda} \geq \boldsymbol{\lambda}'\{\mathbf{I}(\boldsymbol{\theta})\}^{-1}\boldsymbol{\lambda}',$$

i.e.

$$\text{var}(\boldsymbol{\lambda}'\boldsymbol{\theta}^*) \geq \boldsymbol{\lambda}'\{\mathbf{I}(\boldsymbol{\theta})\}^{-1}\boldsymbol{\lambda}. \qquad (3.3.13)$$

(This is a slight abuse of notation in that 'var $(\boldsymbol{\lambda}'\boldsymbol{\theta}^*)$' is intended to denote the sampling variance of $\boldsymbol{\lambda}'\boldsymbol{\theta}^*$.)

The import of the above inequality is this: for any linear combination of the estimates, the sampling variance is not less than the variance of the *same* linear combination of a conceptual set of random variables Z_i, which have the matrix $\mathbf{V}^0 = \{\mathbf{I}(\boldsymbol{\theta})\}^{-1}$ as their dispersion matrix. For example

$$\text{var}(\theta_j^*) \geq \text{var}(Z_j) = V_{jj}^0,$$

$$\text{var}(a\theta_1^* + b\theta_2^*) \geq \text{var}(aZ_1 + bZ_2)$$

$$= a^2 V_{11}^0 + 2ab V_{12}^0 + b^2 V_{12}^0,$$

etc., where V_{ij}^0 is the (i, j) entry in $\mathbf{V}^0 = [\mathbf{I}(\boldsymbol{\theta})]^{-1}$.

EXAMPLE 3.3.13.   *Mean and variance of a Normal distribution.*   Let X be Normally distributed with $E(X) = \theta_1$ and var $(X) = \theta_2$, so that p.d.f. at x is

$$f(x; \theta_1, \theta_2) = (2\pi)^{-1/2} \theta_2^{-1/2} \exp\{-(x - \theta_1)^2/2\theta_2\}.$$

Let θ_1^* and θ_2^* be unbiased estimates of θ_1 and θ_2 respectively, based on a sample $x_1, x_2, \ldots, x_n$ of size n. (Thus we may take $\theta_1^* = \sum_1^n x_r/n = \bar{x}$, and $\theta_2^* = \sum_1^n (x_r - \bar{x})^2/(n-1)$.) We have

$$\log f(X; \theta_1, \theta_2) = \text{constant} - \tfrac{1}{2}\log \theta_2 - (X - \theta_1)^2/2\theta_2,$$

whence

$$\partial^2 \log f / \partial \theta_1^2 = -1/\theta_2,$$

$$\partial^2 \log f / \partial \theta_1 \, \partial \theta_2 = -(X - \theta_1)/\theta_2^2$$

$$\partial^2 \log f / \partial \theta_2^2 = 1/2\theta_2^2 - (X - \theta_1)^2/4\theta_2^3.$$

Taking expectations, we find the information matrix $\mathbf{I}(\theta_1, \theta_2)$ to be

$$-n \begin{pmatrix} -1/\theta_2 & 0 \\ 0 & -1/2\theta_2^2 \end{pmatrix}$$

whence [see I, Equation (6.4.11)]

$$\mathbf{V}^0 = \{\mathbf{I}(\theta_1, \theta_2)\}^{-1} = \begin{pmatrix} \theta_2/n & 0 \\ 0 & 2\theta_2^2/n \end{pmatrix}.$$

The interpretation of the inequality (3.3.13) for this example is therefore as follows: for every choice of constants a, b, and for all unbiased estimates θ_1^* and θ_2^* of θ_1, θ_2,

$$\text{var} \, (a\theta_1^* + b\theta_2^*) \geq (a, b) \begin{pmatrix} \theta_2/n & 0 \\ 0 & 2\theta_2^2/n \end{pmatrix} \begin{pmatrix} a \\ b \end{pmatrix}$$

$$= a^2 \theta_2/n + 2b^2 \theta_2^2/n.$$

In this example it so happens that, for the estimates $\theta_1^* = \bar{x}$ and $\theta_2^* = \sum_1^n (x_r - \bar{x})^2/(n-1)$, $\mathbf{V}^0$ *coincides* with the dispersion matrix of θ_1^*, θ_2^*, so that the inequality (3.3.13) becomes an identity.

EXAMPLE 3.3.14.   *The 2-parameter gamma distribution.*   Suppose X has the 2-parameter gamma distribution [see II, § 11.3.1], with p.d.f. at x given by

$$f(x; \alpha, \beta) = (x^{\alpha-1} e^{-x/\beta})/\beta^\alpha \Gamma(\alpha), \qquad x \geq 0.$$

Using (3.3.12), and the fact that $E(X) = \alpha\beta$, the information matrix based on n observations turns out to be

$$\mathbf{I}(\alpha, \beta) = -n \begin{pmatrix} \psi'(\alpha) & 1/\beta \\ 1/\beta & \alpha/\beta^2 \end{pmatrix}$$

where $\psi'(\alpha) = d^2 \log \Gamma(\alpha)/d\alpha^2$ [see Abramowitz and Stegun (1970)— Bibliography G]. Inverting this, we have

$$\{\mathbf{I}(\alpha, \beta)\}^{-1} = \mathbf{V}^0 = \frac{1}{D} \begin{pmatrix} \alpha/n\beta^2 & -1/n\beta \\ -1/n\beta & \psi'(\alpha)/n \end{pmatrix}$$

where

$$D = \{\alpha\psi'(\alpha) - 1\}/\beta^2.$$

To exemplify, with $\alpha = 2$ and $\beta = 1$ we have $\psi'(2) = 0.045$, $D = 0.29$, and

$$\mathbf{V}^0 = \frac{1}{n}\begin{pmatrix} 6.90 & -3.45 \\ -3.45 & 2.22 \end{pmatrix},$$

whence, for unbiased estimates α^*, β^*, we have

$$\text{var } \alpha^* \geq 6.90/n,$$

$$\text{var } \beta^* \geq 2.22/n,$$

and, for arbitrary a and b,

$$\text{var } (a\alpha^* + b\beta^*) \geq 6.90a^2/n - 7.90ab/n + 2.22b^2/n.$$

3.4. SUFFICIENCY

3.4.1. Definition of Sufficiency

An earlier section (§ 3.3.3) has discussed the concept of efficiency as a measure of the extent to which the sampling variance of an estimate approaches the smallest value that can theoretically be achieved. The concept of *sufficiency* belongs to a similar category of ideas, but is somewhat more profound.

It was discovered by Fisher that in certain cases it is possible to condense into a single statistic the whole of the sample information that is relevant to the parameter being estimated (using the word 'information' here in its everyday sense). Such a statistic is said to be a *sufficient* estimate for the parameter in question.

(The existence of sufficient statistics—even if only in a restricted class of distributions—is of enormous theoretical importance, as exemplified to some extent in the following pages. From the point of view of the practical statistician however it is perhaps less important, since in practice one may be unable to distinguish between two distributions as possible models, one admitting of sufficient statistics and the other not.)

EXAMPLE 3.4.1.   *Sufficiency of observed proportion as estimate of Binomial parameter.*   To illustrate the concept of sufficiency, consider the estimation of the probability p of scoring a 'six' with a biased six-sided die, from data recording the results of n throws of the die. Intuitively one feels that it would be unnecessary to invoke the n *individual* results (i.e. the individual success or failures) and that only the *total* number of successes (or, equivalently, the proportion of successes) is relevant. This intuition is correct, since, as is shown below, the total number of successes is a 'sufficient statistic' for the estimation of the parameter p.

Denote by $x_1, x_2, \ldots, x_n$ the sequence of results of casting the die, where $x_j = 1$ if the jth throw is a success (i.e. a 'six') and $x_j = 0$ otherwise, $j = 1, 2, \ldots, n$. Randomizing [see II, § 3.3], by shaking the die in its box between throws, imposes independence [see II, § 3.6.2]; and we may regard x_j as a realization of the induced random variable X_j, $j = 1, 2, \ldots, n$, where $X_1, X_2, \ldots, X_n$ are i.i.d., with common distribution given by [see II, § 5.3.1]

$$P(X_j = 1) = p, \qquad P(X_j = 0) = 1 - p, \qquad j = 1, 2, \ldots, n,$$

i.e. by

$$P(X_j = y) = p^y (1-p)^{1-y}, \qquad y = 0, 1; \qquad j = 1, 2, \ldots, n. \qquad (3.4.1)$$

Given the observed total number of successes (r_0, say), the conditional joint distribution [see II, § 13.1.4] of the X_j is given by

$$P\left(X_1 = y_1, X_2 = y_2, \ldots, X_n = y_n \Big| \sum_1^n X_j = r_0 \right) \qquad (y_j = 0 \text{ or } 1, \ j = 1, 2, \ldots, n)$$

$$= P\left(X_1 = y_1, X_2 = y_2, \ldots, X_n = y_n, \sum_1^n X_j = r_0 \right) \Big/ P\left(\sum_1^n X_j = r_0 \right). \qquad (3.4.2)$$

Now if $\sum_1^n y_j = r_0$ the numerator in (3.4.2) reduces to $P(X_1 = y_1, X_2 = y_2, \ldots, X_n = y_n)$, since then it is implicit that also $\sum_1^n X_j = r_0$, and this probability is simply

$$P(X_1 = y_1, X_2 = y_2, \ldots, X_n = y_n)$$

$$= \prod_1^n P(X_j = y_j) \qquad \text{by independence [see II, § 4.4]}$$

$$= \prod_{j=1}^n p^{y_j} (1-p)^{1-y_j} \qquad \text{by (3.4.1)}$$

$$= p^{r_0} (1-p)^{n-r_0} \qquad \text{since } \sum_1^n y_j = r_0.$$

If on the other hand $\sum_1^n y_j \neq r_0$ the numerator in (3.4.2) reduces to zero, since it is the probability of an impossible event.

The denominator in (3.4.2) is

$$P\left(\sum_1^n X_j = r_0 \right) = \binom{n}{r_0} p^{r_0} (1-p)^{n-r_0}$$

since $\sum_1^n X_j$ has the Binominal (n, p) distribution [see II, § 5.2.2].

Finally then,

$$P\left(X_1 = y_1, \ldots, X_n = y_n \Big| \sum_1^n X_j = r_0 \right) = \begin{cases} 1 \Big/ \binom{n}{r_0}, & \sum_1^n y_j = r_0 \\ 0, & \text{otherwise.} \end{cases}$$

The important feature of this result is that the conditional distribution of the sample values, given the value of the statistic $\sum_1^n x_j$, *does not depend on p*. Now, once the value of this statistic is known, any further inferences about p that takes into account this knowledge can only be derived from the *conditional*

distribution of the sample values, and, since this does not in any way involve
p, no inference about p is possible from it: i.e. given the total number of
successes, there is no possibility of extracting anything further, relevant to p,
from the data. In this sense the statistic $\sum x_r$ contains all the information about
p that it is possible to extract from the sample. This is what is meant by saying
that the statistic $\sum_1^n x_r$ is *sufficient* for p. [This is of course not to say that the
original individual sample values $x_1, x_2, \ldots, x_n$ are irrelevant to *other* inferences. The discussion in the example was based on the assumption that the
X_i are i.i.d., which is indeed valid for the die. In other cases however the
mutual independence of the X_i might be open to question, and the ordered
sample $(x_1, x_2, \ldots, x_n)$ of individual values would certainly be required for
inferences concerning this lack of independence.]

The arguments showing that $\sum x_r$ is sufficient for p also show that $\frac{1}{2}\sum x_r$,
$a \sum x_r + b$, $\exp \sum x_r$, etc., are each sufficient for p. In fact every function of $\sum x_r$
is sufficient. In the case discussed it is clear on intuitive grounds that an
acceptable function would be $\sum x_r/n$, the observed proportion of successes,
since this is an unbiased estimate of p. In a later section [see § 3.4.3] we give
an objective criterion (the Rao–Blackwell Theorem) for making an appropriate
choice.

We now give a formal definition of sufficiency.

DEFINITION 3.4.1.   *Sufficiency.*   Let the (continuous or discrete) random
variables $(X_1, X_2, \ldots, X_n)$ have p.d.f. at $(x_1, x_2, \ldots, x_n)$ given by
$f_n(x_1, x_2, \ldots, x_n; \theta)$, where θ is a (scalar) parameter, and let $\theta^* =
\theta^*(x_1, x_2, \ldots, x_n)$ be a statistic based on the observations $(x_1, x_2, \ldots, x_n)$. Then
θ^* is *sufficient* for θ if, for every other statistic $\tilde{\theta}(x_1, x_2, \ldots, x_n)$, the conditional
distribution of $\tilde{\theta}$ given θ^* does not involve θ.

In particular θ^* is sufficient for θ if the conditional joint distribution of
$X_1, X_2, \ldots, X_n$, given θ^*, does not involve θ. [See also § 4.13.1(b).]

EXAMPLE 3.4.2.   *Sufficiency of sample mean as estimate of exponential
parameter.*   Let $(x_1, \ldots, x_n)$ be a sample of observations on the exponential
r.v. X, whose p.d.f. at x is

$$f(x; \theta) = \theta\, e^{-\theta x}, \qquad x \geq 0.$$

Then $\bar{x} = \sum_1^n x_j/n$ is sufficient for θ. For, in terms of the i.i.d. variables
$X_1, X_2, \ldots, X_n$, which are statistical copies of X such that x_j is a realization
of X_j $(j = 1, 2, \ldots, n)$, the conditional distribution of the sample, given the
statistic $\bar{x}$, is defined by the conditional p.d.f. at $(u_1, u_2, \ldots, u_n)$ of
$(X_1, X_2, \ldots, X_n)$ conditional on $\sum_1^n X_j = n\bar{x}$.

If $\sum u_j = n\bar{x}$ this is

$$\left\{ \prod_1^n f(u_j; \theta) \right\} \Big/ g(n\bar{x}; \theta) \qquad\qquad (3.4.3)$$

where $g(z; \theta)$ is the p.d.f. of $\sum_1^n X_j$ at z; while, if $\sum u_j \neq n\bar{x}$, it is zero (cf. Example 3.4.1). Now [see II, § 11.3.2]

$$g(z; \theta) = \frac{1}{(n-1)!} \theta^n z^{n-1} e^{-\theta z},$$

while

$$\prod_1^n f(u_j; \theta) = \theta^n e^{-\theta \sum u_j}$$

$$= \theta^n e^{-n\bar{x}\theta}, \qquad \sum u_j = n\bar{x}.$$

Thus (3.4.3) reduces in the non-trivial case to

$$\theta^n e^{-n\bar{x}\theta} \Big/ \left\{ \frac{1}{(n-1)!} \theta^n (n\bar{x})^{n-1} e^{-n\bar{x}\theta} \right\} = (n-1)!/(n\bar{x})^{n-1}.$$

which does not involve θ. It follows that $\bar{x}$ is sufficient for θ.

EXAMPLE 3.4.3 (*continuation*).   It was shown in Example 3.4.2 that, given data $x_1, x_2, \ldots, x_n$ from the p.d.f. $\theta e^{-\theta x}$, the sample mean $\bar{x}$ is a sufficient statistic for θ. It does *not* follow that $\bar{x}$ is in any sense a good estimate of θ. In fact $\bar{x}$ is a completely unacceptable estimate of θ, since it is not even dimensionally correct [see § 3.3.1(a)]: for $E(X) = 1/\theta$, whence $\bar{x}$ has the dimensions of θ^{-1}, not those of θ. What does follow from Example 3.4.2 is that the best possible estimate of θ must be a function of $\bar{x}$. The answer to the question 'What function?' is not provided by the concept of sufficiency but must be supplied by other criteria (such as consistency: see § 3.3.1). In the present case the above remark about dimensions suggests that $1/\bar{x}$ might be an acceptable estimate. In fact the distribution of $Z = \sum_1^n X_j$ is given by (3.4.4), whence the expected value of $1/\bar{X} = n/Z$ is

$$\int_0^\infty (n/z) g(z; \theta) \, dz = n\theta/(n-1)$$

[see II, § 10.4.1]. It follows that $(n-1)/n\bar{x}$ is an unbiased [see § 3.3.2] function of the sufficient statistic $\bar{x}$ and is from this point of view the best possible estimate of θ. (A formal procedure for obtaining an unbiased sufficient statistic is given in section 3.4.3.)

3.4.2.   The Factorization Criterion and the Exponential Family

Examples 3.4.1, 3.4.2 and 3.4.3 demonstrate the direct application of the definition of sufficiency. A simpler approach may be made in terms of the 'factorization criterion' which provides an immediate answer to the question whether a sufficient statistic exists, and identifies it when it does exist. The criterion is as follows.

THEOREM 3.4.1.   *Factorization criterion for sufficiency.   Let the sampling p.d.f. of the observations* $x_1, x_2, \ldots, x_n$ *be* $f_n(x_1, x_2, \ldots, x_n; \theta)$. *A sufficient statistic* $\theta^* = \theta^*(x_1, x_2, \ldots, x_n)$ *for* θ *exists if and only if the function* f_n *may be written in the factorized form*

$$f_n(x_1, x_2, \ldots, x_n; \theta) = g\{\theta^*(x_1, \ldots, x_n), \theta\}h(x_1, x_2, \ldots, x_n), \quad (3.4.4)$$

where the factor $h(\cdot)$ *does not depend on* θ. (*In particular cases the factor* $h(x_1, x_2, \ldots, x_n)$ *may degenerate to a constant.*)

EXAMPLE 3.4.4.   *Factorization criterion and the Bernoulli distribution.*   In Example 3.4.1 the joint distribution of the data is given by the p.d.f.

$$f_n(x_1, x_2, \ldots, x_n; \theta) = \theta^{\Sigma x_j}(1 - \theta)^{n - \Sigma x_j}$$

(replacing p in (3.4.1) *by* θ). This is of the form (3.4.4), with $\theta^* = \sum x_j$, $g(\theta^*, \theta) = \theta^{\theta^*}(1 - \theta)^{n - \theta^*}$, and $h(x_1, \ldots, x_n) = 1$. Hence $\sum x_j$ is sufficient for θ.

In Example 3.4.2 the joint distribution of the data at $(x_1, x_2, \ldots, x_n)$ is

$$f_n(x_1, x_2, \ldots, x_n; \theta) = \theta^n e^{-\theta \Sigma x_j}.$$

This, too, is of the form (3.4.4), with $\theta^* = \sum x_j$, $g(\theta^*, \theta) = \theta^n e^{-\theta \theta^*}$, and $h(x_1, \ldots, x_n) = 1$.

EXAMPLE 3.4.5.   *Factorization criterion and the Normal distribution.*   For the Normal $(\theta, 1)$ distribution the sample p.d.f. at $(x_1, x_2, \ldots, x_n)$ is

$$(2\pi)^{-n/2} \exp\{-\tfrac{1}{2}\sum(x_i - \theta)^2\} = (2\pi)^{-n/2} \exp\{-\tfrac{1}{2}\sum(x_i - \bar{x})^2 - \tfrac{1}{2}n(\bar{x} - \theta)^2\}$$

$$= g(\theta^*; \theta)h(x_1, \ldots, x_n)$$

with $\theta^* = \bar{x}$, $g(\theta^*; \theta) = (2\pi)^{-n/2} \exp\{-\tfrac{1}{2}n(\theta^* - \theta)^2\}$ and $h(x_1, \ldots, x_n) = \exp\{-\tfrac{1}{2}\sum(x_i - \bar{x})^2\}$, whence $\theta^* = \bar{x}$ is sufficient for θ.

Similarly for the Normal distribution with $E(X) = 0$ and $\text{var}(X) = \theta$ we have $f_n(x_1, \ldots, x_n; \theta) = (2\pi\theta)^{-n/2} \exp(-\sum x_i^2/\theta)$, whence $\sum x_i^2$ is a sufficient statistic for the variance θ (and also, therefore, for the standard deviation $\theta^{1/2}$).

EXAMPLE 3.4.6.   *Factorization criterion and the gamma distribution.*   For the one-parameter gamma distribution [see II, § 11.3] with shape parameter θ, for which the p.d.f. at x is $x^{\theta - 1} e^{-x}/\Gamma(\theta)$, $x > 0$, we have

$$f(x_1, x_2, \ldots, x_n; \theta) = \left(\prod_1^n x_j^{\theta - 1}\right) e^{-\Sigma x_j}/\Gamma^n(\theta).$$

Here the factorization (3.4.4) is achieved with $\theta^* = \prod_1^n x_j$, which statistic is therefore sufficient for θ.

It is not required in Theorem 3.4.4 that the x_i be observations on mutually independent and identically distributed variables. In the context of one-parameter distributions, however, it is usually the case that the x_i form a random sample of observations from a single one-parameter distribution, as in Examples 3.4.4, 3.4.5 and 3.4.6. In these circumstances, under conditions of wide generality, a distribution that admits of a sufficient statistic must belong to the *exponential family* of distributions, defined as follows:

DEFINITION 3.4.2. *The exponential family.* The (one-parameter) exponential family (or exponential-type class) of univariate distributions has p.d.f. at x given by

$$f(x, \theta) = \exp\{A(x)B(\theta) + C(x) + D(\theta)\}, \qquad (3.4.5)$$

where $A(x)$, $B(\theta)$, $C(x)$ and $D(\theta)$ are arbitrary functions of the specified arguments, subject only to the restriction that $f(x)$ is a p.d.f., i.e. that $f(x)$ be non-negative and normalized. [The class is sometimes called the Darmois–Pitman–Koopmans class, after its discoverers.]

On applying the factorization criterion of Theorem 3.4.1 to (3.4.5) it will be seen that the sample distribution may be written in the factorized form

$$f_n(x_1, x_2, \ldots, x_n; \theta) = \{\exp B(\theta) \sum A(x_i) + nD(\theta)\} \cdot \{\exp \sum C(x_i)\}$$

whence the statistic $\theta^* = \sum A(x_i)$ is sufficient for θ.

If in addition $\sum A(x_i)/n$ is unbiased for θ, we have an estimate that satisfies the Cramér–Rao inequality (3.3.5): see (3.3.8). The estimate $\sum A(x_i)/n$ is then unbiased, efficient and sufficient.

An example of the exponential family is the p.d.f. of the one-parameter gamma distribution with shape-parameter θ (see Example 3.4.6):

$$x^{\theta-1} e^{-x}/\Gamma(\theta) = \exp\{(\log x)(\theta - 1) - x - \log \Gamma(\theta)\}.$$

This is of the form (3.4.5) with $A(x) = \log x$, $B(\theta) = \theta - 1$, $C(x) = -x$ and $D(\theta) = -\log \Gamma(\theta)$. The sufficient statistic $\theta^* = \sum A(x_i)$ becomes $\sum \log x = \log(\prod x_j)$—or, of course, any transform of this—in agreement with the result obtained in Example 3.4.6 that $\prod x_j$ is sufficient for θ.

Further familiar examples are furnished by the Binomial and the Negative Binomial distribution.

EXAMPLE 3.4.6(a). *The Binomial as a member of the exponential family.* If X is the number of successes in a fixed number n of Bernoulli (θ) trials, the p.d.f. of X at x is

$$f(x; \theta, n) = \binom{n}{x} \theta^x (1 - \theta)^{n-x}, \qquad x = 0, 1, \ldots, n \qquad (3.4.5(a))$$

whence

$$\log f = \log \binom{n}{x} + x \log \frac{\theta}{1 - \theta} + n \log(1 - \theta).$$

This is of the form (3.4.5) with $\log \binom{n}{x} = C(x)$, $x \log\{\theta/(1-\theta)\} = A(x)B(\theta)$, and $n \log (1-\theta) = D(\theta)$. Thus the Binomial is a member of the exponential family, and the statistic $\theta^* = A(x) = x$ is sufficient for θ (c.f. Example 3.4.4).

EXAMPLE 3.4.6(b).   *The Negative Binomial as a member of the exponential family.*

If, in contrast to the situation in Example 3.4.6(a), N is the number of Bernoulli (θ) trials required to achieve a fixed member x of successes, N has the Negative Binomial distribution with

$$P(N = n) = f(n; \theta, x)$$

$$= \binom{n-1}{x-1} \theta^x (1-\theta)^{n-x}, \qquad n = x, x+1, x+2, \ldots \qquad (3.4.5(b))$$

(c.f. Example 3.3.0). Here

$$\log f = \log \binom{n-1}{x-1} + n \log \frac{\theta}{1-\theta} + n \log (1-\theta)$$

and again we see by comparison with (3.4.5) that the p.d.f. belongs to the exponential family and that n is sufficient for θ.

EXAMPLE 3.4.6(c).   *The effect of truncation.*   Suppose, first, that X is a Poisson (θ) variable, and that, in a sample of n observations on r, the value $X = r$ was observed with frequency n_r, for $r = 0, 1, \ldots, k$, where $\sum n_r = n$. The probability of getting this sample is

$$f_n(n_0, n_1, \ldots, n_k; \theta) = \prod_0^k (e^{-\theta}\theta^r/r!)^{n_r}$$

$$= e^{-n\theta} \theta^{\sum_0^k rn_f} \Big/ \prod_0^k (r!)^{n_r}.$$

Since this factorizes in the form (3.4.4), with $\theta^* = \sum_0^k rn_r$, $g(\theta^*, \theta) = e^{-\theta}\theta^{\sum rn_r}$, and $h = 1/\prod (r!)^{n_r}$, it follows that $\sum rn_r$ (the sum of all k observed values) is sufficient for θ.

Suppose now that, instead, the zero values $X = 0$ were unobservable, possibly on account of a fault in the experimental equipment, so that X has a truncated [see II, § 6.7] Poisson distribution with the zero class missing [see II, Example 6.7.2]. The p.d.f. of X is now

$$P(X = x) = \phi(\theta)\theta^r!, \qquad r = 1, 2, \ldots, \phi(\theta) = e^{-\theta}/(1-e^{-\theta}).$$

The probability of the sample is

$$\prod_1^k \{\phi(\theta)\theta^r/r!\}^{n_r} = \{\phi(\theta)\}^n \theta^{\sum_1^k rn_r} \Big/ \prod_1^k (r!)^{n_r}.$$

This, also, factorizes in the form (3.4.4), with $\theta^* = \sum_1^k rn_r$ (the same statistic, as it happens, as in the untruncated case), $g(\theta^*, \theta) = \phi^n(\theta)\theta^{\Sigma_1^k r n_r}$, and $h = 1/\prod_1^k (r!)^{n_r}$. Thus $\sum_1^k r n_r$ is sufficient for θ.

This last example illustrates the general result that, if X (continuous or discrete) has p.d.f. $f(x; \theta) = a'(\theta)\psi(x, \theta)$ and admits of a sufficient statistic for θ, a truncated version of X, say one in which observations can be made only for $a \le X \le b$, also admits of a sufficient statistic for θ. For $f(x, \theta)$ must have the 'exponential family' from (3.4.5). Now the truncated p.d.f. is

$$f_{\text{trunc}}(\chi, \theta) = f(x, \theta)/\lambda(\theta), \qquad a \le x \le b$$

where

$$\lambda(\theta) = \int_a^b f(x, \theta)\, dx$$

and so

$$f_{\text{trunc}}(x, \theta) = \frac{1}{\lambda(\theta)} \exp\{A(x)B(\theta) + C(x) + D(\theta)\}$$

$$= \exp\{A(x)B(\theta) + C(x) + D_1(\theta)\}$$

where

$$D_1(\theta) = D(\theta) - \log \lambda(\theta).$$

Thus $f_{\text{trunc}}(x, \theta)$ belongs to the exponential family, and so admits of a sufficient statistic for θ.

3.4.3. Sufficiency and Minimum-variance Unbiased Estimates

(a) *The Rao–Blackwell Theorem*

The various criteria outlined above enable one to find a sufficient statistic θ^* when such exists, but, as was pointed out in Example 3.4.3, one is still left with the problem of finding a suitable transform of θ^* to enable it to be used as a sensible estimate of θ.

An almost trivial example is provided by the Binomial (n, θ) distribution. If X has this distribution, an observed value x of X is sufficient for θ (see Example 3.4.5(a)), but it is not an acceptable estimate of θ: θ must be between 0 and 1, while, if $n = 20$ (for example), x may be, say, 19. The obvious solution is to take as estimate, not x, but x/n (which is unbiased for θ).

Similar properties are displayed by the Negative Binomial distribution. If N has this distribution, as in Example 3.4.5(b), an observed value n of N is sufficient for θ, but n is unacceptable, for dimensional reasons, as an estimate of θ, as is indicated by the fact that $E(N) = x/\theta$. The step required to obtain an acceptable (e.g. unbiased) transform of x is not quite so obvious as in the Binomial case mentioned above. A little thought however leads one to the estimate $(x - 1)/(n - 1)$, which is unbiased for θ (c.f. Example 3.3.0).

One cannot however rely on always being able to find an ad hoc method like the foregoing. This following theorem provides an objective algorithm for achieving the desired end.

THEOREM 3.4.2   (*The Rao–Blackwell Theorem*). *Suppose* $s = s(x_1, x_2, \ldots, x_n)$ *is a sufficient but biased estimate of* θ, *based on a sample* $(x_1, x_2, \ldots, x_n)$ *of observations on a random variable* X, *and suppose* $u(x_1, x_2, \ldots, x_n)$ *is an unbiased but not sufficient estimate of* θ. *Let* $S = s(X_1, X_2, \ldots, X_n)$ *and* $U = u(X_1, X_2, \ldots, X_n)$, *where* $X_1, X_2, \ldots, X_n$ *are statistical copies of* X. *Then the conditional expectation* [*see* II; § 8.9]

$$\theta^*(s) = E(U|S = s)$$

is an unbiased sufficient estimate of θ, *with*

$$\text{var}\{\theta^*(S)\} \leq \text{var}(U).$$

As a trivial illustration of the application of the theorem, consider a sample $(x_1, x_2, \ldots, x_n)$ of observations on a Bernoulli (θ) variable X. The statistic $s = \sum_1^a x_r$ is sufficient for θ, but not unbiased. The statistic x_1 is unbiased but not sufficient. The Rao–Blackwell statistic derived from these is

$$\theta^*(s) = E\left(x_1 \,\middle|\, \sum_1^n x_r = s \right)$$

$$= \frac{1}{n} E\left(\sum_1^n x_r \,\middle|\, \sum_1^n x_r = s \right) \quad \text{by symmetry}$$

$$= \frac{1}{n} s.$$

This estimate, s/n, is sufficient and unbiased. The following example is more substantial.

EXAMPLE 3.4.7.   *An application of the Rao–Blackwell Theorem.*   Let $(x_1, x_2, \ldots, x_n)$ be a random sample of observations on the r.v. X with p.d.f. $f(x) = \theta^2 x\, e^{-\theta x}$, $x > 0$. (Note that $E(X) = 2/\theta$). From the form of the sample p.d.f. $\prod_1^n f(x_j) = \theta^{2n}(\prod x_j)\, e^{-\theta \Sigma x_j}$ it follows, using Theorem 3.4.1, that $s = s(x_1, x_2, \ldots, x_n) = \sum_1^n x_j$ is sufficient for θ. Since the sampling expectation of this statistic is $2n/\theta$ it is dimensionally unacceptable. On the other hand the estimate $u = u(x_1, x_2, \ldots, x_n) = 1/x_1$ is unbiased for θ, as its sampling expectation is $\int_0^\infty (x^{-1})(\theta^2 x\, e^{-\theta x})\, dx = \theta$. According to Theorem 3.4.2, the estimate $\theta^* = \theta^*(s) = E(U|S = s) = E(X_1^{-1} |\sum_1^n X_r = s)$ is unbiased and sufficient for θ. First note that the conditional distribution of U, given $S = s$, must be parameter-free, from the definition of a sufficient statistic, and so θ^* *is* a statistic. To evaluate it we need the conditional distribution of X_1, given $\sum_1^n X_r = s$. We first note that $f(x)$ is a special case of the gamma distribution, and, from the additive properties of the gamma family [see II, § 11.3.2] we see that the p.d.f.

of $\sum_1^n X_r$ at s is

$$g_n(y) = \theta^{2n} s^{2n-1} e^{-\theta s} / (2n-1)!, \qquad s > 0.$$

To find the conditional distribution of X_1, given $\sum_1^n X_r = s$, we first need the joint p.d.f. $h(x, s)$ of X_1 at x and $\sum_1^n X_r$ at s. This is evidently the same as the joint p.d.f. of X_1 at x and $\sum_2^n X_r$ at $s - x$, and, since X_1 and $\sum_2^n X_r$ are mutually independent, this is

$$h(x, s) = f(x) g_{n-1}(s - x)$$
$$= \theta^{2n} x (s - x)^{2n-3} e^{-\theta s} / (2n-3)!$$

The conditional p.d.f. of X_1 at x, given $\sum_1^n X_r = s$, is

$$f_c(x|s) = h(x, s)/g_n(s) = (2n-1)(2n-2) x (s-x)^{2n-3} / s^{2n-1}, \qquad 0 < x < s.$$

Finally, the required unbiased function of s is

$$\theta^*(s) = E(U|s) = E(X_1^{-1}|s) = \int_0^s x^{-1} f_c(x|s) \, dx$$

$$= \frac{(2n-1)(2n-2)}{s^{2n-1}} \int_0^s (s-x)^{2n-3} \, dx$$

$$= (2n-1)/s = (2n-1)/n\bar{x},$$

where $\bar{x} = s/n$ is the sample mean.

This is evidently a sufficient statistic, since it is a function of the sufficient statistic s. That it is indeed unbiased may readily be verified directly, since

$$E\{\theta^*(S)\} = (2n-1) E(S^{-1})$$

$$= (2n-1) \int_0^\infty s^{-1} g_n(s) \, ds$$

$$= \theta.$$

(b) *Minimum variance unbiased estimates and sufficiency*

Once one has a sufficient estimator S of θ, and any unbiased estimator U, the Rao–Blackwell Theorem shows how to produce an *unbiased* sufficient estimate

$$\theta^*(s) = E(U|S = s)$$

with variance not greater than var (U). From the point of view of concentration (see § 3.3.1(a)), this θ^* is evidently better than U. Suppose one had worked with some other unbiased estimator U_1: would one then have obtained a different—and possibly superior—unbiased sufficient estimate of θ? The answer is No: subject to not very restrictive conditions, the Rao–Blackwell estimate θ^* is unique and is therefore the MVUE of θ.

This follows from a celebrated result of Rao [Rao 1947, 1949: see Bibliography] to the effect that:

if a complete sufficient statistic exists, then every function of it is a uniformly minimal variance unbiased estimate of its expected value.

[A sufficient statistic is said to be complete if no function of it has zero expectation unless, with probability equal to 1, it is identically zero.]

EXAMPLE 3.4.8. *MVUEs of Binomial and Negative Binomial parameters.* We saw in section 3.4.5(a) that, if x is the number of successes in n Bernoulli (θ) trials, x/n is an unbiased sufficient estimate of θ. It can be shown that x is *complete* in the above sense. It follows that x/n is the MVUE of θ.

Similarly, if n is the number of trials required for a fixed number x of successes, $(x-1)/(n-1)$ is the MVUE of θ.

3.4.4. Sufficiency in the Multi-parameter Case

The concept of a sufficient statistic, introduced in section 3.4.1 for a one-parameter family of distributions, may be extended to the case where there are several parameters. The appropriate extension of Definition 3.4.1 is the following:

DEFINITION 3.4.3. *Joint sufficiency.* Let the (continuous or discrete) random variables $X_1, X_2, \ldots, X_n$ have p.d.f. at $(x_1, x_2, \ldots, x_n)$ given by $f_n(x_1, x_2, \ldots, x_n; \theta_1, \theta_2, \ldots, \theta_k)$, where $\theta_1, \theta_2, \ldots, \theta_k$ are parameters. The statistics $\theta_j^*(x_1, \ldots, x_n)$, $j = 1, 2, \ldots, m$ are *jointly sufficient* for the θ_j if, for every other set of m statistics $\tilde{\theta}_j = \tilde{\theta}_j(x_1, \ldots, x_n)$, $j = 1, 2, \ldots, m$, the conditional joint distribution of $\tilde{\theta}_1, \tilde{\theta}_2, \ldots, \tilde{\theta}_m$ given $\theta_1^*, \theta_2^*, \ldots, \theta^*$ does not involve any of the parameters $\theta_1, \theta_2, \ldots, \theta_k$. They are a *minimal* jointly sufficient set if m is the smallest integer for which this statement holds.

In particular the θ_j^* are jointly sufficient for $\theta_1, \theta_2, \ldots, \theta_k$ if the conditional joint distribution of $X_1, X_2, \ldots, X_n$, given the θ_j^*, does not involve any of $\theta_1, \theta_2, \ldots, \theta_k$.

As in the one-parameter case, direct application of the theorem can sometimes be laborious, and it is usually easier to use instead the (equivalent) multiparameter version of the factorization criterion of Theorem 3.4.1. This is as follows:

THEOREM 3.4.3. *Factorization criterion. With*

$$f_n(x_1, x_2, \ldots, x_n; \theta_1, \ldots, \theta_k)$$

defined as in Definition 3.4.3, a set of statistics θ_j^*, $j = 1, 2, \ldots, k$, that are jointly sufficient for the parameters $\theta_1, \ldots, \theta_k$ exists if and only if the function

f_n may be written in the factorized form

$$f_n(x_1, \ldots, x_n; \theta_1, \ldots, \theta_k) = g(\theta_1^*, \theta_2^*, \ldots, \theta_n^*; \theta_1, \theta_2, \ldots, \theta_n) k(x_1, x_2, \ldots, x_n).$$
$$(3.4.6)$$

EXAMPLE 3.4.9.   *Joint sufficiency of $\bar{x}$ and s^2 for Normal μ and σ^2.*   In the case where the X_j are i.i.d. Normal variables with $E(X_j) = \theta_1$, var $(X_j) = \theta_2$, $j = 1, 2, \ldots, n$, the usual estimates of θ_1 and θ_2 are, respectively,

$$\theta_1^* = \bar{x}, \qquad \theta_2^* = \sum_1^n (x_j - \bar{x})^2/(n-1) \qquad (= s^2, \text{say}).$$

It turns out that these estimates are jointly sufficient for θ_1 and θ_2. To see this we note that the sample p.d.f. may be written in the form

$$(2\pi\theta_2)^{-n/2} \exp{-\frac{1}{2\theta_2} \sum (x_j - \theta_1)^2}$$

$$= (2\pi\theta_2)^{-n/2} \exp{-\frac{1}{2\theta_2}\{\sum (x_j - \bar{x})^2 + n(\bar{x} - \theta_1)^2\}}$$

$$= (2\pi\theta_2)^{-n/2} \exp{-\frac{1}{2\theta_2}\{(n-1)s^2 + n(\bar{x} - \theta_1)^2\}}.$$

This is of the form (3.4.6) with $\theta_1^* = \bar{x}$, $\theta_2^* = s^2$, and $h(x_1, x_2, \ldots, x_n) = 1$. It follows that $\bar{x}$ and s^2 are jointly sufficient for θ_1 and θ_2. Using the language of section 3.4.1, we may say that $\bar{x}$ and s^2 together contain all the information about θ_1 and θ_2 that the sample contained.

As in the one-parameter case, every pair of algebraically independent functions of the pair θ_1^*, θ_2^* also provides a set of jointly sufficient estimates. In Example 3.4.8 it is the pair explicitly discussed, namely $\theta_1^* = \bar{x}$, $\theta_2^* = s^2$, that are usually appropriate, since θ_1^* and θ_2^* are, separately, unbiased estimates of θ_1 and θ_2 respectively.

3.5.   PRACTICAL METHODS OF CONSTRUCTING ESTIMATES: AN INTRODUCTION

3.5.1.   Graphical Methods

If the random variable X has a two-parameter c.d.f. of the form $P(X \le x) = F(x; \theta_1, \theta_2) = H\{(x - \theta_1)/\theta_2\}$, where θ_1 and θ_2 are the parameters (θ_1 then being called a *location* parameter and θ_2 a *scale* parameter) it is possible to devise special graph paper, with scales that do not depend on θ_1 or θ_2, on which the graph of $F(x; \theta_1, \theta_2)$ is a straight line [see § 3.2.2(d)]. If a plot on this paper of the empirical c.d.f. [see § 3.2.2(d); § 14.2] of a sample gives a set of points that appears to be scattered around a straight line, this

provides a crude test for the assumption that the sampled distribution belongs
to the family described by the c.d.f. $F(x; \theta_1, \theta_2)$. From a straight line, fitted
to the plotted points by eye, rough estimates of θ_1 and θ_2 may be obtained.

The most commonly occurring families of distributions of this form are the
Normal [see II, § 11.4], the log-Normal [see II, § 11.5], and the Weibull [see
II, § 11.9]. Appropriate graph papers are commercially available. A graphical
method of this kind is often a useful precursor to a more objective analytical
procedure.

EXAMPLE 3.5.1.   *Use of Normal Probability Paper to estimate μ and σ*

The following frequency table summarizes measurements of the heights of
1456 women.

Central height (inches) x_r	Upper boundary of call $x_r + \frac{1}{2}$	Frequency t_r	Cumulative frequency (= no. of women (with height $\leq x_r + \frac{1}{2}$)	Cumulative frequency as % of total
52·5	53	0·5	0·5	0·03
53·5	54	0·5	1	0·07
54·5	55	0	1	0·07
55·5	56	1	2	0·14
56·5	57	5	7	0·48
57·5	58	15	22	1·51
58·5	59	15·5	37·5	2·5
59·5	60	52	89·5	6·1
60·5	61	101	190·5	13·08
61·5	62	150	340·5	23·3
62·5	63	199	539·5	37·05
63·5	64	223	762·5	52·37
64·5	65	215	977·5	67·14
65·5	66	169·5	1147	78·78
66·5	67	151·5	1298·5	89·18
67·5	68	81·5	1380	94·78
68·5	69	40·5	1420·5	97·56
69·5	70	19·5	1440	99·90
70·5	71	10	1450	99·59
71·5	72	5	1455	99·93
72·5	73	0	1455	99·93
73·5	74	1	1456	100

(One of the subjects measured 53 in. exactly, to the accuracy of the measuring
equipment that was used. She was recorded as half a unit in the cell centred
at 52·5 and half in the next cell. Similar explanations account for the other
half units in the frequency columns.)

The empirical c.d.f. (that is, the set of cumulative frequencies, expressed as a percentage of the total number, plotted against the corresponding upper cell boundary) is shown in Figure 3.5.1, and a straight line fitted by eye to the plotted points. The points appear to accord reasonably well with this line, showing that the underlying population (i.e. the population of women's heights) is approximately Normal.

For a Normal distribution the interpretation of the graph may be worked out with the aid of the standard Normal c.d.f., as illustrated in Figure 3.5.2.

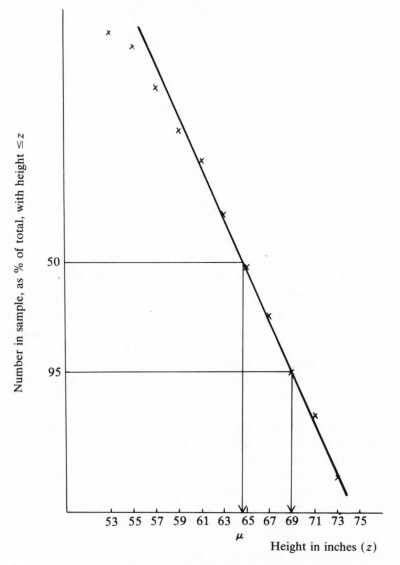

Figure 3.5.1: Empirical c.d.f. plotted on Normal ('Arithmetic') probability graph paper.

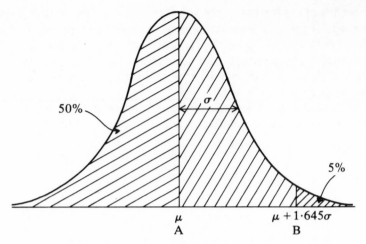

Figure 3.5.2: A sketch of the Normal (μ, σ) p.d.f., showing the 50% point (A) and the 95% point (B).

The 50% point (64·6 in. in our case: see Figure 3.5.1) gives an estimate of μ, and the 95% point (69·0 in.) an estimate of $\mu + 1·645\sigma$.

Thus our graphical treatment verifies that the population sampled is approximately Normal, with parameters $\mu = 64·6$ in. and $\sigma = (69-64·6)/1·645 = 2·67$ in. approximately.

3.5.2. Minimum-variance Unbiased Estimates and the Method of Least Squares

As explained in section 3.3.2, a (restricted) minimum-variance unbiased estimate of a parameter θ based on a sample $(x_1, x_2, \ldots, x_n)$ is a function $h(x_1, x_2, \ldots, x_n; a_1, a_2, \ldots, a_k)$, where the 'coefficients' $a_1, a_2, \ldots, a_k$ are chosen so as to ensure that the sampling expectation of the estimate is identically equal to θ, and, subject to this restraint, the sampling variance is as small as possible. The choice of the functional form of $h(\cdot)$ is usually based on dimensionality arguments. In Example 3.3.4, the parameter θ was the expected value of X, and a linear function was therefore thought to be appropriate. Example 3.3.4 illustrates the use of a function that is linear in the coefficients and quadratic in the observations.

The method is most frequently used in circumstances where it seems reasonable to assume linearity in the coefficients. In such cases an extremely powerful systematization of the method is available. This is the famous 'Principle of Least Squares', described in detail in Chapter 8. (This is one of the oldest estimation methods known: it was used, for example, by Legendre in 1805 and there is a famous exposition of the method by Gauss, dated 1809. See Chapter 15 of Pearson and Kendall (1970)—Bibliography D.) The link between 'Least Squares' and 'Minimum-variance unbiased' is illustrated by the following example. Suppose that, for each of n accurately-known loadings $x_1, x_2, \ldots, x_n$, the corresponding deflections $y_1, y_2, \ldots, y_n$ of a steel beam are observed. It is assumed that the 'levels' x_r of the loadings form part of the predetermined experimental design: they are not 'observations' in the technical sense of being realizations of random variables. On the other hand the deflections y_r *are* 'observations': they are not known in advance, and indeed the whole point of the experiment is to observe and measure them as accurately as possible subject to the limitations imposed by the experimental equipment. Physical considerations suggest that, over the range of loadings considered, and in the absence of error, the deflection $y(x)$ induced by a load x would have the form

$$y(x) = \theta_0 + \theta_1 x + \theta_2 x^2,$$

for appropriate values of the coefficients θ_0, θ_1, θ_2; while in reality the observations satisfy the relations

$$y_r = \theta_0 + \theta_1 x_r + \theta_2 x_r^2 + e_r, \qquad r = 1, 2, \ldots, n,$$

where the e_r represent the *observational errors*. According to the principle of least squares the estimates $\tilde{\theta}_r$ of the θ_r, $r = 0, 1, 2$, are to be chosen as the values of the (unknown) θ_r which minimize the 'sum of squares'

$$\sum_{r=1}^{n} e_r^2 = \sum_{r=1}^{n} (y_r - \theta_0 - \theta_1 x_r - \theta_2 x_r^2)^2. \tag{3.5.1}$$

On the other hand, if we had decided to use minimum-variance unbiased estimates that were linear in the observations y_r, and which involved as coefficients arbitrary functions of the predetermined loading levels x_r, we should have looked for estimates of the form

$$\hat{\theta}_r = a_r + \sum_{s=1}^{n} b_{rs} y_s, \qquad r = 0, 1, 2,$$

with the coefficients a_r, b_{rs} determined by the conditions that (a) $\hat{\theta}_r$ should be an unbiased estimate of θ_r $(r = 0, 1, 2)$, and (b), subject to condition (a), the sampling variance of each of $\hat{\theta}_0$, $\hat{\theta}_1$ and $\hat{\theta}_2$ should be as small as possible.

It turns out for this 'linear' case that (subject to conditions explained below) the 'Least Squares' estimates $\tilde{\theta}_r$ and the 'Minimum-variance Unbiased' estimates $\hat{\theta}_r$ [see § 3.3.2] are *exactly the same*. (The conditions are, briefly, that (i) each of the errors e_r has sampling expectation equal to zero, (ii) the errors all have the same sampling variance, and (iii) the errors are uncorrelated).

The process of minimizing the sum of squares $\sum e_r^2$ is manipulatively straightforward, and simple explicit expressions may be found for the estimates. In our example, it is easily seen that θ_0^*, θ_1^*, θ_2^* are the (unique) solutions of the simultaneous linear equations [see I, § 5.8]:

$$\theta_0^* + \theta_1^* \sum x_r + \theta_2^* \sum x_r^2 = \sum y_r$$

$$\theta_0^* \sum x_r + \theta_1^* \sum x_r^2 + \theta_2^* \sum x_r^3 = \sum x_r y_r$$

$$\theta_0^* \sum x_r^2 + \theta_1^* \sum x_r^3 + \theta_2^* \sum x_r^4 = \sum x_r^2 y_r.$$

Formally, the solution is the estimate-vector $\boldsymbol{\theta}^* = (\theta_0^*, \theta_1^*, \theta_2^*)'$ given by the linear form

$$\boldsymbol{\theta}^* = \mathbf{C}\mathbf{y} \qquad (3.5.2)$$

where $\mathbf{y}' = (y_1, y_2, \ldots, y_n)$ and $\mathbf{C} = \mathbf{B}^{-1}\mathbf{A}'$, with

$$\mathbf{B} = \begin{pmatrix} 1 & \sum x_r & \sum x_r^2 \\ \sum x_r & \sum x_r^2 & \sum x_r^3 \\ \sum x_r^2 & \sum x_r^3 & \sum x_r^4 \end{pmatrix}, \qquad \mathbf{A}' = \begin{pmatrix} 1 & 1 & 1 & \cdots & 1 \\ x_1 & x_2 & x_3 & \cdots & x_n \\ x_1^2 & x_2^2 & x_3^2 & \cdots & x_n^2 \end{pmatrix}.$$

Generalizations of this, and results on the sampling properties of $\boldsymbol{\theta}^*$ are contained in the *Gauss–Markov* Theorem and its applications. These are described in Chapter 8.

3.5.3.   The Method of Moments

Let $(x_1, x_2, \ldots, x_n)$ be a sample of observations on a random variable X whose p.d.f. at x is $f(x; \theta_1, \theta_2, \ldots, \theta_k)$, where the θ_r are unknown parameters. The (population) *moments* of X (about the origin) are the quantities [see § 2.1.2]

$$\mu_r' = E(X^r),$$
$$= h_r(\theta_1, \theta_2, \ldots, \theta_k), \qquad r = 1, 2, \ldots \qquad (3.5.3)$$

say. Here the $h_r(\cdot)$ are known functions of the unknown parameters. The corresponding *sample moments* are the quantities

$$m_r' = \sum_{j=1}^{n} x_j^r / n, \qquad r = 1, 2, \ldots \qquad (3.5.4)$$

The method of moments bases itself on the intuitive notion that the sample moments will be approximately equal to the population moments, and the 'moment estimates' $\theta_r^{(m)}$, $r = 1, 2, \ldots, k$ are obtained by equating the first k population moments to the corresponding sample moments and solving the resulting equations

$$h_r(\theta_1^{(m)}, \theta_2^{(m)}, \ldots, \theta_k^{(m)}) = m_r', \qquad r = 1, 2, \ldots, k.$$

The method is easy to apply and, though lacking a firm theoretical justification, often provides acceptable estimates. It can also produce estimates of very low efficiency. In general the method is inferior to the method of maximum likelihood (see § 3.5.4), which, however, usually requires the numerical solution of non-linear equations. For this reason the Method of Moments was at one time a popular one, but with contemporary computational facilities its raison d'être has largely disappeared. Moment estimates may however serve as useful and readily available first approximations to be used in an iterative procedure for solving the likelihood equations. [See Example 6.4.3].

EXAMPLE 3.5.2. *Estimation of Normal μ and σ by the method of moments.* In the case of the Normal (μ, σ) distribution one has

$$\mu_1' = E(X) = \mu$$

and

$$\mu_2' = E(X^2) = \sigma^2 + \mu^2.$$

The moment equations are then

$$\mu^{(m)} = \sum x_j/n \qquad (= \bar{x}, \text{say})$$

and

$$(\mu^2)^{(m)} + (\sigma^2)^{(m)} = \sum x_j^2/n \qquad (= \overline{x^2}, \text{say}),$$

whence the moment estimates are

$$\mu^{(m)} = \bar{x}, \qquad (\sigma^2)^{(m)} = \overline{x^2} - (\bar{x})^2 = (\sum x_j^2 - \bar{x} \sum x)/n.$$

As will be seen in section 6.4.1, the moment estimates in this example coincide with the estimates provided by the method of maximum likelihood.

EXAMPLE 3.5.3. *Estimation of gamma parameters by moments.* In the case of the gamma distribution with shape parameter α and scale parameter β, with p.d.f.

$$f(x; \alpha, \beta) = x^{\alpha-1} e^{-x/\beta}/\beta^{\alpha} \Gamma(\alpha), \qquad x > 0,$$

the first two moments about the origin are

$$\mu_1' = E(X) = \int_0^{\infty} xf(x; \alpha, \beta) \, dx$$

$$= \alpha\beta$$

and

$$\mu_2' = E(X^2) = \int_0^{\infty} x^2 f(x; \alpha, \beta) \, dx$$

$$= \alpha(\alpha+1)\beta^2.$$

Thus, based on a sample $(x_1, x_2, \ldots, x_n)$, the moment estimates $\alpha^{(m)}$, $\beta^{(m)}$ of α and β are the roots of the equations

$$\alpha\beta = \bar{x} \qquad \left(= \sum_1^n x_j/n\right)$$

$$\alpha(\alpha+1)\beta^2 = \overline{x^2} \qquad \left(= \sum_1^n x_j^2/n\right),$$

viz:

$$\alpha^{(m)} = (\bar{x})^2/\{\overline{x^2}-(\bar{x})^2\}.$$

$$\beta^{(m)} = \{\overline{x^2}-(\bar{x})^2\}/\bar{x}. \tag{3.5.5}$$

[The estimation of these parameters is discussed further in Example 6.4.3, where a numerical illustration is given.]

3.5.4.   The Method of Maximum Likelihood

This, the most widely used and the most powerful method at present available, is described in some detail in Chapter 6. The present section is intended as a brief introduction only.

The method is traced by M. G. Kendall to a paper published by Daniel Bernoulli in 1777. See Pearson and Kendall (1970), Chapter 11—Bibliography D.

We start with a simple example.

EXAMPLE 3.5.4.   *Maximum likelihood estimate of exponential parameter.* Suppose that X is exponentially distributed, with unknown expectation; that is, the p.d.f. of X belongs to the one-parameter family of which a typical member is

$$f(x, \theta) = \begin{cases} \theta\, e^{-\theta x}, & x \geq 0 \\ 0, & \text{otherwise.} \end{cases}$$

The family is generated by allowing θ to run through all possible positive values. The particular value of θ (θ^0 say) appropriate to *our* X is unknown: we shall call it the 'true' value of θ. It is desired to estimate θ^0 from data $(x_1, x_2, \ldots, x_n)$, consisting of independent observations on X. Bearing in mind that the x_r have fixed known values, we construct the function

$$l(\theta) = l(\theta; x_1, \ldots, x_n) = \prod_1^n f(x_i; \theta) = \theta^n\, e^{-\theta\Sigma_1^n x_r}, \qquad \theta > 0, \tag{3.5.6}$$

as a function of the *undetermined variable* θ, in which the given data values serve as *known* and fixed coefficients. This is called the *likelihood function* of the data [see § 4.13.1, § 6.2.1].

In our example, X is a continuous variable. The data $x_1, x_2, \ldots, x_n$ must be regarded as finite (rounded) approximations to the infinite decimals required for the exact specification of real numbers, so that 'x_r' really means 'some value lying in the interval $x_r \pm \frac{1}{2}h$', where h is the size of the measurement grid—1 mm, say, of the x_r are measured to the nearest mm. With small values of h of this order the probability

$$P(X \in x_r \pm \tfrac{1}{2}h)$$

may be replaced, with sufficient accuracy, by

$$hf(x_r; \theta), \qquad r = 1, 2, \ldots, n.$$

The probability of obtaining the observed sample, for a given value of θ, is thus

$$h^n \prod_1^h f(x_r; \theta).$$

Thus, for every fixed value (θ_1, say) of the parameter, the numerical value of the likelihood is proportional to $\prod_1^n f(x_r; \theta_1)$, and we may therefore take the likelihood (which need be specified only up to a multiplicative constant or function of the data not involving θ) to be

$$\prod_1^n f(x_r; \theta)$$

where now the x_r are held fixed and θ is an undetermined variable.

There is of course a fundamental difference between probability and likelihood: in a probability statement, one is concerned with a set of possible outcomes, with a fixed value of θ. In a likelihood statement, on the other hand, the outcome values are fixed, while all possible values of θ are under consideration. Under suitable circumstances, sums of probabilities are probabilities, but sums of likelihoods are not likelihoods, etc.

In spite of these differences there are features in which probabilities and likelihoods resemble each other. Relatively large likelihoods correspond to more plausible values of θ than do relatively small likelihoods, just as large probabilities correspond to more strongly expected outcomes than do small probabilities.

Of two values θ_a and θ_b of θ, θ_a is said to be more likely than θ_b in the sense of being more likely to be near the true value θ^0, if $l(\theta_a)/l(\theta_b) > 1$. The value $\theta_{\max}$ which maximizes the likelihood function has the property that $l(\theta_{\max})/l(\theta_b) > 1$ for every choice of θ_b ($\theta_b \neq \theta_{\max}$) and, in this sense, is the most likely value of θ (on the given data). The method of maximum likelihood takes this maximizing value $\theta_{\max}$ (which depends of course on the data $x_1, x_2, \ldots, x_n$) as the estimate of θ^0. It is called the maximum-likelihood estimate (MLE) of θ^0.

The procedure is illustrated in Figure 3.5.3, which shows the graph of our likelihood function, together with the MLE. In this example the value of $\theta_{\max}$ may be obtained by differentiation, as the relevant root of the *likelihood equation* $dl(\theta)/d\theta = 0$, or, equivalently, of the equation

$$d\{\log l(\theta)\}/d\theta = 0$$

where $l(\theta)$ is given by (3.5.6). Thus, in our example, the likelihood equation reduces to

$$\frac{n}{\theta} - \sum_1^n x_r = 0,$$

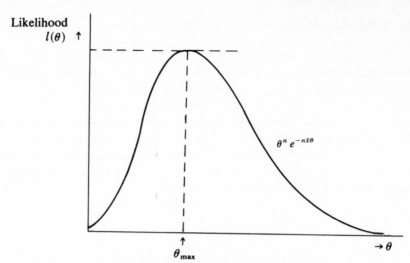

Figure 3.5.3: The likelihood function $l(\theta)$ of Example 3.5.4.

whence

$$\theta_{max} = n/\sum x_r = 1/\bar{x},$$

where $\bar{x}$ is the sample mean.

The above description needs supplementing: strictly speaking the likelihood function should be taken as $al(\theta)$, where a is an arbitrary positive function of the observations and $l(\theta)$ is defined as in (3.5.6). This does not affect the maximizing procedure, since, for every positive a, $al(\theta)$ and $l(\theta)$ attain their maxima at the same value of θ.

In practice, applications of maximum likelihood do not usually make explicit reference to a 'true value θ^0' which picks out the distribution under investigation as the particular member $f(x; \theta^0)$ of the postulated family of p.d.f.'s $f(x, \theta)$, $\theta \in \Omega$ (Ω being the specified 'parameter space': in the example $\Omega = \{\theta: \theta > 0\}$). Instead, (i) one usually speaks (with some abuse of language) of the problem of estimating the parameter θ of the p.d.f. $f(x, \theta)$, meaning by this use of 'θ' the true value θ^0, and (ii) one speaks at the same time of the likelihood function $l(\theta)$, meaning by *this* use of 'θ' a variable whose range is the parameter space Ω.

The maximizing procedure may often be simplified by working with the *log-likelihood* function $\log l(\theta)$ instead of the likelihood, since one then has a sum rather than a product to differentiate, and $\log l(\theta)$ attains its maximum at the same value θ_{max} as does $l(\theta)$. [It must not however be taken for granted that the maximum may in all cases be found by differentiating. For counter-examples see Chapter 6.]

When, as in Example 3.5.3, the likelihood equation has a simple explicit solution, it is possible to investigate the sampling distribution of the estimate by direct means. However it more commonly happens that the solution has to be obtained by an iterative numerical procedure, and a direct examination of the sampling distribution is not then feasible. The general theory (see Chapter 6) provides simple and powerful approximations for use in these cases.

The method readily extends to cases where there are several parameters and the observations are not necessarily i.i.d. (see Chapter 6).

3.5.5.   Normal Linear Models in which Maximum Likelihood and Least Squares Estimates Coincide

A common estimation problem is exemplified by the following: the yield of a chemical process is assumed to depend linearly on the temperature, the duration of cooling, and the quantity of activator present, plus of course the inevitable experimental error. When the *levels* of the *controlled variables* are set at

$$t_r \text{ (temperature)}$$

$$d_r \text{ (duration)}$$

$$a_r \text{ (activation)}$$

it is assumed that the yield (the *response variable*) y_r is a realization of a Normal variable Y_r with

$$E(Y_r) = \theta_0 + \theta_1 t_r + \theta_2 d_r + \theta_3 a_r$$

and

$$\text{var} (Y_r) = \sigma^2.$$

Here the θ's are unknown constants which express the sensitivity of the average yield to changes in the levels, and σ^2 is an unknown but constant variance of the yields.

We suppose that there is one observation for each combination of levels. (In practice there might be several; the present restriction to one is made solely in the interest of simplicity of notation.) On data $y_1, y_2, \ldots, y_n$ the *likelihood* function is proportional to

$$\sigma^{-n} \exp -\frac{1}{2\sigma^2} \Sigma (y_r - \theta_0 - \theta_1 t_r - \theta_2 d_r - \theta_3 a_r)^2$$

and this is a maximum with respect to the θ's when

$$\Sigma (y_r - \theta_0 - \theta_1 t_r - \theta_2 d_r - \theta_3 a_r)^2$$

is a minimum. Thus the *maximum likelihood* estimates of the 'linear' parameters $\theta_0, \theta_1, \theta_2, \theta_3$ coincide in this example with their *least squares* estimates.

E.H.L.

3.6. FURTHER READING

The topics treated in this chapter are basic to the whole of statistical inference, and good discussions of them, at various levels of sophistication, will be found in most general texts on statistics. The following is a selection of recommended texts; Barnett (1982), Fisher (1959), Kalbfleisch (1979), Kendall and Stewart (1973), Volume 2, Rao (1965), Zacks (1971)—all in Bibliography C.

CHAPTER 4

Interval Estimation

4.1. INTRODUCTION: THE PROBLEM

4.1.1. An Intuitive Approach

In earlier sections we have seen that a statistic $t(x_1, x_2, \ldots, x_n)$ which has an acceptable sampling distribution [see § 2.2] may be regarded as an estimate [see § 3.1] of the parameter θ of the p.d.f. $f(x, \theta)$ of a random variable X [see § 3.1]. In a sense, the available information about the unknown parameter θ is contained in the sampling distribution of the estimate, but this information may be difficult to interpret since that distribution itself depends on the unknown value of θ. (The situation is different in the Bayesian approach. See § 15.4.2.)

What is needed is a method of expressing the probable accuracy of the estimate by means which involve only the observed statistic and which do not depend on the unknown true value. An intuitively attractive approach that suggests that this is not an impossible task is as follows:

EXAMPLE 4.1.1. *Sample mean as estimate of expected value.* Suppose that X is Normal $(\theta, 1)$. The sample mean $\bar{x}$ of a sample of size n is a realization of the induced random variable [see Definition 2.2.1] $\bar{X}$ which is Normal $(\theta, 1/\sqrt{n})$. Changes in the value of θ shift the p.d.f. of $\bar{X}$ without changing its shape. Of the possible values of θ and the corresponding density functions, consider the following three (shown in Figure 4.1.1), with the observed value of $\bar{x}$ indicated for each on the same diagram:

In (i) the observed $\bar{x}$ lies in an extremely improbable position and we may say that, with high probability, this putative value (θ_1) of θ may be dismissed. The same applies to θ_3 in (iii). In (ii), on the other hand, the observed $\bar{x}$ lies in a region of high probability density, and the value θ_2 of θ is compatible with $\bar{x}$. There must be a value θ_l of θ between θ_1 and θ_2 and a value θ_u of θ between θ_2 and θ_3 such that values of θ satisfying $\theta_l \le \theta \le \theta_u$ may be regarded as being in some sense plausible, while values of θ lying outside this interval are not. But how plausible? and plausible in what sense? In the Bayesian

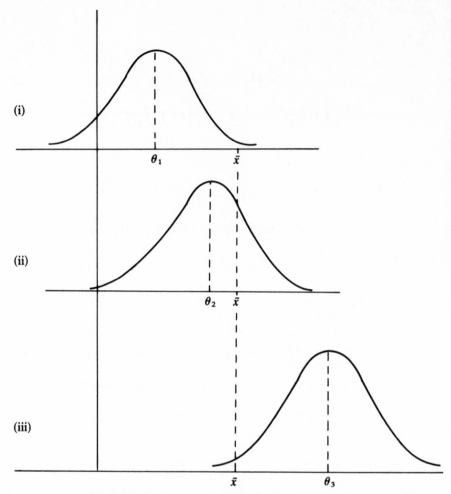

Figure 4.1.1: An observed sample mean $\bar{x}$ in relation to three possible values $\theta_1, \theta_2, \theta_3$ of $E(X)$.

approach there is a probability distribution of values of θ, and we may assess the plausibility of the interval (θ_l, θ_u) as the 'posterior probability' that the random variable θ belongs to the interval. This viewpoint is developed in Chapter 15.

Fisher argued for assessment in terms of the 'fiducial probability' of this interval, but this suffers from the drawback that there is less than a universal understanding of what fiducial probability means. (See, however, Kendall and Stuart (1973), Volume 2, Chapter 21—Bibliography C.].

The generally accepted non-Bayesian solution to the problem is a formalization of the following idea: if the value of θ is θ_l, $\bar{x}$ is a realization of a random variable $\bar{X}$ which is Normal $(\theta_l, 1/\sqrt{n})$, and the greatest distance d that such a realization can plausibly be from θ_l, if it is greater than θ_l, is one such that

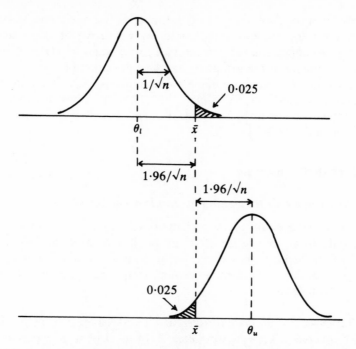

Figure 4.1.2: Lower and upper extremes θ_l and θ_u of plausible locations for $E(X)$, in the light of the observed sample mean $\bar{x}$.

$P(\bar{X} - \theta_l \geq d)$ has a conventionally small value—0·025, say. This fixes d at $1·96/\sqrt{n}$ (see Appendix T4). Thus, in terms of the observed statistic $\bar{x}$,

$$\theta_l = \bar{x} - 1·96/\sqrt{n}.$$

Similarly,

$$\theta_u = \bar{x} + 1·96/\sqrt{n}.$$

This somewhat informal argument leads to the concept of a *confidence interval*, which is explained in Section 4.2.

4.1.2. Standard Error

In Example 4.1.1 we considered the Normal $(\theta, 1)$ distribution. If we had considered, instead, the Normal (θ, σ) distribution [see II, § 11.4.3], with σ known, we should have been led to $\bar{x} \pm 1·96\sigma/\sqrt{n}$ as the intuitively probable interval for θ, since $\sigma/\sqrt{n}$ is the sampling standard deviation of $\bar{x}$ [see Definition 2.3.2]. If, however, the value of σ is not known (and this is usually the case) this interval cannot be evaluated, and it is then tempting to replace the unknown σ by an appropriate *estimate* σ^* of σ, in which case it is plausible to argue that $\bar{x} \pm 2\sigma^*/\sqrt{n}$ is a rough and ready approximation to a 95% interval for θ.

Similarly, if instead of using $\bar{x}$ to estimate θ, we use some statistic $t = t(x_1, x_2, \ldots, x_n)$ to estimate a parameter τ, we arrive at the interval $t \pm 2\sqrt{\{v(t)\}}*$ as an approximate 95% interval for τ, where $\sqrt{\{v(t)\}}*$ denotes an appropriate estimate of the sampling standard deviation of t.

These ideas can of course be made more rigorous and it is the purpose of this chapter to discuss ways of doing this. They do, however, indicate the intuitive appeal of the quantity $\sqrt{\{v(t)\}}*$. This is called the *standard error* of t. [See Definition 2.3.2.]

4.1.3. Probability Intervals

(a) *Probability intervals for continuous random variables*

A fully detailed account of the variability of a random variable X can be obtained only from knowledge of its probability distribution [see II, § 4.3, § 10.1]. It often happens, however, that one needs to find a simpler way of expressing variability. A convenient summarizing form is the *probability interval*, defined as follows.

DEFINITION 4.1.1.   *Probability interval.*   Let X have a continuous distribution which involves a known parameter θ (or a set of such parameters) and let $a = a(\theta)$ and $b = b(\theta)$, with $a < b$, be numbers such that, for given $p(0 < p < 1)$,
$$P(a \le X \le b) = p.$$
or, equivalently,
$$\int_a^b f(x, \theta) \, dx = F(b, \theta) - F(a, \theta) = p$$

where $f(x, \theta)$ and $F(x, \theta)$ denote respectively the p.d.f. and the c.d.f. of X at x. Then the interval (a, b), which depends on the value of θ, is called a probability interval for X with content (or 'size') p, or, equivalently, is a '$100p$% probability interval for X'.

(Note that, because X is a continuous random variable, it would make no difference if the expression $P(a \le X \le b)$ in the definition were replaced by any one of $P(a < X \le b)$, $P(a \le X \le b)$, $P(a < X < b)$.)

The interval (a, b) may be said, colloquially, to 'contain a proportion p of the distribution', in the sense that, in the long run, a proportion p of repeated observations on X would lie in the interval. The concept is illustrated in Figure 4.1.3 for a unimodal distribution.

The intuitive content of the concept is this: if p is large (say 0·95 or 0·99) we are 'practically certain' that any realization of X will lie in the probability interval (a, b). Here a and b are any numbers $(a < b)$ satisfying
$$F(b, \theta) - F(a, \theta) = p$$

where $F(x, \theta)$ denotes the c.d.f. of X.

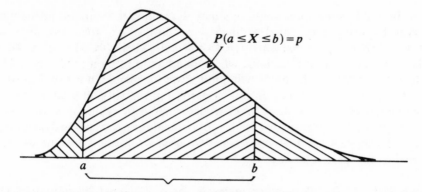

Figure 4.1.3: The interval (a, b) is a $100p\%$ probability interval for the distribution whose p.d.f. is shown. It is a *central* $100p\%$ interval if the tail areas to the left of a and to the right of b are each equal to $\frac{1}{2}(1-p)$.

[If X has a discrete distribution it will in general not be possible to find values of a and b corresponding exactly to a given p, and a modification of the definition is then called for. See Example 4.1.3.]

EXAMPLE 4.1.2. *Probability intervals for the standard Normal distribution.* If X is Normal (μ, σ), so that $U = (X - \mu)/\sigma$ is Normal $(0, 1)$, a 95% probability interval for U is any interval (a', b') such that

$$\Phi(b') - \Phi(a') = 0{\cdot}95$$

where $\Phi(\cdot)$ denotes the standard Normal c.d.f. Examples of such intervals (for U), obtained from published tables of $\Phi(\cdot)$ [see Appendix, Tables T3, T4], are (i) $(-3{\cdot}00, 1{\cdot}66)$, (ii) $(-2{\cdot}50, 1{\cdot}71)$, (iii) $(-1{\cdot}96, 1{\cdot}96)$, (iv) $(-1{\cdot}75, 2{\cdot}33)$. The corresponding intervals for $X (= \mu + \sigma U)$ will be

(i) $(\mu - 3\sigma, \mu + 1{\cdot}66\sigma)$, (length $= 4{\cdot}66\sigma$)

(ii) $(\mu - 2{\cdot}5\sigma, \mu + 1{\cdot}71\sigma)$, (length $= 4{\cdot}21\sigma$)

(iii) $(\mu - 1{\cdot}96\sigma, \mu + 1{\cdot}96\sigma)$, (length $= 3{\cdot}92\sigma$)

(iv) $(\mu - 1{\cdot}75\sigma, \mu + 2{\cdot}33\sigma)$. (length $= 4{\cdot}08\sigma$).

As will be clear from these examples, the specification of the content of a probability interval does not uniquely define its end-points. To achieve uniqueness one must impose some constraints. Most commonly one requires either (a) that the probability interval should be minimal length, or (b) that it should be 'central' or 'symmetrical' in the sense that

$$P(X \le a) = P(X \ge b). \qquad (4.1.1)$$

For a symmetrical unimodal distribution [see II, 10.1.3] the alternative criteria are equivalent to one another. In unimodal unsymmetrical (continuous) cases when the p.d.f. is not monotone, one may determine the shortest probability interval by using the 'equal-ordinates' condition, viz. that, for given p, $F(b, \theta) - F(a, \theta)$ is minimal when $f(b, \theta) = f(a, \theta)$, where $f(x, \theta) = dF(x, \theta)/dx$ is the p.d.f. of X. In applications, where one has to rely on published tables of the c.d.f., tables of the p.d.f. being not always available, it is usual to work with the symmetry condition. In this case, for a probability interval of content p, the values of a and b in (4.1.1) are determined by taking a as the lower $100(1 - \frac{1}{2}p)$ and b as the upper $100(1 - \frac{1}{2}p)$ percentage points of the distribution of X. In Example 4.1.2 the 'symmetrical' interval is (c); this is also the shortest interval in this particular case.

EXAMPLE 4.1.3. *Probability intervals for a J-shaped distribution.* The exponential distribution with p.d.f. $\theta^{-1} \exp(-x/\theta)$ for $x > 0$ [see II, § 10.2.3] has a monotone decreasing ('J-shaped') density and so the 'equal ordinates' condition does not apply. It is clear that for this case, for each given p, the shortest probability interval with content p is $(0, b)$, where b is chosen so that $P(X \le b) = p$, that is, where $1 - \exp(-b/\theta) = p$, so that

$$b = \theta \log \frac{1}{1-p}.$$

The shortest 95% interval is therefore $(0, 2 \cdot 996 \theta)$. The symmetrical probability interval of size $0 \cdot 95$ is the interval (a', b'), where

$$P(X \le a') = 1 - \exp(-a'/\theta) = 0 \cdot 025$$

and

$$P(X \ge b') = \exp(-b'/\theta) = 0 \cdot 025$$

so that $a' = 0 \cdot 078 \theta$, $b' = 3 \cdot 69 \theta$.

(b) *Probability intervals for discrete random variables*

In the case of a continuous random variable X, the interval (a, b) is a $100p\%$ probability interval if

$$P(a \le X \le b) = p,$$

or, equivalently, if

$$P(a < X < b) = p.$$

These are equivalent because, X being continuous, $P(X = a) = P(X = b) = 0$. This is no longer true in general if X is discrete. Suppose for example that X is a Poisson variable, and that a and b are positive integers. Then $P(X = a) > 0$ and $P(X = b) > 0$, so that

$$P(a < X < b) < P(a \le X \le b).$$

To avoid ambiguity therefore, in the discrete case, we say that $[a, b]$ is a $100p\%$ *closed* probability interval for X if

$$P(a \leq X \leq b) = p.$$

(It is sometimes convenient for integer-valued variables to speak equivalently of the *open* interval $(a-1, b+1)$, for which $P(a-1 < X < b+1) = p$.)

A more serious complication that arises with discrete variables is this: for given a and b we can always evaluate $p = P(a \leq X \leq b)$, but for given p we cannot in general find any interval (a, b), much less a 'central' interval (a, b), such that $P(a \leq X \leq b) = p$.

What we *can* do is to find a 'quasi-central' probability interval with probability content *at least* $100p\%$, and as near to $100p\%$ as possible. For a closed interval of this sort with $p = 0.95$ we find from the cumulative distribution tables the value r_l such that

$$P(R < r_l) \leq 0.025$$

while

$$P(R < r_l + 1) > 0.025,$$

and the value r_u such that

$$P(R > r_u) \leq 0.025$$

while

$$P(R > r_u - 1) > 0.025.$$

Then

$$P(r_l \leq R \leq r_u) = 1 - P(R < r_l) - P(R > r_u) \geq 0.95.$$

Care is needed in using tables since these usually—but not universally—give values of $P(R \geq r)$. (This is the case with our Binomial tables given in Appendix T1.) A diagram such as is shown in Figure 4.1.4. may help to sort things out. For the case illustrated, the Binomial (n, p) distribution, with $n = 20$, $p = 0.4$, the quasi-central probability interval with content at least 95% is $(4, 12)$, with actual probability content 96.2%.

4.2. CONFIDENCE INTERVALS AND CONFIDENCE LIMITS

An estimate $\hat{\theta} = t(x_1, x_2, \ldots, x_n)$ of a parameter θ is a realization of the induced random variable [see Definition 2.2.1] $T = t(X_1, X_2, \ldots, X_n)$ whose variability may be specified in terms of its probability distribution, i.e. in terms of the sampling distribution of the estimate. This distribution will of course depend on the unknown value of θ.

One may apply to this distribution the concepts outlined in section 4.1 and obtain a summarizing description of the variability of $\hat{\theta}$ in terms of a probability interval, which, in turn, will depend on θ.

(*R* is Binomial $(20, 0.4)$. See Appendix, Table T1.)

r	$P(R < r)$
1	0.0000
2	0.0005
3	0.0036
4	$0.0160 \ldots P(R < 4) = 0.0160 < 0.0250$
5	$0.0510 \ldots P(R < 5) = 0.0510 > 0.0250$

(too big)   Hence $r_l = 4$

r	$P(> r)$
$\vdots$	$\vdots$
14	0.0016
13	0.0065
12	$0.012 \ldots P(R > 12) = 0.0210 < 0.0250$
11	$0.0565 \ldots P(R > 11) = 0.0565 > 0.0250$

(too small) Hence $r_u = 12$

$$P(r_l \leq R \leq r_u) = P(R = 4) + P(R = 5) + \ldots + P(R = 12)$$

$$= 0.963.$$

r: 0 1 2 3 | 4 5 6 7 8 9 10 11 12 | 13 14 15 ... 20

Prob: 0.0160 | 0.9630 | 0.0210

(the probability interval)

Figure 4.1.4. A quasi-central approximate 95% probability interval of the Binomial $(20, 0.4)$ distribution.

Even when we take into account the fact that the value of θ is unknown, this is not entirely useless information. Usually, however, one requires more direct information about the accuracy of the estimate. One way of doing this would be to formalize the intuitive concept explained in Example 4.1.1, and to construct, if possible, an interval which had a specified probability of containing the unknown value of θ. Since the concept of probability applies only to a random variable and since (from a non-Bayesian viewpoint) θ is not a random variable, it follows that this can be done only if the end-points of the interval are random variables. The resulting interval is called a *confidence interval* and its end-points are called *confidence limits*. Before giving a formal definition, we illustrate with a simple example. (The example is somewhat artificial, as it is based on a Normal distribution in which the variance is supposed to be known, and the expectation θ is the only unknown parameter. Its artificiality is, it is hoped, justified by the simplicity and familiarity of the sampling distribution involved.)

EXAMPLE 4.2.1.   *Confidence interval for Normal μ when σ is known.*   Let $x_1, x_2, \ldots, x_n$ be observations on the random variable X whose distribution is Normal $(\theta, 1)$. Then the statistic $\bar{x} = \sum_1^n x_j / n$ is a realization of a Normal $(\theta, 1/\sqrt{n})$ variable $\bar{X}$. The corresponding 'standardized' variable $U = \sqrt{n}(\bar{X} - \theta)$ is Normal $(0, 1)$. As in Example 4.1.2, we can construct a symmetrical probability interval for U of content $0·95$, say; this is the interval $(-1·96, 1·96)$. We then have

$$0·95 = P(-1·96 \le U \le 1·96)$$
$$= P(-1·96 \le \sqrt{n}(\bar{X} - \theta) \le 1·96)$$
$$= P(\theta - 1·96/\sqrt{n} \le \bar{X} \le \theta + 1·96/\sqrt{n}).$$

Now the proposition

$$\theta - 1·96/\sqrt{n} \le \bar{X} \le \theta + 1·96/\sqrt{n} \qquad (4.2.2)$$

is equivalent to the two propositions

$$\bar{X} \ge \theta - 1·96/\sqrt{n},$$

and

$$\bar{X} \le \theta + 1·96/\sqrt{n},$$

i.e. to the two propositions

$$\theta \le \bar{X} + 1·96/\sqrt{n},$$

and

$$\theta \ge \bar{X} - 1·96/\sqrt{n},$$

which may be combined to give

$$\bar{X} - 1·96/\sqrt{n} \le \theta \le \bar{X} + 1·96/\sqrt{n}. \qquad (4.2.3)$$

Equation (4.2.1) may therefore be written in the 'inverted' form

$$0·95 = P(\bar{X} - 1·96/\sqrt{n} \le \theta \le \bar{X} + 1·96/\sqrt{n}),$$

which has the following meaning: with probability $0·95$, the random interval

$$(\bar{X} - 1·96/\sqrt{n}, \bar{X} + 1·96/\sqrt{n}) \qquad (4.2.4)$$

will include the (unknown) true value of θ. (By a *random interval* is meant an interval whose end-points are random variables). Taking $\bar{x}$, the observed sample mean, to be a realization of $\bar{X}$, we may say that the interval

$$(\bar{x} - 1·96/\sqrt{n}, \bar{x} + 1·96/\sqrt{n}), \qquad (4.2.5)$$

whose endpoints (for given n) are known numbers, is a realization of the random interval (4.2.6). (Compare Example 4.1.2.) This realization is called a *confidence interval for θ, with confidence coefficient* $0·95$, or, more colloquially, a 95% confidence interval for θ. Repetitions of the sampling procedure will generate different sample means $\bar{x}_1, \bar{x}_2, \ldots$ and hence different realizations $(\bar{x}_j - 1·96/\sqrt{n}, \bar{x}_j + 1·96/\sqrt{n})$ of (4.2.6), i.e. numerically different confidence intervals. In the long run, in 95% of such cases, θ will be inside the confidence interval. This is another way of saying that (using this statistic) 95% of the individual confidence intervals will contain the unknown θ. In this sense one may attach a 'confidence level' of 95% to the statement that θ will lie inside the 95% confidence interval generated by a particular sample. The situation is illustrated in Figure 4.2.1.

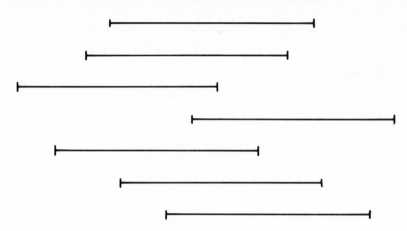

Figure 4.2.1: Examples of confidence intervals for θ generated by samples of size 25 from a Normal $(\theta, 1)$ distribution.

Confidence interval for μ when X is Normal (μ, σ) (with σ known):

The usual notation for the expectation and the standard deviation are μ and σ respectively. If X is Normal (μ, σ), the central 95% confidence interval (4.2.5) for μ becomes

$$(\bar{x} - 1{\cdot}96\sigma/\sqrt{n},\ \bar{x} + 1{\cdot}96\sigma/\sqrt{n}).   \tag{4.2.6}$$

We now give a formal definition of a confidence interval, and, in the following section, a formalization of the procedure employed in Example 4.2.1 for its construction.

DEFINITION 4.2.1.   *Confidence interval, confidence limits.*   A confidence interval for the parameter θ of the distribution of a random variable X, at confidence level $100p\%$, based on a sample $(x_1, x_2, \ldots, x_n)$, is an interval with endpoints $w_1(x_1, x_2, \ldots, x_n)$, $w_2(x_1, x_2, \ldots, x_n)$, realizations of the induced random variables $W_1 = w_1(X_1, X_2, \ldots, X_n)$, $W_2 = w_2(X_1, X_2, \ldots, X_n)$ which are such that

$$P(W_1 \le \theta \le W_2) = p.$$

The end-points w_1, w_2 of the confidence interval are called *confidence limits.* (Here the X_r are statistical copies of X [see Definition 2.2.1].)

As in the case of the probability intervals discussed in section 4.1.1, the intuitive interpretation of the confidence interval is that, if p is large, (say $0{\cdot}95$ or $0{\cdot}99$), it is highly likely that θ does in fact lie in between the given limits.

4.3. CONSTRUCTING A CONFIDENCE INTERVAL BY USING A PIVOT

In Example 4.2.1 the construction of the confidence interval (4.2.4) was based on equation (4.2.1), which expresses a property of the distribution of $\sqrt{n}(\bar{X} - \theta)$ (that is, of the sampling distribution of the quantity $\sqrt{n}(\bar{x} - \theta)$). Because $\sqrt{n}(\bar{X} - \theta)$ has a 'standard' *parameter-free* distribution (namely, in this case, the standard Normal) it was possible to obtain the 95% probability interval $(-1\cdot96, 1\cdot96)$ shown in (4.2.1); and because it involves both (i) a random variable $(\bar{X})$ of which a realization $(\bar{x})$ has been observed and (ii) the unknown parameter θ, it was possible, by inserting the inequalities expressed in (4.2.2), to rewrite them in the logically equivalent 'inverted' form (4.2.3), in which the unknown parameter θ is shown as being included in an interval whose endpoints are defined by observable random variables $\bar{X} \pm 1\cdot96/\sqrt{n}$.

The crucial quantity $\sqrt{n}(\bar{x} - \theta)$ is an example of a *pivot*, which we now define:

DEFINITION 4.3.1.   *Pivot.*   Let $x_1, x_2, \ldots, x_n$ be observations on a random variable X whose distribution involves an unknown parameter θ, and let $h = h(x_1, x_2, \ldots, x_n)$ be a statistic.

The quantity

$$q(h, \theta)$$

is a *pivot* for θ if its sampling distribution is parameter-free.

Once one has a pivot, the procedure is as follows: The pivot $q(h, \theta)$ is a realization of the induced random variable $Q = q(H, \theta)$, where $H = h(X_1, X_2, \ldots, X_n)$. Here the X_j are, as usual, statistical copies of X [see Definition 2.2.1]. Denote the (parameter-free) c.d.f. of Q by

$$G(q) = P(Q \le q),$$

and construct the symmetrical $100p\%$ (e.g. 95%) probability interval (a, b) for Q, i.e. the interval (a, b) defined by

$$G(a) = 1 - G(b) = (1 - p)/2 \tag{4.3.1}$$

$(= 0\cdot025$, say) as in (4.1.1). We then have

$$P\{a \le Q \le b\} = p$$

i.e.

$$P\{a \le q(H, \theta) \le b\} = p \tag{4.3.2}$$

(compare (4.2.1).) It is next necessary to solve the inequalities

$$\left. \begin{array}{l} q(H, \theta) \ge a \\ q(H, \theta) \le b, \end{array} \right\} \tag{4.3.3}$$

(compare (4.2.3)) obtaining, say,

$$\left.\begin{array}{ll} \theta \le w_2(a, H) & (= W_2, \text{say}) \\ \theta \ge w_1(b, H) & (= W_1, \text{say}) \end{array}\right\} \tag{4.3.4}$$

(compare (4.2.4)) which may be combined in the form

$$W_1 \le \theta \le W_2,$$

a proposition which holds with probability p (compare Definition 4.2.1). Thus the $100p\%$ confidence interval for θ is

$$\{w_1(b, h), w_2(a, h)\},$$

where

$$h = h(x_1, x_2, \ldots, x_n),$$

is the statistic on which the construction was based.

EXAMPLE 4.3.1.   *Confidence interval for exponential parameter.*   Let X be exponentially distributed [see II, § 10.2.3], with expectation θ, so that the p.d.f. of X at x is $\theta^{-1} \exp(-x/\theta)$, $x > 0$. From a sample $(x_1, x_2, \ldots, x_n)$ construct the (sufficient [see § 3.4.1]) statistic $h(x_1, x_2, \ldots, x_n) = \sum_1^n x_r$. This is a realization of the random variable $H = \sum_1^n X_r$, where the X_r are statistical copies of X. The distribution of H is of the gamma form [see II, § 11.3], with p.d.f. at h given by

$$\theta^{-n} h^{n-1} e^{-h/\theta}/(n-1)!, \qquad h > 0,$$

from which it may be seen [see II, § 10.7] that the quantity

$$Q = H/\theta$$

has p.d.f. at q given by

$$q^{n-1} e^{-q}/n!, \qquad q > 0.$$

This is parameter-free, whence $q = \sum_1^n x_r/\theta$ is a pivot for θ. We may construct a central 95% probability interval for Q with the aid of tables of the χ^2 distribution, see Appendix T6 since $2Q$ has a χ^2 distribution on $2n$ degrees of freedom. [Compare § 2.5.4(a)(ii).] For example, if $n = 10$, the corresponding χ^2 distribution has 20 degrees of freedom, whence, from the table in Appendix T6, the central 95% probability interval for $2Q$ is the interval $(9 \cdot 591, 34 \cdot 170)$. The corresponding interval for $Q(= H/\theta)$ is therefore $(4 \cdot 795, 17 \cdot 085)$, whence

$$P\{4 \cdot 795 \le H/\theta \le 17 \cdot 085\} = 0 \cdot 95$$

or, equivalently

$$P\{4 \cdot 795\theta \le H \le 17 \cdot 085\theta\} \times 0 \cdot 95 \tag{4.3.5}$$

(compare (4.3.2), with $a = 4 \cdot 795$, $q(H, \theta) = H/\theta$, $b = 17 \cdot 085$, $p = 0 \cdot 95$). Inverting the inequalities

$$H/\theta \geq 4 \cdot 795, \qquad H/\theta \leq 17 \cdot 085$$

(compare (4.3.3)) we obtain

$$\theta \leq H/4 \cdot 795 \ = 0 \cdot 2085 H \quad (= W_2 \text{ in } (4.3.4))$$

$$\theta \geq H/17 \cdot 085 = 0 \cdot 0585 H \quad (= W_1 \text{ in } (4.3.4))$$

which we combine in the form

$$0 \cdot 0585 H \leq \theta \leq 0 \cdot 2085 H, \tag{4.3.6}$$

which holds with probability $0 \cdot 95$. Thus the 95% confidence interval for θ is the interval $(0 \cdot 0585 h, 0 \cdot 2085 h)$, where $h = \sum_1^n x_r = n\bar{x} = 10\bar{x}$ in our case, $\bar{x}$ denoting the observed sample mean. Finally, then, our confidence interval for the unknown expectation θ is $(0 \cdot 585\bar{x}, 2 \cdot 085\bar{x})$. With $\bar{x} = 2 \cdot 1$, for example, the 95% confidence interval for θ would be $(1 \cdot 23, 4 \cdot 38)$.

4.4.   FACTORS AFFECTING THE INTERPRETATION OF A CONFIDENCE INTERVAL AS A MEASURE OF THE PRECISION OF AN ESTIMATE

It is to be noted that the confidence interval for a parameter, obtained from a given sample, is determined by the choice of the 'working statistic' $h(x_1, x_2, \ldots, x_n)$. If this is a sufficient statistic (as in Example 4.3.1) there is no ambiguity: the information in the sample will have been exhaustively extracted. What, however, are we to do when there is no simple sufficient statistic? Intuitively one feels that one ought to use the best statistic available, e.g. an efficient [see Definition 3.3.3] statistic such as the maximum likelihood estimate [see § 3.5.4, § 6.2.2]. This however introduces the possibility of some ambiguity, which is increased if, for reasons of tractability, one works with some other plausible statistic (as is done in Example 4.6.1).

In accepted statistical practice, the choice of working statistic is in fact influenced by tractability, familiarity, the availability of tables, etc. Fisher would regard the stated precision of inferences of this sort as having a lower level of validity than that enjoyed by a probability statement based on a sufficient statistic. Working statisticians appear not to be unduly worried by this difficulty. A possibly consistent viewpoint is to regard the confidence level of the statement that the interval contains θ as representing a probability measured on a reference set of conceptual repetitions of the procedure of sampling and the computation of the confidence interval *based on the stated working statistic.*

Far more straightforward in the interpretation of a confidence interval are such factors as the size of the sample and the magnitude of the confidence level. In example 4.2.1, which dealt with the parameter θ of a Normal $(\theta, 1)$ distribution, the central 95% confidence interval was found to be $(\bar{x} - 1\cdot96/\sqrt{n}, \bar{x} + 1\cdot96/\sqrt{n})$, the length of this interval being $3\cdot92/\sqrt{n}$. If one had worked with a confidence level of 99% the interval would have been $(\bar{x} - 2\cdot58/\sqrt{n}, \bar{x} + 2\cdot58/\sqrt{n})$, the length this time being $5\cdot16/\sqrt{n}$. This illustrates the general result that, if one wishes to increase the probability that one's assertion is correct when one says that the parameter lies in a specified interval, the price one has to pay is that of having a larger interval. On the other hand, a larger sample can be expected to yield a more precise estimate, and the length of the confidence interval for a given confidence level may therefore be expected to diminish with increasing sample size. In the example quoted the length is proportional to $n^{-1/2}$.

The example quoted is unusual in that the length of the confidence interval depends only on the sample size and the confidence coefficient. In general it will depend also on the relevant statistic, as discussed above; thus, in Example 4.3.1, the $100p\%$ confidence interval for the expectation θ of an exponential distribution, based on a sample of size n and the sample mean $\bar{x}$ as relevant statistic, was derived from the probability interval

$$k_p(1, n)\bar{x} \le 2n\bar{x}/\theta \le k_p(2, n)\bar{x},$$

where the confidence limits $k_p(1, n)$, $k_p(2, n)$ are defined in terms of the chi-squared distribution on $2n$ degrees of freedom by the inequalities

$$P\{\chi^2(2n) \le k_p(1, n)\} = \tfrac{1}{2}(1 - p) = P\{\chi^2(2n) \ge k_p(2, n)\}.$$

(Equation (4.3.6) expresses this result for the case $n = 10$, $p = 0\cdot95$.) Thus the confidence limits for θ are $2n\bar{x}/k_p(2, n), 2n\bar{x}/k_p(1, n)$, and the expected length of the confidence interval is $2n\theta\{1/k_p(2, n) - 1/k_p(1, n)\}$.
Some representative values of this, for $p = 0\cdot95$ and $p = 0\cdot99$, and $n = 5, 10, 15, 20$ are as follows:

Sample size n	Confidence level	
	0·95	0·99
5	2·59θ	4·24θ
10	1·50θ	2·39θ
15	1·15θ	1·62θ
20	0·96θ	1·33θ

[The preceding discussion is related to statistical practice. It has to be admitted, however, that, while it may well be true that 'practice makes perfect', it is not invariably the case in statistics that practice *is* perfect. (At least, it is not always perfectly justifiable.) To measure the uncertainty of an estimate

θ^* of a parameter θ by the length of the confidence interval makes sense when θ is a location parameter, since the p.d.f. of the sampled random variable is then of the form $f(x-\theta)$, so that a change in θ from θ_1 to θ_2 shifts the p.d.f. through a distance proportional to $\theta_1 - \theta_2$. When however, θ is not a location parameter, other measures might be preferable. For example, if θ is a scale parameter, the p.d.f. being of the form $f(x/\theta)$, a change in θ from θ_1 to θ_2 affects the argument of f multiplicatively, changing x/θ_1 through the factor θ_2/θ_2, to x/θ_2. This suggests that the magnitude of the uncertainty in θ implied by the confidence interval (θ_2, θ_1) ought to be measured by the ratio θ_1/θ_2 rather than by the length of the interval.]

4.5. CONFIDENCE INTERVALS WHEN THERE ARE SEVERAL PARAMETERS

Among the questions relating to confidence intervals that arise when there are several parameters are the following:

(i) Can we obtain separate confidence intervals for the individual parameters?

(ii) Can we obtain confidence intervals for a combination of the parameters, for example for the sum, difference or ratio of two of them?

(iii) Can we obtain joint (multidimensional) confidence regions for several parameters?

These and related questions are treated in some generality in Chapter 8 in the context of the Analysis of Variance. At an introductory level we give an elementary discussion of (i) and (ii) in the present section; a deeper discussion, touching on (iii), is given in section 4.9.

4.5.1. Individual Confidence Intervals

We first give examples of separate confidence intervals for an individual parameter, taking the variance σ^2 of the Normal (μ, σ) family (Example 4.5.1), the expectation μ of the same family (Example 4.5.2), and each of the three parameters involved in Normal theory linear regression (Example 4.5.3).

EXAMPLE 4.5.1. *Confidence interval for the variance (or the standard deviation) of a Normal (μ, σ) distribution.* Let $(x_1, x_2, \ldots, x_n)$ be a sample from the normal (μ, σ) distribution. Then, with $\bar{x} = \sum_1^n x_r/n$, the statistic $s^2 = \sum_1^n (x_r - \bar{x})^2/(n-1)$, an unbiased estimate of σ^2, is a realization of the random variable $S^2 = \sum_1^n (X_r - \bar{X})^2/(n-1)$, where the X_r are i.i.d. Normal (μ, σ^2) variables and $\bar{X}$ is Normal $(\mu, \sigma/\sqrt{n})$. Then $(n-1)S^2/\sigma^2$ has the $\chi^2(n-1)$ distribution [see § 2.5.4(d)], that is the Chi-squared distribution $n-1$ degrees of freedom [see II, § 11.4.11], for which we may construct the central $100p\%$ probability interval (a, b) by taking a and b to be defined by the following:

$$P(\chi^2(n-1) \le a) = P(\chi^2(n-1) \ge b) = \tfrac{1}{2}(1-p). \qquad (4.5.1)$$

The values of a and b may be read off from the published χ^2 tables [see Appendix T6]. Then, with probability p,

$$a \leq (n-1)S^2/\sigma^2 \leq b,$$

or, equivalently,

$$(n-1)S^2/b \leq \sigma^2 \leq (n-1)S^2/a.$$

This in turn is equivalent to

$$\{(n-1)S^2/b\}^{1/2} \leq \sigma \leq \{(n-1)S^2/a\}.$$

Thus the central $100p\%$ confidence intervals for σ^2 and for σ are as follows:

for

$$\sigma^2: ((n-1)s^2/b, (n-1)s^2/a),$$

for $\hspace{8cm}$ (4.5.2)

$$\sigma: (\{(n-1)s^2/b\}^{1/2}, \{(n-1)s^2/a\}^{1/2}).$$

In this example the existence of a second parameter μ has not affected the argument in any way. [Compare Example 5.8.6.]

The following data (quoted by Fisher (1970)—see Bibliography C) relate to the effectiveness of two sleep-inducing drugs. Each of ten patients took each of two drugs at time intervals large enough to justify the assumption that their reactions to the drugs were mutually independent.

	Additional hours of sleep		Difference
Patient	Drug A	Drug B	y
1	0·7	1·9	1·2
2	−1·6	0·8	2·4
3	−0·2	1·1	1·3
4	−1·2	0·1	1·3
5	−0·1	−0·1	0·0
6	3·4	4·4	1·0
7	3·7	5·5	1·8
8	0·8	1·6	0·8
9	0·0	4·6	4·6
10	2·0	3·4	1·4

Table 4.5.1

The figures in the final column may be regarded as independent observations (y_r) measuring the relative potencies, taken here as the difference between the potencies of the two drugs. We assume the y_r to come from a Normal (μ, σ) distribution, and take as estimate of σ^2 the statistic

$$s^2 = \sum (y_r - \bar{y})^2 / (n - 1)$$

with

$$n = 10, \qquad \bar{y} = \sum y_r / 10 = 1 \cdot 58,$$

$$9s^2 = \sum (y_r - \bar{y})^2 = \sum y_r^2 - 10\bar{y}^2$$

$$= 38 \cdot 58 - 24 \cdot 96 = 13 \cdot 62.$$

Thus the estimates of σ^2 and of σ are

$$s^2 = 1 \cdot 513 \quad \text{and} \quad s = 1 \cdot 230$$

respectively. The central 95% probability interval for χ^2 on 9 d.f. is $(2 \cdot 700, 19 \cdot 023)$, whence the 95% confidence interval for σ^2 is

$$(13 \cdot 62 / 19 \cdot 023, 13 \cdot 62 / 2 \cdot 700),$$

that is

$$(0 \cdot 716, 5 \cdot 044).$$

The corresponding interval for σ is

$$(0 \cdot 846, 2 \cdot 246).$$

EXAMPLE 4.5.2.   Confidence interval for the expectation μ of a Normal distribution.

(a) Population variance (σ^2) known.

This case was dealt with in Example 4.2.1 on data $x_1, x_2, \ldots, x_n$: the 95% confidence interval for μ is $\bar{x} \pm 1 \cdot 96 \sigma / \sqrt{n}$, $\bar{x}$ being the sample mean.

(b) Population variance not known: rough and ready method using standard error.

This merely replaces the unknown σ in (a) above by its estimate s. More generally, suppose θ^* to be an approximately unbiased estimate of a parameter θ where the sampling distribution of θ^* is approximately Normal. Then the 'rough and ready' 95% confidence interval for θ is

$$\theta^* \pm 2 \text{ s.e. } (\theta^*),$$

where s.e. (θ^*) denotes the standard error of θ^* (ie an appropriate estimate of the sampling standard deviation of θ^*).

This has particular relevance in connection with the method of maximum likelihood.

(c) Population variance not known. Exact confidence interval using Student's distribution.

There is a new difficulty here, in that we are looking for a confidence interval for one parameter (μ) when there is a second parameter (σ) whose value is unknown. The problem was solved by Student's brilliant idea of eliminating σ by a process now known as Studentization.

If $(x_1, x_2, \ldots, x_n)$ is a sample of observation on a Normal (μ, σ) population, and we define the statistics

$$\bar{x} = \sum x_j/n, \qquad s^2 = \sum (x_j - \bar{x})^2/(n-1),$$

the quantity $(\bar{x} - \mu)/s$ is invariant under changes in the value of σ: for if σ is changed to σ' it remains true that

$$\frac{n-\mu}{s} = \frac{(x-\mu)/\sigma}{s/\sigma} = \frac{(x-\mu)/\sigma'}{s/\sigma'}.$$

The more convenient quantity

$$t_{n-1} = (\bar{x} - \mu)/(s/\sqrt{n}),$$

called 'Student's t on $n-1$ degrees of freedom' has a parameter-free sampling distribution (see § 2.5.5) and so is a *pivot* for μ.

The distribution is symmetrical, whence, if a is the value exceeded with probability $\frac{1}{2}(1-p)$, the required $100p\%$ central probability interval for μ is the interval

$$(\bar{x} - as/\sqrt{n}, \bar{x} + as/\sqrt{n}).$$

Values of a corresponding to specified values of p may be obtained from the appropriate table (see Appendix T5). (Compare Example 5.8.2.)

If, for example, $n = 10$, there are 9 degrees of freedom, and taking $p = 0.95$, the value of a is 2·262. Using the same data as in Example 4.5.1 we have $\bar{y} = 1.58$, $s^2 = 1.513$, $n = 10$, and $s/\sqrt{n} = 0.389$. Thus the 95% confidence interval for μ is

$$1.58 \pm (2.262)(0.389),$$

i.e.

$$1.58 \pm 0.88 = (0.70, 2.46). \tag{4.5.3}$$

This indicates that there is in all probability a real difference in potency between the two drugs, the extra hours of sleep induced by Drug B as compared with Drug A being about 1·58 hours on average (or, more precisely, 1.58 ± 0.88 hours, at the 95% level). The numerical quantities involved are illustrated in the diagram of Student's distribution shown in Figure 4.5.1.

The intuitive approach used in Example 4.1.1 will be found to apply in this example too, on replacing the Normal p.d.f. of Example 4.1.1 by the p.d.f. of Student's distribution.

Figure 4.5.2 shows the p.d.f. of Student's distribution for 9 degrees of freedom, with various values of $t = (\bar{x} - \mu)/(s/\sqrt{10}) = (1.58 - \mu)/0.389$; i.e. of $\mu = 1.58 - 0.389t$. In that figure the value t_1 of t (and hence the corresponding value of μ, namely $\mu_1 = 5.40 - 0.601t_1$) is too improbable to be accepted. The same applies to t_3. The value of t_2 lies in a region of high probability

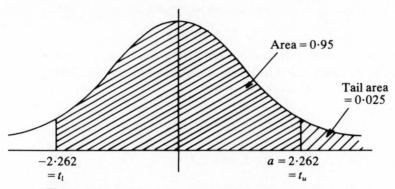

Figure 4.5.1: p.d.f. of Student's distribution on 9 d.f.

density, so that $\mu_2 = 1\cdot58 - 0\cdot389 t_2$ is compatible with the data. The conventional boundaries between acceptable and unacceptable values of μ (at the 95% level) are given by t_l and t_u as in Figure 4.5.1, where t_l is the $0\cdot025$ quantile and t_u the $0\cdot975$ quantile of Student's distribution on 9 degrees of freedom. These are $t_l = -2\cdot262$, $t_u = 2\cdot262$, whence the confidence interval is (μ_l, μ_u), with $\mu_l = 1\cdot58 + 0\cdot389 t_l = 0\cdot70$ and $\mu_u = 1\cdot58 + 0\cdot389 t_u = 2\cdot46$.

EXAMPLE 4.5.3. *Simple linear regression.* If the pair of random variables (X, Y) have a bivariate distribution [see II, § 13.1.1], the conditional expectation $E(Y|X = x) = g(x)$ [see II, § 8.9)] is called the regression of Y on X. If $g(x)$ is linear in x, say $g(x) = \alpha' + \beta' x$, we have 'simple linear regression'. If the observed value of Y corresponding to the predetermined value x_r of X

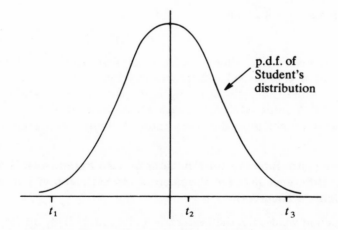

Figure 4.5.2: t_1 represents an implausibly large negative value of $t = (\bar{x} - \mu)/(s/\sqrt{n})$, and so corresponds to a value of μ exceeding $\bar{x}$ by an implausibly large amount; similarly t_3 corresponds to an implausibly low value of μ; while t_2 corresponds to a plausible value of μ.

is y_r, $r = 1, 2, \ldots, n$, it is convenient to write $g(x)$ in the form

$$g(x) = \alpha + \beta(x - \bar{x}),$$

where $\bar{x} = \sum_1^n x_r/n$. If the conditional distribution of Y, given $X = x$, is Normal $(g(x), \sigma)$ for each value of x, the maximum likelihood estimates of α and β [see Example 6.4.4] are, respectively,

$$\hat{\alpha} = \bar{y} \left(= \sum_1^n y_r/n \right),$$

$$\hat{\beta} = \sum_1^n y_r(x_r - \bar{x})/\sum_1^n (x_r - \bar{x})^2$$

and the corresponding unbiased estimate of σ^2 is

$$s^2 = \sum y_r^2 - n\hat{\alpha}^2 - \hat{\beta}^2 \sum_1^n (x_r - \bar{x})^2. \tag{4.5.4}$$

It then turns out that the quantities

$$t_1 = (\hat{\alpha} - \alpha)/s_1 \qquad \text{(where } s_1^2 = s^2/n),$$

$$t_2 = (\hat{\beta} - \beta)/s_2 \qquad \text{(where } s_2^2 = s^2/\sum (x_r - \bar{x})^2),$$

have, independently, Student's distribution [see § 2.5.5] on $n - 2$ degrees of freedom, while $(n-2)s^2/\sigma^2$ has a chi-squared distribution [see II, § 11.4.11] on $n - 2$ degrees of freedom. Thus the central 95% confidence intervals for α, β and σ^2 are, respectively

$$(\hat{\alpha} - t^0 s_1, \qquad \hat{\alpha} + t^0 s_1),$$

$$(\hat{\beta} - t^0 s_2, \qquad \hat{\beta} + t^0 s_2),$$

(compare Example 4.4.2), and

$$((n-2)s^2/c_2, \qquad (n-2)s^2/c_1),$$

(compare Example 4.4.1), where t^0 is the value of Student's t, on $n - 2$ degrees of freedom, that is exceeded with probability $0 \cdot 025$, and c_1 and c_2 are the values of χ^2, on $n - 2$ degrees of freedom, that are exceeded with probability $0 \cdot 975$ and $0 \cdot 025$ respectively. [See also Example 4.5.4 and § 5·8·5. For a numerical example and more detailed treatment of linear regression see § 6.5.]

4.5.2. Confidence Intervals for Functions of Two Parameters, Including the Difference between Parameters and the Ratio of Parameters. (Fieller's Theorem)

In this section we discuss the best-known and most widely applied cases of confidence intervals for a function of two parameters. These are (for Normal-theory models), the following:

(i) the difference $\mu_1 - \mu_2$ between two expectations; (Example 4.5.4);

(ii) the ordinate $\alpha + \beta x_0$ of a linear regression at a determined value x_0 of the controlled or 'independent' variable x (Example 4.5.5);

(iii) the difference between the slopes of two regression lines (Example 4.5.6);

(iv) the ratio of two variances (Example 4.5.7);

(v) the ratio of two parameters that are estimated by linear functions of the data (Fieller's Theorem) (Example 4.5.8).

EXAMPLE 4.5.4.   *Confidence limits for the difference between expectations in two Normal samples with equal variances.*   Let $(x_1, x_2, \ldots, x_{n_1})$ be a sample from the Normal (μ_1, σ) distribution, and $(y_1, y_2, \ldots, y_{n_2})$ a sample from the Normal (μ_2, σ) distribution. The conventional estimates for μ_1 and μ_2 are the respective sample means $\bar{x}$ and $\bar{y}$. Defining s_1^2 and s_2^2 by the usual formulae:

$$s_1^2 = \sum_1^{n_1} (x_r - \bar{x})^2 / (n_1 - 1), \qquad s_2^2 = \sum_1^{n_2} (y_r - \bar{y})^2 / (n_2 - 1),$$

we take as the *pooled* estimate of the common variance the quantity s^2 defined by

$$(n_1 + n_2 - 2)s^2 = (n_1 - 1)s_1^2 + (n_2 - 1)s_2^2. \tag{4.5.6}$$

The appropriate estimate of the sampling variance of $\bar{x}$ is s^2/n_1, and that of $\bar{y}$ is s^2/n_2, so that the estimated variance of $(\bar{x} - \bar{y})$ is $(s^2/n_1 + s^2/n_2)$. Thus

$$t = \{(\bar{x} - \bar{y}) - (\mu_1 - \mu_2)\} \Big/ \left\{ s \sqrt{\left(\frac{1}{n_1} + \frac{1}{n_2} \right)} \right\}$$

is a realization of Student's distribution on $n_1 + n_2 - 2$ degrees of freedom [see Example 2.5.3], whence, as in Example 4.5.2, the interval (a, b) is a central 95% confidence interval for $\mu_1 - \mu_2$, where

$$a = \bar{x} - \bar{y} - t_0 s \sqrt{\left(\frac{1}{n_1} + \frac{1}{n_2} \right)},$$

$$b = \bar{x} - \bar{y} + t_0 s \sqrt{\left(\frac{1}{n_1} + \frac{1}{n_2} \right)}, \tag{4.5.7}$$

t_0 being the 0·975 quantile of Student's distribution on $n_1 + n_2 - 2$ degrees of freedom. [Compare Example 5.8.4.]

We exemplify with the data given in Table 4.5.1, this time however treating the 'Drug A' and 'Drug B' column, for illustrative purposes, as if they had come from different and independent samples of patients. We regard the Drug A figures as values of the x_r, and the Drug B figures as values of the y_r. In applying the formulas given above, we have

$$\bar{x} = 0·75, \qquad \bar{y} = 2·33,$$

with $n_1 = n_2 = 10$, and s^2 given by

$$18s^2 = \sum (x_r - \bar{x})^2 + \sum (y_r - \bar{y})^2$$
$$= (\sum x_r^2 - 10\bar{x}^2) + (\sum y_r^2 - 10\bar{y}^2)$$
$$= (34 \cdot 43 - 5 \cdot 62) + (90 \cdot 37 - 54 \cdot 29)$$
$$= 64 \cdot 88,$$

so that

$$s^2 = 3 \cdot 60 \quad \text{and} \quad s = 1 \cdot 899.$$

Then

$$s\sqrt{\left(\frac{1}{n_1} + \frac{1}{n_2}\right)} = 0 \cdot 849.$$

The 2·5% and 97·5% values of Student's 't' on 18 d.f. are $-2 \cdot 101$ and $+2 \cdot 101$ (see Appendix T5), whence the 95% confidence interval for $\mu_1 - \mu_2$ is

$$-1 \cdot 58 \pm (2 \cdot 101)(0 \cdot 849) = -1 \cdot 58 \pm 1 \cdot 78 = (-3 \cdot 36, 0 \cdot 20).$$

Note that the zero value is covered by this interval: that is, in this analysis, the data are consistent with the hypothesis that $\mu_1 = \mu_2$.

[This example is intended for illustrative purposes only. It assumes that the x_r and the y_r are observed value of independent random variables X and Y, whereas in reality the data actually refer to the effects of different drugs on the same subjects, so that X and Y cannot be presumed to be independent. On the contrary, one would expect X and Y to be positively correlated so that the proper estimate of $\mathrm{var}(X - Y)$ ($= \mathrm{var}(X) - 2\,\mathrm{cov}(X, Y) + \mathrm{var}(Y)$, with $\mathrm{cov}(X, Y) > 0$) would be less than the estimate obtained on the independence assumption. That is the reason why the properly evaluated confidence interval (4.5.3), namely $1 \cdot 58 \pm 0 \cdot 88$, is shorter than that obtained on the independence assumption above, namely $1 \cdot 58 \pm 1 \cdot 78$.]

EXAMPLE 4.5.5.   *Confidence limits for a regression at a specified value of x.*   In the notation of Example 4.5.3, the regression

$$y = \alpha + \beta(x - \bar{x})$$

is estimated at $x = x_0$ by

$$\hat{\alpha} + \hat{\beta}(x_0 - \bar{x}).$$

This is a realization of a Normal variable with expected value

$$\alpha + \beta(x_0 - \bar{x})$$

and variance

$$\sigma^2\left\{\frac{1}{n} + \frac{(x_0 - \bar{x})^2}{\sum (x_r - \bar{x})^2}\right\},$$

where σ^2 is estimated by s^2 as given in (4.5.4). The required 95% confidence interval for $\alpha + \beta(x_0 - \bar{x})$ is then (c.f. Example 4.5.2)

$$\hat{\alpha} + \hat{\beta}(x_0 - \bar{x}) + t_{97.5}\sqrt{v}$$

where

$$v^2 = s^2\left\{\frac{1}{n} + \frac{(x_0 - \bar{x})^2}{\sum(x_r - \bar{x})^2}\right\}$$

is the estimated variance of $\hat{\alpha} + \hat{\beta}(x_0 - \bar{x})$, and $t_{97.5}$ denotes the 97·5% quantile of Student's distribution (on $n - 2$ d.f.)

EXAMPLE 4.5.6. *Confidence limits for the difference between the slopes of two linear regression.* Suppose we have two samples of the kind referred to in Example 4.5.3, one with observations y_r and corresponding values x_r of the 'independent' variable, $r = 1, 2, \ldots, n$, and the other with observations y'_r, independent variable levels x'_r, $r = 1, 2, \ldots, n'$, from which we estimates the linear regressions

$$\alpha_1 + \beta_1(x - \bar{x}), \qquad \alpha_2 + \beta_2(x' - \bar{x}'),$$

obtaining as estimates of the parameters the quantities

$$\hat{\alpha}_1 = a_1, \qquad \hat{\beta} = b_1, \qquad \hat{\alpha}_2 = a_2, \qquad \hat{\beta}_2 = b_2,$$

as the estimates of the corresponding parameters. The sampling expectation of $b_1 - b_2$ is $\beta_1 - \beta_2$, and its sampling variance is $\sigma_1^2/\sum_1 + \sigma_2^2/\sum_2$, where σ_1^2, σ_2^2 are the variances of the two populations, and $\sum_1 = \sum(x_r - \bar{x})^2$, $\sum_2 = \sum(x'_r - \bar{x}')^2$. The sampling variance of $b_1 - b_2$ is estimated as

$$v = s_1^2/\sum_1 + s_2^2/\sum_2$$

where s_1^2 denotes the quantity defined as s^2 in (4.5.4), and s_2^2 the corresponding quantity with x'_r, y'_r replacing x_r, y_r. The estimate v has $n + n' - 2$ degrees of freedom.

Then, as in Example 4.5.2, the 95% confidence interval for $\beta_1 - \beta_2$ is

$$b_1 - b_2 \pm t_{97.5}\sqrt{v}$$

where $t_{97.5}$ is the 97·5% point of Student's distribution on $n + n' - 2$ d.f.

EXAMPLE 4.5.7. *Confidence limits for the ratio of two variances in Normal samples.* The method used in Example 4.5.1 to obtain confidence limits for the variance σ^2 of a Normal population can be extended to find limits for the ratio σ_1^2/σ_2^2 of the variances of two Normal populations. Using the same notation for sample sizes and for s_1^2 and s_2^2 as in Example 4.5.3, we note that

$$(n_1 - 1)s_1^2/\sigma_1^2, \qquad (n_2 - 1)s_2^2/\sigma_2^2 \qquad (4.5.8)$$

are independent realizations of chi-squared variables on $n_1 - 1$ and $n_2 - 1$

degrees of freedom respectively. It follows that the ratio

$$\left(\frac{s_1^2}{\sigma_1^2}\right)\Big/\left(\frac{s_2^2}{\sigma_2^2}\right)$$

is a realization of the 'variance-ratio' distribution [see § 2.5.6] on n_1-1, n_2-1 degrees of freedom, that is the 'F_{ν_1,ν_2} distribution' with $\nu_1 = n_1-1$, $\nu_2 = n_2-1$. Denote by a and b the 0·025 quantile and the 0·975 quantile-i.e. a is the 2·5% point and b the 97·5% point [see II, § 10.3.3]—of this distribution (obtainable from published tables: an illustration is given below. A version of the table is given in the Appendix (T7).) Then we obtain a 95% confidence interval for σ_2^2/σ_1^2 from the inequality

$$a \le \left(\frac{s_1^2}{\sigma_1^2}\right)\Big/\left(\frac{s_2^2}{\sigma_2^2}\right) \le b, \qquad \text{(with probability 0·95),}$$

i.e. from

$$as_2^2/s_1^2 \le \sigma_2^2/\sigma_1^2 \le bs_2^2/s_1^2.$$

Thus the confidence interval for σ_2^2/σ_1^2 is

$$(as_2^2/s_1^2,\ bs_2^2/s_1^2). \qquad (4.5.9)$$

Equivalently, the confidence interval for σ_2/σ_1 is

$$(a^{1/2}s_2/s_1,\ b^{1/2}s_2/s_1) \qquad (4.5.10)$$

and that for σ_1^2/σ_2^2 is

$$(s_1^2/bs_2^2,\ s_1^2/as_2^2). \qquad (4.5.11)$$

If the desired confidence coefficient were, say, 0·99 instead of the value used above, a and b should be taken as the 0·005 quantile and the 0·995 quantile, respectively, of the F distribution.

To illustrate, suppose $n_1 = 20$ and $n_2 = 30$, and that $s_1^2/s_2^2 = 1$. The 95% confidence interval for σ_1^2/σ_2^2 is obtained by taking a and b as the 0·025 and 0·975 quantiles of the $F_{19,29}$ distribution. In published tables (e.g. in the Biometrika tables) b is called the 'upper 2·5% point'. In our case its value is $b = 2·40$. The 'lower 2·5% point' a is not directly tabulated since it coincides with the reciprocal of the upper 2·5% point of $F_{29,19}$ (note the reversal of the degrees of freedom; and similarly for the upper and lower 0·5% points, etc. [see § 2.5.6]). In our case therefore $1/a = 2·23$, $a = 0·448$. Thus our 95% confidence interval for σ_1^2/σ_2^2 is $(0·448, 2·40)$.

Similarly for the data used in Example 4.5.4: we have $n_1 = n_2 = 10$, $9_{s_1}^2 = 28·81$, $9_{s_2}^2 = 36·08$, $1/a = b = 4·03$, whence the 95% confidence interval for σ_1^2/σ_2^2 is $(0·311, 5·04)$, and that for (σ_1/σ_2) is $(0·558, 2·24)$. Note that this interval includes the value 1, so that the data are consistent with the hypothesis of equal variances.

EXAMPLE 4.5.8. *Confidence interval for a ratio of parameters estimated by linear statistics (Fieller's Theorem).* In this problem we are concerned with the ratio $\lambda = \alpha/\beta$, where α and β are parameters relating to two samples of Normally distributed data with common variance. For instance, α and β might be the expected values μ_1 and μ_2 of the parent population, or the slopes of the regression lines; or, in the regression $y = \alpha_1 + \beta_1 x$, λ might be the x-value at which the regression attains the value y_0, since this x-value is $(y_0 - \alpha_1)/\beta_1$, which is of the ratio form specified for λ; etc. We suppose a and b to be unbiased estimates of α and β that are linear functions of the observations, with estimated variances v_{11} and v_{22}, and covariance v_{12}, all these being derived from the combined data of the two samples, with f degrees of freedom.

Consider the quantity $\lambda b - a$. The sampling distribution of this is Normal, with zero expectation, and with variance estimated as

$$v = \lambda^2 v_{22} - 2\lambda v_{12} + v_{11}, \qquad (4.5.12)$$

on f degrees of freedom. The quantity

$$(\lambda b - a)/\sqrt{v} \qquad (4.5.13)$$

has Student's distribution on f degrees of freedom, whence, with probability 0·95

$$(\lambda b - a)^2 \le t_{95}^2 v$$

where t_{95} denotes the 95% point of Student's distribution on f degrees of freedom. The roots λ_1, λ_2 of the quadratic

$$(\lambda b - a)^2 = t_{95}^2 v \qquad (4.5.14)$$

are thus the end-points of the required 95% confidence interval for $\lambda = \alpha/\beta$.

The quadratic in λ is, explicitly,

$$\lambda^2(b^2 - t^2 v_{22}) - 2\lambda(ab - t^2 v_{12}) + (a^2 - t^2 v_{11}) = 0, \qquad (4.5.15)$$

where t stands for t_{95}. (Similarly for t_{99} etc.) This is Fieller's result.

If, as in many applications, $v_{12} = 0$, the roots of the quadratic may be written in the following form, showing the departures of λ_1 and λ_2 from the 'natural' obvious estimate a/b of α/β:

$$\frac{\lambda_1}{\lambda_2} = \frac{a}{b}\left[\frac{1 \pm \sqrt{\{1 - (1 - t^2 v_{22}/b^2)(1 - t^2 v_{11}/a^2)\}}}{1 - t^2 v_{22}/b^2}\right].$$

In the numerical example that follows we also give the values of *approximate* confidence limits obtained from the approximate formula given in Example 2.7.7, namely that the sampling variance of a/b is

$$(b^2 v_{11} + a^2 v_{22})/b^4,$$

so that

$$\frac{\lambda_1}{\lambda_2} \cong \frac{a}{b}\left[1 \pm t\sqrt{\left(\frac{v_{11}}{a^2} + \frac{v_{22}}{b^2}\right)}\right]. \qquad (4.5.16)$$

If, as for instance in section 6.6.6, the ratio under examination is $\lambda = -\alpha/\beta$, where α and β are estimated by a and b respectively, the quantity to be considered is not $\lambda b - a$ but $\lambda b + a$. The sampling expectation is $(-\alpha/\beta)\beta + \alpha$, which is zero as required. The quantity v of (4.5.12) must be replaced by

$$\lambda^2 v_{22} + 2\lambda v_{12} + v_1$$

and the quadratic (4.5.15) by

$$\lambda^2(b^2 - t^2 v_{22}) + 2\lambda(ab - t^2 v_{12}) + (a^2 - t^2 v_{11}) = 0.$$

As a numerical illustration we consider the data used in Example 4.5.4. There we obtained confidence limits for the difference $\mu_1 - \mu_2$ between the expected potencies of the two drugs, measured for each drug in terms of the increase in the number of hours sleep produced. We now consider instead the ratio μ_1/μ_2. In using (4.5.15), therefore, we have

$$a = \hat{\mu}_1 = \bar{y}_1 = 0\cdot75, \qquad b = \hat{\mu}_2 = \bar{y}_2 = 2\cdot33,$$

$$a/b = 0\cdot322,$$

$$v_{11} = v_{22} = s^2/9 = 3\cdot605/9 = 0\cdot401,$$

and

$$v_{12} = 0.$$

Thus

$$v = 0\cdot401(\lambda^2 + 1).$$

The value of t_{95} for 18 d.t. is $2\cdot101$. The quadratic (4.5.14) is

$$(2\cdot33\lambda - 0\cdot75)^2 = (2\cdot101)^2(0\cdot401)(\lambda^2 + 1).$$

The roots are

$$\lambda_1 = -0\cdot270, \qquad \lambda_2 = 1\cdot225.$$

For comparison, the approximation (4.5.16) gives

$$\lambda_1 = -0\cdot28, \qquad \lambda_2 = 0\cdot92.$$

[A further numerical example is given in § 6.6.6.]

4.6.   CONSTRUCTION OF CONFIDENCE INTERVALS WITHOUT USING PIVOTS

In § 4.3 it was explained how to construct a confidence interval, in the case of a single parameter, with the aid of a pivot. A pivot can always be found for a c.d.f. $F(x, \theta)$ that is continuous in x since, for an observation x_i, the sampling distribution of $F(x_i, \theta)$ is the Uniform $(0, 1)$ distribution [see II, § 10.2.1] by the probability integral transform theorem [see II, Theorem

10.7.2]. This is a parameter-free distribution. It follows that any function of the quantities $F(x_1, \theta), \ldots, F(x_n, \theta)$ will be a pivot. In particular, $\prod_1^n F(x_r, \theta)$ is a pivot, with sampling distribution derivable from the fact that $q = -\sum_1^n \log F(x_r, \theta)$ is a realization of the gamma distribution $q^{n-1} e^{-q}/\Gamma(n)$. The inversion of the inequalities (4.3.3) may however present difficulties and an alternative method would be desirable in such cases. Such a method will now be exemplified.

EXAMPLE 4.6.1.   *Confidence limits for the shape parameter of a gamma distribution.*   We suppose our random variable X to have the gamma distribution with shape parameter λ of unknown magnitude, and unit scale parameter, the p.d.f. at x being

$$x^{\lambda-1} e^{-x}/\Gamma(\lambda), \qquad x > 0 \qquad (\theta > 0),$$

so that $E(X) = \lambda$ [see II, § 11.3].

We assume the availability of a sample $x_1, x_2, \ldots, x_n$. As explained earlier the confidence interval (assuming that one can be found) will depend on the choice of working statistic. It is good practice to take as working statistic a *sufficient* estimate of the parameter in question, but in the present case this must be a function of $\prod_1^n x_i$, a somewhat intractable quantity.

We therefore use instead the 'intuitive' statistic $y = \sum_1^n x_r$. This, which is an unbiased estimate of $n\lambda$, is a realization of the random variable Y whose p.d.f. [see II, § 11.3.2] is

$$g(y, \theta) = y^{\theta-1} e^{-y}/\Gamma(\theta), \qquad y > 0 \qquad (\theta > 0),$$

where $\theta = n\lambda$, (so that $E(Y) = \theta = n\lambda$), and whose c.d.f. is

$$P(Y \le y) = G(y, \theta) = \int_0^y z^{\theta-1} e^{-z} \, dz/\Gamma(\theta).$$

For an arbitrary value of θ we can find a central 95% probability interval (y_1, y_2) for Y, by finding numbers y_1 and y_2 such that

$$G(y_1, \theta) = 0 \cdot 025, \qquad G(y_2, \theta) = 0 \cdot 975.$$

[See § 4.1.3.] For example, taking $\theta = 2 \cdot 0$, and $n = 10$, we have

$$G(y_1, 2 \cdot 0) = P(Y \le y_1) = P(2Y \le 2y_1)$$
$$= P\{\chi^2(4) \le 2y_1\},$$

since $2Y$ has the chi-squared distribution on $2\theta = 4$ degrees of freedom (the '$\chi^2(4)$' distribution). [See § 2.5.4(a)(i).] From chi-squared tables [see Appendix (76)] we see that $2y_1 = 0 \cdot 484$, whence $y_1 = 0 \cdot 242$. Similarly $y_2 = 5 \cdot 572$. We can repeat this calculation for other values of θ, obtaining the results shown in Table 4.6.1 (where the values of y_1 and y_2 corresponding to a particular value of θ are labelled $y_1(\theta)$ and $y_2(\theta)$ respectively).

θ	$y_1(\theta)$	$y_2(\theta)$
1	0·0253	3·689
2	0·242	5·572
3	0·618	7·224
4	1·090	8·768
5	1·624	10·242
6	2·202	11·668
7	2·814	13·060
8	3·454	14·422
9	4·116	15·713
10	4·796	17·085
11	5·491	18·390
12	6·200	19·682
13	6·922	20·962
14	7·654	22·230
15	8·396	23·490

Table 4.6.1: Values of $y_1(\theta)$ and $y_2(\theta)$ such that $P\{y_1(\theta) \le Y \le y_2(\theta)\} = 0\cdot95$ where Y has the gamma distribution with shape parameter θ $(= 10\lambda)$.

Thus we have achieved essentially the same position as we had with the aid of a pivot in (4.3.5) in Example 4.3.1, the only difference being that the end-points of the probability interval are defined by tabulated functions instead of being given by explicit formulae.

We now have the problem of inverting the proposition

$$y_1(\theta) \le Y \le y_2(\theta),$$

(which holds with probability 0·95) i.e. of deriving a logically equivalent and therefore equiprobable proposition of the form

$$a_1(Y) \le \theta \le a_2(Y).$$

Both $y_1(\theta)$ and $y_2(\theta)$ are continuous monotonically [see IV, § 2.7] increasing functions of θ, as illustrated in Figure 4.6.1. The proposition '$Y \le y_2(\theta)$' is exemplified by the point (θ_0, Y_0), where $Y_0 \le y_2(\theta_0)$, i.e. where Y_0 lies on or below the curve $y_2(\theta)$. Because the curve is continuous and monotone the function $y_2(\theta)$ is invertible [see IV, § 2.7], i.e. there exists a unique value $\theta_0' = y_2^{-1}(Y_0)$ such that $Y_0 = y_2(\theta_0')$, found by drawing a horizontal through (θ_0, Y_0) and noting the abscissa θ_0' of the point where it meets the curve. Because the curve is monotone increasing, $\theta_0' \le \theta_0$. Thus the proposition '$Y_0 \le y_2(\theta_0)$' is logically equivalent to the 'inverted' proposition

$$\theta_0 \ge \theta_0'$$

where θ_0' $(= \theta_0'(Y_0), \text{ say}) = y_2^{-1}(Y_0)$.

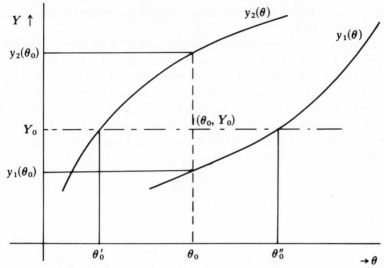

Figure 4.6.1: Graphs of the functions $y_1(\theta)$, $y_2(\theta)$ of Table 4.6.1. Y_0 is a typical value of Y such that the point (θ_0, Y_0), lies between the two curves.. The figure illustrates the equivalence of the propositions

$$y_1(\theta_0) \le Y_0 \le y_2(\theta_0)$$

$$\theta_0' \le \theta_0 \le \theta_0''.$$

Similarly, using the lower curve $y_1(\theta)$, we can see that the proposition '$Y_0 \ge y_1(\theta)$' is logically equivalent to the proposition

$$\theta_0 \le \theta_0''$$

where θ_0'' $(= \theta_0''(Y_0),$ say$) = y_1^{-1}(Y_0)$.

It follows that the proposition

$$y_1(\theta_0) \le Y_0 \le y_2(\theta_0)$$

(which has probability 0.95) is equivalent to the proposition

$$\theta_0'(Y_0) \le \theta_0 \le \theta_0''(Y_0)$$

which, therefore, also has probability 0.95. (See Figure 4.6.1.) This holds for every point (θ, Y) in the interval $(y_1(\theta), y_2(\theta))$, for every value of θ. It follows that, for an observed value y_0 of Y, the interval $(\theta_0'(y_0), \theta_0''(y_0))$ is a 95% confidence interval for θ.

The lower limit $\theta_0' = \theta_0'(y_0)$ is defined by the condition

$$P\{Y(\theta_0') > y_0\} = 0.025$$

where $Y(\theta_0')$ is a gamma variate with shape parameter θ_0', that is θ_0' is the parameter value which makes the upper tail area beyond y_0 take the value 0.025; and the upper limit $\theta_0'' = \theta_0''(y_0)$ is defined by the condition

$$P\{Y(\theta_0'') < y_0\} = 0.025,$$

so that θ_0'' is the parameter value which makes the lower tail area below y_0 take the value 0·025.

For a given y_0, e.g. $y_0 = 6$, we may obtain θ_0' and θ_0'' from tables, or graphically as exemplified by the more accurate version of Figure 4.6.1, drawn by plotting the points tabulated in Table 4.6.1, given in Figure 4.6.2. From this it will be seen that an observed value $y = 6$ (corresponding to an unbiased estimate $\lambda^* = 0·6$ for λ) produces the 95% confidence interval (2·3, 11·7) for $\theta = 10\lambda$, and, correspondingly, the 95% confidence interval (0·23, 1·17) for λ. (The region between the upper and lower curves in Figure 4.6.2 is known as a *confidence belt* or *confidence band*.)

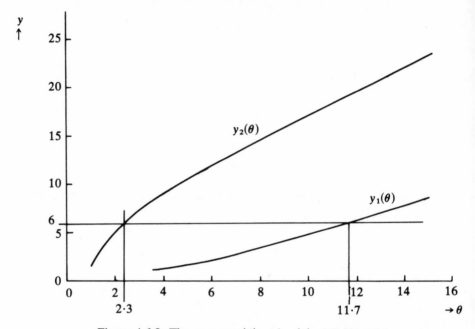

Figure 4.6.2: The curves $y_1(\theta)$ and $y_2(\theta)$ of Table 4.6.1.

The intuitive approach here that corresponds to the discussion preceding Example 4.3.1 is to consider some representative gamma density functions corresponding to a set of θ-values illustrated in Figure 4.6.3. The very small value θ_1 of θ in (i) and the large value θ_3 in (iii) are implausible in the sense that in each case the observed statistic y_1 lies in a region of low probability density, while θ_2 is compatible with y since y lies in a region of high probability density. The boundaries between plausible and implausible values may be fixed at θ_u and θ_l, as in the earlier discussion, where, as indicated in Figure 4.6.4,

$$P(Y \geq y_1; \theta_u) = P(Y \leq y_1; \theta_l) = 0·025.$$

For $y = 6$ we compute θ_l and θ_u by using the fact that $2\theta_l$ is the degrees of

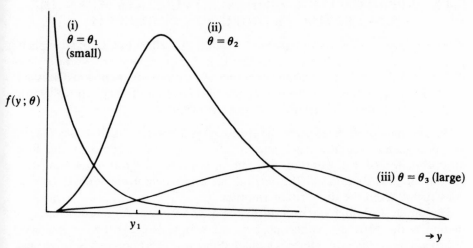

Figure 4.6.3: Gamma densities $f(y; \theta)$ corresponding to various values of the parameter θ.

freedom parameter for the chi-squared distribution whose 0·025 quantile is 12, and $2\theta_u$ its value in the corresponding distribution whose 0·975 quantile is 12. On inspecting the chi-squared table [Appendix (T6)] we see that $2\theta_u$ lies between 23 and 24, and $2\theta_l$ between 4 and 5. Interpolation gives $\theta_l = 2\cdot3$ and $\theta_l = 11\cdot7$, in agreement with the values obtained above.

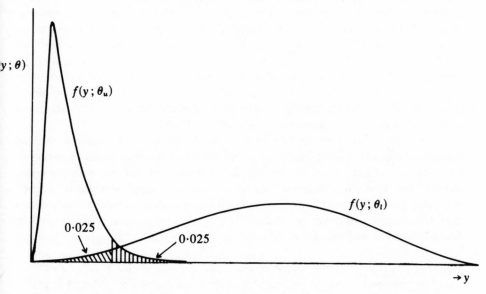

Figure 4.6.4: Gamma densities $f(y; \theta)$ with parameter values θ_l, θ_u respectively.

4.7. APPROXIMATE CONFIDENCE INTERVALS WHEN THE UNDERLYING DISTRIBUTION IS DISCRETE

We now proceed to examples where, in addition to the absence of an explicit pivot, we have the further complication that the distribution is discrete. The difficulties raised by this, and some methods of dealing with them, are discussed in Blyth and Hutchinson (1960), Clopper and Pearson (1934), Crow (1956), Eudey (1949), Pearson (1950), and Stevens (1950).

We confine our discussions to the Binomial and Poisson distributions. There are two distinct approaches, both involving approximations. In the first approach, the central discrete distribution is replaced by an approximating continuous distribution. In the second, an approximate confidence band is constructed from the discrete distribution.

The simplest (though not necessarily the most accurate) continuous approximation is the Normal. If this can be used, simple Normal theory produces approximate confidence intervals with no difficulty. This approach is developed below, for both Binomial and Poisson, in Examples 4.7.1 and 4.7.2.

A more sophisticated continuous family of approximations may be generated by pretending that the discrete random variable actually observed is a quantized form of an underlying continuous distribution; the approximating random variable Z may thus be represented as

$$Z = R + U,$$

where R is the discrete (integer-valued) variable we are really concerned with, and U is a continuous variable, independent of R, whose values are confined to the interval $(-\frac{1}{2}, \frac{1}{2})$. To allot a specified distribution to U (for example, the continuous Uniform) is to specify the distribution of the continuous approximation Z (in terms of course that involve the unknown parameter θ that controls the distribution of R). If, for example, R is Binomial (n, θ), and U is continuous Uniform $(-\frac{1}{2}, \frac{1}{2})$, a five-decimal 'observed' value of Z may be generated by taking the observed value r of R and adding to $r - \frac{1}{2}$ a five-decimal random number, i.e. the number $0 \cdot s_1 s_2 s_3 s_4 s_5$, where s_1, s_2, etc. are 'random digits' [see, e.g. The RAND Corporation (1955) in Bibliography F]. Methods based on this idea have been developed by Tocher (1950).

In the second kind of approximation in which we work with the original unmodified discrete distribution it is possible to obtain confidence intervals that are not based on an approximating distribution but for which we can specify only that the confidence levels are not less than some desired level. We cannot obtain, for example, 95% confidence intervals, but can only assert that the confidence level is *at least* 95%. This approach is developed in Examples 4.7.3 and 4.7.4.

EXAMPLE 4.7.1. *Confidence interval for the parameter θ of a Binomial (n, θ) distribution, using the Normal approximation.* A good continuous approximation to the distribution of a Binomial (n, θ) variable R, for moderate and large values of n, is

$$P(R \le x) = P(X \le x), \qquad 0 \le x \le n \qquad (4.7.1)$$

where X is Normal (μ, σ), with

$$\mu = E(X) = E(R) = n\theta,$$

and

$$\sigma^2 = \text{var}(X) = \text{var}(R) = n\theta/(1-\theta)$$

[see II, § 11.4.7]. That is, R is approximately Normal (μ, σ). It is perhaps more natural to work in terms of R/n, the obvious estimator of θ; this random variable is approximately Normal $(\theta, \sqrt{\theta(1-\theta)/n})$.

(a) *A crude approximation*

For the Normal (μ, σ) distribution of X, the 95% confidence interval for μ based on an observation x is $x \pm 1 \cdot 95\sigma$. The crudest (but widely used) procedure replaces x by the observed value $p = r/n$ of R/n (p is the observed proportion of successes in the sample of n trials) and σ by the estimated value of $\sqrt{\{\theta(1-\theta)/n\}}$, that is by $\sqrt{\{p(1-p)/n\}}$. Then the approximate 95% confidence interval for θ is

$$p \pm 1 \cdot 96\sqrt{\{p(1-p)/n\}}. \qquad (4.7.2)$$

This may without serious further error be replaced by

$$p \pm 2\sqrt{\{p(1-p)/n\}}. \qquad (4.7.3)$$

(b) *A better approximation*

If X is Normal (μ, σ) we may assert with probability $1 - \alpha$ that

$$-a \le (X - \mu)/\sigma \le a$$

where

$$\Phi(a) = 1 - \tfrac{1}{2}\alpha,$$

Φ being, as usual, the standard Normal integral [see Appendix (T3), (T4)]. Thus, with probability $1 - \alpha$,

$$X \le \mu + a\sigma \quad and \quad X \ge \mu - a\sigma.$$

Since R is approximately Normal (μ, σ), with $\mu = n\theta$ and $\sigma^2 = n\theta(1-\theta)$, we may assert with probability *approximately* 95%, that

$$R \le n\theta + a\sqrt{\{n\theta(1-\theta)\}} \quad and \quad R \ge n\theta - a\sqrt{\{n\theta(1-\theta)\}}.$$

To the accuracy of this approximation, the $100(1-\alpha)\%$ confidence interval for θ will then be the interval of θ-values that satisfies the inequalities

$$r \leq n\theta + a\sqrt{\{n\theta(1-\theta)\}} \quad \text{and} \quad r \geq n\theta - a\sqrt{\{n\theta(1-\theta)\}}$$

that is, the interval (θ_1, θ_2), where θ_1 and θ_2 are the roots of the quadratic equation

$$(r - n\theta)^2 = a^2 n\theta(1-\theta). \tag{4.7.4}$$

In terms of p $(= r/n$, the observed proportion of successes), this becomes

$$(1 + a^2/n^2)\theta^2 - (2p + a^2/n)\theta + p^2 = 0.$$

The roots are

$$\frac{\theta_1}{\theta_2} = \frac{p + a^2/2n \pm a\sqrt{\{p(1-p)/n + a^2/4n^2\}}}{1 + a^2/n}. \tag{4.7.5}$$

For the 95% confidence interval ($\alpha = 0{\cdot}05$), $a = 1{\cdot}96 \cong 2$, whence the 95% confidence limits θ_1, θ_2 are given by

$$\theta_1, \theta_2 = \frac{p + 2/n \pm 2\sqrt{\{p(1-p)/n + 1/n^2\}}}{1 + 4/n} \tag{4.7.6}$$

approximately.

For example, if $r = 8$ and $n = 20$, the 95% confidence interval obtained from this procedure is derived from (4.7.4), which in this case is the equation

$$(8 - 20\theta)^2 = (1{\cdot}96)^2 20\theta(1-\theta),$$

whence the confidence interval is $(0{\cdot}216, 0{\cdot}617)$. Replacing $1{\cdot}96$ by 2 would give effectively the same answer. The corresponding values from the crude approximation (4.7.4) in (a) above would be $0{\cdot}4 \pm 2\sqrt{0{\cdot}012} = 0{\cdot}04 \pm 0{\cdot}219 = (0{\cdot}181, 0{\cdot}619)$.

EXAMPLE 4.7.2. *Confidence interval for the parameter of a Poisson distribution, using the Normal approximation.* We suppose that X is a Poisson (θ) variable [see II, § 5.4] and that a sample $(x_1, x_2, \ldots, x_n)$ is available, with sample mean $\bar{x}$. Then, provided θ is not too small, X is approximately Normal $(\theta, \sqrt{\theta})$, and so the sampling distribution of $\bar{x}$ is Normal $(\theta, \sqrt{(\theta/n)})$. Using the kind of argument employed in Example 4.7.1 it will be seen that the approximate $100p\%$ confidence interval for θ is found from the statement that, with probability $0{\cdot}95$,

$$-a \leq \frac{\bar{x} - \theta}{\sqrt{(\theta/n)}} \leq a, \quad (\text{where } a = \Phi^{-1}(\tfrac{1}{2} + \tfrac{1}{2}p))$$

so that the confidence limits θ_1, θ_2 are the root of the equation

$$\frac{\bar{x} - \theta}{\sqrt{(\theta/n)}} = a,$$

i.e. of the quadratic

$$(\bar{x} - \theta)^2 = a^2 \theta / n. \tag{4.7.7}$$

If $p = 0.95$, $a = 1.96 \approx 2$.

As an example consider the data recorded in Table 3.2.3. In 2606 time intervals, each of length 7.5 seconds, the total number of radio-active particles emitted from a certain specimen was 10,070. The mean number per interval is then $\bar{x} = 10,070/2606 = 3.864$, and thus is the Maximum-Likelihood estimate of θ for this sample. The accuracy of this estimate is expressed in terms of the 95% confidence interval obtained from the quadratic (4.7.7);

$$(3.864 - \theta)^2 = (1.96)^2 \theta / 2606,$$

viz. $(\theta_1, \theta_2) = 3.864 \pm 0.049$.

The estimate is evidently highly accurate, as would be expected from such a large sample.

We now move on to a more profound examination of the problems considered in Examples 4.7.1 and 4.7.2, this time taking into account the discreteness of the data.

EXAMPLE 4.7.3. *Confidence interval for the parameter θ of a Binomial (n, θ) distribution, taking discreteness into account.* Let X_θ have the Bernoulli distribution [see II, § 5.2.1] with success parameter θ, so that the p.d.f. of X_θ is

$$P(X_\theta = x) = \theta^x (1 - \theta)^{1-x}, \qquad x = 0, 1; \qquad 0 < \theta < 1.$$

In a sample of size n (i.e. in n Bernoulli trials) the total number of successes $r = \sum_1^n x_i$ is a sufficient statistic for θ, and we shall therefore take this as our working statistic. The corresponding (unbiased) estimate of θ is $\hat{\theta} = r/n$. The statistic r is a realization of the Binomial (n, θ) variable R_θ, for which the p.d.f. is

$$P(R_\theta = r) = \binom{n}{r} \theta^r (1 - \theta)^{n-r}, \qquad r = 0, 1, \ldots, n. \tag{4.7.8}$$

We proceed, as nearly as we can, in the manner exemplified in Example 4.6.1 for a continuous variable.

The statistic r is a realization of the Binomial (n, θ) variable R_θ, which has the p.d.f. given in (4.7.6) [see II, § 5.2.2]. The first task is to construct, as nearly as possible, appropriate central $100p\%$ ($= 95\%$, say) probability intervals for R_θ, for each value of θ ($0 < \theta < 1$). Since R_θ is discrete, central probability intervals of content exactly 95% are unattainable. (See § 4.1.3(b).) Instead, we construct quasi-central open probability intervals $(r_l(\theta), r_u(\theta))$ of content at least 95%, i.e. such that

$$P\{r_l(\theta) < R_\theta < r_u(\theta)\} \geq 0.95; \tag{4.7.9}$$

with

$$P\{R_\theta \le r_l(\theta)\} \le 0.025,$$

where $r_l(\theta)$ is the largest value of r for which $P\{R_\theta \le r\} \le 0.025$, or, equivalently, for which $P(R_\theta > r) \ge 0.975$), and

$$P\{R_\theta \ge r_u(\theta)\} \le 0.025,$$

where $r_u(\theta)$ is the smallest value of r for which $P\{R_\theta \ge r\} \le 0.025$, or equivalently, for which $P(R_\theta < r) \ge 0.975$. [See Example 4.1.4.]

Next, we plot the values of $r_l(\theta)$ and $r_u(\theta)$ as functions of θ, obtaining graphs of the form shown in Figure 4.7.1, and attempt to interpret these in terms of a confidence band.

To see how the functions $r_u(\theta)$ and $r_l(\theta)$ may be constructed, we examine the case where $\theta = 0.45$. From the Binomial tables [see Appendix (T1)] we obtain the following results.

	r	$P(R_{0.45} \le r)$		$P(R_{0.45} \ge r)$
	0	0.00253		1.0000
$\rightarrow$	1	0.02325	$\leftarrow$	
	2	0.09955		$\vdots$
	$\vdots$			
	8	$\vdots$		0.02740
$\rightarrow$	9		$\rightarrow$	0.00451
	10	1.0000		0.00035

We see that

$$r_l(0.45) = 1, \qquad r_u(0.45) = 9.$$

(The arrows in the table draw attention to the probabilities that are as close as possible to 0.025 but do not exceed 0.025.)

Proceeding in this way one may quickly construct the values of $r_u(\theta)$ and $r_l(\theta)$ for each tabulated θ: these are as shown below for $\theta = 0.10(0.05)0.90$.

θ	$r_l(\theta)$	$r_u(\theta)$	θ	$r_l(\theta)$	$r_u(\theta)$
0.10		4	0.55	2	9
0.15		5	0.60	3	10
0.20		6	0.65	3	10
0.25		6	0.70	4	
0.30		7	0.75	5	
0.35		8	0.80	5	
0.40	0	8	0.85	6	
0.45	1	9	0.90	7	
0.50	1	9			

From this coarse tabulation it is not possible to see with any accuracy where the jumps in $r_u(\theta)$ and $r_l(\theta)$ occur, but these may be determined with arbitrarily high accuracy by using more detailed Binomial tables, and interpolating them when necessary. The graphs of $r_u(\theta)$ and $r_l(\theta)$ are as illustrated in Figure 4.7.1.

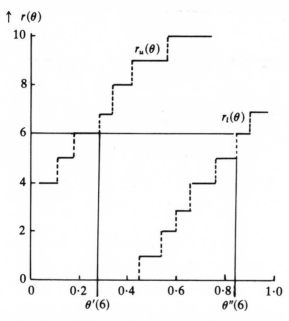

Figure 4.7.1: Graph of $r_u(\theta)$ and $r_l(\theta)$, such that $P\{r_l(\theta) < R_\theta < r_u(\theta)\} \leq 0.95$.

We have now imitated as best we can the construction of the central 95% probability intervals of Example 4.6.1 and the corresponding curves $y_1(\theta)$, $y_2(\theta)$ shown in Figure 4.6.1 and have come up with the discontinuous 'step function' graphs shown in Figure 4.7.1. The next step is to imitate as best we can the argument which showed that the zone between the curves in Figure 4.6.1 was a confidence belt. The argument cannot be exactly duplicated; in the continuous case of Example 4.6.1 it rested in the invertibility of the functions $y_1(\theta)$, $y_2(\theta)$ whereas our step-functions do not correspond to invertible functions.

This, however, is less of a problem than might appear. As will be seen on referring to Figure 4.7.2, for any value of θ, the proposition

$$r_l(\theta) < r < r_u(\theta) \tag{4.7.10}$$

(note the strict inequalities) is equivalent to the proposition

$$\theta'(r) \leq \theta \leq \theta''(r), \tag{4.7.11}$$

where $\theta'(r)$ $(= \theta$, say) is the abscissa of the right-hand end-point of the

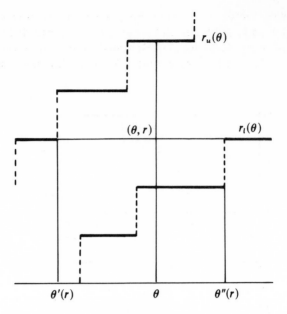

Figure 4.7.2: Illustrating the logical implication $r_l(\theta) < r < r_u(\theta) \Rightarrow \theta'(r) \le \theta \le \theta''(r)$.

horizontal segment ('step') of $r_u(\theta)$ whose height is r, i.e. for which

$$r_u(\theta') = r;$$

and $\theta''(r)$ $(= \theta''$, say) is the abscissa of the left-hand end-point of the step of $r_l(\theta)$ whose height is r, i.e. for which

$$r_l(\theta'') = r.$$

(It should be noted that every possible value of r coincides with the heights of *some* step since, for both, the values are the integers $0, 1, 2, \ldots, n$.) The portions of $\theta'(6)$ and $\theta''(6)$, corresponding to $r = 6$, are shown in Figure 4.7.1.

Since (4.7.10) and (4.7.11) are equivalent propositions (each implying the other) for every realization R_θ of r and every value of θ $(0 < \theta < 1)$, it follows that

$$P\{r_l(\theta) < R_\theta < r_u(\theta)\} = P\{\theta'(R_\theta) \le \theta \le \theta''(R_\theta)\}$$

where $\theta'(R_\theta)$, $\theta''(R_\theta)$ are the random variables induced by $\theta'(r)$, $\theta''(r)$ respectively. Now the probability on the left is not less than $0 \cdot 95$, by construction (and, subject to that, as nearly equal to $0 \cdot 95$ as possible), whence

$$P\{\theta'(R_\theta) \le \theta \le \theta''(R_\theta)\} \ge 0 \cdot 95.$$

Thus, if r is a realization of R_θ, the interval

$$(\theta'(r), \theta''(r)) \tag{4.7.12}$$

is a quasi-central confidence interval for θ at confidence level not less than 0·95 (and as near to 0·95) as possible.

Tables are available for the evaluation of $\theta'(r)$ and $\theta''(r)$ for each value of r and a range of sample sizes. A version of these is given in the chart provided in the Appendix (T10). The chart is scaled for a range of values of r/n where r is the observed value of the Binomial (n, θ) variable R_θ. From this we see, for example, that if $r = 8$ and $n = 20$, so that $r/n = 0·40$, the confidence interval is $(0·19, 0·64)$. (The 95% interval for the same data given by the Normal approximations of Example 4.7.1 were $(0·18, 0·62)$ by the 'crude' method and $(0·22, 0·62)$ by the more accurate method.)

EXAMPLE 4.7.4.   *Confidence interval for the parameter of a Poisson distribution.*   Methods exactly analogous to those explained in Example 4.7.3 may be employed to determine confidence intervals for the parameter of a Poisson distribution [see II, § 5.4].

If the number, R, of occurrences of an event in a given time (or in a given area, volume, etc.) is Poisson (θ), and the total number of occurrences observed in that time is c, confidence intervals at a confidence level of at least $100p\%$ $(p = 0·998, 0·99, 0·98, 0·95, 0·90)$ may be obtained from Table T11 in the Appendix. $(p = 1·2\alpha$ in the table).

4.8. DISTRIBUTION-FREE CONFIDENCE INTERVALS FOR QUANTILES

In the procedures discussed earlier in this chapter, confidence intervals were obtained in terms of the sampling distribution of the working statistic. There are cases however where confidence intervals can be obtained in terms that do not depend on the underlying distribution. In particular, a distribution-free confidence interval, may be obtained for a *quantile* (otherwise called fractile or percentile; see II, § 10.3.3) of a given continuous distribution, the p-quantile of a c.d.f. $F(x, \theta)$ being the value ξ_p defined by the equation

$$F(\xi_p) = p \qquad (0 < p < 1).$$

Let $x_{(1)}, x_{(2)}, \ldots, x_{(n)}$ be the *order statistics* [see II, § 15.1] of a sample of n observations from the given distribution, so that

$$x_{(1)} < x_{(2)} < \ldots < x_{(n)}.$$

The sampling p.d.f. of $x_{(r)}$ at y is

$$g(y) = \{F(y)\}^{r-1}\{1 - F(y)\}^{n-r}f(y)/B(r, n-r+1),$$

where $B(u, \sigma)$ denotes the beta function [see IV, § 10.2].

Now, if $X_{(r)}$ is the random variable of which $x_{(r)}$ is a realization,

$$P\{X_{(r)} \leq \xi_p\} = \int_{-\infty}^{\xi_p} g(y) \, dy,$$

and, on making the substitution $u = F(y)$ and using the fact that $u = F(\xi_p) = p$ when $y = \xi_p$, this reduces to

$$\int_0^p u^{r-1}(1-u)^{n-r}\,du / B(r, n-r+1) = I_p(r, n-r+1),$$

the incomplete beta function ratio. This result is 'distribution-free': that is, it does not depend on the distribution function $F(x)$. From a corresponding property of the joint distribution of the two order statistics $X_{(r)}$ and $X_{(s)}$, with $r < s$, it can similarly be shown that the interval

$$(x_{(r)}, x_{!s)})\quad r < s \tag{4.8.1}$$

is a distribution-free confidence interval for ξ_p with confidence level

$$I_p(r, n-f+1) - I_p(s, n-s+1). \tag{4.8.2}$$

Values of $I_p(u, v)$ may be obtained from published Tables of the Incomplete Beta Function [see Thompson (1941)—Bibliography G] or, alternatively, from cumulative Binomial probability tables, since

$$I_p(a, n-a+1) = \sum_{s=a}^{n} \binom{n}{s} p^a (1-p)^{n-a},$$

the probability that a Binomial (n, p) variable equals or exceeds the value a.

EXAMPLE 4.8.1.   The extreme values $x_{(1)}$ and $x_{(10)}$ of a sample of size 10 are the endpoints of the confidence interval (4.8.1) for the median (the quantile $\xi_{1/2}$), with confidence coefficient (4.8.2), namely

$$I_{1/2}(1, 10) - I_{1/2}(10, 1) = 1 - 2I_{1/2}(10, 1)$$

$$= 1 - 2(\tfrac{1}{2})^{10} = 0 \cdot 998.$$

Similarly the interval $(x_{(2)}, x_{(9)})$ is a confidence interval for the median, with confidence coefficient $I_{1/2}(2, 9) - I_{1/2}(9, 2) = 0 \cdot 979$.

4.9. MULTIPARAMETER CONFIDENCE REGIONS

4.9.1.   Exact Confident Regions

As will be seen from the earlier sections of this chapter, the theory of confidence intervals for the parameter of a one-parameter distribution is in a fairly healthy state. When we come to two-parameter distributions, the one most widely used and studied is the Normal (μ, σ) distribution, and for that, as we have seen in Examples 4.5.1 and 4.5.2, separate confidence intervals for the two parameters are available. These are (μ_l, μ_u) and (σ_l, σ_u) respectively, where

$$\mu_l = \bar{x} - t_{0 \cdot 025}s/\sqrt{n}, \qquad \mu_u = \bar{x} + t_{0 \cdot 975}s/\sqrt{n}$$

$$\sigma_l = (n-1)^{1/2}s/\chi_{0 \cdot 025}, \quad \sigma_u = (n-1)^{1/2}/\chi_{0 \cdot 975},$$

with (taking the sample to be $x_1, x_2, \ldots, x_n$)

$$\bar{x} = \sum_1^n x_r/n, \qquad s^2 = \sum_1^n (x_r - \bar{x})^2/(n-1),$$

t_p = the p-quantile of Student's distribution
   on $n-1$ degrees of freedom, $p = 0\cdot025, 0\cdot975$

and

χ_p^2 = the p-quantile of the chi-squared distribution
   on $n-1$ degrees of freedom, $p = 0\cdot025, 0\cdot975$.

Thus, with confidence coefficient $0\cdot95$, we have

$$\mu_l \le \mu \le \mu_u,$$

and, with the same confidence coefficient,

$$\sigma_l \le \sigma \le \sigma_u.$$

It does not however follow that, with confidence coefficient $(0\cdot95)^2$ we have the joint confidence region

$$\mu_l \le \mu \le \mu_u \quad and \quad \sigma_l \le \sigma \le \sigma_u$$

(a rectangular two-dimensional interval), since the two confidence intervals are based on the random variables induced by $(\bar{x} - \mu)/s$ and s/σ, respectively, and these are not independent.

Thus, even for this simple and familiar distribution, the question of how to construct a joint confidence region at a specified confidence level is not entirely obvious. One should at this stage ask how one ought to define a joint confidence region, what its properties would be, and to what practical use one could put it.

It has to be said at the outset that the theory of such regions is not in anything like as complete a state as the corresponding single-parameter theory. As to how one might sensibly define a two-dimensional confidence set, it might turn out for some random variable Y, with parameters α and β, estimated as α^* and β^*, that, with 95% probability, and for all α, β,

$$(A^* - \alpha)^2 + (B^* - \beta)^2 \le 1,$$

A^* and B^* being the random variables induced by α^* and β^* (so that α^* is a realization of A^*, and β^* of B^*). The realization of this, namely

$$(\alpha^* - \alpha)^2 + (\beta^* - \beta)^2 \le 1$$

would clearly be a 95% confidence region for (α, β): since the confidence level is 95% that the circle in the (α, β) plane with unit radius and centre α^*, β^* will contain the true values of α and β.

More generally, for a random variable Y whose distribution involves two parameters α, β, an acceptable 95% confidence region for α and β, based on estimates α^*, β^*, might be the set of points on and inside a closed curve $C(\alpha^*, \beta^*)$ in the (α, β) plane, such that

$$P\{(\alpha, \beta) \in C(A^*, B^*)\} = 0\cdot95.$$

That such regions exist in certain cases will be demonstrated below (see Example 4.9.1), where we construct a joint confidence region for the parameters of a Normal (μ, σ) distribution. Before proceeding to the details of that example, however, let us pause to consider how such a region could be used in practice.

The primary question that arises in multiparameter situations is how to assess the reliability of a given combination of the parameters, preferably (in the present context) in the form of a confidence interval for that combination. In the case of two parameters $(\alpha, \beta, \text{say})$ the simplest 'combination' is the parameter α itself (or, equivalently, β itself). We immediately encounter the disconcerting result that it is in general not possible to construct a confidence interval for α, with known confidence coefficient, from a knowledge of the joint 95% confidence region of α and β. Consider for example the situation portrayed in Figure 4.9.1, where the region contained in the closed curve C

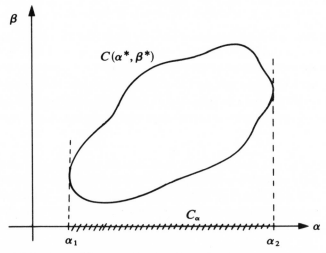

Figure 4.9.1: A confidence region for the pair of parameters (α, β).

is supposed to be a 95% confidence region for (α, β). The 'obvious' 95% confidence interval for α would be the interval (α_1, α_2) formed by the projection C_α the curve C on the axis of α. In fact, however, from the definition of a confidence region, the proposition that (α, β) lies in the set C is equivalent to one or more propositions involving inequalities relating A^*, B^*, α, β, the (joint) probability of these propositions being $0\cdot95$. These inequalities are satisfied by every point (α, β) inside C. That (α, β) should lie inside C implies that α lies inside the projection C_α. It follows from this logical implication that

$$P\{\alpha \in C_\alpha\} \ge 0.95$$

[see II, Theorem 3.4.5]. Thus, while C_α is a confidence interval for α, its confidence level is unknown: all we can tell is that this level is at least 95%.

If we cannot readily construct an exact confidence interval for α, or for β, from the joint confidence region (except in the 'inequality' sense described above), we cannot reasonably hope to construct exact intervals for combinations such as $\alpha + 2\beta$, α/β, etc. Methods giving confidence intervals at 'conservative' levels (based on inequalities) have however been developed. See, e.g. Scheffé, H: (1953) and (1970).

We now proceed to exemplify a two-dimensional confidence region.

EXAMPLE 4.9.1.   *A 95% confidence region for (μ, σ).*   Consider the quantities $(\bar{X} - \mu)/\sigma$, S^2/σ^2, where, as usual, $\bar{X}$ is the random variable induced by the statistic $\bar{x}$, and S^2 that induced by s^2 (using the same notation as above). Then $(\bar{X} - \mu)/(\sigma/\sqrt{n})$ is Normal $(0, 1)$, and $(n - 1)S^2/\sigma^2$ has the chi-squared distribution on $n - 1$ degrees of freedom. Furthermore, $\bar{X}$ and S are statistically independent. [See Theorem 2.5.2].

Now, if u_0 is the 0.975 quantile of the standard Normal distribution,

$$P\{-u_0 \le n^{1/2}(\bar{X} - \mu)/\sigma \le u_0\} = 0.95 \qquad (4.9.1)$$

so that

$$P\{\mu - u_0\sigma/\sqrt{n} \le \bar{X}, \ \mu + u_0\sigma/\sqrt{n} \ge \bar{X}\} = 0.95.$$

It follows that the probability is 0.95 that the wedge-shaped region shown in the figure (Figure 4.9.2) contains the point (μ, σ):

This wedge-shaped region is therefore a confidence region for (μ, σ), with confidence level 0.95. As it is unbounded, it is by itself of no practical utility.

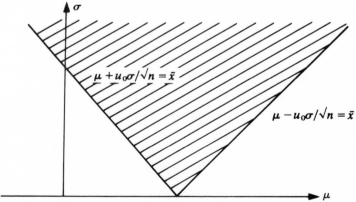

Figure 4.9.2: The hatched wedge-shaped region is an unbounded confidence region for the pair (μ, σ).

Similarly we have

$$P\{a^2 \le (n-1)S^2/\sigma^2 \le b^2\} = 0{\cdot}95, \qquad (4.9.2)$$

where $a^2 = \chi^2_{0 \cdot 025}(n-1)$ and $b^2 = \chi^2_{0 \cdot 975}(n-1)$ are respectively the $0 \cdot 025$ quantile and the $0 \cdot 975$ quantile of the chi-squared distribution on $n-1$ degrees of freedom, so that

$$P\{(n-1)^{1/2}S/b \le \sigma \le (n-1)^{1/2}S/a\} = 0{\cdot}95,$$

whence, in the μ, σ plane, the slab-shaped region shown in Figure 4.9.3 must have probability $0 \cdot 95$ of containing σ. This is an unbounded two-dimensional confidence region for σ.

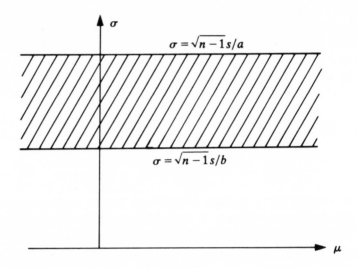

Figure 4.9.3: The hatched slab-shaped region is an unbounded confidence region for σ.

Since $\bar{X}$ and S are statistically independent, the propositions involved in (4.9.1) and (4.9.2) are independent, whence it follows that

$$P\{-u_0 \le n^{1/2}(\bar{X}-\mu) \le u_0 \quad \text{and} \quad a^2 \le (n-1)S^2/\sigma^2 \le b^2\} = (0{\cdot}95)^2,$$

that is, with probability $(0 \cdot 95)^2$,

$$\left. \begin{array}{c} \mu - u_0\sigma/\sqrt{n} \le \bar{X}, \qquad \mu + u_0\sigma/\sqrt{n} \ge \bar{X}, \\[2mm] \text{and} \\[2mm] (n-1)^{1/2}S/b \le \sigma \le (n-1)^{1/2}S/a. \end{array} \right\} \qquad (4.9.3)$$

This is equivalent to the proposition, which therefore has probability $(0 \cdot 95)^2 = 0 \cdot 9025$, that the random region $\mathscr{C}(\bar{X}, S)$ defined by (4.9.3) contains the point (μ, σ). If we replace $\bar{X}$ and S by their observed values $\bar{x}$ and s, the resulting region $\mathscr{C}(\bar{x}, s)$ is therefore a 90·25% confidence region for the pair (μ, σ). The region is illustrated in Figure 4.9.4.

If we wish to end up with a confidence coefficient of, say, 0·95 for the two dimensional region, we repeat the computations taking μ_0 to be the 0·9875 quantile of the standard Normal distribution, viz. 2·24, and (taking $n = 11$) $a = 2·7$, $b = 22·6$, respectively the 0·0125 quantile and the 0·9875 quantile of the chi-squared distribution on 10 degrees of freedom.

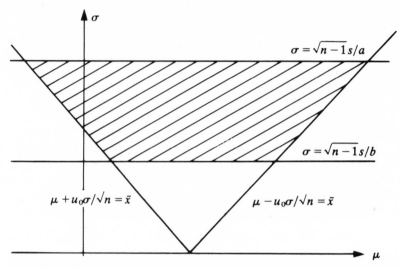

Figure 4.9.4: The hatched region is the intersection of the hatched regions of Figure 4.9.2 and Figure 4.9.3. It is a closed confidence region for the pair (μ, σ).

(To see where the figure 0·9875 comes from, we note that we must replace the probability level of 0·95 by $p = \sqrt{0·95} = 0·975$ in each of (4.9.1) and (4.9.2). To achieve these levels the appropriate quantiles of the Normal and Chi-squared distributions should be $\frac{1}{2}(1-p)$ and $\frac{1}{2}(1+p)$ as indicated in Figure 4.9.5).

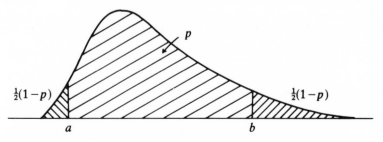

Figure 4.9.5: A central interval of content p is defined by (a, b) where a is the $\frac{1}{2}(1-p)$ quantile and b is the $\{p + \frac{1}{2}(1-p)\} = \frac{1}{2}(1+p)$ quantile.

4.9.2. Elliptical Confidence Regions for the Expectation Vector of a Bivariate Normal Distribution; and Approximate Confidence Regions for a Pair of Maximum Likelihood Estimates

In the theory of estimation, the maximum-likelihood estimate $\hat{\theta}$ of a parameter θ is found in many cases to be approximately Normal $(\theta, \sigma_\theta^2)$, where the approximate sampling variance σ_θ^2 is obtained from the data [see § 6.2.5]. As a very rough-and-ready procedure one may then argue that, $(\hat{\theta} - \theta)/\sigma_\theta$, being standard Normal, will with probability 0·95 lie in the interval $\pm 1\cdot 96$, whence

$$\hat{\theta} \pm 1\cdot 96 \sigma_\theta$$

is an approximate 95% confidence interval for θ.

When there are two parameters, say

$$\boldsymbol{\theta} = (\theta_1, \theta_2),$$

the corresponding approximations regard the Maximum-Likelihood estimate $\hat{\boldsymbol{\theta}} = (\hat{\theta}_1, \hat{\theta}_2)$ as having a bivariate Normal distribution (approximately) with expectation vector $\boldsymbol{\theta}$ and dispersion matrix $\mathbf{V}$ obtainable from the data [see § 6.2.5(h)].

It is then desirable to obtain a 95% confidence region, however rough-and-ready, for $\boldsymbol{\theta}$. We may argue as follows: if

$$\mathbf{V} = \begin{pmatrix} v_1 & c \\ c & v_2 \end{pmatrix},$$

so that the estimated variance of $\hat{\theta}_1$ is v_1, and that of $\hat{\theta}_2$ is v_2, we may of course construct separate approximate 95% confidence intervals

$$\hat{\theta}_1 \pm 1\cdot 96 v_1^{1/2}, \quad \hat{\theta}_2 \pm 1\cdot 96 v_2^{1/2} \tag{4.9.4}$$

for θ_1 and θ_2. Any one of these is satisfactory, but (as pointed out in section 4.9.1) this does not mean that with probability $(0\cdot 95)^2$ the point (θ_1, θ_2) will lie in the rectangle indicated by (4.9.4). A confidence statement involving both θ_1 and θ_2 can however be made as follows. The qualification '(approximately)' applies where $\hat{\theta}_1$ and $\hat{\theta}_2$ are maximum likelihood estimates whose joint sampling distribution is not exactly Bivariate Normal. Where $\hat{\theta}_1$ and $\hat{\theta}_2$ are Bivariate Normal the qualification should be disregarded. For any vector $\mathbf{a}' = (a_1, a_2)$, the random vector

$$a_1 \hat{\theta}_1 + a_2 \hat{\theta}_2$$

is Normal with expectation

$$a_1 \theta_1 + a_2 \theta_2$$

and variance

$$\mathbf{a}'\mathbf{V}\mathbf{a} = a_1^2 v_1 + 2 a_1 a_2 c + a_2^2 v_2,$$

so that the 95% confidence interval for

$$a_1\theta_1 + a_2\theta_2$$

is

$$a_1\hat{\theta}_1 + a_2\hat{\theta}_2 \pm 1\cdot 96\sqrt{(a_1^2 v_1 + 2a_1 a_2 c + a_2^2 v)}. \qquad (4.9.5)$$

An alternative approach which enables one to find a two-dimensional confidence set for θ is to argue that the two-dimensional analogue of the central
95% probability interval $\pm 1\cdot 96\sigma$ of the univariate Normal $(0, \sigma^2)$ distribution
is the area C contained in the (elliptical) contour of constant probability
density defined such that

$$\iint\limits_C f(x_1, x_2)\, dx_1\, dx_2 = 0\cdot 95,$$

where f denotes the p.d.f. of the bivariate Normal $(\mathbf{0}, \mathbf{V})$ distribution. Then,
with probability $0\cdot 95$,

$$\hat{\boldsymbol{\theta}} - \boldsymbol{\theta} \in C$$

so that the 95% confidence region for θ is the ellipse '$C + \hat{\boldsymbol{\theta}}$', that is the ellipse
obtained by translating C so that its centre is at $\hat{\boldsymbol{\theta}}$.

To find the ellipse C proceed as follows:
The p.d.f. of the bivariate Normal $(\mathbf{0}, \mathbf{V})$ distribution [see II, § 13.4.6(i)] is

$$(2\pi)^{-1/2}|\mathbf{V}|^{-1/2} \exp\left(-\tfrac{1}{2}\mathbf{x}'\mathbf{V}^{-1}\mathbf{x}\right).$$

The contour C has equation

$$\mathbf{x}'\mathbf{V}^{-1}\mathbf{x} = b$$

say, where b is to be determined. Write $\mathbf{V} = \boldsymbol{\Sigma}\boldsymbol{\Sigma}'$ and $\mathbf{Y} = \boldsymbol{\Sigma}^{-1}\mathbf{X}$, so that $\mathbf{Y}$ is
bivariate Normal $(\mathbf{0}, \mathbf{I})$ [see II, Example 13.4.8]. Under this transformation
C becomes the circle

$$\mathbf{y}'\mathbf{y} = b, \qquad (\text{or } y_1^2 + y_2^2 = b),$$

where

$$\mathbf{y} = \boldsymbol{\Sigma}^{-1}\mathbf{x}.$$

We thus have to determine b from the condition

$$\frac{1}{2\pi} \iint\limits_{y_1^2 + y_2^2 \leq b} e^{-(y_1^2 + y_2^2)/2}\, dy_1\, dy_2 = 0\cdot 95.$$

The integral is readily evaluated if one transforms to polar coordinates [see
IV, Example 6.2.3]. The above then reduces to

$$1 - e^{-b/2} = 0\cdot 95,$$

whence

$$b = 5 \cdot 991.$$

Then the 95% confidence region for (θ_1, θ_2) is the region contained by the ellipse

$$(\mathbf{x} - \hat{\boldsymbol{\theta}})' \mathbf{V}^{-1} (\mathbf{x} - \hat{\boldsymbol{\theta}}) = 5 \cdot 991,$$

where

$$\mathbf{x}' = (\theta_1, \theta_2).$$

(Similarly for a 99% confidence interval. We then have to solve the equation $1 - e^{-b/2} = 0 \cdot 99$, whence $b = 9 \cdot 210$.) Thus if

$$\mathbf{V} = \begin{pmatrix} v_1 & c \\ c & v_2 \end{pmatrix} = \begin{pmatrix} \sigma_1^2 & \rho \sigma_1 \sigma_2 \\ \rho \sigma_1 \sigma_2 & \sigma_2^2 \end{pmatrix},$$

so that [see I, § 6.4(iii)]

$$\mathbf{V}^{-1} = \frac{1}{1 - \rho^2} \begin{pmatrix} 1/\sigma_1^2 & -\rho/\sigma_1 \sigma_2 \\ -\rho/\sigma_1 \sigma_2 & 1/\sigma_2^2 \end{pmatrix},$$

the equation is

$$\frac{(x_1 - \hat{\theta}_1)^2}{\sigma_1^2} - \frac{2\rho}{\sigma_1 \sigma_2} (x_1 - \hat{\theta}_1)(x_2 - \hat{\theta}_2) + \frac{(x_2 - \hat{\theta}_2)^2}{\sigma_2^2} = 5 \cdot 991(1 - \rho^2). \quad (4.9.6)$$

This region is an accurate (though not unique) 95% confidence region for $\boldsymbol{\theta}$ when the parameters $(\rho, \sigma_1, \sigma_2)$ are known exactly. When they are only known approximately, of course, the confidence region is also only approximately defined.

In the case of linear regression with independent Normally distributed errors, there is a straightforward procedure for constructing exact confidence ellipsoids for the parameters (or any subset of them). This is described in section 8.3.2.

4.10.   CONFIDENCE INTERVAL OBTAINABLE FROM A LARGE SAMPLE, USING THE LIKELIHOOD FUNCTION

When the sample is sufficiently large certain simplifying approximations become possible.

4.10.1.   The Likelihood Function

Let X have p.d.f. $f(x, \theta)$. From a random sample $(x_1, x_2, \cdots, x_n)$ of observations on X one constructs the likelihood function [see § 6.2.1]

$$l(\theta; x_1, x_2, \ldots, x_n) = \prod_1^n f(x_r, \theta). \quad (4.10.1)$$

The log-likelihood function is $\sum_1^n \log f(x_r, \theta)$ and its derivative with respect to θ [see IV, § 4.3] is

$$z(\theta; x_1, x_2, \ldots, x_n) = \frac{\partial \log l}{\partial \theta} = \sum_1^n \frac{\partial}{\partial \theta} \log f(x_r, \theta). \qquad (4.10.2)$$

The sampling distribution of this is the distribution of the induced random variable

$$Z = z(\theta; X_1, X_2, \ldots, X_n) = \sum_1^n \frac{\partial}{\partial \theta} \log f(X_r, \theta) \qquad (4.10.3)$$

where the X_r are i.i.d. statistical copies of X. The random variables $\partial \log f(X_r, \theta)/\partial \theta$, $r = 1, 2, \ldots, n$ are obviously i.i.d., and, by the Central Limit Theorem [see II, § 17.3], their sum Z is approximately Normally distributed when n is large.

To utilize this fact we need expressions for $E(Z)$ and var (Z). The first of these is zero, since

$$E\{\partial \log f(X_r, \theta)/\partial \theta\} = \int_{-\infty}^{\infty} \left\{\frac{\partial}{\partial \theta} \log f(x, \theta)\right\} f(x, \theta)\, dx$$

$$= \int_{-\infty}^{\infty} \frac{\partial f(x, \theta)}{\partial \theta}\, dx$$

$$= \frac{\partial}{\partial \theta} \int_{-\infty}^{\infty} f(x, \theta)\, dx$$

provided $f(x, \theta)$ satisfies appropriate regularity conditions [see IV, § 4.7]. Since $\int_{-\infty}^{\infty} f(x, \theta)\, dx = 1$, it follows that $\partial \log f(X_r, \theta)/\partial \theta$ has expectation equal to zero, and therefore that $E(Z) = 0$.

The variance of $\partial \log f(X_r, \theta)/\partial \theta$ is then

$$E\{\partial \log f(X_r, \theta)/\partial \theta\}^2 = \int_{-\infty}^{\infty} \left\{\frac{\partial \log f(x, \theta)}{\partial \theta}\right\}^2 f(x, \theta)\, dx. \qquad (4.10.4)$$

Now

$$\left(\frac{\partial \log f}{\partial \theta}\right)^2 = \frac{1}{f^2}\left(\frac{\partial f}{\partial \theta}\right)^2,$$

while

$$\frac{\partial^2 \log f}{\partial \theta^2} = -\frac{1}{f^2}\left(\frac{\partial f}{\partial \theta}\right)^2 + \frac{1}{f}\frac{\partial^2 f}{\partial \theta^2},$$

and so

$$\left(\frac{\partial \log f}{\partial \theta}\right)^2 = \frac{1}{f}\frac{\partial^2 f}{\partial \theta^2} - \frac{\partial^2 \log f}{\partial \theta^2},$$

and the integral on the right in (4.10.4) becomes

$$\int_{-\infty}^{\infty} \frac{\partial^2 f}{\partial \theta^2} \, dx - \int_{-\infty}^{\infty} \left\{ \frac{\partial^2 \log f(x, \theta)}{\partial \theta} \right\} f(x, \theta) \, dx,$$

Subject to suitable regularity conditions (cf. § 3.3.3(c)) the first of these two integrals is $(\partial^2/\partial\theta^2) \int_{-\infty}^{\infty} f(x, \theta) \, d\theta$. This reduces to zero, since $\int_{-\infty}^{\infty} f(x, \theta) \, d\theta = 1$. Finally, then, from (4.10.4),

$$\text{var}\,\{\partial \log f(X_r, \theta)/\partial\theta\} = -\int_{-\infty}^{\infty} \left\{ \frac{\partial^2 \log f(x, \theta)}{\partial \theta^2} \right\} f(x, \theta) \, dx$$

$$= -E\{\partial^2 \log f(X_r, \theta)/\partial\theta^2\}, \qquad r = 1, 2, \ldots, n,$$

and, so, from (4.10.3),

$$\text{var}\, Z = -\sum_{r=1}^{n} E\{\partial^2 \log f(X_r, \theta)/\partial\theta^2\};$$

i.e.

$$\text{var}\, Z = I_n(\theta), \tag{4.10.5}$$

say, where

$$I_n(\theta) = -nE\{\partial^2 \log f(X, \theta)/\partial\theta^2\} \tag{4.10.6}$$

is the 'amount of information' in the sample (c.f. (3.3.6)).

Thus Z, defined by (4.9.2), is approximately Normal with zero expectation and with variance $I_n(\theta)$ given by (4.10.6).

4.10.2. Approximate Confidence Intervals, using the Derivative of the Log-Likelihood

It follows from the results obtained in § 4.10.1 that, writing $\omega^2 = I_n(\theta)$,

$$P(-1 \cdot 96\omega \le Z \le 1 \cdot 96\omega) \doteq 0 \cdot 95,$$

$$P(-2 \cdot 576\omega \le Z \le 2 \cdot 576\omega) \doteq 0 \cdot 99, \tag{4.10.7}$$

etc., from which one may construct approximate confidence intervals for θ with confidence coefficient 0·95, 0·99 etc., as illustrated in the following example.

EXAMPLE 4.10.1.   *Confidence interval for Binomial θ by the Likelihood method.*   Consider the random variable X having the Bernoulli distribution with success-parameter θ [see II, § 5.2.1]. The p.d.f. of X at x is

$$f(x, \theta) = \theta^x (1 - \theta)^{1-x}, \qquad x = 0, 1 \qquad (0 \le \theta \le 1).$$

In terms of a sample $(x_1, x_2, \ldots, x_n)$ the likelihood function (4.9.1) is

$$l(\theta; x_1, \ldots, x_n) = \prod_1^n \theta^{x_i}(1-\theta)^{1-x_i}$$

$$= \theta^r(1-\theta)^{n-r}, \qquad r = \sum_1^n x_i.$$

The derivative (4.10.2) of the log-likelihood function is thus

$$z(\theta; x_1, \ldots, x_n) = \frac{\partial}{\partial\theta} \log\{\theta^r(1-\theta)^{n-r}\} = \frac{\partial}{\partial\theta}\{r\log\theta + (n-r)\log(1-\theta)\}$$

$$= \frac{r}{\theta} - \frac{n-r}{1-\theta} = \frac{r-n\theta}{\theta(1-\theta)}, \qquad \left(r = \sum_1^n x_j\right).$$

This is a realization of the random variable

$$Z = (R - n\theta)/\theta(1-\theta) \qquad (4.10.8)$$

(compare (4.10.3)). It may be verified that $E(Z) = 0$: this follows from the fact that, R being Binomial (n, θ), $E(R) = n\theta$.

Next we need $I_n(\theta)$. We have

$$\log f(X, \theta) = X \log\theta + (1-X)\log(1-\theta)$$

whence

$$\frac{\partial^2}{\partial\theta^2} \log f(X, \theta) = -\frac{X}{\theta^2} - \frac{(1-X)}{(1-\theta)^2}.$$

Since $E(X) = \theta$, we have

$$-E\left\{\frac{\partial^2 \log f(X, \theta)}{\partial\theta^2}\right\} = \frac{1}{\theta} + \frac{1}{1-\theta} = \frac{1}{\theta(1-\theta)}$$

so that, by (4.10.5)

$$\text{var}(Z) = I_n(\theta) = n/\theta(1-\theta).$$

The accuracy of this may be independently verified from (4.9.8): since R is binomial (n, θ), $\text{var}(R) = n\theta(1-\theta)$, whence

$$\text{var}(Z) = n\theta(1-\theta)/\{\theta(1-\theta)\}^2 = n/\theta(1-\theta).$$

To set up a 95% confidence interval for θ, then, we use (4.10.7): with probability 0·95, approximately,

$$\frac{-1\cdot96\sqrt{n}}{\sqrt{\{\theta(1-\theta)\}}} \le \frac{R-n\theta}{\theta(1-\theta)} \le \frac{1\cdot96\sqrt{n}}{\sqrt{\{\theta(1-\theta)\}}},$$

i.e.

$$(R - n\theta)^2 \le 3\cdot842n\theta(1-\theta).$$

Thus, from a sample $(x_1, x_2, \ldots, x_n)$, with $\Sigma x_r = r$ (the total number of 'successes') the 95% confidence interval obtained by this procedure is the interval (θ_l, θ_u), where θ_l, θ_u are, respectively, the smaller and the larger roots of the quadratic

$$(r - n\theta)^2 = 3 \cdot 842 n\theta(1 - \theta).$$

For example, if $r = 6$ and $n = 20$ (so that the conventional estimate of θ would be $r/n = 0 \cdot 30$) the quadratic becomes

$$(6 - 20\theta)^2 = (3 \cdot 842)(20)\theta(1 - \theta)$$

whence

$$\theta_l = 0 \cdot 138, \qquad \theta_u = 0 \cdot 526, \tag{4.10.9}$$

to the degree of approximation implied by the assumption of Normality.

[The approximate Normality of the Binomial variable R was in fact the first special case of the Central Limit Theorem to be discovered [see II, § 11.4.7]. A conventional crude version of the above calculation would say that R is approximately Normal with expectation 20θ and variance $20\theta(1 - \theta)$. Replacing θ in the variance expression by its conventional estimate $r/n(=0 \cdot 3)$ gives the variance as $4 \cdot 2$. The argument would then continue: R is approximately Normal $(20\theta, 2 \cdot 05)$, whence the 95% confidence limits for θ are $r \pm (1 \cdot 96)(2 \cdot 05)/20$, i.e. $0 \cdot 30 \pm 0 \cdot 20$, i.e. $(0 \cdot 10, 0 \cdot 50)$. (Compare (4.10.9)).

These large-sample approximations may be compared with the exact value $(0 \cdot 12, 0 \cdot 54)$ of the $100p\%$ confidence interval for θ, with $r = 6$ and $n = 20$, obtained in Example 4.7.1, where, however, it will be recalled, instead of $p = 0 \cdot 95$ we had $p \geq 0 \cdot 95$.

4.10.3   Confidence Intervals Obtained with the Aid of an (Approximately) Normalizing Transformation

It sometimes happens that a statistic which has an 'awkward' sampling distribution may be transformed into one with a more amenable distribution. The best-known example of this is the sample correlation coefficient as an estimate of the bivariate Normal correlation coefficient. [See § 2.7.3(b)].

EXAMPLE 4.10.2.   *Approximate confidence interval for a correlation coefficient.*   Let ρ be the correlation coefficient of the bivariate Normal pair (X, Y). On a sample $(x_1, y_1), (x_2, x_2), \ldots, (x_n, y_n)$, the conventional estimate of ρ is

$$r = \left\{ \sum_{j=1}^{n} (x_j - \bar{x})(y_j - \bar{y}) \right\} \Big/ \left\{ \sum_{s=1}^{n} (x_s - \bar{x})^2 \sum_{t=1}^{n} (y_t - \bar{y})^2 \right\}^{1/2}.$$

To obtain a confidence interval for ρ from an observed value of r we use the fact that, if n is large (say, $n > 50$), the quantity given by

$$z = \tfrac{1}{2} \log \left(\frac{1+r}{1-r} \right)$$

is to a good approximation a realization of a Normal variable Z with $E(Z) = \tfrac{1}{2} \log \{(1+\rho)/(1-\rho)\}$ and $\text{var}(Z) = 1/(n-3)$. Accepting this approximation, it follows that, with 95% probability,

$$Z - 1 \cdot 96 / \sqrt{(n-3)} \le \tfrac{1}{2} \log \frac{1+\rho}{1-\rho} \le Z + 1 \cdot 96 / \sqrt{(n-3)}.$$

Solving these inequalities for ρ, the corresponding 95% confidence interval for ρ is found to be $\{(a-1)/(a+1), (b-1)/(b+1)\}$, where

$$a = \left(\frac{1+r}{1-r} \right) e^{-3 \cdot 92 / \sqrt{(n-3)}}$$

and

$$b = \left(\frac{1+r}{1-r} \right) e^{3 \cdot 92 / \sqrt{(n-3)}}.$$

For example, if $r = 0 \cdot 3$ and $n = 55$, $a = 1 \cdot 07$ and $b = 3 \cdot 20$, whence the 95% confidence interval for ρ is $(0 \cdot 03, 0 \cdot 52)$. [See also Example 5.2.2].

4.11. A CONFIDENCE BAND FOR AN UNKNOWN CONTINUOUS CUMULATIVE DISTRIBUTION FUNCTION

4.11.1. The Empirical c.d.f.

The c.d.f. $F(x)$ of a continuous random variable X is defined as

$$F(x) = P(X \le x), \quad -\infty < x < \infty.$$

From a sample of n observations on X the obvious sample analogue of $F(x)$ is the function $F_n(x)$ defined by:

$$nF_n(x) = \text{number of observations not exceeding } x$$
$$= \text{fr}(X \le x), \tag{4.11.1}$$

say, where 'fr($\mathscr{A}$)' stands for 'the frequency with which the proposition $\mathscr{A}$ is verified in the sample'. This is called the empirical c.d.f. (or the empirical distribution function). An equivalent alternative version is as follows:

DEFINITION 4.11.1. *Order statistics of a sample.* Let $x_{(1)}, x_{(2)}, \ldots, x_{(n)}$ (with $x_{(1)} < x_{(2)} < \ldots < x_{(n)}$) be the order statistics [see II, Chapter 15] of a sample of n observations on a continuous random variable X. The *empirical c.d.f.* of X is

$$F_n(x) = \begin{cases} 0, x < x_{(1)} \\ k/n, x_{(k)} \leq x < x_{(k+1)}, \\ 1, x \geq x_{(n)}. \end{cases} \quad k = 1, 2, \ldots, n-1 \quad (4.11.2)$$

EXAMPLE 4.11.1.   *Order statistics.*   The following set of four-decimal random numbers [see II, § 5.1] is a random sample of ten observations, recorded with 4-figure accuracy, on the continuous uniform $(0, 1)$ distribution [see II, § 10.2.1]:

0·4754	0·0083
0·7591	0·0666
0·5566	0·1330
0·5435	0·8572
0·0392	0·6566

Rearranging the data in increasing order of magnitude we see that the order statistics are

r	$x_{(r)}$	r	$x_{(r)}$
1	0·0083	6	0·5435
2	0·0392	7	0·5566
3	0·0666	8	0·6566
4	0·1330	9	0·7591
5	0·4754	10	0·8572

The empirical c.d.f. $F_{10}(x)$ is thus the function tabulated in Table 4.11.1. The graph of this function is shown in Figure 4.11.1. It is a step function [see IV, Definition 4.9.4] in which the horizontal segments are to be interpreted as intervals closed on the left and open on the right [see I, § 2.6.3]. For comparison the c.d.f. $F(x)$ is also shown, viz. the function

$$F(x) = \begin{cases} 0, & x < 0 \\ x, & 0 \leq x \leq 1 \\ 1, & x > 1. \end{cases}$$

Values of x	$F_{10}(x)$
$x < 0\cdot0083$	0
$0\cdot0083 \leq x < 0\cdot0392$	$0\cdot1$
$0\cdot0392 \leq x < 0\cdot0666$	$0\cdot2$
$0\cdot0660 \leq x < 0\cdot1330$	$0\cdot3$
$0\cdot1330 \leq x < 0\cdot4754$	$0\cdot4$
$0\cdot4754 \leq x < 0\cdot5435$	$0\cdot5$
$0\cdot5435 \leq x < 0\cdot5566$	$0\cdot6$
$0\cdot5566 \leq x < 0\cdot6566$	$0\cdot7$
$0\cdot6566 \leq x < 0\cdot7591$	$0\cdot8$
$0\cdot7591 \leq x < 0\cdot8572$	$0\cdot9$
$x > 0\cdot8572$	$1\cdot0$

Table 4.11.1

4.11.2   The Kolmogorov–Smirnov Distance Between the c.d.f. and the Empirical c.d.f.

For the purposes of constructing a confidence band for an unknown c.d.f. $F(x)$ the most useful measure of closeness of the empirical c.d.f. $F_n(x)$ to the actual c.d.f. $F(x)$ is the Kolmogorov–Smirnov statistic d_n given by

$$d_n = d_n(x_{(1)}, x_{(2)}, \ldots, x_{(n)}) = \sup_x |F_n(x) - F(x)|, \qquad (4.11.3)$$

where 'sup' denotes the supremum, or least upper bound [see I, § 2.6.3].

In a given sample this statistic is the magnitude of the largest vertical distance between $F_n(x)$ and $F(x)$. In Figure 4.11.1 it occurs at $x = 0\cdot1330$, where $F_n(x) = 0\cdot4000$ and $F(x) = 0\cdot1330$, so that $d_n = 0\cdot2670$.

This statistic d_n is a realization of the random variable

$$D_n = d_n(X_{(1)}, X_{(2)}, \ldots, X_{(n)}) \qquad (4.11.4)$$

where the $X_{(n)}$ are the 'order-statistic' random variables reduced by the $x_{(n)}$.

[There is a notational problem here. It is standard practice to denote a c.d.f. by a capital letter such as F, as in the usage $F(x)$ employed above. This conflicts with the convention of reserving capital Latin letters for random variables, but of course no ambiguity is possible. The picture is not quite so clear with the empirical c.d.f. Here also it is standard practice to use a notation such as our $F_n(x)$, as defined in (4.11.2) and used in (4.11.3). For each x, this $F_n(x)$

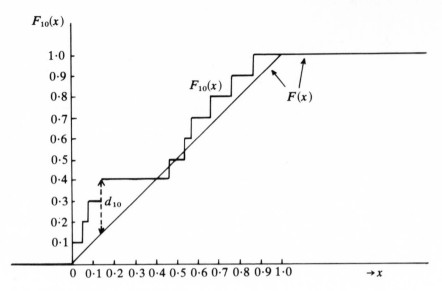

Figure 4.11.1: Graph of the empirical c.d.f. $F_{10}(x)$ of the sample in Example 4.11.1.

(capital letter notwithstanding) is a realization of the random variable

$$F_n^{(r.v.)}(x) = \begin{cases} 0, & x < X_{(1)} \\ k/n, & X_{(k)} \le x < X_{(k+1)}, \qquad k = 1, 2, \ldots, n-1 \\ 1, & x \ge X_{(n)} \end{cases} \qquad (4.11.4)$$

(compare (4.11.2)).

Here we have been forced to use the ungainly notation $F_n^{(r.v.)}(x)$ to distinguish the random variable from its realization $F_n(x)$. Needless to say, in practice nobody ever uses the symbol $F_n^{(r.v.)}(x)$; the same symbol $F_n(x)$ is used for the statistic (4.11.2) and the random variable of which it is a realization, leaving the context to indicate which is meant.]

4.11.3.   The Sampling Distribution of the Kolmogorov–Smirnov Statistic; Confidence Limits for the c.d.f.

It turns out that the sampling distribution of d_n (see Definition 4.11.1) does not depend on the sampled c.d.f. $F(x)$. To see how this comes about, consider the transformation from X to $Y = F(X)$. The c.d.f. of Y at $y = F(x)$ is $G(y)$ given by

$$G(y) = P(Y \le y) = P\{F(x) \le F(x)\}$$

$$= P(X \le x)$$

$$= F(x) = y, \qquad \text{since } F \text{ is monotone increasing,}$$

while the empirical c.d.f. of Y is the function $G_n(y)$ given by

$$nF_n(x) = fr(X \leq x) = fr\{F(X) \leq F(x)\}$$
$$= fr(Y \leq y) = nG_n(y),$$

whence

$$F_n(x) - F(x) = G_n(y) - y.$$

Now $F_n(x)$ is a realization of a function defined on the order statistics $x_{(r)}$ of a random sample of n observations on the variable X whose c.d.f. is $F(x)$. It follows that $G_n(y)$ is a realization of the corresponding function defined on the order statistics $y_{(r)}$ of the random variable $Y = F(X)$, which has the continuous uniform $(0, 1)$ distribution. (See II: Theorem 10.7.2). Thus the sampling distribution of $G_n(y) - y$ is determined by the properties of the uniform $(0, 1)$ distribution, and does not in any way depend on $F(x)$. We have

$$P(D_n \leq d) = P\{\sup_x |F_n(x) - F(x)| \leq d\}$$

$$= P\{\sup_y |G_n(y) - y| \leq d\}$$

$$= K_n(d) \tag{4.11.5}$$

say. This distribution has been tabulated [see, e.g. Owen (1962) or Harter and Owen (1970), Vol. 1, both in Bibliography G]. For the purposes of constructing 95% confidence limits one needs, for each value of n, the value of d such that $K_n(d) = 0.95$. These values are available (see Table 4.11.2) and likewise for 99%, 98%, 90% etc. For $n = 10$, for instance, the 95% value of d is 0.409, so that

$$P(D_{10} \leq 0.409) = 0.95.$$

EXAMPLE 4.11.2.   *Confidence band for a c.d.f.*   It follows that, in a sample of size $n = 10$,

$$0.95 = P(D_{10} \leq 0.409)$$

$$= P(\sup_x |F_{10}(x) - F(x)| \leq 0.409)$$

$$= P\{|F_{10}(x) - F(x)| \leq 0.409 \text{ for all } x\}$$

$$= P\{F_{10}(x) - 0.409 \leq F(x) \leq F_{10} + 0.409 \text{ for all } x\}.$$

Hence for all x, the inequalities

$$F_{10}(x) - 0.409 \leq F(x) \leq F_{10}(x) + 0.409 \tag{4.11.6}$$

define a 95% confidence interval for the c.d.f. $F(x)$, assumed to be unknown. Since however we also have

$$0 \leq F(x) \leq 1$$

we may improve (4.11.6) to the form

$$l_n(x) \le F(x) \le u_n(x) \tag{4.11.7}$$

where

and

$$\left. \begin{aligned} l_n(x) &= \max\,(0,\, F_n(x) - 0{\cdot}409) \\ u_n(x) &= \min\,(1,\, F_n(x) + 0{\cdot}409). \end{aligned} \right\} \tag{4.11.8}$$

In the sample of size 10 discussed in Example 4.11.1 the empirical c.d.f. $F_n(x)$ is tabulated in Table 4.11.1. It follows that the confidence band given by (4.11.8) is as given in Table 4.11.2.

	95% Confidence limits for $F(x)$	
Values of x	$l_{10}(x)$	$u_{10}(x)$
$x < 0$	0	0
$0 \qquad \le x < 0{\cdot}0083$	0	0·409
$0{\cdot}0083 \le x < 0{\cdot}0392$	0	0·509
$0{\cdot}0392 \le x < 0{\cdot}0666$	0	0·609
$0{\cdot}0666 \le x < 0{\cdot}1330$	0	0·709
$0{\cdot}1330 \le x < 0{\cdot}4754$	0	0·809
$0{\cdot}4754 \le x < 0{\cdot}5435$	0·091	0·909
$0{\cdot}5435 \le x < 0{\cdot}5566$	0·191	1
$0{\cdot}5566 \le x < 0{\cdot}6566$	0·291	1
$0{\cdot}6566 \le x < 0{\cdot}7591$	0·391	1
$0{\cdot}6591 \le x < 0{\cdot}8572$	0·491	1
$0{\cdot}8572 \le x < 1$	0·591	1
$x \ge 1$	1	1

Table 4.11.2: The confidence band (4.11.8)

That the confidence band is so hopelessly wide is of course an illustration of the impossibility of extracting very precise limits for a distribution function from a small sample. [The crucial value, 0·409 for $n = 10$, diminishes to the more informative 0·294 when $n = 20$, to 0·242 when $n = 30$, to 0·210 when $n = 40$, and to $1{\cdot}36/\sqrt{n}$ (approximately) for $n > 40$; these all refer to a 95% confidence level.]

The asymptotic distribution of D_n as $n \to \infty$ (in practice this means for $n > 40$) has the comparatively simple form

$$P(D_n < z/\sqrt{n}) = H(z)$$

where

$$H(z) = 1 - 2 \sum_{s=1}^{\infty} (-1)^{s-1}\, e^{-2s^2 z^2}, \qquad z > 0. \tag{4.11.9}$$

Some representative values of $H(z)$ (given in percentile form) are listed in Table 4.11.3:

$H(z)$	0·99	0·98	0·95	0·90	0·85	0·80
z	1·63	1·52	1·36	1·22	1·14	1·07

Table 4.11.3: Percentage points for the asymptotic Kolmogorov–Smirnov distribution (4.11.9).

In a sample of size 100, for instance, the 95% confidence band will be defined by the inequalities

$$\max \{0, F_{100}(x) - 0\cdot136\} < F(x) < \min \{1, F_{100}(x) + 0\cdot136\}.$$

It will be seen that, to construct a 95% confidence belt whose greatest width is 0·1, one would need a sample of size n, where

$$1\cdot36/\sqrt{n} = 0\cdot05$$

i.e.

$$n = 740 \text{ approx.}$$

4.12. TOLERANCE INTERVALS

A problem somewhat akin to that of finding confidence intervals is that of finding bounds between which a given proportion (e.g. 99%) of the sampled population may be asserted to lie. Such an assertion, naturally, can only be made 'with probability p' where $p = 0\cdot95$ or $0\cdot99$, etc. For example, from a sample of men's heights (assumed to be Normally distributed) in a given population, it may be required to find bounds x_1 and x_2 such that one may assert, with probability 0·95, that 99% of the men in the population have heights between x_1 and x_2. Such bounds are called *tolerance limits*. [See Wilks (1961)—Bibliography C].

To formalize this somewhat, we define $100p\%$ tolerance limits for a Normal (μ, σ) population, at probability level β. If μ and σ were known we could easily find a central interval (a_1, a_2) containing a fraction p of the population. The end points a_1 and a_2 satisfy the relation

$$\int_{a_1}^{a_2} f(x; \mu, \sigma) \, dx = p \qquad (4.12.1)$$

where $f(x; \mu, \sigma)$ denotes the Normal (μ, σ) p.d.f. at x. It is easy to find a_1 and a_2 from the Normal tables. For example, if $p = 0\cdot95$,

$$a_1 = \mu - 1\cdot96\sigma, \qquad a_2 = \mu + 1\cdot96\sigma.$$

When μ and σ are not known the best we can do, based on a sample $x_1, x_2, \ldots, x_n$, is to replace $\mu \pm 1\cdot96\sigma$ by, say $\bar{x} \pm \lambda s$, where, as usual, $\bar{x} = \sum_1^n x_r/n$

and $(n-1)s^2 = \sum_1^n (x_r - \bar{x})^2$; and λ is a constant to be determined. Of course we could not then assert that

$$\int_{b_1}^{b_2} f(x; \mu, \sigma)\, dx = p \qquad\qquad (4.12.2)$$

since b_1 and b_2 are statistics. They are realizations of random variables $B_1 = \bar{X} - \lambda S$, $B_2 = \bar{X} + \lambda S$, where $\bar{X}$ and S are the random variables induced by $\bar{x}$ and s. The assertion (4.12.2) must therefore be regarded as a realization of the proposition

$$\int_{\bar{X}-\lambda S}^{\bar{X}+\lambda S} f(x; \mu, \sigma)\, dx = p.$$

Tolerance interval theory is concerned with the related proposition

$$\int_{\bar{X}-\lambda S}^{\bar{X}+\lambda S} f(x; \mu, \sigma)\, dx \geq p.$$

We cannot guarantee the truth of this proposition, but we can require it to have a specified probability, β say:

$$P\left\{ \int_{\bar{X}-\lambda S}^{\bar{X}+\lambda S} f(x; \mu, \sigma)\, dx \geq p \right\} = \beta. \qquad\qquad (4.12.3)$$

If we can find a value of λ that satisfies (4.12.3), the interval $(\bar{x} - \lambda s, \bar{x} + \lambda s)$ is a $100p\%$ tolerance interval, at probability level β.

Such values of λ can in fact be found. A table of values of λ is available for various sample sizes, values of β and values of p.

4.13.   LIKELIHOOD INTERVALS

4.13.1.   Likelihood

(a) *Likelihood and log-likelihood: definitions and examples*

An example of a likelihood function was given in section 3.3.4, together with an interpretation which suggested an intuitive justification for the maximum-likelihood method of finding an estimate of an unknown parameter [see also § 4.10.1]. The generally accepted objective justification for this method is that the sampling distribution of the maximum-likelihood estimate has statistically desirable properties.

It is possible, howev⌐., to develop a calculus of likelihood estimates and their accuracies by methods that do not involve the concept of sampling distributions. The present section provides a brief introduction to this viewpoint. In the present section we shall be concerned with the behaviour of the likelihood function $l(\theta; x_1, x_2, \ldots, x_n)$ *as a function of* θ, taking the data $(x_1, x_2, \ldots, x_n)$ as fixed. The viewpoint to be adopted is that the concept of *probability* is appropriate to a situation where observations are going to be made and we are interested in the probabilities of various possible sets of values that might be observed, taking the parameter θ as fixed (even if unknown); while the concept of *likelihood* is appropriate to a situation where data have already been obtained, and all conceivable values of θ are to be considered in the light of the data.

We might emphasize at this stage that, although likelihood values for particular data sets are numerically equal (or proportional) to the corresponding probability (or probability density) values, *likelihoods are not probabilities* and have quite different properties.

It is convenient to introduce here a more general definition of likelihood than that given in (4.10.1), as follows:

DEFINITION 4.13.1.   *Likelihood.*   Suppose we have a sample of observations $(x_1, x_2, \ldots, x_n)$, where, for $r = 1, 2, \ldots, n$, x_r is a realization of the random variable X_r. Let

$g(x_1, x_2, \ldots, x_n; \theta)$

$$= \begin{cases} P(X_1 = x_1, X_2 = x_2, \ldots, X_n = x_n), & X \text{ discrete} \\ \text{joint p.d.f. of } (X_1, X_2, \ldots, X_n) \text{ at } (x_1, x_2, \ldots, x_n), & X \text{ continuous.} \end{cases}$$

(4.13.1)

Here θ denotes the (scalar) parameter of the joint distribution of the X_r, if that distribution involves only one parameter; in the multiparameter case, θ is a vector. The likelihood function of θ on these data is defined as

$$l(\theta) = l(\theta; x_1, \ldots, x_n) = ag(x_1, x_2, \ldots, x_n; \theta), \qquad \theta \in \Omega, \quad (4.13.2)$$

where Ω denotes the parameter space, that is set of possible values of θ, and $a = a(x_1, x_2, \ldots, x_n)$ is an arbitrary constant (which may depend on the observations x_r). Here the x_r have fixed values, so that the likelihood is a function of θ, not of $x_1, x_2, \ldots, x_n$.

[No significance attaches to the absolute values of a likelihood. We shall be concerned only with *comparisons* of the values of the likelihood function at various values of θ, these comparisons being made in terms of ratios: thus $l(\theta_1)$ is compared with $l(\theta_2)$ in terms of the ratio $l(\theta_1)/l(\theta_2)$, which is of course independent of the arbitrary multiplier a in (4.13.2). In practice (4.13.2) is often replaced by the equivalent

$$l(\theta) \propto g(x_1, x_2, \ldots, x_n; \theta). \tag{4.13.3}$$

If the x_r are i.i.d. [see § 1.4.2(i)], the likelihood function takes the simpler form

$$l(\theta) = l(\theta; x_1, x_2, \ldots, x_n) \propto \prod_1^n f(x_i; \theta) \tag{4.13.4}$$

where $f(x, \theta)$ is the p.d.f. of X at x (including the case where X is discrete, when $f(x; \theta) = P(X = x)$.)

The logarithm of a likelihood function is often more amenable than the likelihood function itself and so one frequently works with the *log-likelihood* function $\log l(\theta)$. Note that, if $l(\theta)$ has a maximum, $\log l(\theta)$ also has a maximum at the same θ-value.

EXAMPLE 4.13.1.   *The Poisson likelihood.*   Suppose X is Poisson (θ), so that

$$P(X = x) = e^{-\theta}\theta^x/x!, \qquad x = 0, 1, \ldots \qquad (\theta > 0).$$

On a sample $(x_1, x_2, \ldots, x_n)$, the likelihood function is

$$l(\theta) = l(\theta; x_1, x_2, \ldots, x_n) \propto \prod_{y=1}^{n} e^{-\theta}\theta^{x_j}/x_j!$$

$$= e^{-n\theta}\theta^{n\bar{x}}/\prod x_j! \qquad \left(n\bar{x} = \sum_{1}^{n} x_j\right)$$

$$\propto e^{-n\theta}\theta^{n\bar{x}}, \qquad \theta > 0.$$

The graph of $l(\theta)$ is illustrated in Figure 4.13.1. Note that, although X is discrete, $l(\theta)$ is a continuous function of θ.

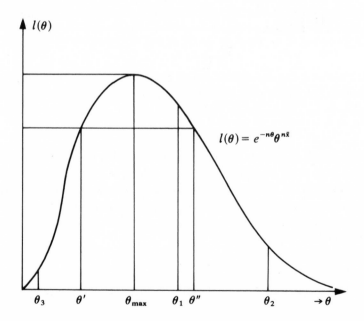

Figure 4.13.1: The graph of a likelihood function, $l(\theta)$. θ_{max} is the value of θ which maximizes $l(\theta)$, θ', θ'' are a pair of values having equal likelihoods.

The likelihood function shown in Figure 4.13.1 has a unique maximum. This occurs at θ_{max}, where

$$dl(\theta)/d\theta]_{\theta=\theta_{max}} = 0 = [d \log l(\theta)/d\theta]_{\theta=\theta_{max}}$$

i.e. where

$$(d/d\theta)(-n\theta + n\bar{x} \log \theta) = 0,$$

so that

$$\theta_{max} = \bar{x}.$$

The maximum value of $l(\theta)$ is proportional to $e^{-n\bar{x}}(\bar{x})^{n\bar{x}}$, and the likelihood of any other value of θ may be compared with this by examining the ratio

$$l(\theta)/l(\theta_{max}) = e^{-n(\theta-\bar{x})}(\theta/\bar{x})^{n\bar{x}}.$$

Invariance.   Suppose we are concerned with the Normal distribution with zero expectation and with variance σ^2, or, equivalently, with standard deviation σ. Ought we to work with the likelihood of σ or with the likelihood of σ^2? Does it matter? One of the attractive properties of likelihood is that it does *not* matter: that is, on a given sample, the likelihoods of σ and of σ^2 are the same (see Example 4.13.2). This is an illustration of the property of *invariance*.

EXAMPLE 4.13.2.   *The likelihood of a function of θ. Invariance.*   On data $(x_1, x_2, \ldots, x_n)$, the likelihood of the standard deviation of the Normal $(0, \sigma)$ distribution is

$$l_1(\sigma) = (2\pi)^{-n/2}\sigma^{-n} \exp\left\{\sum_1^n x_j^2/2(\sigma)^2\right\},$$

while that of the variance $v(=\sigma^2)$ is

$$l_2(v) = (2\pi)^{-n/2}(v)^{-n/2} \exp\left[-\left\{\sum_1^n x_j^2/2v\right\}\right].$$

Clearly these are identical.

In this example σ^2 is a one-to-one function of σ (there being no '$\sigma = \pm\sqrt{\sigma^2}$' ambiguity: σ is by definition positive, so σ is given uniquely by $\sigma = +\sqrt{\sigma^2}$). Clearly the argument works for any one-to-one function: if the likelihood of θ on given data is $l_1(\theta)$, and if $\phi = h(\theta)$, where $h(\cdot)$ is one-to-one, the likelihood of ϕ is

$$l_2(\phi) = l_1\{h^{-1}(\phi)\}.$$

(b) *Likelihood and sufficiency*

The whole of the information about θ that is contained in the sample $x_1, x_2, \ldots, x_n$ ($=x$, say) is expressed in the likelihood function $l(\theta; x)$, and all samples which provide the same likelihood function are equally informative. If, for every pair θ_1, θ_2 of putative values of θ, the ratio $l(\theta_1; x)/l(\theta_2; x)$ is a function of a set of statistics $\theta_1^*(x)$, $\theta_2^*(x), \ldots, \theta_s^*(x)$ only, and cannot be expressed in terms of a smaller number of such statistics, the set $\theta_1^*, \theta_2^*, \ldots, \theta_s^*$ is (minimally) sufficient for θ. [See § 3.4].

Thus if, for example, X has the Poisson (θ) distribution, the likelihood $l(\theta; \mathbf{x})$ on a random sample $(x_1, x_2, \ldots, x_n)$ of observations on X is $l(\theta) = e^{-\theta}\theta^{n\bar{x}}/\prod x_j!$, and

$$\frac{l(\theta_1)}{l(\theta_2)} = e^{-(\theta_1 - \theta_2)}\left(\frac{\theta_1}{\theta_2}\right)^{n\bar{x}},$$

whence $\bar{x}$ is sufficient for θ. In the case of a random sample $(x_1, x_2, \ldots, x_n)$ from the Normal (μ, σ) distribution, with θ now denoting the vector (μ, σ), we have

$$\frac{l(\mu_1, \sigma_1)}{l(\mu_2, \sigma_2)} = \left(\frac{\sigma_2}{\sigma_1}\right)\exp\left[-\frac{1}{2}\left\{\left(\frac{1}{\sigma_1^2} - \frac{1}{\sigma_2^2}\right)\sum x_i^2 - 2\left(\frac{\mu_1}{\sigma_1} - \frac{\mu_2}{\sigma_2}\right)\sum x_i + \left(\frac{\mu_1^2}{\sigma_1^2} - \frac{\mu_2^2}{\sigma_2^2}\right)\right\}\right],$$

whence the pair of statistics $\sum x_i, \sum x_i^2$ are a (minimal) jointly sufficient set for $\mathbf{\theta}$.

4.13.2.   Plausible Values and Likelihood Intervals

(a) *Equiplausible values of θ*

Two values θ', θ'' such that

$$l(\theta'; x_1, \ldots, x_n) = l(\theta''; x_1, \ldots, x_n)$$

are regarded, in the light of the data $(x_1, x_2, \ldots, x_n)$, as equally plausible (or equally implausible) approximations to the unknown true value of θ: for discrete data, the equality of the likelihoods is equivalent to the equation

$$P(X_1 = x_1, \ldots, x_n; \theta') = P(X_1 = x_1, \ldots, x_n = x_n; \theta'').$$

According to this the probability of obtaining the sample actually observed if the true value of θ were θ_1' is equal to the probability of obtaining it if the true value of θ were θ_1''. (See θ' and θ'' in Figure 4.13.1).

This justification for regarding θ' and θ'' as equiplausible can be extended, by a limiting argument, to apply also when the X_r are continuous.

(b) *One value of θ more plausible than another*

If

$$l(\theta_1; x_1, x_2, \ldots, x_n) > l(\theta_2; x_1, x_2, \ldots, x_n),$$

θ_1 is regarded, in the light of the data $(x_1, x_2, \ldots, x_n)$, as being more plausible than θ as an approximation to the unknown true value of θ, since the probability (or probability density) of the data would have a greater value if the true value of θ were θ_1 than if it were θ_2. Thus, in Figure 4.13.1, θ_1 is more plausible than θ_2, and θ_2 more plausible than θ_3. On this argument the most plausible value of all is θ_{max}. (This is the maximum-likelihood estimate of θ. See Chapter 6.)

(c) *Conventionally implausible values of θ: likelihood intervals*

The concepts introduced in the preceding paragraphs lead to the idea that a value of θ such that $l(\theta)/l(\theta_{max})$ is not much smaller than unity is not much less plausible than θ_{max}, while a value of θ such that $l(\theta)/l(\theta_{max})$ is very much less than unity is very much less plausible than θ_{max}. We might for example take as a conventional standard of implusibility the following:

any value of θ such that

$$l(\theta)/l(\theta_{max}) < 0 \cdot 10$$

is to be regarded as conventionally implausible at the 10% likelihood level. Similarly for the 12·5% level, etc. The interval (θ_l, θ_u) such that

$$l(\theta)/l(\theta_{max}) \geq 0 \cdot 10, \quad \theta_l \leq \theta \leq \theta_u$$

would then be regarded as an interval within which, at the 10% likelihood level, every value of θ was a plausible approximation to the unknown true value of θ. Formally, we make following definition.

DEFINITION 4.13.2. *Likelihood intervals (One-parameter case).* For data $(x_1, x_2, \ldots, x_n)$ whose joint sampling distribution depends on a single parameter θ, denote the likelihood function of θ by $l(\theta)$. If there are two values θ_l and θ_u such that

$$l(\theta)/l(\theta_{max}) \geq 0 \cdot 10$$

for

$$\theta_l \leq \theta \leq \theta_u,$$

the interval (θ_l, θ_u) is called a 10% likelihood interval for θ on the given sample. (Similarly for intervals at other percentage levels.)

If one is working with the log-likelihood it should be noted that $\log l(\theta)$ and $l(\theta)$ attain their maxima at the same value θ_{max}. In terms of the log likelihood, the end-points θ_l and θ_u of the 0·01 likelihood interval are of course the roots of the equation

$$\log l(\theta) = \log l(\theta_{max}) + \log (0 \cdot 01).$$

Note that, in the case of a discrete variable which admits of a continuum of possible parameter values, as for instance in the Poisson distribution, the calculation of the likelihood interval does not present the difficulties that arise in such cases with confidence intervals. This is illustrated in Example 4.13.3.

EXAMPLE 4.13.3.   *Likelihood interval for a Poisson parameter.*   In Example 4.13.1 the likelihood function was

$$l(\theta) \propto e^{-n\theta} \theta^{n\bar{x}}$$

with

$$\theta_{\max} = \bar{x}.$$

The endpoints θ_l, θ_u of the 10% likelihood interval are then the appropriate roots of the equation

$$e^{-n(\theta-\bar{x})}(\theta/\bar{x})^{n\bar{x}} = 0\cdot 10.$$

For example, with $n = 10$ and $\bar{x} = 2\cdot 4$ we have to solve the equation

$$e^{-10(\theta-2\cdot 4)}(\theta/2\cdot 4)^{24} = 0\cdot 10,$$

or, taking logs,

$$-10\theta + 24 + 24 \log \theta - 24 \log 2\cdot 4 = \log 0\cdot 10,$$

that is

$$10\theta - 24 \log \theta = 5\cdot 29.$$

The roots (obtained by numerical approximation) are approximately $1\cdot 5$ and $3\cdot 6$. Thus from this sample the *most* plausible value of θ is $2\cdot 4$, any value between $1\cdot 5$ and $3\cdot 6$ is conventionally regarded as plausible, while any value less than $1\cdot 5$ or greater than $3\cdot 6$ is regarded as implausible (at the 10% level).

Not all likelihood functions have a horizontal tangent at their maximum value, nor do they all allow of two values θ_l and θ_u satisying the requirements of Definition 4.13.2. Consider the following example:

EXAMPLE 4.13.4.   *Likelihood function for an extremity of a uniform distribution.*   Suppose that X is uniformly distributed on the interval $(0, \theta)$, so that the p.d.f. of X at x is

$$f(x; \theta) = \begin{cases} 1/\theta, & 0 \leq x \leq \theta \\ 0, & \text{otherwise.} \end{cases}$$

On a sample $(x_1, x_2, \ldots, x_n)$ the likelihood function is

$$l(\theta) = \begin{cases} 1/\theta^n, & 0 \leq x_1 \leq \theta, 0 \leq x_2 \leq \theta, \ldots, 0 \leq x_n \leq \theta, \\ 0, & \text{otherwise,} \end{cases}$$

i.e.

$$l(\theta) = \begin{cases} 1/\theta^n, & 0 \le x_{(n)} \le \theta \quad \text{(i.e. } \theta \ge x_{(n)}); \\ 0, & \text{otherwise} \end{cases}$$

where $x_{(n)}$ is the largest observation. The graph of this function is shown in Figure 4.13.2. In this example the likelihood achieves its maximum value l_{max} when $\theta = x_{(n)}$, a point at which the curve does not have a horizontal tangent.

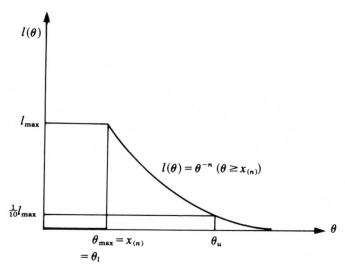

Figure 4.13.2: Likelihood function for the upper extremity of a uniform distribution.

In such a case we would set $\theta_l = x_{(n)}$ and define θ_u (at the 10% level) as the value of θ satisfying the equation

$$(x_{(n)}/\theta)^n = 0 \cdot 10$$

i.e.

$$\theta_u = 10^{1/n} x_{(n)}.$$

For example, if $x_{(n)} = 1 \cdot 8$ and $n = 17$, $\theta_u = 2 \cdot 06$.

The evaluation of likelihood intervals does not require the use of the possibly elaborate sampling distribution required by confidence intervals. The following example illustrates a case where the evaluation of a confidence interval would require considerable computational effort.

EXAMPLE 4.13.5.   *Poisson distribution with zero value missing.*   Suppose X has the truncated Poisson distribution with no zero class, so that the p.d.f. is

$$P(X = r) = \left(\frac{e^{-\theta}}{1 - e^{-\theta}} \right) \theta^r / r!, \qquad r = 1, 2, \dots$$

(see II, § 6.7). The likelihood function is proportional to

$$l_1(\theta) = \left(\frac{e^{-\theta}}{1 - e^{-\theta}}\right)^n \theta^s, \qquad \theta > 0, \tag{4.13.3}$$

where s is the sum of the n observed values of X. (This is a sufficient statistic for θ; see § 3.4.) A few minutes with a pocket calculator produces a table of values of the logarithm of (4.13.3) which can be used to sketch the graph of the log likelihood function. Table 4.13.1 and Figure 4.13.3 illustrate the case

θ	$-\log l_1(\theta)$	θ	$-\log l_1(\theta)$
0·9	11·78	1·8	8·80
1·0	10·83	1·9	9·09
1·1	10·09	2·0	9·37
1·2	9·54	2·1	9·71
1·3	9·14	2·2	10·11
1·4	8·88	2·3	10·57
1·5	8·73	2·4	11·08
1·6	8·69	2·5	11·63
1·7	8·74	2·6	12·24

Table 4.13.1

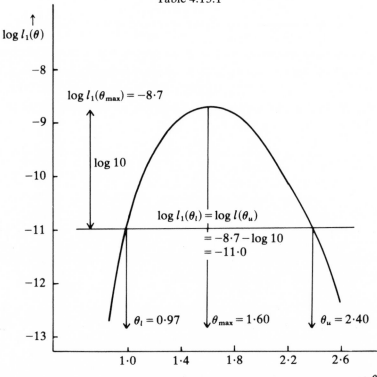

Figure 4.13.3: The log likelihood function $l_1(\theta)$ of Example 4.13.4.

where $n = 20$ and $s = 40$: it will be seen that $\theta_{max} = 1 \cdot 60$ and that the 10% likelihood interval for θ is $(0 \cdot 99, 2 \cdot 39)$.

(d) *Likelihood intervals for θ and for $g(\theta)$*

It follows from the invariance property illustrated in Example 4.13.2 that, if (θ_l, θ_u) is the $100p\%$ likelihood interval for θ on given data, the $100p\%$ likelihood interval for $\phi = g(\theta)$ on the same data is (ϕ_l, ϕ_u) where $\phi_l = g(\theta_l)$ and $\phi_u = g(\theta_u)$. (Here $g(\theta)$ represents any one-to-one transform of θ). Thus, if in Example 4.13.2, the 10% likelihood interval for the standard deviation σ is $(1.2, 2.2)$, the corresponding interval for the variance σ^2 is $(1.44, 4.84)$.

4.13.3. The Two-parameter Case

The definition of likelihood given in Definition 4.13.1 continues to apply when there are several parameters. The situation will be adequately dealt with if we discuss the case of two parameters, θ_1 and θ_2 say. The symbol 'θ' of Definition 4.13.1 is then to be interpreted as the vector (θ_1, θ_2).

EXAMPLE 4.13.6. *The two-parameter Normal distribution.* Let X be Normal with expectation μ and variance v. The likelihood $l(\mu, v)$ of (μ, v) on data $x_1, x_2, \ldots, x_n$ is proportional to

$$v^{-n/2} \exp\left\{-\sum_1^n (x_j - \mu)^2 / 2v\right\}$$

$$= v^{-n/2} \exp\left[-\frac{1}{2v}\left\{\sum (x_j - \bar{x})^2 + n(\bar{x} - \mu)^2\right\}\right]$$

$$= v^{-n/2} \exp -\frac{1}{2v}\{a + n(\mu - \bar{x})^2\}$$

where

$$a = \sum (x_j - \bar{x})^2 = (n-1)s^2$$

and

$$\bar{x} = \sum_1^n x_j / n.$$

Take the case where $n = 10$, $\bar{x} = 10$, $a = 20$. Then we may take the log likelihood to be

$$\log l(\mu, v) = -\tfrac{1}{2}n \log v - \frac{a}{2v} - \frac{n}{2v}(\mu - \bar{x})^2$$

$$= -5 \log v - \frac{10}{v} - \frac{5(\mu - 10)^2}{v}.$$

Figure 6.2.4 shows a sketch of the contours of this surface. Figure 6.2.5 shows a perspective view of the surface.

The analogue for this two-parameter case of the 10% likelihood interval of section 4.13.2(c) is the 10% likelihood contour. This is labelled '$0 \cdot 1 l_{max}$' in Figure 4.13.4. The log-likelihood level of this contour is of course $\log l_{max} - \log_e 10$. On the conventional 10% level any pair of values (μ, v) inside this contour are regarded as plausible, any pair outside it as implausible.

Separate 10% likelihood intervals (μ_l, μ_u) for μ and (v_l, v_u) for are as indicated in sketch in Figure 4.13.4.

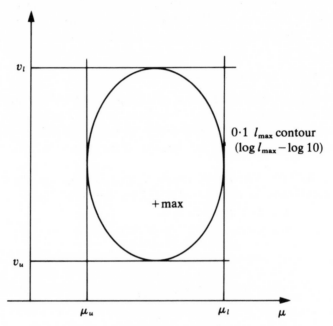

Figure 4.13.4: Separate (one-dimensional) confidence intervals for μ and v [see Example 4.13.6].

4.14. BAYESIAN INTERVALS

From the Bayesian point of view (see Chapter 15) a person's assessment of the probability of a proposition is a measure of his degree of rational belief in the validity of that proposition. An allocation of probabilities to each interval in a partition of the set Ω of possible values of a parameter is equivalent to regarding the parameter as a random variable, in the sense that probability statements can be made about it, most conveniently in terms of its probability distribution. Such an allocation made in the light of prior information and judgement, before the experiment under discussion is performed, leads to a 'prior' distribution, with p.d.f. $g_1(\theta)$ say. Data $(x_1, x_2, \ldots, x_n)$ yielded by the experiment then leads to the likelihood function $l(\theta; x_1, x_2, \ldots, x_n)$, defined as in section 4.13.1, and the observer's updated beliefs about θ are then expressed in terms of the 'posterior' p.d.f. $g_2(\theta|x_1, x_2, \ldots, x_n)$ which is the conditional distribution [see II, § 6.5] of θ given the data. By Bayes' Theorem (see Chapter 15) we are led to the result that

$$g_2(\theta|x_1, x_2, \ldots, x_n) = Ag_1(\theta)l(\theta; x_1, x_2, \ldots, x_n), \quad \theta \in \Omega$$

where A represents the appropriate normalizing constant.

We therefore have a probability distribution of possible values of θ. We may choose as our best estimate of 'the' value of θ the modal value θ_m (the value which maximizes the posterior p.d.f.), or the expected value $\int \theta g_2(\theta|x_1, x_2, \ldots, x_n) \, d\theta$, or a value which minimizes an appropriate penalty function, etc., as explained in Chapter 15. The uncertainty in our estimate is then appropriately expressed in terms of a probability interval (see § 4.1.3).

E.H.L.

4.15. FURTHER READING AND REFERENCES

The texts cited in section 3.6 are again recommended here. For uses of the likelihood function in inference see especially Kalbfleisch (1979), and, for the Bayesian approach, see Lindley (1965)—both in Bibliography C.

References for this chapter are given below.

Blyth, C. R. and Hutchinson, D. W. (1960). Tables of Neyman-shortest Unbiased Confidence Intervals for the Binomial Parameter, *Biometrika* **47**, 381.

Blyth, C. R. and Hutchinson, D. W. (1961). Tables of Neyman-shortest Unbiased Confidence Intervals for the Poisson Parameter, *Biometrika* **48**, 191.

Clopper, C. J. and Pearson, E. S. (1934). The Use of Confidence or Fiducial Limits Illustrated in the Case of a Binomial, *Biometrika* **26**, 404.

Crow, E. L. and Gardner, R. S. (1959). Table of Confidence Limits for the Expectation of a Poisson Variables, *Biometrika* **46**, 441.

Eudey, M. W. (1949). On the Treatment of Discontinuous Random Variables, *Technical Report*, No. 13, Statistical Laboratory, University of California, Berkeley.

Garwood, F. (1936). Fiducial Limits for the Poisson Distribution, *Biometrika* **28**, 437.

Crow, E. L. (1956). Table for Determining Confidence Intervals for a Proposition in Binomial Sampling, *Biometrika* **43**, 423.

Pearson, E. S. (1950). On Questions Raised by the Combination of Tests Based on Discontinuous Distributions, *Biometrika* **37**, 383.

Scheffé, H. (1953). A Method for Judging all Contrasts in the Analysis of Variance, *Biometrika* **40**, 87.

Scheffé, H. (1970). Multiple Testing Versus Multiple Estimation. Improper Confidence Sets . . . , *Anals. of Math Statistics* **41**, 1.

Stevens, W. L. (1950). Fiducial Limits of the Parameter of a Discontinuous Random Variable, *Biometrika* **37**, 117.

Tocher, K. D. (1950). Extension of Neyman–Pearson Theory of Testing Hypothesis to Discontinuous Variables, *Biometrika* **37**, 130.

CHAPTER 5

Statistical Tests

5.1. WHAT IS MEANT BY A SIGNIFICANCE TEST?

Tests of significance (significance tests, or tests of hypotheses, or, simply *tests*) are possibly the most primitive but certainly the most widely used statistical procedures. The relevant literature is vast, and anything approaching a complete survey in a book of this nature would be impracticable. We content ourselves with an introduction to the main concepts, some illustrative examples, and a treatment of the most commonly used tests. Other tests of the same general kind are described in later chapters as the appropriate topics occur: for example the tests usually employed in connection with analyses of variance are dealt with in Chapters 8 and 10. Sequential tests, distribution-free tests and tests of goodness of fit are mentioned only briefly in the present chapter since later chapters are devoted specifically to these topics.

A significance test is a procedure designed to help the statistician to find a sensible answer to a question such as the following:

In two samples of steel, one produced by Method A and the other by Method B, the mean tensile strengths are unequal. Does this indicate that the two methods really produce steels of unequal strengths, or are the observed differences simply attributable to sampling fluctuations?

In this example, the question at issue is whether one variety of steel is superior to another. Similarly we may be interested in the question whether one alleged antidote to influenza is superior to another, whether a non-smoking personal history is superior to a cigarette-smoking history as a precursor to freedom from lung cancer, whether one fertilizer is superior to another in promoting heavy crops of vegetables, and so on.

The following section is devoted to a discussion of simple tests of this general kind.

5.2. INTRODUCTION TO TESTS INVOLVING A SIMPLE NULL HYPOTHESIS IN DISCRETE DISTRIBUTIONS

In this section we introduce the main concepts involved in the simpler tests involving discrete distributions. (Here the author wishes to acknowledge his indebtedness to the excellent treatment of the topic in Kalbfleisch (1979)—(see Bibliography C).)

Continuous distributions are discussed in sections 5.2.5 and 5.8.

5.2.1. A Two-sided (Binomial) Test: Ingredients, Procedure and Interpretation

The following example, which describes a particular simple test, illustrates the general approach and the main concepts employed. In particular it introduces the key concepts of the *significance set* and the *significance level*. Consider an experiment to investigate whether, in a given English city which has a large population of West Indian origin, the proportion of male births amongst children born to West Indian families differs from the British average of 52%. The available data consist of a list, in date order, of all single births in West Indian families during a given year.

(a) *Probability model*

The first ingredient in formulating the test is an appropriate probability model. We shall adopt the simplest possible one, namely one in which the births are regarded as mutually independent Bernoulli trials [see II, § 5.2.1], in each of which the probability of a boy is the same: p, say. For the purposes of this test, this model will be accepted as part of the underlying framework, which is not in itself directly questioned in the test. (Aspects of it which might be open to question, such as the possibility that boy babies are more probable to one age-group of mothers than another, might themselves form the subject of a separate test, but the model is not in question as far as the present test is concerned.)

To formalize the model, let x_r denote the sex of the rth child born in the sequence, with $x_r = 1$ for a boy, $x_r = 0$ for a girl, so that $\sum_1^n x_r$ denotes the total number of boys in the sample. Then, for $r = 1, 2, \ldots, n$, x_r is a realization of the random variable X_r which has the Bernoulli distribution [see II, § 5.2.1]

$$P(X_r = x_r) = p^{x_r}(1-p)^{1-x_r}, \qquad x_r = 0, 1,$$

whence the joint distribution of the data is

$$P(X_r = x_r, r = 1, 2, \ldots, n) = \prod_{r=1}^{n} p^{x_r}(1-p)^{1-x_r}$$

$$= p^{\Sigma x_r}(1-p)^{n-\Sigma x_r}.$$

(b) *Condensing the data; the test statistic*

It is not convenient to have to deal simultaneously with n separate pieces of information. They must be combined into a single statistic, and we shall replace the original probability model of (a) above by a condensed version, namely the sampling distribution of this statistic. The most effective condensation is in terms of a statistic that is *sufficient* for the parameter (p) of interest, since, with such a statistic, no information is lost in the condensation [see § 3.4.1]. In our case the appropriate sufficient statistic is $b_0 = \sum x_r$, the observed total number of boys. Its sampling distribution, namely the distribution of the induced random variable B (of which b_0 is a realization,) is [see II, § 5.2.2]

$$P(B = b) = \binom{n}{b} p^b (1-p)^{n-b}, \qquad b = 0, 1, \ldots, n. \qquad (5.2.1)$$

(c) *Null hypothesis, null distribution*

The question to be answered is this: is the value of p different from the 'British' value of $0\cdot52$? It is preferable to turn the question around and to frame it thus: are the data consistent with the assumption that $p = 0\cdot52$? To answer this question we assume as a working hypothesis that p *does* equal $0\cdot52$. This assumption is called the null hypothesis, denoted by

$$H\colon p = 0\cdot52. \qquad (5.2.2)$$

The joint distribution of the X_r implied by this assumption is obtained by inserting this value of p in (5.2.1), giving the *null distribution* of the X_r, or the 'distribution of the X_r on H', namely

$$P(X_r = x_r, r = 1, 2, \ldots, n | H) = (0\cdot52)^{\sum x_r} (0\cdot48)^{n - \sum x_r}, \qquad \sum x_r = 0, 1, \ldots, n.$$

The *null distribution of the test statistic* is given by the form taken by (5.2.1) when p has the null value $0\cdot52$, namely

$$P(B = b | H) = \binom{n}{b} (0\cdot52)^b (0\cdot48)^{n-b}, \qquad b = 0, 1, \ldots, n \qquad (5.2.3)$$

with $n = 20$ in our case.

The idea underlying the test is this: if the null hypothesis attributes to the data a sufficiently high degree of plausibility (in some sense to be defined), it is regarded as being consistent with the data; if not, it is inconsistent with the data—that is the data differ *significantly* from the hypothesis. What is and what is not a 'sufficiently high degree of plausibility' is discussed below (see paragraphs (e) and (f).)

In the present example, the null hypothesis is *simple*: it completely specifies the value of the parameter. (There is only one parameter involved in this example. In more elaborate tests there might be several parameters [see § 8.3.3]. A *simple* null hypothesis would specify the values of all of them.)

[An example of a test in which the null hypothesis is '*composite*' (i.e. not simple) is the following. Among n_1 births to mothers aged 20–25, b_1 were boys, while n_2 births to mothers aged 30–35 produced b_2 boys. It is required to test whether the proportions b_1/n_1 and b_2/n_2 are significantly different. The null hypothesis in this case would be that the probability of a boy baby is the same in each of the two groups. The value of this common probability is however not specified by the null hypothesis, which is therefore not 'simple'. This test and similar ones are discussed in section 5.4.1.]

(d) *Alternative hypothesis*

The object of the test is to see whether the data may reasonably be regarded as consistent with the null hypothesis, or whether, instead, they are so discordant with it as to discredit it. It is important to know, in general terms, what kinds of discrepancies are regarded as relevant. In the present example the so-called alternative hypothesis against which H may be thought to be competing is

$$H': p \neq 0.52.$$

Thus H would be discredited by data showing a proportion of boys much larger than 0.52 or much smaller than 0.52. For this reason the test is described as *two-sided*. (An example of a *one-sided* test is given below in section 5.2.3.)

(e) *Compatibility of sample with H*

The original question concerning the compatibility of the n ($= 20$) observations with the null hypothesis (5.2.2) may now be replaced by the equivalent question of the compatibility of the observed value b_0 ($=5$) with the null distribution (5.2.3). That distribution is unimodal, with a region of high probability near the centre and regions of low probability in the tails. If the observed value of B lies in the high-probability region, which is to be expected if H is in fact correct, the conclusion to be drawn is that the sample offers no evidence against H: it is consistent with H. If however the observed value b_0 is sufficiently extreme, so extreme as to be improbable on H, that fact is to be regarded as evidence against H.

The argument here resembles the familiar contrapositive argument of Aristotelian logic, according to which, if $\mathscr{A}$ implies $\mathscr{B}$, then not-$\mathscr{B}$ implies not-$\mathscr{A}$, for any propositions $\mathscr{A}$ and $\mathscr{B}$. The statistical version of this is that, if $\mathscr{B}$ is a probable consequence of $\mathscr{A}$, then not-$\mathscr{A}$ is a probable consequence of not-$\mathscr{B}$. Take $\mathscr{A}$ to mean 'H is correct' and $\mathscr{B}$ to mean 'the observed value of b will probably be near the mode of the null distribution'. Then the 'statistical contrapositive' says that, if the observed value of b is not near the mode of the null distribution, H is probably false.

The question is, how extreme a value must b have in order to provide strong evidence against H?. We note from the nature of the original question that, as was remarked earlier, evidence against H would be provided by either an extremely large value of b_0 (near to n) or an extremely small value (near to zero): the test has to be 'two-sided'. Precisely how large or small, however, is not obvious.

(f) *Significance set, significance level (significance probability). Critical region*

There are a number of appealing approaches to the question of how significant is a given value of b_0 as evidence against H. A first attempt might be to regard b_0 as significant in this sense if $P(B = b_0|H)$ [see § 1.4.2 (vi) for this notation] is small. This however suffers from the difficulty that, if the sample is large enough, $P(B = b|H)$ will necessarily be small, whatever the value of b. It is therefore necessary to replace the simple point probability $P(B = b|H)$ by an intuitively equivalent measure that is standardized in some way so as to be free of this defect.

There are various possible ways of doing this. The usual method is to base one's judgement on the probability allotted by H to a particular *set* of possible values of the test statistic B, the set being chosen in such a way that this probability is small if H is correct. The set in question consists of all values that are in a prescribed sense (see below) at least as 'extreme' as the value b_0 actually observed. This set is called the *significance set* $G(b_0)$, and the criterion used to measure the significance to be attributed to b_0 as evidence against H is the *significance level SL*—or *SL*(b_0)—defined as the probability content of the significance set, *computed on the assumption that the null hypothesis H is correct*, viz.:

$$SL(b_0) = P\{B \in G(b_0)|H\}. \tag{5.2.4}$$

The significance level, as here defined, is also known as the *significance probability* of the samples, to distinguish it from a related concept in Neyman and Pearson's decision-theoretic approach to the testing of hypotheses outlined in section 5.12.

The general idea, which we shall develop, is to regard the sample as being consistent with the null hypothesis H if the significance probability is large in a defined sense, and to regard the sample as providing evidence against H if the probability is small [see § 5.2.2].

Critical region

It must be said here that practising statisticians frequently do not evaluate the significance set and the corresponding *significance level of their data* but, instead, specify a *conventional* significance set which, if it had actually been observed, would have had a sufficiently low significance level α (say 0·02, for example) to provide a conventionally high strength $(1 - \alpha)$ of evidence [see § 5.2.1(h)]) against the null hypothesis. This conventional significance set is called a *critical region of size* α. Instead of stating the actual significance level of their sample, adherents of this practice state whether or not their test statistic falls into the critical region or not. If it does, the sample is said to be *significant* at the α level and the null hypothesis is said to be *rejected* at the α level; if it does not, the sample is said to be 'not significant at the α level'.

This approach is not followed in the remainder of the present chapter, but is outlined in more detail in section 5.12.

What values are at least as extreme as b_0?

The definition of the significance set becomes meaningful only when the phrase 'at least as extreme as' is defined. To see that this is not quite trivial, suppose that b_0 is smaller than expected in H. For instance, with $p = 0.52$ and $n = 20$, the expected value would be 10.4, and an observed value of $b = 5$ would be smaller than expected. Possible values as extreme as or more extreme than 5 (in the sense of being smaller, that is 'in the lower tail') are the values 5, 4, 3, 2, 1, 0; but what is the corresponding set in the upper tail? How, that is to say, can we define an observation b', greater than the expected value 10.4, that is at least as extreme (in the sense of a large observation) as b is (in the sense of a small observation)? The following methods are available:

(f_1) *Distance ordering*

In this approach a 'large' value b' $(> E(B|H))$ and a 'small' value b $(< E(B|H))$ are equally significant if they are *equidistant* from $E(B|H)$; values that are further than either from $E(B|H)$ are of course more significant. (Here $E(B|H)$ denotes the expectation of B on H; that is, the expected value of the distribution (5.2.3).) This solves the problem of the two tails by defining the significance set generated by an observation b_0 as

$$G(b_0) = \{b : |b - E(B|H)| \geq |b_0 - E(B|H)|\},$$

so that the significance level of b_0 is

$$SL(b_0) = P\{B \in G(b_0)|H\}$$

$$= P[\{|B - E(B|H)| \geq |b_0 - E(B|H)|\}|H]. \qquad (5.2.5)$$

The points in the distribution of B to be included in this computation are shown in Figure 5.2.1.

Thus, in the case where there are 5 boys among 20 births, and where the null distribution of B is binomial $(20, 0.52)$, we have $E(B|H) = 10.4$, and the significance level of the data is

$$SL = P[\{|B - 10.4| \geq |5 - 10.4|\}|H]$$

$$= P[\{|B - 10.4| \geq 5.4\}|H]$$

$$= P\{B \leq 5|H\} + P\{B \geq 15.8|H\}$$

$$= P\{B \leq 5|H\} + P\{B \geq 16|H\}. \qquad \text{(see Figure 5.2.1)}$$

From tables of the Binomial distribution [see Appendix (T1)] we find $SL = 0.023$.

(f_2) *Probability ordering.*

We suppose initially that b_0, the observed value of B_1 is 'small' in the sense that $b_0 < E(B|H.)$ (Thus, in the example under consideration (see Figure 5.2.1) we have $b_0 = 5$ and $E(B|H) = 10.4$, so $b_0 = 5$ is 'small'.) The *probability ordering*

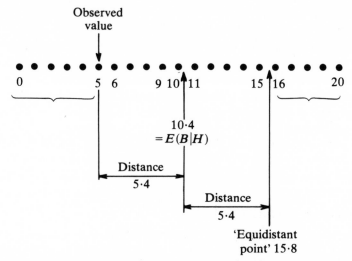

Figure 5.2.1: On H, B is Binomial $(20, 0\cdot52)$, so that $E(B|H) = 10\cdot4$. The observed value, 5, is at a "distance" of $10\cdot4 - 5 = 5\cdot4$ below $E(B|H)$. The "equidistant" point above $E(B|H)$ is $10\cdot4 + 5\cdot4 = 15\cdot8$. The nearest possible realization of B that is at least as extreme as is 5 is 16. The set of points at least as extreme as the observed value is the set

$$(0, 1, \ldots, 5) \cup (16, 17, \ldots, 20)$$

approach seeks to pair with b_0 an equiprobable 'large' value b_0', where 'large' means that $b_0' > E(B|H)$, and 'equiprobable' means

$$P(B = b_0'|H) = P(B = b_0|H).$$

It may well happen however that there is no possible value b_0' that is exactly equiprobable. Thus, in our example with $b_0 = 5$, where the distribution of B on H is Binomial $(20, 0\cdot52)$, we have the following situation:

b_0 (in lower tail)	b_0' (in upper tail)
$P(B = b_0|H) = 0\cdot00975$	$P(B = 15|H) = 0\cdot02171 \ (>0\cdot00975)$ $P(B = 16|H) = 0\cdot00735 \ (<0\cdot00975)$

Thus 15 is two small a value to be equiprobable with b_0, 16 is too large (see Figure 5.2.2.) In such a case we modify the requirement that b_0' be equiprobable (on H) with b_0 to the requirement that b_0' is the smallest integer for which

$$P(B = b_0'|H) < P(B = b_0|H).$$

In the example this leads to $b_0' = 16$.

The significance set generated by the observed b_0 is

$$G(b_0) = \{b : b \le b_0 \text{ or } b \ge b_0'\}$$

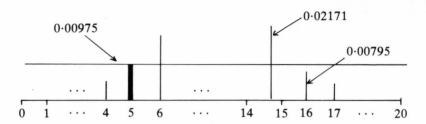

Figure 5.2.2: Part of the p.d.f. of the Binomial $(20, 0\cdot52)$ distribution, for which $P(B = 5) = 0\cdot00975$.

and the significance level of b_0 is

$$SL = P(B \le b_0|H) + P(B \ge b_0'|H)$$

In our example, with 5 boys among 20 births, the significance level is

$$P(B \le 5|H) + P(B \ge 16|H) \quad = 0\cdot023.$$

[The procedure we have described applies when the data value b_0 is 'small'. Obvious modifications apply if b_0 is 'large'.]

In this example the SL value is exactly the same whether one uses the 'distance' definition or the 'smallest probabilities' definition. The two definitions will in fact always give equal results when the null distribution is symmetrical, and nearly equal results when the null distribution is nearly symmetrical, but will differ when the null distribution is markedly skew. Probability ordering is then preferable.

(f_3) *Likelihood-ratio ordering*

For the test statistic b_0, which is a realization of the Binomial (n, p) random variable B, the *likelihood* of p [see § 4.13.1] is proportional to

$$l(p) = p^{b_0}(1 - p)^{n - b_0}.$$

In our case, with $b_0 = 5$ and $n = 20$, this is

$$l(p) = p^5(1 - p)^{15}$$

and, on the hypothesis H that $p = p_H = 0\cdot52$, this is

$$l_H = p_H^5(1 - p_H)^{15} = (0\cdot52)^5(0\cdot48)^{15}.$$

The greatest value of $l(p)$ when p is allowed to vary over its permitted range $(0 \le p \le 1)$, while b_0 is kept fixed, is attained when p takes its 'maximum likelihood' value $\hat{p} = b_0/n$ $(=5/20 = 0\cdot25$ in our example). This maximum value is

$$l_{max} = (0\cdot25)^5(0\cdot75)^{15}.$$

The ratio

$$\lambda = l_H / l_{\max} \tag{5.2.8}$$

is called the *likelihood ratio statistic*. In our example its value is

$$\lambda = (0\cdot52/0\cdot25)^5 (0\cdot48/0\cdot75)^{15}$$

$$= 0\cdot048.$$

For an arbitrary value b of B the value of the likelihood ratio statistic would be

$$\lambda = \lambda(b) \quad \text{say}$$

$$= p_H^b (1 - p_H)^{n-b} \Big/ \left(\frac{b}{n}\right)^b \left(\frac{n-b}{n}\right)^{n-b}$$

$$= (np_H/b)^b \{(n - np_H)/(n - b)\}^{n-b}$$

so that, in our case, with $n = 20$ and $p_H = 0\cdot52$, we have

$$\lambda(b) = \left(\frac{10\cdot4}{b}\right)^b \left(\frac{9\cdot6}{20-b}\right)^{20-b}.$$

On likelihood-ratio ordering, b is 'more extreme' than b_0 if

$$\lambda(b) < \lambda(b_0),$$

whence the significance set is

$$G(b_0) = \{b : \lambda(b) \le \lambda(b_0)\}$$

and the significance level is

$$SL(b_0) = P\{B \in G(b_0)\}.$$

(This is a formalization of the idea that one expects λ to be large, that is, close to unity, when the hypothesis H is correct, and small when it is false.)

For convenience of computation it is usual to replace $\lambda(b)$ by

$$d(b) = -2 \ln \lambda(b)$$

in which case the significance set is

$$G(b_0) = \{b : d(b) \ge d(b_0)\}.$$

In our example the possible values of b and the corresponding values of $d(b)$ are as follows, with

$$d(b) = 2b \ln (b/np_H) + 2(n - b) \ln \{n - b)/(n - np_H)\}$$

$$= 2b \ln (b/10\cdot4) + 2(20 - b) \ln \left(\frac{20-b}{9\cdot6}\right).$$

b	$d(b)$	b	$d(b)$	b	$d(b)$
0	29	7	2·3	14	2·7
1	21	8	1·2	15	4·5
2	16	9	0·4	16	6·8
3	12	10	0·03	17	10
4	8·7	11	0·07	18	14
5	6·1	12	0·5	19	18
6	4·0	13	1·4	20	26

The observed value b_0 is 5, with $d(5) = 6·1$. The smaller values 4, 3, 2, 1, 0 have d-values exceeding 6·1, as do the values 16, 17, 18, 19, 20. Thus the significance set is

$$G(5) = \{0, 1, 2, 3, 4, 5, 16, 17, 18, 19, 20\}$$

$$= \{b : b \le 5 \text{ or } b \ge 16\}.$$

The significance level is

$$SL(5) = P\{B \le 5 | H\} + P\{B \ge 16 | H\}.$$

The significance set coincides with that obtained with probability ordering and the test therefore gives the same significance level, i.e. 0·023. This is typical for simple tests of this sort. The likelihood-ratio method really comes into its own in more elaborate situations, especially those involving more than one parameter. (See § 5.5).

(g) *Interpretation of the significance level. Strength of evidence*

In our numerical example (5 boys among 20 births) we found the SL to be 0·023. How is this to be assessed in making a judgement as to whether or not the data are consistent with the null hypothesis h of (5.2.2), that the proportion of boys among all births in the population sampled is equal to the 'British' figure of 0·52? The answer must be, disappointingly perhaps, that this is largely a matter of convention. As a guide to the intuition however we may use the following arguments. [see § 5.3.] If the null hypothesis H is correct, the observed value of the test statistic is unlikely to be very different from its expected value. It might of course happen by chance, even when H is correct, that the test statistic in a particular case *is* very different from expectation, in which case the significance level will be small. The probability of this, however, is small. In fact, for any α, the probability of obtaining a significance level as small as α is exactly equal to α. More precisely [see § 5.3], when H is correct,

$$P(SL \le \alpha) = \alpha. \tag{5.2.9}$$

Thus for example the chance that the *SL* will be no larger than 0·001 when *H* is correct is only one in a thousand. This is a very small probability. It seems reasonable therefore to regard a significance level of 0·001 as providing strong evidence against *H*. For this kind of reason it has become accepted practice to interpret significance levels in accordance with Table 5.2.1 given below. In the light of this it will be seen that the significance level of 2·3% obtained in our (fictitious) numerical example of 5 boys out of 20 is low enough to cast substantial doubt on the null hypothesis.

If instead the number of boys in the sample had been 7, the significance level, using the 'equal distance' criterion of section f_1, would have been

$$P[\{B-10\cdot4|\geq|7-10\cdot4|\}|H] = P\{|B-10\cdot4|\geq 3\cdot4\}|H\}$$
$$= P(B\leq 7|H) + P(B\geq 13\cdot8|H)$$
$$= P(B\leq 7|H) + P(B\geq 14|H)$$
$$= 0\cdot178.$$

Such a large *SL* would be interpreted as meaning that the data were consistent with the null hypothesis.

(h) *Strength of evidence*

It is particularly to be noted that the *smaller* the *SL* value, the *stronger* is the evidence against *H*. It would perhaps be more convenient if we adopted a direct rather than an inverse measure of the evidence against *H*. Inconvenient or not, however, the significance level is too deeply entrenched to be abandoned. Nevertheless we shall on occasion use a direct measure to be called the *strength of the evidence against H*. This is the complement of the significance level:

strength of evidence against the null hypothesis *H*

$$= 1 - \text{significance level.} \tag{5.2.10}$$

A significance level near zero means a strength of nearly unity—very strong evidence against *H*. A significance level near unity means a strength near zero—very weak evidence against *H*, indicating in fact that the sample is consistent with *H*.

5.2.2. Conventional Interpretation of Significance Levels; Practices used in Specifying Significance Levels; Critical Region

The conventional interpretation of significance levels is given in Table 5.2.1. The concept is further discussed in section 5.3. To say that the significance levels and their interpretations quoted in this table are 'conventional' is not to say that they are arbitrary modes of statistical expression; on the contrary, they provide a generally accepted set of interpolates between, on the one hand, evidence so strong as virtually to kill off any possibility that the null hypothesis is true, and, on the other hand, the absence of any evidence at all against that hypothesis.

SL	Interpretation
$>0\cdot10$	Data consistent with H
$\approx0\cdot05$	Possibly significant. Some doubt cast on the truth of H
$\approx0\cdot02$	Significant. Rather strong evidence against H
$\approx0\cdot01$	Highly significant. H is almost certainly invalidated.

Table 5.2.1: Conventional interpretation of significance levels ('*SL*').

Inverse tables

The exact evaluation of a significance level depends on the possession of detailed information about the cumulative distribution function of the test statistic 'on' the null hypothesis ('on' here meaning 'computed on the assumption of the truth of': this useful abbreviation is widely used.) Given the exigencies of publishing, the amount of detail actually available in tables is often less than the user might wish. The most common space-saving device of table makers is to use *inverse tabulation*, in terms of 'percentage points' [see § 1.4.2(vii)].

The practical effect of having to use inverse tables is that, unless accurate interpolation is possible, one is unable to find the exact significance level of one's data and has instead to make do with inequalities such as that the significance level lies between 2·5% and 5%. This is not as vague as it sounds, as in most cases a rough interpolation enables one to deduce an approximate value (e.g. $SL \approx 3\%$) that is nevertheless sufficiently accurate for many purposes.

'*Data significant at level p*'

Unfortunately it has to be said that some users of statistical methods go further, and, instead of reporting that their significance level lies between 2·5% and 5%, record only that the significance level is less than 5%. For this the locution '*the data are significant at the 5% level*' is used; and similarly for other levels. It cannot be too strongly emphasised that this piece of jargon means '*the significance level of the data is at most* $0\cdot05$'. [see also § 5·10].

Critical region

A related and almost universally used convention is to characterize a test by specifying the value that the relevant statistic would have to take to achieve the precise significance level of $100p\%$ (e.g., for $p = 1$, 2·5, 5). The set of values that would form the significance set (see § 5.2.1(f)) for that particular observation is then called the $100p\%$ '*critical region*' or 'rejection region of the test'. [See § 5·10 for further discussion.]

5.2.3. A One-sided (Binomial) Test

There is a village where, over the year 1980–81, a large proportion of babies born were girls: 25 girls out of 35 single births. This was considered unusual, and various possible explanations were made in the press. One of these was the suggestion that the environment was rich in cadmium, possibly in the form of microscopic dust from a nearby stone quarry. The effect of a man's exposure to cadmium is to increase the probability that any child he fathers will be a girl. To test the cadmium hypothesis it would be necessary to carry out a significance test. As in section § 5.1.2 a reasonable underlying probability model is one which regards single births as mutually independent trials in each of which the probability of a girl is the same. The appropriate null hypothesis H_0 states that $p = 0·48$ (the national average). The alternative hypothesis is

$$H_1 : p > 0·48$$

The question whether p is significantly *less* than 0·48 is of no interest in this example.

With a probability p for girls, the probability of there being a birth sequence $y_1, y_2, \ldots, y_n$, with $y_r = 1$ for a girl and $y_r = 0$ for a boy, in a sample of n births, is given by

$$P(Y_r = y_r, r = 1, 2, \ldots, n) = \prod_{r=1}^{n} p^{y_r}(1-p)^{1-y_r}$$

$$= p^{\Sigma y_r}(1-p)^{n-\Sigma_r}$$

$$= p^g(1-p)^{n-g}, \qquad g = 0, 1, \ldots, n$$

where Y_r is the random variable whose realization is y_r, and $g = \Sigma y_r$ is the total number of girls in the sample. The appropriate sufficient statistic which condenses the data is g, and its sampling distribution [see II, § 5.2.2] is

$$P(G = g) = \binom{n}{g} p^g (1-p)^{n-g}, \qquad g = 0, 1, \ldots, n.$$

The null distribution of G is

$$P(G = g | H_0) = \binom{n}{g}(0·48)^g (0·52)^{n-g}.$$

As we are interested only in the question whether or not p exceeds 0·48, and as this question arises only in connection with samples in which the number of girls is not less than the expected value of G on H_0 (i.e. $0·48n$), the significance level of an observed value of g will be simply

$$SL = P\{G \geq g | H_0\}.$$

There is no lower tail to bother about, and we avoid the difficulties of coping with the second tail, described in section 5.2.2(f).

For the values of n and of g quoted above (viz. $n = 27, g = 35$) the significance level is

$$\sum_{r=25}^{35} \binom{35}{r} (0.48)^r (0.52)^{35-r} = 0.004.$$

This is a very small probability indeed, and the data must be regarded as highly significant. The evidence against H_0 is strong. (This does not of course necessarily establish the "cadmium' hypothesis.)

5.2.4.   Tests with Poisson Distributions

One-sided and two-sided tests to detect the possible departure of a Poisson parameter from its hypothesized value do not differ in principle from the corresponding binomial cases described in sections 5.2.2 and 5.2.4. Consider for example an experiment in which it was desired to test whether or not in a batch of nominally identical manufactured articles the number that were defective in a particular respect was significantly greater than the required norm of 2%, when a sample of 100 articles taken from the batch yielded 4 defectives. On the null hypothesis that the proportion of defectives is in fact 2%, the number D of defectives in the sample will have a hypergeometric distribution which is adequately approximated [see II, § 5.5] by the Poisson distribution

$$P(D = r) = e^{-\lambda} \lambda / r!, \qquad r = 0, 1, \ldots.$$

where $\lambda = 2$.

Here a one-sided test is called for, and the significance level of the data is

$$SL = P(D \geq 4 | H_0)$$

$$= P\{D \geq 4 | D \text{ is a Poisson (2) variable}\}$$

$$= 0.143$$

(from the Poisson tables [see Appendix (T2)]). This is a large probability; the result is not significant: the data are consistent with the 2% hypothesis.

5.2.5.   Tests with Continuous Distributions

We have so far discussed only tests for parameters in discrete distributions. Similar methods are used in continuous distributions.

EXAMPLE 5.2.1.   *The significance of a correlation coefficient.*   Denote by r the sample correlation coefficient [see § 2.5.7] obtained from a sample of n pairs of observations $(x_1, y_1), \ldots, (x_n, y_n)$ assumed to come from a bivariate

Normal distribution [see II, § 13.4.6(i)] with unknown correlation coefficient ρ:

$$r = \{\textstyle\sum x_j y_j - n\bar{x}\bar{y}\} / \sqrt{[(\textstyle\sum x_j^2 - n\bar{x}^2)(\textstyle\sum y_j^2 - n\bar{y}^2)]},$$

where $\bar{x} = \sum x_j/n$, $\bar{y} = \sum y_j/n$. If it is desired to test whether the observed value of r indicates that the data really are correlated, the appropriate null hypothesis is

$$H : \rho = 0.$$

A sufficiently large absolute value of r would tend to discredit the null hypothesis. The question 'how large?' is easier to answer in terms of the transform

$$t = \frac{r}{\sqrt{(1-r^2)}} \sqrt{(n-2)} \qquad (5.2.11)$$

since, on the null hypothesis, the sampling distribution of this is Student's distribution on $n-2$ degrees of freedom [see § 2.7.5]. A large absolute value of r corresponds to a large absolute value of t, and, since the sampling distribution of t is symmetrical about the value 0 [see § 2.5.5], the arguments adduced in section 5.2.1(f_2) adapted to a continuous distribution lead to the following *tail-area* definition of the significance level:

$$SL = P(T \geq |t|) + P(T \leq -|t|)$$

$$= 2P(T > |t|),$$

where T has Student's distribution on $n-2$ degrees of freedom and t is computed in accordance with (5.2.11) from the observed sample correlation coefficient r. For example, Fisher quotes the sample correlation coefficient between the annual wheat yield and the autumn rainfall over 20 years in East Anglia as $r = -0.629$. The corresponding value of t (calculated from (5.2.11) with $n = 20$) is -3.433. The significance level is $2P(T_{18} \geq 3.433)$, where the subscript 18 indicates the number of degress of freedom: see Figure 5.2.3.

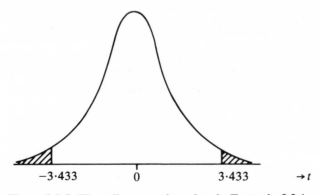

Figure 5.2.3: The tail areas referred to in Example 5.2.1.

Unfortunately the commonly available tables of the cumulative distribution function of Student's distribution are of the 'inverse table' kind [see § 5.2.2] that do not enable one readily to evaluate this probability. Instead they tabulate values of t that would correspond to predetermined values of the tail area. For example, Fisher's table (in '*Statistical Methods for Research Workers*) gives values of t corresponding to SL values of $0 \cdot 01$, $0 \cdot 02$, $0 \cdot 05$, $0 \cdot 1$, $0 \cdot 2(0 \cdot 1)0 \cdot 9$. For 18 d.f. the nearest value tabulated is $2 \cdot 878$, corresponding to an SL of $0 \cdot 01$. It follows that the t-value of $3 \cdot 433$ corresponds to an SL of less than $0 \cdot 01$. This is certainly significant [see Table 5.2.1]: the null hypothesis is strongly discredited; and the existence of a correlation may be taken as established. In this example no great harm has been done by our inability to find the significance level (other than in terms of the inequality: $SL < 0 \cdot 01$). If however the observed sample correlation coefficient r had been $0 \cdot 468$, corresponding to $t = 2 \cdot 25$, the table would merely have told us that the SL lay between $0 \cdot 05$ (corresponding to $t = 2 \cdot 101$) and $0 \cdot 02$ (corresponding to $t = 2 \cdot 552$). A result of this sort might ordinarily be reported as 'significant at the 5% level, not significant at the 2% level'. It should always be understood that this kind of circumlocution is solely a consequence of the structure of published tables and is equivalent to saying that the SL is roughly $0 \cdot 03$ or $0 \cdot 04$ (making such interpolations as the tables permit).

[A further point to be noted by users of published tables is that some versions (as, for instance, Fisher's mentioned above) assume that one is performing a two-sided test and give the corresponding SL, namely $2P(T \geq |t|)$, while others give the single-tail area $P(T \geq |t|)$. The user has to be sure he knows precisely what the table is talking about.]

Tests on Normal samples deserve a separate section, and are discussed in section 5.8. However the following example is worth giving separately.

EXAMPLE 5.2.2.   *Significance of a difference between sample correlation coefficients.*   Suppose that two samples, of sizes n_1 and n_2 respectively, assumed to come from bivariate Normal [see II, § 13.4.6(i)] populations, provide sample correlation coefficients [see § 2.5.7] r_1 and r_2, with $r_1 \neq r_2$. Does this indicate that the population values ρ_1, ρ_2 of the correlation coefficients are unequal? The appropriate null hypothesis is

$$H : \rho_1 = \rho_2,$$

and the question is whether $|r_1 - r_2|$ is large enough to discredit this hypothesis. We again have recourse to a transformation: it is known that

$$z_1 = \tfrac{1}{2} \log \frac{1 + r_1}{1 - r_1}$$

and

$$z_2 = \tfrac{1}{2} \log \frac{1 + r_2}{1 - r_2}$$

are, to a good approximation, realizations of Normal variables with expectations

$$\zeta_1 = \tfrac{1}{2}\log\frac{1+\rho_1}{1-\rho_1},$$

$$\zeta_2 = \tfrac{1}{2}\log\frac{1+\rho_2}{1-\rho_2}$$

and variances $1/(n_1-3)$, $1/(n_2-3)$ respectively [see § 2.7.3(b)]. On H, therefore, with $\rho_1 = \rho_2$, it follows that $z_1 - z_2$ is a realization of a Normal variable with zero expectation and with variance equal to $\omega^2 = \{(1/n_1-3)+(1/n_2-3)\}$ approximately [see II, § 9.2], whence $(z_1 - z_2)/\omega$ may be regarded as an observation on a standard Normal variable U [see II, § 11.4.1]. Large values of $|z_1 - z_2|/\omega$ will correspond to improbable 'tail' values of this distribution, which will discredit H_0. The significance level, for given values z_1 and z_2, is

$$SL = P(U > |z_1 - z_2|/\omega) + P(U < -|z_1 - z_2|/\omega)$$

$$= 2P(U > |z_1 - z_2|/\omega)$$

by symmetry. For example, with $r_1 = 0 \cdot 3$, $n_1 = 10$ and $r_2 = 0 \cdot 6$, $n_2 = 15$, we have $z_1 = 0 \cdot 31$, $z_2 = 0 \cdot 69$, and $\omega^2 = (1/7) + (1/12) = 0 \cdot 226$, so that $\omega = 0 \cdot 475$, and $|z_1 - z_2|/\omega = 0 \cdot 8$, whence

$$SL = 2P(U > 0 \cdot 8)$$

$$= 0 \cdot 42$$

from the Normal tables [see Appendix (T4)]. This is a large probability. The difference between r_1 and r_2 is not significant. There is no evidence against the null hypothesis.

5.2.6. Choosing a Test Statistic

The following is an example in which the choice of test statistic is more explicit than in Examples 5.2.1 and 5.2.2.

EXAMPLE 5.2.3. *Test of the parameter of an exponential distribution.* Suppose we have n realizations $x_1, \ldots, x_n$ of an exponential variable X [see II, § 11.2] with p.d.f. $f(x) = \theta^{-1} e^{-x/\theta}$ ($x > 0$), and it is desired to test the null hypothesis H that $\theta = \theta_0$ against the one-sided alternative that $\theta > \theta_0$. The likelihood function is $\theta^{-n} e^{-\Sigma x_i/\theta}$, whence $s = \sum x_i$ is a sufficient statistic for θ [see § 4.13.1]. This suggests that s or an appropriate transform of s would be a suitable text statistic. The sampling expectation of s is $n\theta$, whence s has the same dimensions as θ, and s/n ($= \bar{x}$, the sample mean) is a good estimate of θ: large values of s/n are unlikely on H, but more likely if the alternative hypothesis holds, whence $\bar{x}$ is a suitable test statistic.

To work out the significance level we need the sampling distribution of $\bar{x}$. The sampling distribution of $s = \sum x_i$ is [see § 2.4]

$$g(s) = s^{n-1} e^{-s/\theta} / \theta^n (n-1)!, \qquad s > 0.$$

Tables of the corresponding c.d.f. are not readily available, but the quantity $z = 2s/\theta$ has p.d.f.

$$h(z) = (\tfrac{1}{2}z)^{n-1} e^{-z/2} / 2\Gamma(n)$$

which is the chi-squared distribution on $2n$ degrees of freedom [see II, § 11.2.2, § 11.4.11], the c.d.f. of which is widely tabulated. [Here $\Gamma(n) = (n-1)!$]. Then the *SL* of an observed sample mean $\bar{x} = s/n$ is

$$SL = P(\bar{X} \geq \bar{x} | H)$$

where $\bar{X}$ denotes the r.v. induced by the statistic $\bar{x}$

$$= P(S \geq s | H)$$

(where $S = n\bar{X}$)

$$= P(\chi^2_{2n} \geq 2n\bar{x}/\theta_0)$$

where $\chi^2_{2n} = 2S/\theta_0$ has the chi-squared distribution on $2n$ degrees of freedom.

Similar principles apply, mutatis mutandis, when the alternative is $\theta < \theta_0$. Suppose for example that the sum of 18 lifetimes (assumed to be exponentially distributed) of electric lamp bulbs of nominal life 100 hours turned out to be 1500 hours. Here the null value is $\theta_0 = 100$ and the significance level is

$$P(\chi^2_{36} \leq 3000/100) = P(\chi^2_{36} \leq 30).$$

Standard chi-squared tables [see Appendix (T6)] give

$$P(\chi^2_{36} \geq 28 \cdot 7) = 0 \cdot 80, \qquad P(\chi^2_{36} \geq 31 \cdot 1) = 0 \cdot 70 \qquad (5.2.11)$$

whence

$$0 \cdot 70 \leq P(\chi^2_{36} \geq 30) \leq 0 \cdot 80$$

so that

$$0 \cdot 20 \leq P(\chi^2_{36} \leq 30) \leq 0 \cdot 30.$$

In particular, the significance level exceeds $0 \cdot 20$, whence the results are certainly not significant. The data do not provide evidence against the hypothesis that the mean lifetime is 100 hours.

[In this example the inequalities (5.2.11) sufficed. Linear interpolation would have given $P(\chi^2_{36} \leq 30) \approx 0 \cdot 75$. If however a more precise result were required it would be necessary to use more detailed tables or to use an appropriate transformation in terms of a variable for which more detailed tables are readily available. The best known of these, for $\nu > 30$ [see § 2.7.3(c)], gives the following good approximation in terms of the Standard Normal variable U:

$$P(\chi^2_\nu \geq k) \approx P(U \geq u)$$

where

$$u = \frac{(k/\nu)^{1/3} - 1 + 2/9\nu}{\sqrt{(2/9\nu)}}. \tag{5.1.12}$$

For $\nu = 40$ and $k = 59 \cdot 3$ (for example) we find $u = 1 \cdot 956$, whence the significance level given by the approximation (5.2.12) is $0 \cdot 026$. The correct value is $0 \cdot 025$.]

5.3.   CRITERIA FOR A TEST

The main qualitative requirement of a test is that it should be unlikely to discredit a true hypothesis, but that it should have a high probability of discrediting a false hypothesis. Up till now we have relied on intuition to justify the assumption that our tests do actually behave in this way. In this section we discuss an objective justification of the procedure, and show how in a specified sense one test may be superior to another.

5.3.1.   Sensitivity Function

To avoid obscuring the principles by mathematical detail we restrict attention to a particular simple case, namely that in which we have a vector $x = (x_1, x_2, \ldots, x_n)$ of independent observations x_r on a Normal $(\theta, 1)$ variable, where we wish to test the null hypothesis

$$H(0): \theta = 0 \tag{5.3.1}$$

against the one-sided alternative that $\theta > 0$. In this example we take as test statistic the sample mean $\bar{x}_0$ and as significance set $G(\bar{x}_0)$ the set of possible values of the sample mean that exceed our observed value $\bar{x}_0$:

$$G(\bar{x}_0) = \left\{ \mathbf{x} : \sum_1^n x_r \geq n\bar{x}_0 \right\} \tag{5.3.2}$$

where $\mathbf{x}$ is the sample vector. The significance level (z_0, say) of the observed statistic $\bar{x}_0$ is, as in (5.2.4),

$$z_0 = SL(\bar{x}_0) = P\{\bar{X} \in G(\bar{x}_0)|H(0)\}$$
$$= P\{\bar{X} \geq \bar{x}_0 | \bar{X} \text{ is Normal } (0, 1/\sqrt{n})\}. \tag{5.3.3}$$

Here $\bar{X}$ denotes the random variable induced by the sample mean: this is Normal $(\theta, n^{-1/2})$, where $\theta = E(X)$; and, on $H(0)$, we take $\theta = 0$. We therefore have

$$z_0 = SL(\bar{x}_0) = 1 - \Phi(\bar{x}_0\sqrt{n}) \tag{5.3.4}$$

where Φ, as usual, denotes the standard Normal c.d.f. [see II, § 11.4.1]. This significance level z_0 is entirely determined by the statistic $\bar{x}_0\sqrt{n}$, and so is itself a statistic; we emphasise this, when necessary, by calling it the *significance level statistic.* It is a realization of the *significance level random variable* $Z = SL(\bar{X})$, that is

$$Z = SL(\bar{X}) = 1 - \Phi(\bar{X}\sqrt{n}). \tag{5.3.5}$$

It cannot be too strongly emphasised that, while the significance level z_0 of an observed statistic $\bar{x}_0$ is, by definition, a probability computed on the assumption of the validity of the null hypothesis

$$H(0) : E(X) = 0$$

[see (5.3.3)], the sampling *distribution* of z_0—that is the distribution of Z—depends on the *actual* (unknown) value (θ; say) of $E(X)$. This distribution (in cumulative form) is

$$P(Z \le z | \theta) \qquad (0 \le z \le 1)$$

$$= Q(z, \theta),$$

say.

When $\theta = 0$, that is when the null hypothesis is correct, we have

$$Q(z, 0) = P(Z \le z | 0)$$

$$= P\{1 - \Phi(\bar{X}\sqrt{n}) \le z | E(X) = 0\} \qquad \text{[see (5.3.5)]}$$

$$= P\{\Phi(\bar{X}\sqrt{n}) \ge 1 - z | E(X) = 0\}$$

$$= P\{\bar{X}\sqrt{n} \ge u(z) | E(X) = 0\}$$

(where $\Phi\{u(z)\} = 1 - z$)

$$= 1 - \Phi\{u(z)\}$$

(since $\bar{X}\sqrt{n}$ is a standard Normal variable when $E(X) = 0$)

$$= z, \qquad 0 \le z \le 1. \tag{5.3.6}$$

Thus, when $H(0)$ is correct, the sampling distribution of the significance level is Uniform $(0, 1)$; and the chance of observing a *small* significance level z (say $\le 0 \cdot 01$) is correspondingly small (one in a hundred, for the case quoted). Thus there is only a very small chance of obtaining strong evidence against $H(0)$ when $H(0)$ is in fact correct.

A test of the kind described does therefore satisfy the first criterion stated in the opening lines of section 5.3. What about the second criterion, that a test should have a high probability of discrediting a false hypothesis? To answer this, we evaluate the sampling distribution of the significance level for an undetermined value of θ. The probability that our observed significance level will be z, or less, when the true value of $E(X)$ is θ, is

$$Q(z, \theta) = P(Z \le z | \theta)$$

$$= P\{1 - \Phi(\bar{x}\sqrt{n}) \le z | \theta\}$$

$$= P\{\bar{X}\sqrt{n} \ge \Phi^{-1}(1 - z) | \theta\} \qquad \text{(using (5.3.5))}$$

(since Φ is a monotone increasing function)

$$= 1 - \Phi\{\Phi^{-1}(1-z) - \theta\sqrt{n}\} \tag{5.3.7}$$

since $\bar{X}\sqrt{n} - \theta\sqrt{n}$ is a standard Normal variable.

To interpret this it is helpful to consider values of $Q(z, \theta)$ for a fixed value (z_0 say) of Z, and for all possible relevant values of θ. (In this 'one-sided' example the only relevant values of θ are non-negative ones.) If we take this fixed z_0 to be the significance level of the observed sample mean $\bar{x}_0$, we have

$$z_0 = SL(\bar{x}_0) = P\{\bar{X} \geq \bar{x}_0 | \bar{X} \text{ Normal } (0, 1/\sqrt{n})\}$$

(where n is the sample size)

$$= 1 - \Phi(\bar{x}_0\sqrt{n}),$$

and the value of $Q(z_0, \theta)$ as given in (5.3.7) then reduces to

$$Q(z_0, \theta) = 1 - \Phi\{(\bar{x}_0 - \theta)\sqrt{n}\}, \tag{5.3.8}$$

with

$$Q(z_0, 0) = 1 - \Phi(\bar{x}\sqrt{n})$$

$$= z_0$$

by (5.3.6).

It will be shown below that the values of this function $Q(z_0, \theta)$ measure the extent to which the test procedure is capable of detecting departures of θ from the hypothesized value 0, that is for discriminating between one value of θ and another. For this reason it is called the *sensitivity function* of the test. (It is in fact numerically the same as the *power function* of the Neyman–Pearson theory of tests, [see § 5.10] but has a somewhat different interpretation)

To illustrate with a numerical example, suppose that a Normal sample of size $n = 20$ gave a sample mean $\bar{x}_0 = 0 \cdot 458$. It is known that the parent distribution has unit variance. The expected value (θ) is of unknown value, but, according to the null hypothesis, $\theta = 0$. The significance level of the sample with respect to this hypothesis is

$$z_0 = SL(\bar{x}_0) = P(\bar{X} \geq 0 \cdot 458)\bar{X} \text{ Normal } (\theta, 1/\sqrt{20})$$

$$= 1 - \Phi(0 \cdot 458 \times 4 \cdot 472)$$

(since $\sqrt{20} = 4 \cdot 472$)

$$= 1 - \Phi(2 \cdot 048)$$

$$= 0 \cdot 02.$$

Some representative values of $Q(0 \cdot 02, \theta)$, calculated as

$$Q(0 \cdot 02, \theta) = 1 - \Phi\{(0 \cdot 458 - \theta)4 \cdot 472\}$$

as in (5.3.8) are as follows:

θ	0	0·2	0·4	0·6	0·8	1·0
$Q(0·02, \theta)$	0·020	0·125	0·398	0·737	0·937	0·992

$$(5.3.9)$$

The following points emerge:

(i) $Q(0·02, 0)$ is equal to the significance level, 0·02. This is a 'small' value. The meaning is that, if the null hypothesis were correct (i.e. if it were true that $\theta = 0$) the probability of obtaining a significance level as small as 0·02 or smaller would be precisely 0·02. Equivalently, the chance of obtaining strong evidence against $H(0)$ (i.e. evidence of strength 0·98) is small (0·02, in fact): The test procedure is *unlikely to produce strong evidence against a valid hypothesis*.

(ii) $Q(0·02, 1·0) = 0·992$. This is the (extremely large) probability of obtaining a significance level of 0·02 when the null hypothesis is far from being correct (θ is 1·0, while $H(0)$ alleges that it is zero). A significance level of 0·02 would be taken as providing strong evidence against the hypothesis. Thus we see that the test is *almost certain to provide strong evidence against* $H(0)$ *when* $H(0)$ *is far from the truth*.

(iii) The same applies, less forcefully, to the situation that arises when $H(0)$, although false, is nearer the truth than in (ii); for example when $\theta = 0·6$.

(iv) When $H(0)$ is still closer to the truth, e.g. when $\theta = 0·4$, the procedure is somewhat indecisive, having only a (roughly) 40% chance of providing evidence against $H(0)$ of strength 0·98; and of course, a 60% chance of failure to produce such evidence.

A test which has a high chance of producing strong evidence against a hypothesis that is 'only slightly false', and which is unlikely to produce such evidence against a strictly correct hypothesis is clearly a *sensitive* test for discriminating between values of θ near to zero. Thus the function $Q(z_0, \theta)$ may be regarded as a *sensitivity function* for the test.

That all tests are not equally sensitive may be demonstrated by the test in which the test statistic is the mean of the first and last observations in the sample ($=\tilde{x}_0$, say), all other observations being ignored. This test will clearly have the same properties as that in which the test statistic is the sample mean, as in the earlier example, with a sample size of $n = 2$. The sensitivity function for this will therefore be

$$Q'(z_0, \theta) = 1 - \Phi(\tilde{x}_0\sqrt{2} - \theta\sqrt{2}) \qquad (5.3.10)$$

by (5.3.8). The values of $Q'(z_0, \theta)$, for the same value $z_0 = 0·02$ of the significance level as considered earlier in (5.3.9), are shown below:

θ	0	0·2	0·4	0·6	0·8	1·0	2·0	3·0
$Q'(0·02, \theta)$	0·020	0·039	0·069	0·115	0·180	0·264	0·118	0·014

This test has the same significance level (0·02) as that based on the sample size of 20, whose sensitivity function Q is tabulated in (5.3.9), but the values of Q' are less than the corresponding values of Q for every positive value of θ, showing that a false hypothesis, whatever its degree of falsity, is more likely to be discredited by the first test than by the second (which only uses part of the data). Similar results hold for all values of the significance level z. (The fact that $Q(z, \theta) < Q(z', \theta)$ for negative values of θ does not affect the argument, since the only competing values of θ that are regarded as relevant in this one-sided test are positive ones.) The forms of the sensitivity functions $Q(z, \theta)$, with a fixed common value of the significance level z for the two tests are as illustrated in Fig. 5.3.1. The sensitivity function $Q(z, \theta)$ of a test, with $z = \alpha$, has the same values as (but a different interpretation from) the α-level power function $W(\theta)$ of Neyman–Pearson Theory. [See § 5.10.]

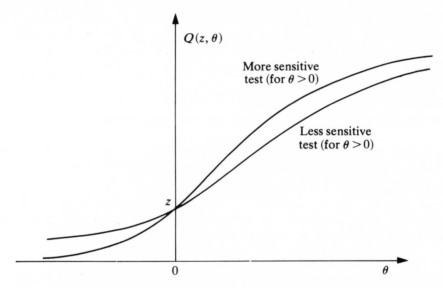

Figure 5.3.1: Sensitivity functions for a one-sided test.

5.3.2. Sensitivity Function of the One-tailed Test when σ is Unknown, in a Sample from a Normal (θ, σ) Distribution

The exposition given in section 5.3.1 was intended to explain the principles involved in sensitivity analysis without obscuring them by unfamiliar mathematics. This was made possible by dealing with a test of a hypothesis specifying the expected value θ in a Normal $(\theta, 1)$ distribution, the only distribution involved in the calculations being the Normal. We now briefly outline the corresponding calculations for the more realistic case where the parent-distribution is Normal (θ, σ), the value of σ being unknown.

Define $X_1, X_2, \ldots, X_n$ as independent random variables each of which is Normal (θ, σ), with $\mathbf{x} = (X_1, X_2, \ldots, X_n)$, $\bar{X} = \sum_1^n X_r/n$, and $S^2 = \sum_1^n (X_r - \bar{X})^2/(n-1)$; and define

$$T' = \bar{X} n^{1/2}/S.$$

We are concerned with the hypothesis

$$H(0): \quad \theta = 0,$$

a hypothesis that says nothing about σ.

Let $x_1, x_2, \ldots, x_r$ denote typical realizations of the X_r, and let $\mathbf{x} = (x_1, x_2, \ldots, x_r)$, $\bar{x} = \sum_1^n x_r/n$, $s^2 = \sum_1^n (x_r - \bar{x})^2/(n-1)$, and $t' = \bar{x} n^{1/2}/s$. Suppose the values of the observation vector and the corresponding statistics in our actual sample to be $\mathbf{x}_0$, $\bar{x}_0$, s_0^2 and $t_0' = \bar{x}_0 n^{1/2}/s_0$. Then $\bar{x}_0$ is our estimate of θ, and s_0^2 our estimate of σ^2; and t_0' our value of Student's ratio. When $\theta = 0$ this latter is a realization of Student's distribution on $n-1$ d.f.; its value will be much more likely to be near 0 than to be large (provided $\theta = 0$) and we shall regard large values of t' as tending to discredit the null hypothesis.

We therefore take as significance region the set

$$G(t_0') = \{\mathbf{x} : t' \geq t_0'\}.$$

The significance level (SL) of our sample vector $\mathbf{x}_0$ is then

$$z_0 = P\{T' \in G(t_0') | \theta = 0\}$$

$$= P\{T' \geq t_0' | T' \text{ is Student } (n-1)\}$$

$$= 1 - \Psi_{n-1}(t_0'),$$

where $\Psi_{n-1}(t)$ denotes the c.d.f. at t of Student's distribution in $n-1$ d.f. The significance level random variable Z is then

$$Z_1 = 1 - \Psi_{n-1}(T')$$

(cf. (5.3.5)) and the sensitivity function is

$$Q(z, \theta) = P(Z \leq z | \theta)$$

$$= P\{\Psi_{n-1}(T') \geq 1 - z | \theta\}$$

$$= P\{T' \geq \Psi_{n-1}^{-1}(1 - z) | \theta\}.$$

Now $T' = \bar{X} n^{1/2}/S$. When $E(X) = \theta$, the r.v. $(\bar{X} - \theta) n^{1/2}/S$ has Student's distribution on $n-1$ d.f., whence T' has the *non-central* Student distribution on $n-1$ d.t. with non-centrality parameter $\lambda = \lambda(\theta, n)$ given by

$$\lambda(\theta) = -\theta n^{1/2}/\sigma.$$

[See § 2.8.3.] Denote the c.d.f. at w of this distribution by the symbol

$$H_{n-1}\{w, \lambda(\theta)\},$$

so that

$$Q(z, \theta) = 1 - H_{n-1}\{\Psi_{n-1}^{-1}(1-z), \lambda(\theta)\}$$

(compare (5.3.7)). When $\theta = 0$ this becomes

$$Q(z_0, 0) = 1 - \Psi_{n-1}\{\Psi_{n-1}^{-1}(1-z_0)\},$$

since $\lambda(0) = 0$, and the non-central Student distribution with c.d.f. $H_{n-1}\{w, \lambda(\theta)\}$ becomes the ordinary 'central' Student distribution with c.d.f. $\Psi_{n-1}(w)$. Thus

$$Q(z_0, 0) = 1 - (1 - z_0)$$

$$= z_0,$$

so that $Q(z_0, 0)$ is the observed significance level, as in the earlier example (cf. (5.3.6)).

For example, if $\bar{x}_0 = 0.458$, $s_0 = 1.00$, $n = 20$, and $t'_0 = 0.458\sqrt{(20)} = 2.048$, the significance level is

$$P\{T > 2.048 | T \text{ is Student (19)}\} = 0.022$$

(from tables). This is a small value: the probability of getting such a small value if $H(0)$ were true is 0.022, and equivalently, the strength of the evidence against $H(0)$ is 0.978. The result is therefore an unlikely consequence of $H(0)$, which hypothesis has strong evidence against it and is therefore regarded as being fairly firmly discredited.

The probability of obtaining evidence of strength $1 - z$ against $H(0)$, when $E(X) = \theta$, is equal to the probability of obtaining a significance level of z or less; that probability is given by $Q(z, \theta)$, numerical values of which may be obtained from the appropriate tables of the non-central Student distribution with $n - 1$ d.f. and parameter λ. Qualitatively the function resembles the functions illustrated in Figure 5.3.1.

5.3.3.   Sensitivity Function of a Two-tailed Test

In section 5.3.1 we considered a sample $(x_1, x_2, \ldots, x_n)$ from a Normal $(\theta, 1)$ population and discussed the significance of the sample mean $\bar{x}$ in relation to the null hypothesis that $\theta = 0$, the alternative being the 'one-sided' hypothesis that $\theta > 0$.

If the allowed alternatives had included the possibility that $\theta < 0$ as well as $\theta > 0$ we should have had a 'two-sided' test. We now proceed on the assumption that the alternative hypothesis is in fact this two-sided one.

In this case the significance level statistic corresponding to (5.3.3) is

$$z_0 = SL(\bar{x}_0) = 2P\{\bar{x} \geq |\bar{x}_0| \,\big|\, X \text{ Normal } (0, 1)\}$$

$$= 2\{1 - \Phi(|\bar{x}_0|\sqrt{n})\}. \tag{5.3.11}$$

The cumulative sampling distribution of z_0, when $\theta = 0$, is

$$Q(z, 0) = P(Z \le z | \theta = 0)$$

where Z is the random variable induced by z_0; that is,

$$Z = 2\{1 - \Phi(|\bar{X}|\sqrt{n})\}.$$

Thus

$$Q(z, 0) = P[2\{1 - \Phi(|\bar{X}|\sqrt{n})\} \le z | \theta = 0]$$
$$= P\{\Phi(|\bar{X}|\sqrt{n}) \ge 1 - \tfrac{1}{2}z | \theta = 0\}$$
$$= P\{|\bar{X}|\sqrt{n} \ge \Phi^{-1}(1 - \tfrac{1}{2}z) | \theta = 0\}$$
$$= 2P\{\bar{X}\sqrt{n} \ge \Phi^{-1}(1 - \tfrac{1}{2}z) | \theta = 0\}$$

(taking $\bar{x}$ as positive, for definiteness)

$$= 2[1 - \Phi\{\Phi^{-1}(1 - \tfrac{1}{2}z)\}]$$

(since $\bar{X}\sqrt{n}$ is Normal $(\theta, 1)$)

$$= z, \qquad (0 \le z \le 1). \tag{5.3.12}$$

This is the same result as was obtained in the single-tailed test in section 5.3.1: when the null hypothesis is correct, the sampling distribution of the significance level is Uniform $(0, 1)$.

The corresponding sampling distribution when θ is not necessarily zero is

$$P\{Z \le z, \theta\} = Q(z, \theta)$$
$$= P[2\{1 - \Phi(|\bar{X}|\sqrt{n})\} \le z | \theta]$$
$$= P\{|\bar{X}|\sqrt{n} \ge \Phi^{-1}(1 - \tfrac{1}{2}z) | \theta\}$$
$$= P\{\bar{X}\sqrt{n} \ge \Phi^{-1}(1 - \tfrac{1}{2}z) | \theta\} + P\{\bar{X}\sqrt{n} \le -\Phi^{-1}(1 - \tfrac{1}{2}z) | \theta\}$$
$$= 1 - \Phi\{\Phi^{-1}(1 - \tfrac{1}{2}z) - \theta\sqrt{n}\}$$
$$+ 1 - \Phi\{\Phi^{-1}(1 - \tfrac{1}{2}z) + \theta\sqrt{n}\}$$

since, where $E(X) = \theta$, $\bar{X}\sqrt{n}$ is Normal $(\theta\sqrt{n}, 1)$. If now, as in the one-sided case, we fix the value of z in $Q(z, \theta)$ at $z = z_0$, the significance level, for which we have

$$1 - \tfrac{1}{2}z_0 = \Phi(|\bar{x}_0|\sqrt{n})$$

(5.3.11), so that

$$\Phi^{-1}(1 - \tfrac{1}{2}z_0) = |\bar{x}_0|\sqrt{n},$$

the function $Q(z_0, \theta)$ reduces

$$Q(z_0, \theta) = 2 - \Phi\{(|\bar{x}_0| - \theta)\sqrt{n}\} - \Phi\{(|\bar{x}_0| + \theta)\sqrt{n}\}. \tag{5.3.13}$$

This is the analogue of (5.3.8) for a two-tailed test. The graph of $Q(z, \theta)$

against θ for a fixed z is exemplified in Figure 5.3.2. This shows that the probability of obtaining strong evidence against $H_0 : \theta = 0$, when $E(X = \theta)$, increases with $|\theta|$. A corresponding analysis for the case where the parent variance is unknown may be carried out with the aid of the non-central Student distribution, as in section 5.3.2.

[The z-level sensitivity function has the same values as, but a different interpretation from, the 'power function' discussed in section 5.10.]

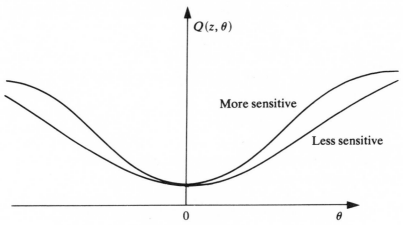

Figure 5.3.2: Sensitivity functions for a two-sided test.

5.4. TESTS INVOLVING A COMPOSITE NULL HYPOTHESIS

The examples discussed in the preceding section have been confined to situations in which, for a given probability model, the null hypothesis completely specified the null distribution. However, situations commonly arise in which this is not the case. The null hypothesis is then said to be *composite*. Examples arise both with one-parameter probability models and with multi-parameter models. An example of the former is the comparison of two Binomial proportions, with the null hypotheses that the two Binomial parameters p_1 and p_2 are equal. An example of the multiparameter kind is the null hypothesis that the expectation of a Normal distribution is zero, the standard deviation being unknown and not specified by the hypothesis (an example of a 'nuisance parameter').

5.4.1. Conditional Tests: Equality of Binomial Proportions; Ratio of Poisson Parameters

The standard method of dealing with the difficulties raised by the composite nature of a null hypothesis is to work with a suitably *conditioned* version of the null distribution. This is best explained in terms of examples.

EXAMPLE 5.4.1.   *Test for the equality of binomial proportions.*   In an investigation to see whether the probability of giving birth to a male child depended on the age of the mother it was found that in a sample of m single births to mothers aged between 20 and 25 the number of boy babies was x_0, while in a corresponding sample of size n of mothers aged between 30 and 35 the number was y_0. [c.f. § 5.2.1(c).] Take as underlying probability model the Binomial (m, p_1) distribution [see II, § 5.2.2] for the boys born to the younger mothers and the independent [see II, § 6.6] Binomial (n, p_2) for the others, so that the density of the joint sampling distribution of x_0 and y_0 at (x, y) is

$$\binom{m}{x}\binom{n}{y}p_1^x(1-p_1)^{m-x}p_x^y(1-p_2)^{n-y}.$$

On the null hypothesis H that $p_1 = p_2$ $(= p$, say) this becomes

$$f(x, y) = \binom{m}{x}\binom{n}{y}p^{x+y}(1-p)^{m+n-x-y}.$$

Clearly $x + y$ is a sufficient statistic for p [see Example 3.4.1], whence the conditional joint distribution of x and y given $x + y = s$ does not depend on p. In fact this conditional density is

$$g(x, y|s) = \frac{f(x, y)}{P(S = s)}, \qquad (x + y = s)$$

where S is the random variable induced by s.

Since (on H) S is Binomial $(m + n, p)$ we have

$$P(S = s) = \binom{m+n}{s}p^s(1-p)^{m+n-s},$$

whence

$$g(x, y|s) = \frac{\binom{m}{x}\binom{n}{y}}{\binom{m+n}{s}}, \qquad (x + y = s).$$

Here $y = s - x$, so that this is in reality a univariate distribution, which we may write as

$$h(x|s) = \frac{\binom{m}{x}\binom{n}{s-x}}{\binom{m+n}{s}}, \qquad x = 0, 1, \ldots, \min(m, s). \tag{5.4.1}$$

This parameter-free (Hypergeometric [see II, § 5.3]) distribution is the required *conditioned* version of the null distribution. It might be helpful to note that this is equal to the probability of obtaining x defective items in a sample of size s taken without replacement from a collection containing m defective items and n non-defective items. Tables of this distribution (the hypergeometric) are available [see, e.g., Liebermann and Owen (1961)— Bibliography G].

The Binomial approximation

The Hypergeometric distribution involves three parameters, namely m, n and s, so tables are necessarily bulky. If s is small compared with m and n, the probabilities do not differ greatly from the results that would have been obtained from sampling *with* replacement, in which case $n(x|s)$ may be approximated by $b(x; s, p)$, the Binomial (s, p) probability, with $p = m/(m+n)$; that is

$$h(x|s) \approx b(x; s, p) = \binom{s}{x} p^x (1-p)^{s-x}, \qquad x = 0, 1, \ldots, s, \qquad p = m/(m+n)$$

(5.4.2)

provided

$$s \ll \min(m, n).$$
(5.4.3)

These probabilities may readily be evaluated with the aid of the recursion

$$b(x+1; s, p) = \left(\frac{s-x}{x+1}\right)\left(\frac{p}{1-p}\right) b(x; s, p), \qquad x = 0, 1, \ldots, s-1,$$

(5.4.4)

or alternatively obtained from the relatively compact Binomial tables [see Appendix (T1)].

An example of the Hypergeometric distribution (5.4.1) to which the Binomial approximation (5.4.2) does not apply is the case where $m = 15$, $n = 20$, $s = 17$. The values of $h(x|17)$ are tabulated in Table 5.4.1.

x	$h(x)$	x	$h(x)$	x	$h(x)$
0	0·0000	6	0·1852	12	0·0016
1	0·0000	7	0·2620	13	0·0001
2	0·0004	8	0·2381	14	0·0000
3	0·0039	9	0·1389	15	0·0000
4	0·0233	10	0·0513		
5	0·0834	11	0·0117	Total	0·9997

(The total differs slightly from 1·000 because of rounding errors.)

Table 5.4.1: Values of $h(x) = h(x|17)$

The observed value $x_0 = 10$ lies in the upper tail, and the probability of values as extreme or more so in this direction is the sum $h(10) + h(11) + \ldots + h(15) = 0·0647$. Using 'probability ordering' as in section 5.2.1 (f_2) we see that the probability of values as extreme as x_0 or more so in the opposite direction (the lower tail) is the sum $h(4) + h(5) + \ldots + h(0)$, since $h(4)$ is the largest value, working from below, that does not exceed $h(10)$. This lower tail sum is 0·0276. The significance level is therefore

$$SL = 0 \cdot 0647 + 0 \cdot 0276$$

$$= 0 \cdot 0923.$$

This is a fairly large probability and it follows that the results are not significant; that is, the data do not invalidate the null hypotheses that $p_1 = p_2$.

EXAMPLE 5.4.2.   *Test for equality of Poisson parameters.*   Just as in the case discussed in the above example (Example 5.4.1), a null hypothesis that postulates the equality of two unknown Poisson parameters is composite, and may be tested by using a conditioned null distribution.

Suppose we have observations $x'_1, x'_2, \ldots, x'_m$ on a Poisson variable X [see II, § 5.4] and $y'_1, y'_2, \ldots, y'_n$ on another Poisson variable Y. Denote the respective (unknown) parameters by θ_1 and θ_2. It is required to test the null hypothesis H that $\theta_1 = \theta_2$ ($= \theta$, say). The underlying probability model is that the joint sampling distribution of the data has density at $x_1, x_2, \ldots, x_m, y_1, y_2, \ldots, y_n$ given by

$$e^{-m\theta_1 - n\theta_2} \theta_1^{\Sigma x_i} \theta_2^{\Sigma y_j} / (\Pi x_i!)(\Pi y_j!).$$

The null hypothesis is

$$H: \theta_1 = \theta_2 \ (= \theta, \text{ say}),$$

which is to be tested against the alternative that $\theta_1 \neq \theta_2$. We now have to choose a test statistic (cf. § 5.2.1(d)): in this case, a pair of test statistics. Since

$$s_1 = \Sigma \, x_i \quad \text{and} \quad s_2 = \Sigma \, y_j$$

are sufficient [see § 3.4], respectively, for θ_1 and θ_2, the data may be condensed into the statistics

$$s'_1 = \Sigma \, x'_i, \qquad s'_2 = \Sigma \, y'_j.$$

The joint sampling distribution of the data may therefore be replaced by the simpler but equivalent joint sampling distribution of s_1 and s_2. Since these are sums of Poisson variables they are themselves Poisson variables [see Table 2.4.1], with parameters $m\theta_1, n\theta_2$ respectively. Their joint sampling distribution at s_1, s_2 is given by

$$e^{-m\theta_1 - n\theta_2} (m\theta_1)^{s_1}(n\theta_2)^{s_2} / s_1! \, s_2!.$$

On the null hypothesis H (that $\theta_1 = \theta_2 = \theta$), this reduces to the null distribution

$$f(s_1, s_2) \, e^{-(m+n)\theta} m^{s_1} n^{s_2} \theta^{s_1 + s_2} / s_1! \, s_2!.$$

Now for the conditioning: we observe that $s = s_1 + s_2$ is sufficient for θ in this null distribution, whence we may obtain a θ-free conditional distribution of S_1 and S_2, given $S_1 + S_2$, as

$$P(S_1 = s_1, S_2 = s_2 | S_1 + S_2 = s).$$

Since (on H) $S_1 + S_2$ is Poisson, with parameter $(m + n)\theta$, the conditioned null joint distribution of S_1 and S_2 is

$$g(s_1, s_2 | s) = \frac{m^{s_1} n^{s_2}}{s_1! \, s_2!} \bigg/ \frac{(m+n)^s}{s!}, \qquad s = s_1 + s_2,$$

Since $s_2 = s - s_1$ this is in reality a univariate distribution. Its density may be written in the form

$$h(s_1 | s) = \frac{s!}{s_1! \, s_2!} \left(\frac{m}{m+n}\right)^{s_1} \left(\frac{n}{m+n}\right)^{s_2}, \qquad s_2 = s - s_1$$

$$= \binom{s}{s_1} p^{s_1} (1-p)^{s - s_1}, \qquad p = m/(m+n). \tag{5.4.5}$$

This is the Binomial (s, p) distribution where $p = m/(m+n)$ is known. To test the significance of the observed value $s_1' = \sum x_i'$ with respect to the null hypothesis we have to see whether s_1' lies in a region of relatively high probability or whether it lies in one or other of the low-probability regions (i.e. in one of the tails). If the former, the data would be regarded as being consistent with the hypothesis that $\theta_1 = \theta_2$; if the latter, the hypothesis would be more or less strongly discredited. The procedure is precisely that followed in section 5.2.1.

For example suppose the x-data to be counts of radio-active particles emitted by specimen A in m intervals each of 10 seconds duration and the y-data to arise similarly in n intervals from specimen B, with

$$m = 20, \qquad \sum x_i' = s_1' = 15, \qquad n = 30, \qquad \sum y_i' = s_2' = 45. \tag{5.4.6}$$

The conditioned null sampling density of s_1' at s_1 is

$$h(s_1 | 50) = \binom{50}{s_1} p^{s_1} (1-p)^{50 - s_1}, \qquad s_1 = 0, 1, \ldots, 50,$$

with $p = m/(m+n) = 20/50 = 0.4$. The observed value $s_1' = 15$ lies in the lower tail of this distribution. Since the distribution is not very skew, the (two-sided) significance level computed with probability ordering will be nearly equal to that computed with distance ordering (cf. § 5.2.1(f_1) and (f_2)), whence

$$SL = P(S_1 \le 15 | s = 50) + P(S_1 \ge 25 | s = 50),$$

(15 and 25 being equidistant from the expected value of $0.4 \times 50 = 20$). Using the Binomial tables for $n = 50$, $p = 0.4$, this gives

$$SL = 0.19.$$

The data are not significant. The null hypothesis is not discredited.

EXAMPLE 5.4.3.   *Test for specified ratio of Poisson parameters.*   The principles utilized in Example 5.4.2 may be applied to test whether the data are consistent with the hypothesis that the two Poisson parameters are in a specified ratio $k:1$. If the null hypothesis is

$$H : \theta_1 = k\theta_2,$$

or, equivalently,

$$H : \theta_1 = k\theta, \qquad \theta_2 = \theta$$

where k is a specified multiplier, the null distribution (5.4.3) becomes

$$f(s_1, s_2) = e^{-(km+n)\theta} (km)^{s_1} n^{s_2} \theta^{s_1+s_2} / s_1! s_2!,$$

the sampling distribution of $s = s_1 + s_2$ is Poisson with parameter $(km+n)\theta$, and the conditioned null joint distribution (5.4.4) becomes

$$g(s_1, s_2 | s) = \frac{(km)^{s_1} n^{s_2}}{s_1! s_2!} \bigg/ \frac{(km+n)^s}{s!}, \qquad s = s_1 + s_2,$$

reducing to

$$h(s_1 s_2) = \binom{s}{s_1} p^{s_1}(1-p)^{s-s_1}, \qquad p = km/(km+n). \qquad (5.4.7)$$

For example, with the data (5.4.6) of Example 5.4.2, and $k = 1 \cdot 5$, we have $p = 30/60 = 0 \cdot 5$. The significance level becomes

$$SL = P(S_1 \le 15 | s = 50) + P(S_1 \ge 35 | s = 50),$$

(15 and 35 being equidistant from the expected value of $0 \cdot 5 \times 50 = 25$), whence

$$SL = 0 \cdot 003.$$

This is a very small probability: the data are highly significant; the hypothesis that $\theta_1 = 1 \cdot 5\theta_2$ is severely discredited.

[Note on the choice of an appropriate null hypothesis: In this example the data suggest that θ_1 is about half of θ_2. If we were concerned with the estimation of θ_1 and θ_2, maximum likelihood [see § 3.5.4 and Example 6.3.3] would in fact have led us to the estimates $\hat{\theta}_1 = s_1'/m = 0 \cdot 75$, $\hat{\theta}_2 = s_2'/n = 1 \cdot 5$, so that $\hat{\theta}_2 = 2\hat{\theta}_1$. There would not be much point in looking at the data, seeing that $\hat{\theta}_2 = 2\hat{\theta}_1$, and *on this basis* formulating and testing the hypothesis that $\theta_2 = 2\theta_1$. The result of such a test would clearly be 'not significant', which in the circumstances would merely mean that the data were consistent with the hypothesis that had been suggested by the data themselves. This is equivalent to saying that data are consistent with a hypothesis that is consistent with the data: not a particularly profound inference. It would similarly be possible, after carefully examining the data, to formulate a hypothesis which was inconsistent with the data. The situation is quite different when the null hypothesis has been formulated *before* the data have been examined, since in this case the consistency or inconsistency of the data with the hypothesis leads to a real inference in which one has learned something new.

These remarks illustrate the principle that, to carry out a test of significance, the null hypothesis should be formulated without reference to the data used in the test.]

5.4.2.   Tests for Independence in 2×2 Contingency Tables; Fisher's 'Exact Test'

A contingency table is a bivariate or cross-classified frequency table, as in the following example, which classifies a given number (n) of students according to their performance (a) in an end-of-term examination and (b) in term-time tests.

		Examination score (%)				
		<30	30–49	50–69	≥70	Total
	Poor	n_{11}	n_{12}	n_{13}	n_{14}	r_1
Test score	Moderate	n_{21}	n_{22}	n_{23}	n_{24}	r_2
	Good	n_{31}	n_{32}	n_{33}	n_{34}	r_3
	Total	c_1	c_2	c_3	c_4	n

Here, for example n_{23} is the number who scored 50–69 in the examination and were rated 'moderate' in the tests. The table also gives the row totals r_1, r_2, r_3 and the column totals c_1, c_2, c_3, c_4, as well as the grand total n. As there are three rows and four columns of data (ignoring the 'total' row and column for this purpose) it is a '3×4' contingency table.

The simplest such tables are 2×2 *tables* [see also § 7.4.1]. These may arise in various ways, of which perhaps the most important are those exemplified below:

(a) 'Cross-classification' (e.g. schoolchildren classified according to hair colour and degree of freckling):

	Fair haired	Dark haired	Totals
Heavily freckled	n_{11}	n_{12}	r_1
Lightly freckled	n_{21}	n_{22}	r_2
Totals	c_1	c_2	n

(b) 'Two treatments' (e.g. patients given medical treatment for rheumatism):

	Pain alleviated	Pain not alleviated	
Given aspirin	n_{11}	n_{12}	r_1
Given placebo	n_{21}	n_{22}	r_2
Totals	c_1	c_2	n

(c) 'Two populations' (e.g. each of two 'populations', one consisting of the inhabitants of Exeter and the other those of Edinburgh, is sampled for possession of the B-negative blood group):

	B-neg	Other	
Sample from Exeter	n_{11}	n_{12}	r_1
Sample from Edinburgh	n_{21}	n_{22}	r_2
	c_1	c_2	n

In particular cases it may not always be obvious that a set of data belongs to one of those types rather than another. Fortunately the method of analysis is the same in all three cases.

The question of interest is whether or not the row attributes are independent of the column attributes. Thus, in (a), is hair colour related to the degree of freckling, or has hair colour nothing to do with freckling (i.e. independent of it)?; in (b) does aspirin alleviate rheumatic pains, or, on the contrary, does it make no difference to the pain whether aspirin or the placebo was administered?; in (c) do the two populations contain different proportions of B-negative blood-group carriers, or are the proportions equal?

In the third of these examples we are asking whether the proportion p_1 of B-negatives in one population (of which the sample value is n_{11} out of r_1) differs from the corresponding proportion p_2 in the other population (where the sample value is n_{21} out of r_2). Provided the sample sizes r_1 and r_2 are reasonably small compared with the population sizes so that the binomial model applies to each of the two samples, the question is whether or not two binomial parameters p_1, p_2 are significantly different. The (conditional) test for this has been described in Example 5.4.1.

The 'two-treatments' case (b) may be tested in the same way; that is, one tests whether proportion n_{11}/r_1 of alleviations among the aspirin-treated patients differs significantly from the proportion n_{21}/r_2 among those given the placebo.

The true cross classification case (a) is, in principle, different, but it turns out to be amenable to precisely the same kind of analysis. To see this, let us adopt a more general notation in which the table in section (a) is replaced by the following (where '$\bar{A}$' means 'not-A', etc.):

	B	$\bar{B}$		
A	n_{11}	n_{12}	r	(5.4.8)
$\bar{A}$	n_{21}	n_{22}	$n-r$	
	c	$n-c$	n	

Here n_{12}, for example denotes the frequency of '$A\bar{B}$' individuals, that is those possessing attribution A and $\bar{B}$. Suppose the probability of a randomly chosen individual being an 'A' is $P(A) = \alpha$, and the probability that he is a B is $P(B) = \beta$, so that the marginal probabilities in the bivariate distribution [see II, § 6.3] are as shown:

	B	$\bar{B}$	
A			α
$\bar{A}$			$1-\alpha$
	β	$1-\beta$	

The appropriate null hypothesis (independence of hair colour and freckling) is that A and B are statistically independent [see II, § 3.5] so that

$$P(AB) = P(A)P(B) = \alpha\beta$$

(which, of course, implies that $P(A\bar{B}) = P(A)P(\bar{B}) = \alpha(1-\beta)$, etc.). Thus, on the null hypothesis, the bivariate probabilities associated with the four cells of the 2×2 table are

	B	$\bar{B}$
A	$p_{11} = \alpha\beta$	$p_{12} = \alpha(1-\beta)$
$\bar{A}$	$p_{21} = (1-\alpha)\beta$	$p_{22} = (1-\alpha)(1-\beta)$

In samples of size n, from a population specified by these probabilities, the joint distribution of the cell frequencies n_{11}, n_{12}, n_{21}, n_{22} is the Multinomial [see II, § 6.4.2] with p.d.f. given by:

$$f(n_{11}, n_{12}, n_{21}, n_{22}) = \frac{n!}{n_{11}! \, n_{12}! \, n_{21}! \, n_{22}!} \, p_{11}^{n_{11}} p_{12}^{n_{12}} p_{21}^{n_{21}} p_{22}^{n_{22}},$$

$$(n_{11} + n_{12} + n_{21} + n_{22} = n)$$

$$= \frac{n!}{n_{11}! \, n_{12}! \, n_{21}! \, n_{22}!} \, \alpha^r (1-\alpha)^{n-r} \beta^c (1-\beta)^{n-c}.$$

Clearly r is sufficient for α, and c for β. (Compare Example 5.4.1). We may therefore obtain a parameter-free conditional distribution by conditioning on r and c, i.e. as

$$g(n_{11}, n_{12}, n_{21}, n_{22} | r, c) = \frac{f(n_{11}, n_{12}, n_{21}, n_{22})}{h(r, c)},$$

where $h(r, c)$ is the joint sampling density of r and c. Since, on H_0, A and B are independent,

$$h(r, c) = \binom{n}{r} \alpha^r (1+\alpha)^{n-r} \cdot \binom{n}{c} \beta^c (1-\beta)^{n-c}.$$

Thus the parameter-free conditional null distribution is

$$g(n_{11}, n_{12}, n_{21}, n_{22} | r, c) = \frac{n!}{n_{11}! \, n_{12}! \, n_{21}! \, n_{22}!} \bigg/ \binom{n}{r}\binom{n}{c}, \qquad (5.4.9)$$

$$n_{11} + n_{12} = r, \quad n_{21} + n_{22} = c, \quad r + c = b.$$

This is in reality a univariate distribution which (taking n_{11} as the variable) may be written in terms of the notation displayed in (5.4.8) as

$$g(n_{11} | r, c) = \binom{r}{n_{11}}\binom{n-r}{c-n_{11}} \bigg/ \binom{n}{c}, \qquad n_{11} = 0, 1, \ldots, \min(r, c),$$

$$(5.4.10)$$

and we test the null hypothesis by computing the significance level of the observed value of n_{11} in this Hypergeometric distribution. The test has therefore reduced to exactly the same procedure as in the comparison of two binomial probabilities described in Example 5.4.1.

The test, which is due to R. A. Fisher, is often called 'Fisher's exact test' to distinguish it from an arithmetically simpler approximate procedure based on the χ^2 distribution, described in section 7.4.1. Fisher's own treatment of the test may be found in Fisher (1970) section 21.02 (see Bibliography C).

The comparison of two Binomial probabilities discussed in Example 5.4.1 called for a two-tailed test. A conditional test for independence in 2×2 tables as described in this section may, similarly, call for a two-sided treatment; it might, however, call for a one-sided treatment. Case (c) above ('two populations') is one in which a two-sided test is required, since the null hypothesis of equal populations would be contradicted not only by the proportion of B-negatives being higher in Exeter than in Edinburgh but also by the proportion being higher in Edinburgh than in Exeter.

In the situation exemplified by Case (b) above ('two treatments') however, the 'effect', if any, is one-directional, since data tending to discredit the null hypothesis would be those in which n_{11} was as small as or *smaller* than the number expected on the null hypothesis. For this, therefore the significance level would be the single 'tail'

$$\sum_{\substack{s \\ s \le n_{11}}} g(s|r, c)$$

where g is defined as in (5.4.10).

EXAMPLE 5.4.4.   *Effectiveness of an anti-cholera injection.*   An example of the kind of data calling for a one-sided test is provided by the following data on the numbers of members of a certain community who were attacked by cholera after having been inoculated with an anticholera drug, and the numbers not so inoculated who had been attacked (Greenwood and Yule's data):

	Attacked	Not attacked	Total
Inoculated	5	1625	1630
Not inoculated	11	1022	1033
Total	16	2647	2663

What we are interested in is the effectiveness of the inoculation in preventing infection, and therefore in the degree of smallness of the inoculated/attacked frequency $n_{11} = 5$. The notation required, as shown in (5.4.8), is

$u_{11} = 5$	$n_{12} = 1625$	$r = 1630$
$n_{21} = 11$	$n_{22} = 1022$	$n - r = 1033$
$c = 16$	$n - c = 2647$	$n = 2663$

and the significance level, with respect to the null hypothesis that the inoculation is useless, is, by (5.4.1),

$$\sum_{s \le n_{11}} \binom{r}{s}\binom{n-r}{c-s} \Big/ \binom{n}{c} = \sum_{s \le 5} \binom{1630}{s}\binom{1033}{16-s} \Big/ \binom{2663}{16}.$$

The Hypergeometric distribution involved here satisfies the conditions given in (5.4.3) for the application of the Binomial approximation (5.4.2), so that the significance level becomes, approximately,

$$\sum_{s \leq 5} \binom{16}{s} p^s q^{16-s}, \qquad p = 1630/2663 = 0.6129$$

$$q = 1 - p = 0.3871.$$

One evaluates

$$b_5 = \binom{16}{5} p^5 q^{11} = 0.0112,$$

and then finds

$$b_4 = (5/12)(q/p) b_5 \text{ etc.}$$

as indicated in (5.4.4), resulting in a significance level of

$$SL = 0.015.$$

This is a small enough value (about 1 in 70) to discredit the null hypothesis. The conclusion is that the attack-rate among those who had been inoculated is significantly less than it is among the others.

[See Example 7.4.1 for an application of the χ^2 test to the same data.]

EXAMPLE 5.4.5.   *Criminality among criminals' twins.*   A famous example of a 2×2 table is the following one, quoted by Fisher (1970) (see Bibliography C). The sample consists of 30 criminals each of whom had a twin brother. The 30 were cross-classified, one way according to the nature of the twinning (13 identical and 17 non-identical) and, the other way, according to the criminality or otherwise of the brother (12 convicted, 15 not):

	Brother convicted	Brother not convicted	
Identical twin	10	3	13
Non-identical twin	2	15	17
	12	18	30

(Reprinted with permission of Macmillan Publishing Company from *Statistical Methods for Research Workers*, 14th edition, by Sir Ronald A. Fisher. Copyright © 1970 University of Adelaide.)

On the null hypothesis, criminality is not significantly more frequent among identical twin brothers of criminals than among non-identical twin brothers. Is this supported by the data? Here again we need the one-tailed version of the conditional test. The null conditional distribution (5.3.10) becomes

$$g(n_{11}|r, c) = \binom{13}{n_{11}}\binom{17}{12-n_{11}} \Big/ \binom{30}{12}, \qquad n_{11} = 0, 1, \ldots, 12,$$

and the significance level is the sum of the probabilities allocated by this distribution to values of n_{11} that are not less than the observed value, viz. to the values 10, 11, 12. Thus

$$SL = \left\{\binom{13}{10}\binom{17}{2} + \binom{13}{11}\binom{17}{1} + \binom{13}{12}\binom{17}{0}\right\} \Big/ \binom{30}{12}$$

$$= 0\cdot0005.$$

The data are thus very highly significant; the null hypothesis is overwhelmingly discredited.

Labour-saving in the analysis of 2×2 tables

The computations exemplified above may readily be carried out with the aid of a small calculator. Two alternatives are however available. One is to use Finney's 'Tables for testing significance in a 2×2 contingency table' [Pearson and Hartley (1966) Vol. 1, Table 38—see Bibliography G]. The other is to use the χ^2 approximation explained in Chapter 7 [see § 7.4.1]. Finney's tables do not of course give the precise significance level for every possible 2×2 table, but, for sample sizes up to $n = 60$, enable one to tell whether the SL is less than 0·005, between 0·005 and 0·01, between 0·01 and 0·05, or greater than 0·05; with somewhat less detail for n up to 80.

Conditional tests for $a\times b$ tables

The methods explained above for the analysis of 2×2 tables may be extended to $a\times b$ tables. An account of the exact conditional will be found in Kalbfleisch (1979) II—see Bibliography C. The χ^2 approximation, which often suffices, is outlined in Chapter 7.

5.5.   TESTS INVOLVING MORE THAN ONE PARAMETER. GENERALIZED LIKELIHOOD RATIO TESTS

The likelihood ratio test for a simple hypothesis was described in section 5.2.1 (f_3). A generalization of that procedure is particularly useful when there are several parameters and the null hypotheses specifies the value(s) of only a subset of them. We first give a formal definition of the procedure and then illustrate it with examples.

Suppose that the probability model for the data involves m parameters denoted by $\theta = (\theta_1, \theta_2, \ldots, \theta_m)$. Let Ω denote the 'unrestricted parameter space', that is the set of all allowable values of θ; and suppose the null hypothesis specifies the values of k of the parameters ($k < m$), but says nothing about the others. We denote this by

$$H : \theta \in \Omega_H \qquad (\Omega_H \subset \Omega). \qquad (5.5.1)$$

It will be convenient to call Ω_H the restricted parameter space. (In particular cases θ might be a scalar, and Ω_H might contain a single point only.) Denote the likelihood of θ [see § 4.1.3.1] on the data vector x by

$$l(\theta; x), \qquad (5.5.2)$$

its greatest value for variations of θ when $\theta \in \Omega_H$ (x remaining constant) by

$$l_H = \max_{\theta \in \Omega_H} l(\theta; x), \qquad (5.5.3)$$

and its greatest value for unrestricted variations of θ over Ω by

$$l_{max} = \max_{\theta \in \Omega} l(\theta; x). \qquad (5.5.4)$$

The generalized likelihood ratio is defined as

$$\lambda = \lambda(x) = l_H / l_{max}, \qquad (5.5.5)$$

and the significance set for testing H when the data vector is x_0 is

$$G(x_0) = \{x : \lambda(x) \le \lambda(x_0)\}$$
$$= \{x : d(x) \ge d(\bar{x}_0)\}$$

where

$$d(x) = -2 \ln \lambda(x), \qquad (5.5.6)$$

this latter quantity often being more convenient to compute and manipulate than $\lambda(x)$ itself. The significance level is

$$SL(x_0) = P(X \in G(x_0)|H). \qquad (5.5.7)$$

We illustrate with an important example.

EXAMPLE 5.5.1.   *Student's t-test.*   In this example it is required to test the expectation of a Normal distribution when the standard deviation is unknown. The data vector is

$$x = (x_1, x_2, \ldots, x_n)$$

where the x_i are realizations of a Normal (μ, σ) variable, so that the likelihood function (5.5.2) is proportional to

$$l(\mu, \sigma; x) = \sigma^{-n} \exp -\frac{1}{2\sigma^2} \Sigma (x_i - \mu)^2.$$

The unrestricted parameter space is

$$\Omega : -\infty < \mu < \infty; \qquad \sigma > 0.$$

Suppose the null hypothesis H specifies the value of μ as μ_0, leaving σ unspecified, so that (5.5.1) becomes

$$H: \mu = \mu_0,$$

whence the restricted parameter space Ω_H is

$$\Omega_H: \mu = \mu_0, \qquad \sigma > 0.$$

In the restricted case the maximizing value $\tilde{\sigma}$ of σ is given by

$$\tilde{\sigma}^2 = \sum (x_i - \mu_0)^2 / n,$$

whence

$$l_H = \max_{\sigma > 0} l(\mu, \sigma; x)$$

$$= l(\mu_0, \tilde{\sigma}; x) = \tilde{\sigma}^{-n} e^{-1/2n}.$$

In the unrestricted case the maximizing values $\hat{\mu}$ of μ and $\hat{\sigma}$ of σ are given by

$$\hat{\mu} = \bar{x} = \sum x_r / n$$

and

$$\hat{\sigma}^2 = \sum (x_r - \bar{x})^2 / n,$$

whence

$$l_{max} = \max_{\substack{-\infty < \mu < \infty \\ \sigma > 0}} l(\mu, \sigma; x)$$

$$= l(\hat{\mu}, \hat{\sigma}; x) = \hat{\sigma}^{-n} e^{-1/2n}.$$

It follows that

$$\lambda = (\tilde{\sigma}/\hat{\sigma})^{-n}.$$

Now

$$(\tilde{\sigma}/\hat{\sigma})^2 = \sum (x_i - \mu_0)^2 / \sum (x_i - \bar{x})^2$$

$$= 1 + n(\bar{x} - \mu_0)^2 / \sum (x_i - \bar{x})^2.$$

In terms of

$$s^2 = \sum (x_i - \bar{x})^2 / (n - 1)$$

this becomes

$$1 + \{t(x)\}^2 / (n - 1)$$

where

$$t(x) = n^{1/2}(\bar{x} - \mu_0)/s \qquad\qquad (5.5.8)$$

is Student's statistic. (Note that s^2 is the usual estimate of the parameter σ^2 in this situation.)

The significance set G is the set of all possible n-ples $(x_1, x_2, \ldots, x_n)$ for which the value of $\lambda = (\tilde{\sigma}/\hat{\sigma})^{-n}$ would be less than the value calculated from the actual observations, i.e. for which the value of $|t(x)|$ would be greater than that observed. Thus the significance level is

$$SL(t) = P\{|T| \geq |t| \,|\, H\}$$

where t is the observed value of Student's statistic (5.5.8) and T is the corresponding random variable. The distribution of T has a standard form which is widely tabulated [see Appendix (T5)] under the title 'Student's distribution on $n-1$ degrees of freedom'. [See § 2.5.5]. [The test is intuitively appealing. It differs from what would be done if σ had a known value (see § 5.8.1) only in replacing the unknown σ by its estimate $\hat{\sigma}$ and using the sampling distribution implied by this replacement (see § 5.8.2).]

For a numerical illustration of the t-test (the 1-sided version) see Examples 5.8.1 and 5.8.2.

EXAMPLE 5.5.2.   *Likelihood ratio tests for regression parameters.*   A generalization of Example 5.5.1, which is of fundamental importance in statistical theory, arises in the theory of 'linear regression'. The data vector is again denoted by $x = (x_1, x_2, \ldots, x_n)$, but this time the observation x_r is supposed to come from a Normal distribution with expectation

$$E(X_r) = a_{r1}\theta_1 + a_{r2}\theta_2 + \ldots + a_{rm}\theta_m, \qquad r = 1, 2, \ldots, n \qquad (m < n) \tag{5.5.9}$$

where the a_{rs} are given constants and the θ_s are unknown parameters (called 'regression coefficients'), and with variance σ^2 (also unknown but the same for all observations). It is required to test a hypothesis that specifies the values of a subset of the regression coefficients.

For example the data might represent tensile strength measurements at various temperatures, with the probability model giving the expected strength at temperature c (in degrees Centigrade) as $\theta_1 + c\theta_2$, so that θ_1 denotes the expected strength at freezing point and θ_2 the expected increase per degree; in this case (5.5.9) simplifies to

$$E(X_r) = \theta_1 + c_r\theta_2, \qquad r = 1, 2, \ldots, n$$

(i.e. $a_{r1} = 0$, $a_{r2} = c_r$, $a_{r3} = \ldots = a_{rm} = 0$, $r = 1, 2, \ldots, n$), where c_r denotes the temperature at which X_r was observed, yielding the observed strength x_r. An example of the kind of hypothesis to be tested might be

$$H: \theta_2 = 0$$

(c.f. (5.5.1)). For the situation envisaged in (5.5.9), let the unrestricted parameter space be

$$\Omega: \begin{cases} -\infty < \theta_r < \infty, & r = 1, 2, \ldots, n, \\ \sigma > 0, \end{cases}$$

and let us suppose the null hypothesis (5.5.1) is

$$H: \theta_1 = \theta_2 = \ldots = \theta_k = 0, \qquad (k < m) \qquad (5.5.10)$$

so that the restricted parameter space Ω_H is

$$\Omega_H : \begin{cases} \theta_r = 0, & r = 1, 2, \ldots, k; \\ -\infty < \theta_2 < \infty, & s = k+1, k+2, \ldots, m; \\ \sigma > 0. \end{cases}$$

The calculations involved in evaluating the likelihood ratio statistic are similar to those carried out in Example 5.5.1. Instead of Student's ratio t, the test statistic that emerges is Fisher's ratio F, viz.

$$F = \frac{(n-m)(S_H - S_\Omega)}{(m-k)S_\Omega} \qquad (5.5.11)$$

where S_H is the minimum of the 'restricted sum of squares'

$$\sum_r (x_r - a_{r1}\theta_1 - \ldots a_{rk}\theta_k)^2$$

and S_Ω is the minimum of the 'unrestricted sum of squares'

$$\sum_r (x_r - a_{r1}\theta_1 - \ldots - a_{rk}\theta_k - \ldots - a_{rm}\theta_m)^2.$$

It turns out that, on H, the sampling distribution of the statistic F is the standard 'F distribution on $(n-m)$ and $(m-k)$ degrees of freedom', usually called $F_{n-m, m-k}$ in tables, and the significance level is the (widely tabulated) probability (on H) of exceeding the observed value of F. [See § 2.5.6.]

(The statistic (5.5.11) is always denoted by the letter F, and this usage forces us to depart from the usual convention in which a capital letter denotes a random variable and the corresponding small letter a realization of it.)

Several illustrations are provided in Chapter 8.

5.6. LARGE SAMPLE APPROXIMATION TO THE SIGNIFICANCE LEVEL OF A LIKELIHOOD RATIO TEST

In the test situation described in section 5.5, the sampling distribution on H of the test statistic

$$d(x) = -2 \ln \lambda(x),$$

of (5.5.6) is approximately the chi-squared distribution on k degrees of freedom, where k is the number of parameters specified by the null hypothesis. (This result assumes that the probability model satisfies certain regularity conditions.)

The approximation can in practice be surprisingly good even for samples of modest size, as is shown in the following example.

EXAMPLE 5.6.1. 2×2 *Contingency table.*   Consider a 2×2 cross-classified contingency table (cf. § 5.4.2) with cell frequencies a, b, c and d, row and column totals r_1, r_2, c_1, c_2 and sample size n; as in the following table:

a	b	r_1
c	d	r_2
c_1	c_2	n

Denote the corresponding probabilities by

p_{11}	p_{12}	p_{10}
p_{21}	p_{22}	p_{20}
p_{01}	p_{02}	1

where $p_{10} = p_{11} + p_{12}$, $p_{20} = p_{21} + p_{22}$, $p_{01} = p_{11} + p_{21}$, and $p_{02} = p_{12} + p_{22}$, and $p_{10} + p_{20} = p_{01} + p_{02} = 1$. These probabilities may be written in terms of the three parameters α, β, θ, as follows:

$\alpha\beta + \theta$	$\alpha(1-\beta) - \theta$	α
$(1-\alpha)\beta - \theta$	$(1-\alpha)(1-\beta) + \theta$	$1-\alpha$
β	$1-\beta$	1

The likelihood function [see § 4.13] is proportional to

$$l(\alpha, \beta, \theta) = (\alpha\beta + \theta)^a \{\alpha(1-\beta) - \theta\}^b \{(1-\alpha)\beta - \theta\}^c \{(1-\alpha)(1-\beta) + \theta\}^d.$$

$$(5.6.1)$$

Suppose we wish to test the hypothesis H that the two categorizations are mutually independent, that is that the probability of getting an observation the (r, s) cell $(r, s = 1, 2)$ is the product $p_r q_s$ where p_r is the sum of the two cell probabilities in the rth row and q_s the corresponding sum for the sth column. This is equivalent to

$$H: \theta = 0.$$

The greatest value of $l(\alpha, \beta, \theta)$ for variations of α, β, θ subject to H is

$$l_H = l(\tilde{\alpha}, \tilde{\beta}, 0)$$

where

$$\tilde{\alpha} = r_1/n, \qquad \tilde{\beta} = c_1/n,$$

so that

$$l_H = (r_1c_1/n^2)^a(r_1c_2/n^2)^b(r_2c_1/n^2)^c(r_2c_2/n^2)^d;$$

while the greatest value, with no restrictions, is

$$l_{\max} = l(\hat{\alpha}_1\hat{\beta}, \hat{\theta})$$

where

$$\hat{\alpha}\hat{\beta} + \hat{\theta} = a/n, \qquad \hat{\alpha}(1-\hat{\beta}) - \hat{\theta} = b/n,$$

$$(1-\hat{\alpha})\hat{\beta} - \hat{\theta} = c/n, \qquad (1-\alpha)(1-\hat{\beta}) + \hat{\theta} = d/n,$$

so that

$$l_{\max} = (a/n)^a(b/n)^b(c/n)^c(d/n)^d.$$

Then the likelihood ratio statistic (5.5.5) is

$$\lambda = l_H/l_{\max}$$

$$= \left(\frac{r_1c_1}{na}\right)^a\left(\frac{r_1c_2}{nb}\right)^b\left(\frac{r_2c_1}{nc}\right)^c\left(\frac{r_2c_2}{nd}\right)^d.$$

For the criminality-in-twins data quoted in Example 5.4.4, the table was

$a = 10$	$b = 3$	$13 = r_1$
$c = 2$	$d = 15$	$17 = r_2$
$c_1 = 12$	$c_2 = 18$	$30 = n$

$$r_1c_1 = 5\cdot2, \qquad r_1c_2 = 7\cdot8,$$

$$r_2c_1 = 6\cdot8, \qquad r_2c_2 = 10\cdot2,$$

whence

$$\lambda = \left(\frac{5\cdot2}{10}\right)^{10}\left(\frac{7\cdot8}{3}\right)^3\left(\frac{6\cdot8}{2}\right)^2\left(\frac{10\cdot2}{15}\right)^{15}$$

and

$$-2\ln\lambda = 14.$$

Since the null hypothesis specified the value of only one parameter, the asymptotic theorem (if it could be taken as applying to such a small sample) would tell us that, on H, this value 14 was to be regarded as a realization of the χ^2 distribution on 1 degree of freedom. The significance level, on this interpretation, would therefore be

$$SL = P(\chi_1^2 \geq 14) = P(U^2 \geq 14)$$

where U is a standard Normal variable (since U^2 has the χ_1^2 distribution). The significance level is therefore

$$SL = P(|U| \geq 3 \cdot 7)$$

$$= 0 \cdot 0002.$$

The significance level obtained in the conditional test in Example 5.4.4 was 0·0004. The large sample likelihood ratio approximation is therefore in error by only 0·0002.

5.7. RANDOMIZATION TESTS

In the analysis of experimental results it is usual to postulate a probability model and to conduct the analysis in terms of the parameters of this model. This is however not, strictly, necessary in a properly designed experiment. The following example outlines Fisher's treatment of a celebrated experiment of Charles Darwin. [See Fisher (1951) Chapter III.]

EXAMPLE 5.7.1. *Comparison of two 'treatments'.* In his book *Statistical Methods and Scientific Inference* Fisher discusses a famous experiment carried out by Charles Darwin on the comparative growth-rates of seeds of plants of the same variety, one set·of seeds having been produced by cross-fertilization and one by self-fertilization. Fifteen pairs of seeds were grown under comparable conditions, in 15 pots, each pot having a pair of seeds of which one was 'crossed' and the other 'selfed'. For each pair of the resulting plants the excess in height of the 'crossed' over the 'selfed' plant was recorded, with the following results:

49	23	56
−67	28	24
8	41	75
16	14	60
6	29	−40 (Total 314)

Table 5.7.1: Differences in height
(in eighths of an inch).

Since in thirteen of the fifteen comparisons the cross-fertilized plant was taller than the self-fertilized one, there is clearly evidence in support of the contention that cross-fertilization produces taller plants than does self-fertilization. The problem is to decide how strong this evidence is.

Fisher's argument went like this: each pot had two seeds, one selfed and one crossed, placed diametrically opposite each other, one on the east and one on the west, say. Strictly speaking no conclusions can be drawn unless we can assume that the allocation of seed, selfed or crossed, to a particular location was made randomly, e.g. by tossing a coin. Let us suppose that this was done.

We now tentatively adopt the null hypothesis H that there are *no* systematic differences in growth potential as between selfed and crossed seeds. The observed differences must then be attributed to environmental factors such as soil fertility, environmental atmosphere conditions, sun, light, etc.

In one experiment involving 15 pairs of plants, the observed differences $z_i = x_i - y_i$ (where x_i = height of cross-fertilized plant in ith pot, and y_i = height of self-fertilized plant in the same pot) were as shown in Table 5.7.1.

It is proposed to take the mean of these differences (or equivalently, their sum) as the relevant statistic.

The actual disposition of seeds (selfed on the east and crossed on the west or vice versa) was one of the 2^{15} possible arrangements, all equally likely because of randomization. The arrangement that differs from the one actually adopted only in respect of the first pot would, on the null hypothesis, have given results differing from those obtained only in replacing the value 49 for the first pot by -49, and the value of $\sum z_i$ would therefore have been 116. In a similar way one can compute the $2^{15} = 32,768$ equally likely values of $\sum z_i$, one for each possible arrangement. On carrying out the computation it is found that in 863 cases the value of the sum would be as large as or larger than that actually observed, so that the significance level is

$$P\{\text{sum of deviations} \geq 314\} = \frac{863}{32,768}$$

$$= 0 \cdot 026,$$

thus showing (to quote Fisher) 'a consistent advantage of cross-fertilized seed . . .: since only once in [about] 40 trials [i.e. with probability $0 \cdot 026$] would a chance deviation have been observed both so large and in the right direction'.

5.8. STANDARD TESTS WITH THE NORMAL DISTRIBUTION MODEL

In this section we describe some of the tests that are commonly used when the underlying probability model is Normal. In each case the significance level may be interpreted in accordance with Table 5.2.1.

5.8.1. The Significance of a Sample Mean when the Variance is Known

A sample $x = (x_1, x_2, \ldots, x_n)$ is drawn from a population, assumed to be Normal, in which the expectation μ is unknown but the variance is known to equal σ_0^2 [see II, § 11.4.3]. (That the variance is known is almost always an unrealistic assumption. The purpose of describing the procedure is primarily as an introduction to the more realistic t-test of section 5.8.2). To test whether the data are in accord with the null hypothesis H that $\mu = \mu_0$ (e.g. $\mu_0 = 0$), form the statistic

$$u = (\bar{x} - \mu_0)/(\sigma_0/\sqrt{n}), \tag{5.8.1}$$

where $\bar{x}$ is the sample mean $(x_1 + \ldots + x_n)/n$. This is a realization of the r.v. $U = (\bar{X} - \mu_0)/(\sigma_0\sqrt{n})$ [see § 2.5.3(b)], where $n\bar{X} = X_1 + \ldots + X_n$, the X_r being independent Normal (μ, σ_0) variables, whence $\bar{X}$ is Normal $(\mu, \sigma_0/\sqrt{n})$ and U is Normal $(\mu - \mu_0, 1)$.

If the competing hypothesis is one-sided, that is

$$H: \mu > \mu_0,$$

the significance level of the observed value of the statistic u is

$$SL = P(U \geq u)$$

$$= 1 - \Phi(u).$$

In the case however where the competing hypothesis is two-sided, viz.

$$\mu \neq \mu_0,$$

the significance level is

$$SL = P(U \geq u \text{ or } U \leq -u)$$

$$= 2\{1 - \Phi(u)\}.$$

(Φ denotes the standard Normal integral. Some published versions (see for example Appendix T3) of the Φ-tables tabulate the function $\Phi(u)$ against u, others tabulate $1 - \Phi(u)$ against u.)

5.8.2.   The Significance of a Mean when the Variance is Unknown: Student's *t*-Test

The sample $x = (x_1, x_2, \ldots, x_n)$ is drawn from a population, assumed to be Normal, with unknown parameters, the expectation being denoted by μ and the variance by σ^2. The null hypothesis is

$$H: \mu = \mu_0.$$

The procedure is to evaluate the following statistics:

(i) Sample mean: $\bar{x} = \sum_1^n x_r/n$.
(ii) Sample variance: $s^2 = \{\sum_1^n x_r^2 - n\bar{x}^2\}/(n-1)$
$$= \{\sum_1^n x_r^2 - \bar{x}\sum_1^n x_r\}/(n-1),$$

and then to construct 'Student's ratio':

$$t = (\bar{x} - \mu)/(s/\sqrt{n}). \tag{5.8.2}$$

(Compare (5.8.1). In (5.8.2) the unknown σ_0 has been replaced by its estimate s. See also Example 4.5.2.)

On H, t is a realization of Student's random variable T_{n-1} with $n-1$ degrees of freedom ('d.f.').

If the competing hypothesis is the one-sided one

$$H': \mu > \mu_0$$

(so that values of $\bar{x}$ less than μ_0 are of no interest or are regarded as virtually impossible, and the value of t obtained from (5.8.2) is virtually certain to be positive), the significance level is

$$SL = P(T_{n-1} \geq t). \qquad (5.8.3)$$

If the competing hypothesis is one-sided in the other direction, viz.

$$H': \mu < \mu_0,$$

the significance level is

$$SL = P(T_{n-1} \leq t), \qquad (5.8.4)$$

t having in this case a negative value. In the two-sided case, where the competing hypothesis is

$$\mu \neq \mu_0$$

the significance level is

$$SL = 2P(T_{n-1} \geq |t|). \qquad (5.8.5)$$

Published tables do not in fact tabulate $p_{n-1}(t) = P(T_{n-1} \geq t)$ against t. They tabulate t against $p_{n-1}(t)$ [see Appendix (T5)]. The significance level then has to be found by interpolation. [See section headed 'Inverse Tables' in § 5.2.2]. A complication which demands care from the user is that some versions tabulate, instead t against $2p_{n-1}(t)$.

The most important case of the test has $\mu_0 = 0$, which occurs when testing for the existence of an 'effect' in meaningfully paired sets of data, as in the following example.

EXAMPLE 5.8.1. *Matched pairs.* Each of n specimens of wire is cut into two pieces, one of which (chosen at random) has its tensile strength measured at a fixed low temperature and the other at a fixed high temperature. It is desired to test whether this temperature difference affects the tensile strength. The data are:

Specimen No.	1	2	...	n
Low-temperature measurement	x_1	x_2	...	x_n
High-temperature measurement	y_1	y_2	...	y_n
Difference	z_1	z_2	...	z_n

where $z_r = x_r - y_r$, $r = 1, 2, \ldots, n$. By analysing the z's, rather than the x's and y's separately, one is considering only comparisons between pieces of the same original specimen and so eliminating irrelevant differences that might exist between the specimens themselves. The z's are regarded as realizations of a Normal variable which, on the null hypothesis that there is no temperature effect on the strength of the wire, has zero expectation (so that here $\mu_0 = 0$), and unknown variance. Evaluate

$$\bar{z} = \sum_1^n z_r/n,$$

$$s^2 = \left(\sum_1^n z_r^2 - n\bar{z}^2\right)\Big/(n-1),$$

and

$$t = \bar{z}/(s/\sqrt{n}).$$

Carry out the t-test (one-sided or two-sided, whichever is appropriate: if for example it was known on technical grounds that the effect, if one existed, would be a decrease of strength with increase of temperature, the one-sided test using (5.8.4) would be the one to use) with $n-1$ degrees of freedom. With $n = 10$, for example and $t = -2\cdot95$, the (one-sided) test of the null hypothesis that there is no effect, against the hypothesis that there is a negative effect, gives the significance level

$$SL = P(T_9 \le -2\cdot95)$$

$$= P(T_9 \ge 2\cdot95)$$

$$= 0\cdot002$$

[see Appendix (T5)].

This provides strong evidence against the null hypothesis. (Note on the role of the Normality of the underlying probability model: The t-test is *robust* against moderate departures from Normality; that is, it is rather insensitive to such departures, so that, in small samples where no strong evidence of the nature of the underlying distribution is available, it is fairly safe to use the t-test.)

EXAMPLE 5.8.2.   *Darwin's data.*   If we analyse Darwin's experiment (see Example 5.7.1) from this standpoint, we regard the 15 height differences in Table 5.7.1 as 15 independent realizations $x_1, x_2, \ldots, x_{15}$ of a Normal (μ, σ) variable, where σ is unknown, and, on the null hypothesis, $\mu = 0$. To test this hypothesis we compute

$$t = \bar{x}/(s/\sqrt{15})$$

where

$$x = 314/15 = 20\cdot933$$

and

$$14s^2 = \sum x_i^2 - 15\bar{x}^2$$

$$= 25814 - 6573,$$

$$s^2 = 1374\cdot35,$$

$$s = 37\cdot07$$

$$s/\sqrt{15} = 9\cdot572,$$

so that

$$t = 2 \cdot 187.$$

On 14 d.f. the significance level for a one-sided test, using the t-tables in Appendix T5, is $0 \cdot 025$, a small enough value to discredit the null hypothesis. The experiment therefore provides fairly strong evidence to support the hypothesis that cross-fertilization produces taller plants than does self-fertilization.

It might be noted that the significance level obtained here is very close to the value $(0 \cdot 026)$ obtained by Fisher's randomization analysis (Example 5.7.1).

5.8.3.   The Fisher–Behrens Test for the Significance of the Difference between Two Means

Here we have two samples, $(x_1, x_2, \ldots, x_m)$ and $(y_1, y_2, \ldots, y_n)$, of sizes m and n respectively, from Normal populations with respective expectation μ_1, μ_2 and unknown variances σ_1^2, σ_2^2. It is desired to test the null hypothesis that $\mu_1 = \mu_2$ without making the assumption that $\sigma_1^2 = \sigma_2^2$.

Evaluate the sample means

$$\bar{x} = \sum_1^m x_r / m, \qquad \bar{y} = \sum_1^n y_r / n,$$

and the sample variances

$$\left. \begin{aligned} s_1^2 &= \left(\sum_1^m x_r^2 - m\bar{x}^2 \right) / (m - 1), \\ s_2^2 &= \left(\sum_1^n y_r^2 - n\bar{y} \right)^2 / (n - 1). \end{aligned} \right\} \tag{5.8.6}$$

Then s_1^2/m and s_2^2/n are the sample variances of $\bar{x}$, $\bar{y}$ respectively, and $s_1^2/m + s_2^2/n$ that of $\bar{x} - \bar{y}$ [see II, § 9.2.5].

Evaluate the *Fisher–Behrens statistic*

$$d = (\bar{x} - \bar{y}) / \sqrt{\left(\frac{s_1^2}{m} + \frac{s_2^2}{n} \right)}.$$

The significance level of this test statistic depends not only on m and n but also on the ratio s_1^2/s_2^2, expressed in the tables through

$$\theta = \tan^{-1} \left(\frac{s_1 / \sqrt{m}}{s_2 / \sqrt{n}} \right).$$

This significance level is obtainable by interpolation from Sukhatme's table (Table VI in Fisher and Yates (1974)—Bibliography G), which gives the 1% and 5% values of d for m and n equal to 6, 8, 12, 24, and ∞, and for $\theta = 0°$, 15°, 30°, 45°, 60°, 75° and 90°.

[Warning: The notation employed in the Fisher and Yates tables uses s_1^2 to denote what we call s_1^2/m, and s_2^2 for our s_2^2/n.]

For example, with $m = 20$, $\bar{x} = 29 \cdot 233$, $s_1^2 = 5 \cdot 62$ and $n = 10$, $\bar{y} = 27 \cdot 562$, $s_2^2 = 2 \cdot 19$, we have $s_1^2/m + s_2^2/n = 0 \cdot 500$, whence

$$d = 2 \cdot 364,$$

and

$$\theta = \text{arc tan} \, (1 \cdot 133)$$

$$= 48 \cdot 6°.$$

The relevant entries in the table give:

5% *points*

	n	$\theta = 45°$	$\theta = 60°$
$m = 12$	8	2·229	2·262
	12	2·167	2·169
$m = 24$	8	2·175	2·236
	12	2·112	2·142

1% *points*

	n		
$m = 12$	8	2·083	3·192
	12	2·954	2·978
$m = 24$	8	2·988	3·158
	12	2·853	2·938

(Reproduced by permission of Longman Group Ltd., from *Structural Tables for Biological, Agricultural and Medical Research*, by R. A. Fisher and F. Yates, 1974.)

Linear interpolation gives the following values for our sample, ($m = 20$, $n = 10$ and $\theta = 48 \cdot 6°$):

5% point: 2·235

1% point: 2·971.

Since our observed value, 2·364, lies between these two values, its significance level is between 0·01 and 0·05; about 0·04.

This test is not often used in practice because, unless the unknown variances σ_1^2 and σ_2^2 are severely unequal, the assumption that they are in fact equal gives results close enough to the Fisher–Behrens test, with very much less labour. [See § 5.8.4.]

5.8.4.   The *t*-Test for the Significance of the Difference between Two Means (Variances Being Equal)

Let $(x_1, x_2, \ldots, x_m)$ and $(y_1, y_2, \ldots, y_n)$ be samples from Normal distributions with unknown parameters (μ_1, σ), (μ_2, σ) respectively. Note that the unknown variances are assumed to be equal, with common value σ^2. To test the null hypothesis that $\mu_1 = \mu_2$, proceed as follows. Evaluate the sample means

$$\bar{x} = \sum_1^m x_r/m, \qquad \bar{y} = \sum_1^n y_r/n,$$

the sample 'sums of squares',

$$(m-1)s_1^2 = \left(\sum_1^m x_r^2 - m\bar{x}^2 \right)$$

$$(n-1)s_2^2 = \left(\sum_1^n y_r^2 - n\bar{y}^2 \right)$$

(*c.f.* (5.8.6)), and the pooled estimate of the common variance:

$$s^2 = \{(m-1)s_1^2 + (n-1)s_2^2\}/(m+n-2). \tag{5.8.7}$$

Then the estimate of the sampling variance of $\bar{x} - \bar{y}$ is

$$s^2 \left(\frac{1}{m} + \frac{1}{n} \right) = \frac{m+n}{mn} s^2,$$

so that the standard error of $\bar{x} - \bar{y}$ is

$$s \sqrt{\left(\frac{1}{m} + \frac{1}{n} \right)} = s\sqrt{\{(m+n)/mn\}}.$$

This motivates the next step, which is to evaluate

$$t = \frac{(\bar{x} - \bar{y})}{s} \sqrt{\left(\frac{mn}{m+n} \right)}. \tag{5.8.8}$$

On the null hypothesis this has Student's distribution [see § 2.5.5] on $m+n-2$ degrees of freedom. The significance level may be obtained from the *t*-tables as in section 5.8.2. [see also Example 4.5.5].

For example, with the data analysed by the Fisher–Behrens test in section 5.8.3, but with the assumption now of a common variance, we have

$$m = 20, \qquad \bar{x} = 29 \cdot 233, \qquad s_1^2 = 5 \cdot 62$$

$$n = 10, \qquad \bar{y} = 27 \cdot 562, \qquad s_2^2 = 2 \cdot 19,$$

whence, by (5.8.7), the pooled variance estimate is s^2, where

$$28s^2 = (19 \times 5 \cdot 62) + (9 \times 2 \cdot 19).$$

The pooled estimate of the common variance σ^2 is $s^2 = 4 \cdot 5175$ and that of the difference between the means is

$$s^2\left(\frac{1}{20}+\frac{1}{10}\right) = 0 \cdot 678 = (0 \cdot 823)^2.$$

Thus our value of t is

$$t = \frac{1 \cdot 671}{0 \cdot 823} = 2 \cdot 030.$$

The (two-sided) significance level of this, on 28 d.f., is $0 \cdot 051$. (Note that this value is close to the value of 'about $0 \cdot 04$' obtained by the Fisher–Behrens test in section 5.8.3).

5.8.5.   The t-Test for the Significance of a Regression Coefficient in Simple Linear Regression

Suppose that we measure the height x and the weight y of each of n individuals, obtaining the data $(x_1, y_1), (x_2, y_2), \ldots, (x_n, y_n)$; or the height x of a father and the height y of his eldest adult son; or the temperature x at which steel has been annealed and the tensile strength y of the resulting material; and so on. The observations $(x_r, y_r), r = 1, 2, \ldots, n$, may be regarded as independent realizations of the pair of random variables (X, Y). It is frequently useful to investigate either or both of the conditional expectations $E(Y|X = x), E(X|Y = y)$. These are, respectively, the regression of Y on X and the regression of X on Y. In many cases of practical interest one, or both, is a linear function. In the case of the regression of heights of fathers (Y) or heights of sons (X), for example, the regression

$$E(Y|X = x)$$

is (to an acceptable approximation) a linear function of x, say

$$E(Y|X = x) = \alpha' + \beta' x$$

[see Examples 4.5.3 and 4.5.4.] A plot of y_r against x_r, $r = 1, 2, \ldots, n$, would give a 'belt' of points indicating a linear trend. To estimate α' and β' it is more convenient to represent the equation of the line as

$$\alpha + \beta(x - \bar{x})$$

where $\bar{x} = \sum_1^n x_r/n$, so that $\beta' = \beta$ and $\alpha' = \alpha - \beta\bar{x}$. Estimation of α and β by least squares (see Chapter 8) leads to the estimate

$$a = \bar{y}\left(= \sum_1^n y_r/n \right) \tag{5.8.9}$$

for α, and

$$b = \sum y_r(x_r - \bar{x}) \bigg/ \sum (x_r - \bar{x})^2 \tag{5.8.10}$$

for β. If the conditional distribution of Y, given $X = x$, may be taken as Normal (μ, σ) for each x, a and b are realizations of independent Normal variables with parameters $(\alpha, \sigma/\sqrt{n})$ and $(\beta, \sigma/\sqrt{\sum_1^n (x_r - \bar{x})^2})$ respectively. The appropriate estimate of σ^2 is now

$$s^2 = \sum \{y_r - a - b(x_r - \bar{x})\}^2 / (n-2)$$
$$= \{\sum y_r^2 - n\bar{y}'^2 - b^2(\sum x_r^2 - n\bar{x}^2)\}/(n-2) \qquad (5.8.11)$$

since then

$$\frac{a - \alpha}{s/\sqrt{n}} \qquad (5.8.12)$$

and

$$\frac{b - \beta}{s/\sqrt{\sum (x_r - \bar{x})^2}} \qquad (5.8.13)$$

are independent realizations of Student's distribution, on $n-2$ degrees of freedom in each case, so that significance tests intended to answer questions such as the following:

does a differ significantly from a postulated value α_0 of α?

does b differ significantly from a postulated value β_0 of β?

may be constructed with the aid of Student's distribution. For a numerical example and further details see section 6.5.

EXAMPLE 5.8.3. *Significance of the slope parameter.* A common application is to the question: are the values of y affected at all by the values of x? that is, is $\beta = 0$? In other words, does the observed values of b differ from zero merely as a result of sampling fluctuations? The appropriate null hypothesis is that $\beta = 0$. Then (compare (5.8.13))

$$t = \frac{b\sqrt{\sum (x_r - \bar{x})^2}}{s} \qquad (5.8.14)$$

is a realization of Student's ratio on $n-2$ d.f. If the alternative hypothesis is that $\beta \neq 0$, the two-sided t-test is required, the significance level of the sample is

$$2P(T \geq |t|)$$

where T has Student's distribution on $n-2$ d.f., and t is computed from (5.8.14), the value of s in this expression being calculated from (5.8.11). The significance of the estimate b is then interpreted in terms of this value in accordance with Table 5.1.2.

The significance of a with respect to the null hypothesis that $\alpha = \alpha_0$ may be similarly and independently tested against the hypothesis that (e.g.) $\alpha \neq \alpha_0$

by evaluating

$$t = (a - \alpha_0)/(s/\sqrt{n}).$$

On the null hypothesis this has Student's distribution on $n-2$ degrees of freedom (c.f. § 5.8.2).

5.8.6.   Test for the Equality of Two Variances

Let $(x_1, x_2, \ldots, x_m)$ be a sample of observations on the random variable X which is Normal (μ_1, σ_1), and $(y_1, y_2, \ldots, y_n)$ a sample of observations on the random variable Y which is Normal (μ_2, σ_2). Are the data consistent with the null hypothesis H_0 that $\sigma_1^2 = \sigma_2^2$ $(= \sigma^2$, say)?
Compute the following quantities:

$$\bar{x} = \sum_1^m x_r/m,$$

$$s_1^2 = \sum_1^m (x_r - \bar{x})^2/(m-1) = \left\{ \sum_1^m x_r^2 - \bar{x} \sum_1^m x_r \right\}/(m-1),$$

$$\bar{y} = \sum_1^n y_r/n,$$

$$s_2^2 = \sum_1^n (y_r - \bar{y})^2/(n-1) = \left\{ \sum_1^n y_r^2 - \bar{y} \sum_1^n y_r \right\}/(n-1).$$

Then, if H_0 holds,

$$(m-1)s_1^2/\sigma^2 \quad \text{and} \quad (n-1)s_2^2/\sigma^2,$$

are independently distributed as $\chi^2(m-1)$ and $\chi^2(n-1)$ respectively [see § 2.5.4(a)] (where, as usual, '$\chi^2(\nu)$' refers to the chi-squared distribution on ν degrees of freedom). It follows that the 'variance ratio' [see § 2.5.6]

$$s_1^2/s_2^2$$

is a realization of a random variable which has the 'F-distribution' on $m-1$ and $n-1$ degrees of freedom, which we abbreviate to 'the $F(m-1, n-1)$ distribution'. [see also Example 4.5.7.]

Case (i): *The alternative hypothesis is that $\sigma_1^2 > \sigma_2^2$*

On the null hypothesis, the observed value of the ratio s_1^2/s_2^2 is likely to be close to unity, and large values of the ratio will be unlikely; whereas on the alternative hypothesis large values of the ratio are more likely to happen. The appropriate significance set is therefore the upper tail of the F-distribution lying beyond the observed value of s_1^2/s_2^2, and the significance level is

$$SL = P\{F(m-1, n-1) \geq s_1^2/s_2^2\}$$

(On this alternative hypothesis, values of s_1^2/s_2^2 that are ≤ 1 are regarded as being consistent with the null hypothesis.) A curve showing the p.d.f. of the F distribution is given in Figure 2.5.2.

Case (ii): *The alternative hypothesis is that $\sigma_1^2 < \sigma_2^2$.*

Here the observed value of the ratio s_1^2/s_2^2 is likely to be small on the alternative hypothesis, so that the significance level is

$$SL = P\{F(m-1, n-1) \leq s_1^2/s_2^2\}.$$

Published tables do not usually give these lower tail distributions. We may however recover case (i) by interchanging m and n, and s_1^2 and s_2^2, whence the significance level is

$$SL = P\{F(n-1, m-1) \geq s_2^2/s_1^2\}.$$

Case (iii): *The alternative is the two sided hypothesis that $\sigma_1^2 \neq \sigma_2^2$.*

In this case the significance level of the observed ratio is

$$SL = \begin{cases} 2P\{F(m-1, n-1) \geq s_1^2/s_2^2\} & \text{if } s_1^2 > s_2^2 \\ 2P\{F(n-1, m-1) \geq s_2^2/s_1^2\} & s_1^2 < s_2^2 \end{cases}.$$

EXAMPLE 5.8.4.   *Equality of two variances.*   In sections 5.8.3 and 5.8.4 we considered samples of sizes $m = 20$ and $n = 10$ with sample variances $s_1^2 = 5 \cdot 62$ and $s_2^2 = 2 \cdot 19$, so that

$$s_1^2/s_2^2 = 2 \cdot 57.$$

The significance level of this ratio in relation to the null hypothesis of equal variances against the two-sided alternative of unequal variances is thus

$$SL = 2P\{F(19, 9) \geq 2 \cdot 57\}.$$

Tables of the F-distribution [for an example see Appendix T7] do not tabulate the SL against observed values of s_1^2/s_2^2; they tabulate values of the statistic $r = s'^2/s''^2$ against values of $P(F \geq r)$, where r is the larger of s_2^2/s_2^2 and s_2^2/s_1^2. The relevant section of one of the better (i.e. more detailed) tabulations of the distribution gives the information shown in Table 5.8.1 below for $F(15, 9)$ and $F(20, 9)$. The final column of this Table gives in parentheses the interpolated values of $F(19, 9)$, these being the values required by our example. (These latter values are not included in the published tables.)

It will be seen that our value of the variance ratio, which is $2 \cdot 57$ with degrees of freedom equal to 19 for the numerator and 9 for the denominator, lies between the $0 \cdot 100$ value and the $0 \cdot 050$ value; it corresponds to a probability value of about $0 \cdot 080$, so that its significance level, being twice the entry, is $0 \cdot 16$. This is a high value, and the data must be judged 'not significant' i.e. consistent with the null hypothesis.

[It must be pointed out that this test is not robust: it is sensitive to departures from Normality.]

Degrees of freedom of denominator	Degrees of freedom of numerator		
9	15	20	(19)
Probability of a larger value			
0·100	2·34	2·30	(2·31)
0·050	3·01	2·94	(2·95)
0·025	3·77	3·67	(3·69)
0·010	4·96	4·81	(4·83)
0·005	6·03	5·83	(5·87)

Table 5.8.1: Values of s'^2/s''^2 that have the stated significance level.

5.8.7. Tests for the Equality of Several Means: Introduction to the Analysis of Variance

(a) *Introduction*

In section 5.8.4 the problem discussed was that of assessing the significance of the difference between the means of two samples. The underlying probability model was that the two sampled populations were both Normal, with common variance, but with possibly differing expectations, and the question being asked was whether the data were consistent with the hypothesis that these expectations were in reality equal. In applications the two samples would be measurements on otherwise comparable entities which had been given different 'treatments', and the difference, if any, between the expectations would be attributed to a difference between the 'effects' of the treatments. The measurements might for example be yields of wheat, the two treatments corresponding to the application of different fertilizers, one to the field from which one sample came, another to the field from which the other sample came.

What if one wished to compare three (or more) treatments? One way might be to compare them in pairs, using the method of section 5.8.4 for each pair. This is rather cumbersome and unsatisfactory (not all the pairs being independent) and one would prefer a generalization of the two-sample procedure to enable one to deal with the question:'Are the three (or more) expectations equal?

(b) *Comparison of two means as analysis of variance*

A direct extension of the argument of section 5.8.4 is not at first sight obvious. To see what changes are needed to enable one to make the required generalization, consider the following alternative way of looking at the analysis in section 5.8.4: First get rid of the square root in (5.8.8) by squaring. We obtain the expression $mn(\bar{x}-\bar{y})^2/(m+n)s^2$. Now consider the relation between the numerator and the denominator. To express this we use the notation

$$z_1 = x_1, \qquad z_2 = x_2, \ldots, z_m = x_m$$

$$z_{m+1} = y_1, \qquad z_{m+2} = y_2, \ldots, z_{m+n} = y_n,$$

so that the vector $(z_1, z_2, \ldots, z_{m+1})$ represents the two samples thrown together as in the partitioned [see I, § 6.6] vector $\mathbf{z}' = (\mathbf{x}' \mid \mathbf{y}')$. Let

$$\bar{z} = \sum_1^{m+n} z_r/(m+n) = (m\bar{x} + n\bar{y})/(m+n).$$

This quantity is called the *grand mean*. The following is an algebraic identity:

$$\sum_1^{m+n} (z_r - \bar{z})^2 = \underbrace{\sum_1^m (x_r - \bar{x})^2 + \sum_1^n (y_r - \bar{y})^2}_{} + \frac{mn}{m+n}(\bar{x}-\bar{y})^2$$

$$= \qquad\qquad w \qquad\qquad + \quad b \qquad\qquad\qquad (5.8.15)$$

say. The first term on the right, called the 'sum of squares of the x's' is proportional to the variability of the members of the first sample. (It is in fact $m-1$) times the sample variance.) The second term on the right, the sum of the squares of the y's, has a similar meaning with respect to the second sample. These two terms taken together form a measure of the variability 'within' the samples; their sum, w, is called the '*within-samples* sum of squares'.

The last term on the right, b, is clearly a measure of difference between the samples, and is called the '*between-samples* sum of squares'. Finally, the term on the left, called the 'total sum of squares' is a measure of the variability of the entire set of data, which is analysed by the identity into a 'within-samples' component and a 'between-samples' component.

If the two treatments under investigation have in reality no differential effects, the apparent difference between them, as indicated by b, must be due solely to the same kind of random fluctuations that cause the within-samples variations indicated by w. In fact it is not difficult to see that their sampling expectations are then

$$\text{for } w : (m+n-2)\sigma^2, \qquad \text{for } b : \sigma^2,$$

where σ^2 is the common variance of the observations.

If however there is a real difference between the populations, their expectations differing by an amount δ, the sampling expectation of B increases to

$$\sigma^2 + \frac{mn}{m+n}\delta^2.$$

Since σ^2 is unknown, the appropriate procedure is to evaluate the ratio b/w, or, more conveniently, $(m+n-2)b/w$. Numerator and denominator will then have equal expectations on the null hypothesis of no differential effect, but the expectation of the numerator will be greater than that of the denominator if there is a real differential effect.

More precisely, one considers the sampling distribution of the quantity $(m+n-2)b/w$. On the null hypothesis this is the $F(1, m+n-2)$ distribution. The significance level of the data with respect to the null hypothesis of no differential effect against the hypothesis of a real effect is

$$SL = P\{F(1, m+n-2) > (m+n-2)b_0/w_0\} \qquad (5.8.16)$$

where b_0 and w_0 denote the observed values of the quantities b and w defined according to (5.8.15).

Comparison of this with (5.8.8) will show that the square of the quantity t defined in that expression is identically equal to $(m+n-2)b/w$, their explicit common value being

$$\frac{(m+n-2)mn}{(m+n)} \frac{(\bar{x}-\bar{y})^2}{\sum(x_r-\bar{x})^2+\sum(y_r-\bar{y})^2}.$$

Further, the distribution of the square root of the random variable $F(1, m+n-2)$ is Student's distribution on $m+n-2$ df. Thus (5.8.16) gives exactly the same significance level as does the t-test of (5.8.8).

What, then, is the point of this analysis? The answer is that it is this method, in terms of the 'sums of squares identity' (5.5.15), that points the way to the required generalization to three or more samples. Before proceeding we summarize the results of the two-sample analysis in terms of the following table:

Source of variation	Sum of squares	df
Between samples	$\dfrac{mn}{m+n}(\bar{x}-\bar{y})^2$	1
Within samples	$\sum(x_r-\bar{x})^2+\sum(y_r-\bar{y})^2$	$m+n-2$
Total	$\sum(z_r-\bar{z})^2$	$m+n-1$

This method of presenting the arithmetical values of the various terms on the identity (5.8.15), together with their divisors ('d.f.' = degrees of freedom) and the interpretation given in the first column, is called an *analysis of variance table*.

We have presented the argument at some length because it is fundamental not only for the comparison of 3 or more means, discussed in (c) below, but because the analysis of variance has a very wide field of application in statistics in general. [see Chapters 8, 10.]

(c) *The case of k samples*

Suppose we now have k samples, corresponding respectively to k treatments, of sizes $n_1, n_2, \ldots, n_k$, as in Table 5.8.2 (where the samples are displayed in 'data-vector' format).

Treatment no	1	2	...	k	
Data	x_{11} x_{21} $\vdots$ $x_{n_1 1}$	x_{12} x_{22} $\vdots$ $x_{n_2 2}$		x_{1k} x_{2k} $\vdots$ $x_{n_k k}$	
Total	$C_1 = \sum_r x_{r1}$	$C_2 = \sum_r x_{r2}$	...	$C_k = \sum_r x_{rk}$	$\sum_r \sum_s x_{rs} = G$
Number of observations	n_1	n_2	...	n_k	n
Mean	$c_1 = x._1$	$c_2 = x._2$	...	$c_k = x._k$	$x.. = g$

Table 5.8.2.

The sums-of-squares identity is

$$\sum_r \sum_s (x_{rs} - x..)^2 = \sum_r \sum_s (x_{rs} - x._s)^2 + \sum_s n_s (x._s - x..)^2, \qquad (5.8.17)$$

viz

$$\text{Total sum of squares} = \left(\begin{array}{c}\text{within-sample}\\\text{sum of squares}\end{array}\right) + \left(\begin{array}{c}\text{between samples}\\\text{sum of squares}\end{array}\right)$$

$$= \quad w \quad + \quad b,$$

say. The sampling expectation of w is $(n-k)\sigma^2$ where now $n = \sum n_r = $ total number of observations. The sampling expectation of b is $(k-1)\sigma^2$ if the null hypothesis holds (i.e. if the k treatments produce no real differential effects), but greater than this if the null hypothesis is false (i.e. if at least some of the treatments produce real differential effects). Thus the quantity $f = \{b/(k-1)\}/\{w/(n-k)\}$ is an appropriate test statistic, large values of which (i.e. values 'significantly' exceeding unity) indicate a real treatment effect. More precisely, the sampling distribution of the statistic is the $F(k-1, n-k)$ distribution, and the significance level of the data in relation to the null-hypothesis of 'no treatment effects' is

$$SL = P\{F(k, n-k) > f\}$$

It is convenient to lay out the arithmetic in the form of an analysis of variance table as follows

Source of variation	Sum of squares	d.f.	Test statistic
Between samples	b	$k-1$	$f = \dfrac{b}{k-1} \bigg/ \dfrac{w}{n-k}$
Within samples	w	$n-k$	
Total	$\sum_r \sum_s x_{rs}^2 - nx_{..}^2$	$n-1$	Distribution: $F(k-1, n-k)$

Table 5.8.3: Within and between analysis of variance.

The 'total' sum of squares is evaluated from the expression given in the table. The value of b is evaluated as

$$b = \sum_s n_s x_{.s}^2 - nx_{..}^2,$$

and the value of w found by subtracting b from the total sum of squares.

In practice the arithmetic is less onerous than at first sight appears. One needs to evaluate the column totals $C_1, C_2, \ldots, C_k$, the grand total G, the column means $c_1, c_2, \ldots, c_k$ (where $c_j = C_j/n_j$, $j = 1, 2, \ldots, k$), the grand means $g = G/n$, and the sum of squares, S say, of all n observations. Then the

$$\text{'total sum of squares'} = S - Gg$$

and

$$b = \text{'between-samples sum of squares'}$$

$$= C_1 c_1 + C_2 c_2 + \ldots + C_k c_k - Gg$$

and

$$w = \text{'within-samples sum of squares'}$$

$$= S - (C_1 c_1 + C_2 c_2 + \ldots + C_k c_k).$$

Finally one computes $f = \{b/(k-1)\}/(w/(n-k)\}$ and finds the significance level from the $F(k-1, n-k)$ distribution as

$$SL = P\{F(k-1, n-k) > f\}.$$

Equivalently, the data might be presented in a frequency table format, as in Table 5.8.4. Here n_{ij} denotes the frequency of the response x_i to treatment number j:

Treatment No.		1	2	$\ldots$	k
Response	x_1 x_2 $\vdots$ x_a	n_{11} n_{21} $\vdots$ n_{a1}	n_{12} n_{22} $\vdots$ n_{a2}	$\ldots$	n_{1k} n_{2k} $\vdots$ n_{ak}
Total (i.e. sum of all observations)		$C_1 = \sum_i n_i x_i$	$C_2 = \sum_i n_{i2} x_i$	$\ldots$ $C_k = \sum_i n_{ik} x_i$	$G = \sum_i \sum_j n_{ij} x_i$
Number of observations		n_1	n_2	$\ldots$ n_k	n
Mean		$c_1 = x._1$	$c_2 = x._2$	$\ldots$ $c_k = x._k$	$g = x..$

Table 5.8.4: Notation for k samples, frequency tables format.

Here $x._1 = \sum n_{i1} x_i / n_i$, etc.: and $x.. = \sum \sum n_{ij} x_i / n$. In this version the sum of squares identity becomes

$$\sum_i \sum_j n_{ij} (x_i - x..)^2 = \sum_i \sum_j n_{ij} (x_i - x._j)^2 + \sum_j n_j (x._j - x..)^2$$

$$= \qquad w \qquad + \qquad b. \qquad (5.8.18)$$

The sum of squares of all n observations is

$$S = \sum_i \sum_j n_{ij} x_i^2 ;$$

the column totals C_j and means c_j, and the grand total G and grand mean g are as indicated in the table; and the values of b and w and of the significance level (in relation do the hypothesis that the treatments produce no differential effects) are then calculated as in the data-vector format given earlier.

EXAMPLE 5.8.5.   *Test for treatment effect.*   Table 6.5.3 presents an illustrative set of data in frequency table form. Here the 'treatments' are temperatures: treatment number j is denoted by x_j. The responses, denoted by x_1, x_2, etc. in the account given above, are called $y_1, y_2, \ldots, y_{25}$ in the table referred to. The column totals C_j and means c_j are given by:

j	1	2	3	4	5
C_j	409·32	189·76	206·84	32·00	−208·21
c_j	4·548	3·514	2·492	0·320	−2·420

j	6	7	8	9
C_j	$-222 \cdot 90$	$-633 \cdot 35$	$-652 \cdot 68$	$-408 \cdot 79$
c_j	$-1 \cdot 627$	$-4 \cdot 623$	$-6 \cdot 660$	$-7 \cdot 713$

G and g are, respectively, $-1264 \cdot 13$ and $-1 \cdot 536$, while the value of S is $18143 \cdot 70$. The analysis of variance table is therefore as follows:

Source of Variation	Sum of Squares	d.f.	Mean square
Between arrays (b)	12,370	$k-1=8$	$1546 \cdot 25 \ = \dfrac{b}{k-1}$
Within arrays (w)	3,832	$n-k=814$	$4 \cdot 708 = \dfrac{w}{n-k}$
Total	16,202	$n-1=822$	

Table 5.8.5: Analysis of variance table.

The variance ratio is therefore 329. The probability of obtaining such a large (or larger) value of the F distribution is of course vanishingly small, whence the Hypothesis that there are no significant differences among the column means is shown to be untenable (as was of course obvious from the beginning).

5.9. TESTS FOR NORMALITY

As the concept of Normality plays such a large rôle in so many statistical procedures, it is rather important to be able to verify that one's sample does not differ significantly from the Normal form, and that the sampling distributions of the statistics used are Normal.

The following dictum of Fisher's may serve to put the problem in perspective: 'Departures from Normal form, unless very strongly marked, can only be detected in large samples; conversely, they make little difference to statistical tests on other questions'. Of course, as far as their sampling distributions are concerned, some statistics are more sensitive than others to departures from Normality. Bartlett's test for the equality of variances, outlined in section 5.10, is a case in point.

Use of probability paper

Whilst the investigation of such sensitivity questions can be difficult, it is quite easy to test whether or not a given sample can reasonably be regarded as having come from a Normal distribution. A first test can quickly be effected by plotting the sample c.d.f. on probability graph paper [see § 3.2.2(d)]. The points of a sample c.d.f. from a Normal sample will lie on or near a straight line, while if the population is markedly non-Normal they will approximate a curve of some kind. Example 3.5.1 shows the computation of a sample c.d.f., and Figure 3.5.1 shows the plot on probability paper.

A more objective procedure is described below.

Test for Normality in terms of skewness and kurtosis

The coefficient of skewness (γ_1, say) of a random variable X is

$$\text{skew}(X) = \gamma_1 = E(X - \mu)^3 / \{\text{var}(X)\}^{3/2}$$

[see II, § 9·10], where

$$\mu = E(X).$$

In the notation of section 2.1 this is

$$\gamma_1 = \mu_3 / \mu_2^{3/2}$$

[see (2.1.5), (2.1.6)]. In a symmetrical distribution such as the Normal, this equals zero. The corresponding sample skewness is

$$m_3 / m_2^{3/2},$$

where, in a sample of size n,

$$m_r = \sum_j f_j (x_j - \bar{x})^r / n, \qquad r = 2, 3, \ldots$$

[see (2.1.8)]. Fisher recommends a slight modification of this as an estimate of γ_1, namely

$$g_1 = k_3 / k_2^{3/2}$$

where

$$k_3 = m_3 \bigg/ \left(1 - \frac{1}{n}\right)\left(1 - \frac{2}{n}\right)$$

and

$$k_2 = m_2 \bigg/ \left(\frac{1-1}{n}\right).$$

Here m_2 and m_3 may be obtained from the uncentred moments m_2', m_3' by the use of (2.1.9).

Similarly the kurtosis (the standardized fourth moment) of a random variable X is

$$\gamma_2 = \mu_4 / \mu_2^2,$$

the corresponding sample quantity is m_4/m_2^2, and the recommended estimate is

$$g_2 = k_4/k_2^2,$$

where

$$k_4 = m_4 \bigg/ \left(1 - \frac{2}{n+1}\right)\left(1 - \frac{2}{n}\right)\left(1 - \frac{3}{n}\right) - 3m_2^2 \bigg/ \left(1 - \frac{2}{n}\right)\left(1 - \frac{3}{n}\right).$$

For a Normal distribution the sampling distributions of g_1 and g_2 may be taken to be Normal, with zero expectation in each case, and with variances given by

$$6n(n-1)/(n+3)(n+1)(n-2),$$

$$24n(n-1)^2/(n+5)(n+3)(n-2)(n-3)$$

respectively. The test for Normality based on these results is illustrated in Example 5.9.1.

EXAMPLE 5.9.1.   *Sample skewness and kintosis.*   For the following rainfall data [see Fisher (1970). (Reprinted with permission of Macmillan Publishing Company from *Statistical Methods for Research Workers*, 14th edition, by Sir Ronald A. Fisher. Copyright © 1970 University of Adelaide.) § 14 Table 3—(Bibliography C)] the plot of the sample c.d.f. on probability paper is shown in Figure 5.9.1

Rainfall (inches)	Frequency	Rainfall	Frequency
16	1	28	4
17	0	29	8
18	0	30	9
19	3	31	6
20	2	32	7
21	3	33	4
22	0	34	4
23	3	35	4
24	2	36	3
25	12	37	3
26	4	38	0
27	7	39	1
			$n = 90$

The data lead to the following values for the sample moments:

$$m_1' = 28 \cdot 62, \qquad m_2 = 23 \cdot 013, \qquad m_3 = -24 \cdot 909, \qquad m_4 = 1403 \cdot 656,$$

whence

$$k_2 = 23 \cdot 2715, \qquad k_3 = -25 \cdot 761, \qquad k_4 = -162 \cdot 487;$$

$$g_1 = -0 \cdot 231, \quad \text{with standard error } \pm 0 \cdot 254$$

$$g_2 = -0 \cdot 302, \quad \text{with standard error } \pm 0 \cdot 503.$$

Since both g_1 and g_2 are smaller in magnitude than their respective standard error, neither is significant: the data must be judged to be consistent with the hypothesis of Normality.

Test for Normality using χ^2

The χ^2 test is described in Chapter 7 and the application of the procedure to the data of Example 5.9.1 is given in Example 7.4.1.

5.10.   TEST FOR THE EQUALITY OF k VARIANCES (BARTLETT'S TEST)

Suppose we have k samples, one from each of k Normal distributions, as for example in the following array:

Sample No.

	1	2	...	k
Data	y_{11}	y_{12}		y_{1k}
	y_{21}	y_{22}		y_{2k}
	$\vdots$	$\vdots$	...	$\vdots$
	$y_{n_1 1}$	$y_{n_2 2}$		$y_{n_k k}$
Mean	$y_{\cdot 1}$	$y_{\cdot 2}$		$y_{\cdot k}$

To test whether the k sampled populations have equal variances, we may in the first instance use a graphical method: the sample c.d.f.s, plotted in probability paper [see § 5.9.1] will approximate to parallel lines if the distributions concerned are Normals with equal variances. (For an example of this, see Figure 6.5.3.)

The more objective procedure of Bartlett's test is as follows: for each sample compute the usual variance estimate

$$s_j^2 = \sum_r (y_{rj} - y_{\cdot j})^2 / (n_j - 1), \qquad j = 1, 2, \ldots, k$$

and also the pooled estimate

$$s^2 = \sum_{j=1}^{k} (n_j - 1)s_j^2 \Big/ \sum_{j=1}^{k} (n_j - 1).$$

Bartlett's test criterion for the quality of the variances is

$$w = \{\sum (n_j - 1)\} \log s^2 - \sum (n_j - 1) \log s_j^2$$

(logarithms, as elsewhere in this book, being to base e). Good approximations to the significance levels are obtainable by using the following transformations due to Box. Calculate

$$A = \frac{1}{3(k-1)} \left\{ \sum \left(\frac{1}{n_j - 1}\right) - \frac{1}{\sum (n_j - 1)} \right\}$$

$$f_1 = k - 1$$

$$f_2 = (k+1)/A^2$$

$$b = f_2/(1 - A + 2/f_2).$$

Then the sampling distribution of

$$f_2 w / f_1 (b - w)$$

is (approximately) the F-distribution on f_1 and f_2 degrees of freedom.

EXAMPLE 5.10.1. *Equality of three variances.* We apply the test to the following three samples.

Sample No. j	d.f. $n_j - 1$	s_j^2	$\log s_j^2$
1	81	58·57	4·070
2	44	76·84	4·342
3	13	79·67	4·378

$$
\begin{array}{ccc}
\sum (n_j - 1) & \sum (n_j - 1)s_j^2 & \sum (n_j - 1) \log s_j^2 \\
= 138 & = 9160{\cdot}84 & = 577{\cdot}632
\end{array}
$$

$$s^2 = 66{\cdot}383 \qquad \{\sum (n_j - 1)\} \log s^2 = 578{\cdot}971$$

$$n = 1{\cdot}339$$

$$A = \tfrac{1}{6}(0{\cdot}1120 - 0{\cdot}0072) = 0{\cdot}0175$$

$$f_1 = 2$$

$$f_2 = 13061$$

$$b = 13291$$

$$f_2 w / f_1 (b - w) = 0{\cdot}66.$$

On the 'equal variances' hypothesis this is a realization of an F distribution on 2 and 13061 degrees of freedom. Now, for large values of ν_2, the F distribution on ν_1 and ν_2 degrees of freedom is well approximated by the χ^2 distribution on ν_1 d.f., taking $\nu_1 F$ as the observed χ^2 [see § 2.5.6]. Thus, on the null hypothesis, $\nu_1 F = 2 \times 0 \cdot 66 = 1 \cdot 32$ is a realization of a χ^2 variable on 2 d.f., where the significance level is about $0 \cdot 6$. This is a very large probability, and the conclusion is that the data are consistent with the hypothesis of equal variances.

It has to be pointed out that this test is not robust: it is intended for Normal distributions and it is not accurate for markedly non-Normal distributions. The (approximate) Normality of the original samples must therefore be established before applying the test, and even so caution is required in the interpretation of significance levels in the $0 \cdot 01$ to $0 \cdot 07$ region. In the example considered above the significance level is high enough to justify the inference that the variances are equal.

5.11.   COMBINING THE RESULTS OF SEVERAL TESTS

Suppose that, in k independent experiments, significance levels $q_1, q_2, \ldots, q_k$ had been obtained with respect to a hypothesis H. If some or all of the significance levels were indecisive, it is possible that the entire set of data, considered as relating to single enlarged experiment, might give a more informative result. Consider, for example, three separate experiments to test whether a certain additive increased the crushing strength of concrete, where in each case a t-test was carried out to test the hypothesis that the additive was ineffective, with significance levels $0 \cdot 051$, $0 \cdot 038$ and $0 \cdot 046$. We may combine these significance levels by the following argument. The sampling distribution, on the null hypothesis, of an observed significance level s is uniform $(0, 1)$, [see Equation (5.3.6)] whence that of $-2 \log s$ is the chi-squared distribution on 2 degrees of freedom [see § 2.7.6]. It follows from the closure property of the chi-squared family [see § 2.5.4(a)] that, if $s_1, s_2, \ldots, s_k$ are independent significance levels, the quantity

$$q = -2 \sum_1^n \log s_r$$

is distributed as chi-squared on $2n$ degrees of freedom. The probability lying beyond q in the tail of this distribution is then the required 'combined' significance level. In our example we had

Test No.	$SL = s$	$-2 \ln s$	d.f.
1	0·051	5·952	2
2	0·038	6·540	2
3	0·046	6·158	2
Total		18·650	6

The probability of exceeding 18·650 in the chi-squared distribution on 6 d.f. is 0·005. The three indecisive significance levels therefore combine to give the highly significant level of 0·005: the hypothesis that the additive is ineffective is strongly discredited.

The test is due to Fisher [see Fisher (1970) § 2.1.1—Bibliography C].

5.12. NEYMAN–PEARSON THEORY

5.12.1. Sampling Inspection Schemes

In routine and repetitive industrial production inspection for quality, each day's production or each load of goods delivered might be sampled for quality, and the whole load 'sentenced' on the result of inspection of the sample. An example of a very simple procedure of this sort would be: take 100 items, inspect, and count the number of defective items d in the sample. If $d \le c$ (where c is a given 'acceptance number', e.g. $c = 2$), 'accept' the load; otherwise 'reject' it. ['Rejecting' the load might mean literally rejecting it,·i.e. sending it back to the suppliers, or it might be a code word for subjecting it to further treatment, qualifying for a lower price, etc. etc.] In the long run this procedure will accept most high-quality loads (but will reject some of them) and reject most low quality loads (but will accept some). Suppose, to be definite, that the sampled load is large compared with the sample size, so that the number of defectives in a sample is adequately represented as binomial (n, θ) variable, [see II, § 5.2.2] θ being the (unknown) proportion of defectives in the load; then the probability of accepting a load whose quality is θ is

$$\mathscr{A}(\theta) = P\{R \le c; R \text{ bin } (n, \theta)\}$$

$$= \sum_{0}^{c} \binom{n}{r} \theta^r (1-\theta)^{n-r}.$$

This function, called the *acceptance function*, or the *operating characteristic* of the procedure, has a graph of the kind illustrated in Figure 5.12.1. It will be seen that, if θ_1 is a value of the quality parameter θ such that a load with $\theta \le \theta_1$ would be regarded as 'good', the probability of accepting a good load is at least $\mathscr{A}(\theta_1)$. Equivalently, the probability of rejecting a good load is at most $1 - \mathscr{A}(\theta_1)$. The magnitude of this probability can be adjusted to as small a value (α, say) as desired, by suitably choosing n and c. The values chosen for n and c fix the curve, from which it is seen that the probability of accepting a poor-quality load (i.e. one with large θ) is small: just how small, of course, depends on θ, n and c.

5.12.2. The Neyman–Pearson Theory of Tests

The Neyman–Pearson theory ('N–P theory') of tests uses language and concepts derived from repetitive 'accept or reject' decision processes like that

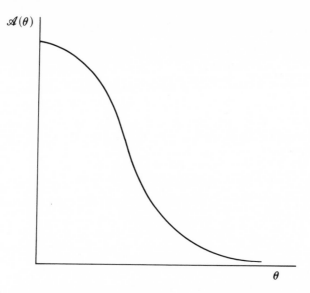

Figure 5.12.1: The operating characteristic function of a sampling inspection scheme.

outlined in section 5.12.1. As in section 5.3, the data vector x_0 arises from a family of distributions indexed by a parameter vector $\theta(\theta \in \Omega)$, this family forming the probability model that underlies the sampling procedure. The null hypothesis $H(\theta_0)$ assumed for the present to be a simple one, postulates that the value of the parameter vector θ in the sampled distribution is θ_0. This is in competition with an 'alternative hypothesis' $H(\theta)$ according to which the parameter value θ is a member of some set Ω' (a specified subset of Ω.) As in § 5.3, a test statistic $s(x_0)$ is chosen. We denote by S the corresponding random variable, so that $s(x_0)$ is a realization of S. Unlike the earlier treatment in 5.3, which utilizes a 'significance set' $G(x_0)$ with significance level

$$SL(x_0) = P\{S \in G(x_0)|H(\theta_0)\} \tag{5.12.1}$$

defined in terms of the observed statistic x_0, the N–P theory makes use of an appropriately selected subset $C(\alpha)$ of the observation space called the *critical region*, with the property that

$$P\{S \in C(\alpha)|H(\theta_0)\} = \alpha \tag{5.12.2}$$

where α is a *predetermined* constant (e.g. $\alpha = 0\cdot01$), called the level of significance or *size* of the test, (or of the critical region). If the observed value x_0 of X falls in this predetermined critical region, the null hypothesis is 'rejected', and the alternative hypothesis 'accepted'; otherwise $H(\theta_0)$ is 'accepted', and the alternative 'rejected'. Like the significance set of section 5.3, the critical region is so chosen that the probability allotted to it by $H(\theta_0)$ is small, while that allotted to it by $H(\theta)$ is large when θ is sufficiently remote from θ_0. We have

$$P\{\text{reject } H(\theta_0) \text{ when } H(\theta_0) \text{ is true}\}$$

$$= P\{S \in C(\alpha)|H(\theta_0)\}$$

$$= \alpha,$$

the predetermined level of significance. This is also known as the 'risk of Type I error'. Similarly

$$P\{\text{reject } H(\theta_0) \text{ when } H(\theta) \text{ is true, } \theta \in \Omega'\}$$

$$= P\{S \in C(\alpha)|H(\theta)\}$$

$$= w(\alpha; \theta), \text{ say.}$$

The quantity $1 - w(\alpha; \theta)$ is known as the 'risk of Type II error' corresponding to the parameter value θ. For a given α the function $w(\alpha; \theta)$ is called the α-level *power function* of the test. As was remarked in section 5.3 it has the same numerical values as the sentitivity function $G(\alpha; \theta)$ of that section, although its interpretation is different. In the approach to testing described in section 5.3 the purpose of a statistical test is seen as the assessment of the strength of the evidence provided by the test statistic against $H(0)$, so as to enable the statistician to make a provisional, tentative informed judgement as to whether or not $H(0)$ is still tenable in the light of the data. The N–P approach, on the other hand, is to make a *decision*: the null hypothesis is to be either *rejected* or *accepted*. In the approach of section 5.3 the significance level is a function of the observed statistic ($\bar{x}_0$, say). It is then natural to compare the values of the sensitivity function $G(\bar{x}_0, \theta)$ for different values of θ but with the same value of the statistic $\bar{x}_0$, namely the value actually observed. In the N–P theory the level of significance α (the *size* of the test) is fixed in advance (at 1% or 5%, say) and we compare values of the power function $w(\alpha, \theta)$ for various values of θ, with the same predetermined values of α.

As an illustration consider the same example that was used in section 5.3, namely that in which the probability model of the sampling process is Normal $(\theta, 1)$, and the test statistic the mean $\bar{x}$ of the n sample values. Take as the alternative hypothesis the 'one-sided' hypothesis

$$H'(\theta): \theta > 0,$$

and as critical region the upper tail region of the sampling distribution of $\bar{X}$ on $H(\theta_0)$ consisting of all x-values such that

$$\bar{x} \geq c(\alpha)$$

where

$$P\{\bar{X} \geq c(\alpha)|H(0)\} = \alpha.$$

Now this is equal to

$$P\{\bar{X} \ge c(\alpha)|\bar{X} \text{ is Normal } (0, 1/\sqrt{n})\} = 1 - \Phi\{\sqrt{n}c(\alpha)\}$$

whence

$$\sqrt{n}c(\alpha) = \Phi^{-1}(1-\alpha).$$

The power function is

$$
\begin{aligned}
w(\alpha; \theta) &= P\{\text{reject } H(0)|H(\theta)\} \\
&= P\{\bar{X} \ge c(\alpha) \text{ when } \bar{X} \text{ is Normal } (\theta, 1/\sqrt{n})\} \\
&= 1 - \Phi\{\sqrt{n}c(\alpha) - \theta\sqrt{n}\} \\
&= 1 - \Phi\{\Phi^{-1}(1-\alpha) - \theta\sqrt{n}\}
\end{aligned}
$$

which is the same function as the sensitivity function (5.3.7). The function is illustrated qualitatively in Figure 5.12.2.

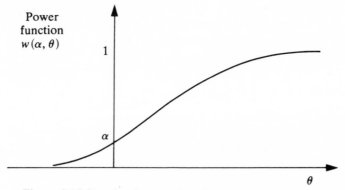

Figure 5.12.2: α-level power function of a one-sided test.

Of two tests of size α, one is regarded as being superior to the other at the parameter value θ if its power at this value of θ is greater than that of the other. A test which is more powerful than all other tests of the same size, for all values of θ in Ω', is called a *uniformly most powerful test.*

The N–P theory appears to have the strong advantage over the section 5.3 approach that it is able to quantify the probability of drawing a false conclusion, i.e. of wrongly rejecting $H(0)$, as a function of the true value θ of the parameter. This, however, requires that one should accept the theory in the strict sense of specifying the significance level in advance and literally rejecting or accepting the null hypothesis, not on the basis of the actual observed value of the test statistic but on the basis of whether or not that observed value falls anywhere in the predetermined critical region. Of course there are circumstances in which this is done. For most one-off tests, however, standard statistical practice is to proceed somewhat tentatively, recording the significance level of the observed value of the statistic as defined in section 5.3 and using the complement of this as a measure of the strength of the evidence against $H(\theta_0)$. This procedure is in conformity with the following modification of the canonical version of the N–P theory: instead of working with a single critical region $C(\alpha)$ of fixed size α(e.g. $\alpha = 0\cdot02$), have in mind a nested array of critical regions of different sizes, e.g. all sizes between $0\cdot001$ and $0\cdot10$; let α_0 denote the size of the critical region which just contains the observed value of the test statistic—i.e. such that this observed value lies in $C(\alpha)$ for $\alpha \le \alpha_0$ but does not lie in $C(\alpha)$ for $\alpha > \alpha_0$. Then $H(\theta_0)$ is 'just' rejected at level α_0, which is equivalent to saying that the statistic has significance level α_0 in the sense of section 5.3. Interpret the phrase '$H(\theta_0)$ is rejected, at level α' as 'the probability of achieving a significance level as small as α_0 (or smaller), if $H(\theta_0)$ were true, would be α_0', whence $1 - \alpha_0$ may be taken as a measure of the strength of the evidence against the null hypothesis.

The Neyman–Pearson lemma

The jewel in the crown of the N–P theory is the Neyman–Pearson lemma, according to which the optimal procedure for testing the simple null hypothesis $H(\theta_0)$ against the simple alternative $H(\theta_1)$ is that of the likelihood ratio test of section 5.2. This is optimal in the sense of having power at θ_1 at least as large as that of any other test that could be formulated.

Composite hypotheses

The above is the merest sketch of the N–P theory, which has extensive ramifications dealing with, amongst other matters, composite hypotheses, and special devices called randomized tests for assimilating discrete probability models into the general theory.

E.H.L.

5.13. FURTHER READING AND REFERENCES

The subject-matter of this chapter is a fundamental part of the theory and practice of statistical inference and is discussed in one way or another in all general texts on statistical theory. The texts cited in section 3.6 are particularly recommended. Kalbfleisch, especially, gives a clear account of tests involving discrete data and of conditional tests; Fisher's (mainly non-mathematical) book is an authoritative source of much of contemporary statistical procedure, excluding Neyman–Pearson theory, which is, however, fully dealt with in the other sources cited. The basic paper on this is Neyman (1937). A work devoted solely to the theory of tests is Lehmann (1959)—details below.

REFERENCES

Fisher, R. A. (1951). *The Design of Experiments*, sixth edition, Oliver and Boyd.
Neyman, J. (1937). Outline of a Theory of Statistical Estimation Based on the Classical Theory of Probability, *Phil. Trans. A* **236**, 333.
Lehmann, E. L. (1959). *Testing Statistical Hypotheses*, Wiley.

CHAPTER 6

Maximum-likelihood Estimates

6.1. INTRODUCTION

As explained in earlier sections (e.g. Chapters 1 and 3) the main idea underlying all statistical analyses is to regard the data values as realizations of random variables. There are procedures in existence which do not require any assumptions about the distribution(s) of these random variables: see Chapter 14; the greater part of statistical practice is however not conducted on 'distribution-free' lines, and one normally postulates a probability model which specifies the family of distributions involved. In simple cases this family may be implied by the sampling procedure; more generally it has to be chosen, with some degree of arbitrariness, from the available families that appear to be compatible with the data.

An example of the 'simple case' arises in sampling inspection, where a sample of fixed size is taken from a finite collection of articles containing an unknown number (θ, say) of 'defective' articles. Subject to reasonable sampling precautions, the probability distribution of the number of defectives in the sample will unambiguously belong to the hypergeometric family [see II, § 5.3]. The p.d.f. will depend on the unknown θ as a parameter. The object of the exercise will be to obtain from the sample an appropriate approximation to the value of θ (this is called an *estimate* of θ) and an acceptable description of the accuracy of this approximation.

As an example of the second category consider the analysis of the total annual flows past a gauging station on a given river. These are non-negative and are, typically, positively skewed [see II, § 9.10], with no ascertainable upper bound. A plausible probability distribution might then be a member of the two-parameter gamma family [see II, § 11.3]; or of the log-Normal family [see II, § 11.5]. In either case, there would be two parameters whose values would have to be estimated from the sample. In cases of this kind it is of course also necessary to have some means of checking that the postulated distribution, with the estimated values of the parameters, does in fact adequately represent the data (see Chapter 7).

The problem of finding a suitable statistic to be used as the estimate of an unknown parameter is sometimes referred to as '*point estimation*', and the problem of describing its accuracy as '*interval estimation*'. These are not very helpful terms but they are part of statisticians' jargon. The estimation procedure to be discussed in the present chapter, the *method of maximum likelihood*, provides both a technique for choosing the statistics to be used as estimates of the relevant parameters and a method of assessing their accuracy.

The concept of likelihood is explained in section 6.2.1, and the method of finding an estimate by maximizing this function is explained in section 6.2.2 and motivated in section 6.2.3. The principal properties and practical procedures are outlined in section 6.2.5. Simple examples are given in sections 6.3 and 6.4. Substantial nontrivial numerical analyses of two important examples follow: linear regression in section 6.5 and dosage-mortality in section 6.6.

6.2. THE METHOD OF MAXIMUM LIKELIHOOD

6.2.1. The Likelihood and Log-likelihood Functions

Data may be presented in a variety of ways, of which the most common are the 'data vector' and the 'frequency table' methods. The data vector x,

$$x = (x_1, x_2, \ldots, x_n)$$

is simply a list of the data values x_r recorded in the order in which they occurred. (Each value x_r may itself be a vector: for example, x_r might consist of the two numbers (u_r, v_r), where u_r and v_r are the observed responses (response A and response B) of the r-th individual in the samples.) If the data values are necessarily integers, as in *counting* experiments, and if the ordering of the sample values is not regarded as important, the data vector may conveniently be abbreviated to a *frequency table* [see § 3.2.2], a one-dimensional version of which would be as follows:

Observed value	1	2	...	s	...	k	Total
Frequency	$n(1)$	$n(2)$	...	$n(s)$	...	$n(k)$	n

$$(6.2.1)$$

(or similarly with observed values starting at 0).

This records the information that observations having magnitude s occurred $n(s)$ times in the sample ($s = 0, 1, \ldots, k$), that the largest observed magnitude was k, and that the sample size was $n = \sum_{s=1}^{k} n(s)$.

If the data values are realizations of *continuous* random variables, it might be that the observations are recorded with a conventionally high degree of accuracy, in which case one may use the data vector method of presentation. Alternatively the data may be recorded as a *grouped* frequency table:

Observed value	$x_r \pm \frac{1}{2}h$
Frequency	$n(r)$

$(r = 1, \ldots, k)$ $(6.2.2)$

Here $n(r)$ denotes the frequency with which observations fall into the 'cell' defined by the 'cell boundaries' $x_r - \frac{1}{2}h$, $x_r + \frac{1}{2}h$, h being the so-called grouping interval. (In some tables the size of the grouping interval itself changes with x_r. See section 3.2.2.) In grouped frequency tables of this kind some information has been sacrificed. The effect of this on estimation procedures will be discussed in section 6.5.

Likelihood and log-likelihood

The maximum-likelihood estimate is obtained from the *likelihood function* [see also § 3.5.4 and § 4.13.1]. To understand what this is we consider first the data-vector version. Suppose, then, that we have a sample of n one-dimensional observations $x_1, x_2, \ldots, x_n$, forming the data vector $x = (x_1, x_2, \ldots, x_n)$. This is regarded as a realization of the random vector $X = (X_1, X_2, \ldots, X_n)$, of which the p.d.f. at x (i.e. the joint p.d.f. of $X_1, X_2, \ldots, X_n$ at $x_1, x_2, \ldots, x_n$ respectively), called the *sample* p.d.f., is here denoted by

$$g(x; \theta) = g(x_1, x_2, \ldots, x_n; \theta_1, \theta_2, \ldots, \theta_p), \qquad \theta_1, \ldots, \theta_p \in \Omega, \quad (6.2.3)$$

where $\theta_1, \theta_2, \ldots, \theta_p$ denote the unknown parameters, the 'parameter space' Ω is the set of their permitted values, and g is supposed to have known functional form. The *likelihood* of θ, on the data x, is defined as

$$l(\theta; x) = a(x)g(x; \theta), \qquad \theta \in \Omega \quad (6.2.4)$$

where $a(x)$ is an arbitrary multiplier that may depend on the data and on known constants but not on the parameters θ_r. The appearance of an arbitrary factor in an entity of such importance might be regarded as surprising: since however the likelihood only appears in contexts involving the ratio of one likelihood value $l(\theta_1; x)$ to another, $l(\theta_2; x)$, with the *same* data vector x, the factor $a(x)$ is indeed irrelevant. Its purpose is to simplify the functional form of the likelihood by removing 'uninformative' factors that do not contain θ.

The likelihood is therefore a function of the parameters $\theta_1, \theta_2, \ldots, \theta_p$ and it is of importance to study the way in which it changes when the values of the θ_r vary; the data vector x is held constant in such manipulations. (This is the exact opposite of what happens in manipulations with the p.d.f., where computations are carried out to evaluate the probabilities associated with, for example, intervals of x-values, during which calculations the parameter vector θ remains fixed.)

It is often convenient to work with the logarithm (usually the *natural* logarithm) of the likelihood. This is called the *log-likelihood*.

The simplest and most common sampling procedure is one in which the X_r may be regarded as independent [see II, § 6.6.2] and identically distributed ('i.i.d.'). In this case the sample p.d.f. is

$$g(x; \theta) = \prod_{r=1}^{n} f(x_r, \theta) \quad (6.2.5)$$

where f is the common p.d.f. of the X_r. We exemplify below for (a) the one-parameter case (Examples 6.2.1, 6.2.2 and 6.2.3) and (b) the two-parameter case (Example 6.2.4) with independent observations. We also give, in Example 6.2.5, a two-parameter likelihood function arising from non-independent (Markovian) data.

EXAMPLE 6.2.1. *Likelihood function for the Poisson parameter.* Suppose the X_r are i.i.d. [see § 1.4.2(i)] Poisson (θ) variables [see II, § 5.4]. The sample p.d.f. (6.2.5) becomes

$$\prod_{r=1}^{n} (e^{-\theta}\theta^{x_r}/x_r!) = e^{-n\theta}\theta^{n\bar{x}}/\prod x_r!.$$

In forming the likelihood function (6.2.4) from this it would be natural to choose $a(x) = \prod x_r!$, so as to remove the uninformative factor $1/\prod x_r!$. The likelihood would then be taken as

$$l(\theta; x) = e^{-n\theta}\theta^{n\bar{x}}, \qquad \theta > 0. \tag{6.2.6}$$

The graph of this function of θ is shown in Figure 4.13.1.

EXAMPLE 6.2.2. *Likelihood function for Normal expectation (variance known).* If X is Normal $(\theta, 1)$ the joint sampling p.d.f. of n independent observations at $(x_1, x_2, \ldots, x_n)$ is

$$(2\pi)^{-n/2} \exp -\tfrac{1}{2}\sum_{1}^{n} (x_r - \theta)^2, \qquad n < x < \infty.$$

This expression is proportional to the likelihood of θ based on a sample $(x_1, x_2, \ldots, x_n)$. The likelihood function may be taken to be

$$\exp -\tfrac{1}{2}\sum_{1}^{n} (x_r - \theta)^2.$$

Since

$$\sum_{1}^{n} (x_r - \theta)^2 = \sum_{1}^{n} (x_r - \bar{x})^2 + n(\theta - \bar{x})^2$$

the likelihood function is proportional to

$$l(\theta) = \exp -\tfrac{1}{2}n(\theta - \bar{x})^2,$$

a function whose graph is shown in Figure 6.2.1.

The graph of the equivalent log-likelihood function

$$\log l(\theta) = -\tfrac{1}{2}n(\theta - \bar{x})^2$$

is shown in Figure 6.2.2.

EXAMPLE 6.2.3. *Likelihood function for Normal variance (when the expectation is known).* If X is Normally distributed with zero expectation and with

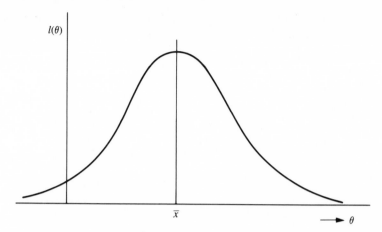

Figure 6.2.1: Likelihood function of θ based on data from a Normal $(\theta, 1)$ distribution (see Example 6.2.2).

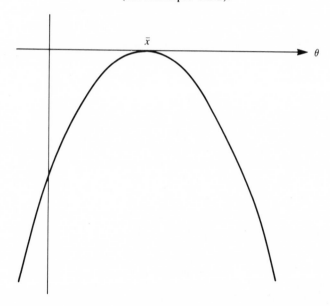

Figure 6.2.2: The log-likelihood function $-\frac{1}{2}n(\theta - \bar{x})^2$ based on data from a Normal $(\theta, 1)$ distribution.

standard deviation θ, the p.d.f. at x is

$$(2\pi)^{-1/2}\theta^{-1} \exp(-x^2/2\theta^2), \qquad \alpha < x < \infty$$

and the sample p.d.f. at $(x_1, x_2, \ldots, x_n)$ is

$$(2\pi)^{-n/2}\theta^{-n} \exp\left(-\sum_1^n x_j^2/2\theta^2\right).$$

The likelihood of θ on a sample $(x_1, x_2, \ldots, x_n)$, on dropping the uninformative multiplicative constant $(2\pi)^{-n/2}$, may be taken as

$$l(\theta) = \theta^{-n} \exp(-c/2\theta^2), \qquad \theta > 0, \left(c = \sum_1^n x_j^2\right) \qquad (6.2.7)$$

and the log-likelihood as

$$\log l(\theta) = -n \log \theta - c/2\theta^2.$$

The graph of $(6.2.7)$ is illustrated in Figure 6.2.3.

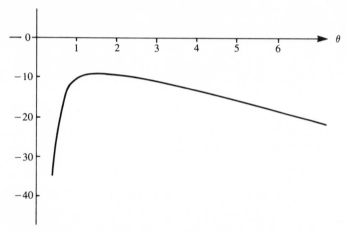

Figure 6.2.3: The log of the likelihood function $l(\theta) = \theta^{-n} \exp - \sum_1^n x_r^2/2\theta^2$ $(n = 10, \sum x_r^2 = 20)$ [see Example 6.2.3].

EXAMPLE 6.2.4.   *The two-parameter Normal distribution.*   If X is Normal (μ, σ), the likelihood function on data $(x_1, x_2, \ldots, x_n)$ is proportional to

$$\sigma^{-n} \exp\left\{-\sum_1^n (x_r - \mu)^2/2\sigma^2\right\}, \qquad -\infty < \mu < \infty, \qquad \sigma > 0. \qquad (6.2.8)$$

Replacing $\sum_1^n (x_j - \mu)^2$ by the equivalent expression $\sum_1^n (x_j - \bar{x})^2 + n(\mu - \bar{x})^2$, where $\bar{x}$ denotes the sample mean, it will be seen that we may write the likelihood function as

$$l(\mu, \sigma) = \sigma^{-n} \exp[-\{a + n(\mu - \bar{x})\}/2\sigma^2]$$

where

$$a = \sum_1^n (x_j - \bar{x})^2,$$

and, equivalently, the log-likelihood function as

$$\log l = \text{constant} - n \log \sigma - \{a + n(\mu - \bar{x})^2\}/2\sigma^2$$

where we may conveniently take the constant to be zero. Another version of this, in terms of the variance $\tau(=\sigma^2)$ is

$$\log l = -\tfrac{1}{2}n \log \tau - \{a + n(\mu - \bar{x})^2\}/2\tau.$$

With the following data, for example:

$$n = 10, \qquad \bar{x} = 10, \qquad a = 20,$$

this becomes

$$\log l = -5 \log \tau - \frac{10}{\tau} - \frac{5(\mu - 10)^2}{\tau}.$$

A rough idea of the form of this surface may be obtained by evaluating $\log l$ on a grid of (μ, τ) values and sketching the contours by 'visual interpolation'. This is illustrated in Figure 6.2.4. A perspective view of the surface is shown in Figure 6.2.5.

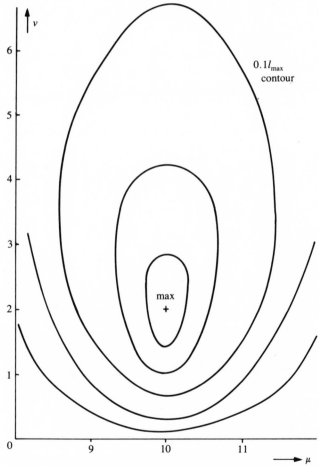

Figure 6.2.4: Contours of the log-likelihood surface of Example 4.13.6 (not to scale).

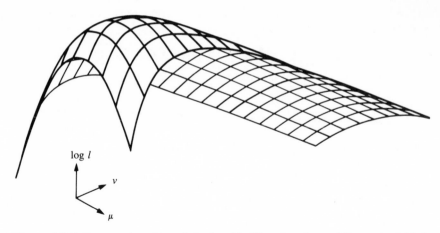

Figure 6.2.5: Perspective view of the log-likelihood surface of Example 4.13.6.

EXAMPLE 6.2.5.   *Markovian data.*   Consider a sequence $(t_1, t_2, \ldots, t_n)$ of 'indicators' showing whether the greatest height of the water level of a given river at a specified location in the rth year of record exceeded a given norm ($t_r = 1$) or failed to do so ($t_r = 0$), $r = 1, 2, \ldots, n$, it being assumed that the t_r are realizations of random variables $T_1, T_2, \ldots, T_n$ which form a segment of a Markov chain [see II, § 19.3] with transition probability matrix

$$
\begin{array}{c|cc}
 & 0 & 1 \\
\hline
r \quad 0 & 1-\alpha & \alpha \\
1 & \beta & 1-\beta
\end{array}
\qquad 0 < \alpha < 1, \qquad 0 < \beta < 1.
$$

The probability that $T_r = t_r$, $r = 1, 2, \ldots, n$ is

$$P(T_1 = t_1, T_2 = t_2, \ldots, T_n = t_n)$$

$$= P(T_1 = t_1) \prod_{r=1}^{n-1} P(T_{r+1} = t_{r+1} | T_r = t_r)$$

$$= \pi(\alpha, \beta)(1-\alpha)^{n(0,0)} \alpha^{n(0,1)} \beta^{n(1,0)} (1-\beta)^{n(1,1)}$$

where, for i and $j = 0$ or 1, $n(i, j)$ denotes the number of transitions in the sample from state i to state j, and $\pi(\alpha, \beta) = P(T_1 = t_1)$. Thus, for example, in the data sequence

$$\overline{0 \; 1} \; 1 \quad 0 \; 0 \; \overline{0 \; 1} \quad 0 \; 0 \; \overline{0 \; 1} \; 1 \; 1 \; 0,$$

for which $n = 14$, there are three transitions from 0 to 1 (marked $\overline{0 \; 1}$) so that $n(0, 1) = 3$. Similarly $n(0, 0) = 4$, $n(1, 0) = 3$, and $n(1, 1) = 3$, so that we have the following transition frequency table for $n(r, s)$:

	$s=0$	$s=1$
$r=0$	4	3
$r=1$	3	3

The (absolute) probability $P(T_1 = t_1)$ may be taken from the limiting equili-
brium distribution [see II, § 19.6]

$$P(T_1 = 0) = \beta/(\alpha + \beta), \qquad P(T_1 = 1) = \alpha/(\alpha + \beta)$$

whence

$$P(T_1 = t) = \alpha^t \beta^{1-t}/(\alpha + \beta), \qquad t = 0, 1.$$

In the data sequence shown, therefore, with $T_1 = 0$, the likelihood function is

$$l(\alpha, \beta) = \frac{\beta}{\alpha + \beta}(1-\alpha)^4 \alpha^3 \beta^3 (1-\beta)^3, \qquad 0 < \alpha < 1, \qquad 0 < \beta < 1.$$

In a larger sample showing five floods in a 50-year record, starting with a
non-flood $(T_1 = 0)$, the transition frequency table was found to be

		s	
		0	1
r	0	41	3
	2	3	2

The likelihood function for this sample is

$$l(\alpha, \beta) = (1-\alpha)^{41} \alpha^3 \beta^4 (1-\beta)^2/(\alpha + \beta). \qquad (6.2.9)$$

The contours of the surface represented by this function are shown in Figure
6.2.6, the contours of $l = 0.5, 1, 2, 4$ and 5×10^{-7} being as indicated. [Continued
in Example 6.2.6a.]

Likelihood evaluated from a frequency table

If the sample data are recorded in the frequency-table form (6.2.1), the
frequencies $n(r)$ are realizations of random variables $N(r)$, the joint distribution of which is multinomial [see II, § 6.4.2], and (6.2.5) has to be replaced by

$$P\{N(r) = n(s), s = 1, 2, \ldots, k\}$$

$$= \left\{ n! \Big/ \prod_{s=1}^{k} n(s)! \right\} \prod_{s=1}^{k} \{P(x_s - \tfrac{1}{2}h < X \leqslant x_s + \tfrac{1}{2}h)\}^{n(s)} \qquad (6.2.10)$$

where the p.d.f. of X is the common p.d.f. of the X_r, namely $f(x, \theta)$. As far
as the likelihood is concerned the combinatorical factor is irrelevant, and we

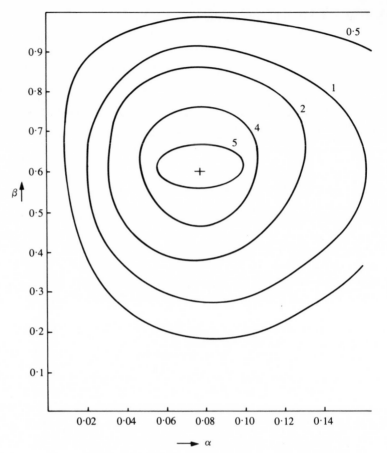

Figure 6.2.6: Contours of the likelihood surface (6.2.9):
$$l(\alpha, \beta) = (1-\alpha)^{41}\alpha^3\beta^4(1-\beta)^2/(\alpha+\beta), \ 0<\alpha<1, \ 0<\beta<1.$$

may take the likelihood as being proportional to

$$\prod\{P(x_s-\tfrac{1}{2}h<X\leq x_s+\tfrac{1}{2}h)^{n(s)}.$$

In the case of discrete random variables this reduces to

$$l(\theta; x) = \prod_{s=1}^{k} \{f(x_s, \theta)\}^{n(s)}, \qquad \theta \in \Omega, \tag{6.2.11}$$

the obvious frequency-table version of (6.2.5). Thus for example, in the case of a frequency table of Poisson (θ) data, the likelihood (6.2.11) becomes

$$\prod_{s=1}^{k} (e^{-\theta}\theta^s/s!)^{n(s)} = e^{-n\theta}\theta^{n\bar{x}}/\prod(s!)^{n(s)}$$

[compare Example 6.2.1]. In practice one would remove the uninformative factor $1/\prod (s!)^{n(s)}$, so that the likelihood function would become

$$l(\theta; x) = e^{-n\theta}\theta^{n\bar{x}},$$

coinciding exactly with the data-vector version discussed earlier (in Example 6.2.1).

If fundamentally continuous data have been grouped into a frequency table as in (6.2.2), the likelihood becomes

$$l(\theta; x) = \prod_{s=1}^{k} \{F(x_s + \tfrac{1}{2}h, \theta) - F(x_s - \tfrac{1}{2}h, \theta)\}^{n(s)}, \qquad (6.2.12)$$

where $F(x, \theta)$ is the common p.d.f. of the X_r.

The additional complications introduced by this effect are discussed later (see § 6.5.2).

6.2.2.   The Maximum-likelihood Estimate (M.L.E.)

We consider first the one-parameter case. The likelihood function $l(\theta; x)$, as defined above, is a function of the parameter value (θ, say) and the data x. The *maximum-likelihood estimate* ('M.L.E.') of θ is the value ($\hat{\theta}$, say) which maximizes $l(\theta; x)$. In most cases of practical interest there is a unique maximizing value for $\theta \in \Omega$ (see Figures 6.2.1 to 6.2.6). This maximizing value $\hat{\theta}$ (which may be a vector) is of course a function of the data x: it is a statistic, and we may on occasion emphasize its dependence on x by writing it as $\hat{\theta}(x)$. For example, in the case discussed above where the x_r are independent realizations of a Poisson (θ) variable, the likelihood of θ is $e^{-n\theta}\theta^{n\bar{x}}$ ($\theta > 0$), where $\bar{x}$ is the sample mean, and this is readily seen to have a unique maximum (see Figure 6.2.1) at $\bar{x}$. The M.L.E. of θ is thus $\hat{\theta}(x) = \bar{x}$.

The statistical properties of $\hat{\theta}(x)$ are those of the induced random variable $\hat{\theta}(X)$ [see § 11.4.2(v)] of which it is a realization. In the Poisson case just mentioned, this induced random variable is just $\bar{X} = \sum_1^n X_r/n$, where the X_r are independent Poisson (θ) variables. It follows that $E\{\hat{\theta}(X)\} = E(\bar{X}) = \theta$, so that $\hat{\theta}$ is unbiased [see § 3.3.2], that var $\{\hat{\theta}(X)\} = \theta/n$, and that the sampling distribution of $n\hat{\theta}$ ($= \sum_1^n x_r$) is Poisson ($n\theta$) [see Table 2.4.1]. The accuracy of $\hat{\theta}$ might be described in various ways, most satisfactorily perhaps by constructing 95% confidence limits for θ in terms of $\hat{\theta}$ (see Chapter 4).

The two-parameter case is illustrated in Example 6.2.6.

EXAMPLE 6.2.6.   *M.L.E. of parameters of the Normal distribution.*   In Example 6.2.4, the likelihood surface (that is the surface representing the function $l(\mu, \sigma)$ of (6.2.8)) is shaped like a smooth mountain with a single summit [see Figures 6.2.4 and 6.2.5]. At the summit, of which the coordinates are, say, $(\hat{\mu}, \hat{\sigma})$, the tangent plane is horizontal, whence, writing l for $l(\mu, \sigma)$,

$$\partial l/\partial \mu = 0 \quad \text{and} \quad \partial l/\partial \sigma = 0$$

at $(\hat{\mu}, \hat{\sigma})$ [see IV, § 5.2]. These are the likelihood equations, whose solution $(\hat{\mu}, \hat{\sigma})$ locates the maximum. Since

$$\frac{\partial \log l}{\partial \mu} = \frac{1}{\mu} \frac{\partial l}{\partial \mu},$$

and similarly for $\partial \log l/\partial \sigma$, the likelihood equations may be replaced by the equivalent pair

$$\partial \log l/\partial \mu = 0, \qquad \partial \log l/\partial \sigma = 0.$$

The differentiations are usually easier to carry out on the *log-likelihood* function $\log l$ than on the likelihood itself.

We have, from (6.2.8),

$$\log l = -n \log \sigma - \sum_1^n (x_j - \mu)^2/2\sigma^2$$

so that

$$\partial \log l/\partial \mu = -\sum_1^n (x_j - \mu)/\sigma^2,$$

$$\partial \log l/\partial \sigma = -n/\sigma + \sum_1^n (x_j - \mu)^2/\sigma^3.$$

Equating these expressions to zero we obtain the estimates $\hat{\mu}, \hat{\sigma}$ given by

$$\hat{\mu} = \bar{x} \quad \text{(the sample mean)}$$

and

$$\hat{\sigma} = \left\{ \sum_1^n (x_j - \bar{x})^2/n \right\}^{1/2}.$$

In the sample, the values are found to be

$$\hat{\mu} = 2, \qquad \hat{\sigma} = 1 \cdot 41.$$

One might equally well have taken μ and $v = \sigma^2$ as the parameters, instead of μ and σ. The likelihood function of μ and v is obtainable from $l(\mu, \sigma)$ on replacing σ by $v^{1/2}$, and the maximizing conditions

$$\partial l/\partial \mu = 0, \qquad \partial l/\partial v = 0$$

lead to the estimation

$$\hat{\mu} = \bar{x}, \qquad \hat{v} = \sum (x_j - \bar{x})^2/n$$

so that $(\hat{\sigma})^2 = \hat{v}$: the M.L.E. of σ^2 is the square of the M.L.E. of σ. This 'invariant' property is of general application, as we shall see.

It is of course not always the case that the likelihood equations have an explicit analytic solution. Indeed, the existence of such a solution is the

exception. Usually the equations have to be solved numerically. The following
is an example.

EXAMPLE 6.2.7.   *Numerical evaluation of M.L.E.s of two parameters.*   The
likelihood function (6.2.9) of Example 6.2.5 leads to the following equations
for the M.L.E.:

$$\frac{3}{\alpha} - \frac{41}{1-\alpha} = \frac{1}{\alpha+\beta}$$

$$\frac{4}{\beta} - \frac{2}{1-\beta} = \frac{1}{a+\beta}.$$

A reasonable first approximation α_1, β_1, may be obtained by comparing the
transition probability matrix with this transition frequency matrix.
   These are, respectively,

		Total				Total
$1-\alpha$	α	1		41	3	44
β	$1-\beta$	1	and	3	2	5

whence we take α_1 to be $3/44 = 0·068$ and β_1 to be $3/5 = 0·600$. The use of
iterative methods, or, equivalently, a fine-scale tabulation of $l(\alpha, \beta)$ in the
neighbourhood of the point $(0·068, 0·600)$, lead to the M.L.E.

$$\hat{\alpha} = 0·066, \qquad \hat{\beta} = 0·607.$$

[Continued in Example 6.4.4.]

6.2.3.   M.L.E.: Intuitive Justification

   'Nothing is easier' (said Fisher) 'than to invent methods of estimation.' He
went on to say that the important thing was to distinguish satisfactory methods
from unsatisfactory ones. There is a very simple argument for preferring the
method of maximum likelihood. The process of evaluating the M.L.E. $\hat{\theta}$ is
equivalent to that of examining every member of the prescribed family $f(x, \theta)$
of distributions, for $\theta \in \Omega$, and selecting the one, namely $f(x, \hat{\theta})$, which allots
maximal probability (or probability density) to the data values actually
obtained. Any value of θ other than $\hat{\theta}$, with $l(\theta; x) < l(\hat{\theta}; x)$, would have
allocated a smaller probability to the data and so would be less plausible, to
an extent indicated by the *relative likelihood* $l(\theta; x)/l(\hat{\theta}; x)$.
   The accuracy of the estimate must depend to some extent on the data-values
and to some extent on the nature of the sampled distribution. If these conspire
to produce a likelihood function with a rather flat maximum, θ-values that
may be far from $\hat{\theta}$ will have a relative likelihood approaching unity, whence
$\hat{\theta}$ must be regarded as somewhat lacking in precision when the likelihood
function has small curvature at the maximum; and conversely a large curvature

is intuitively associated with a comparatively precise estimate. Since the curvature [see IV, Definition 6.1.4] of the graph of $l(\theta)$ at the maximum (where $dl/d\theta = 0$) is equal to $d^2l/d\theta^2$, one would expect this quantity to be a useful measure of accuracy. As will be seen, the logarithm of the likelihood is more relevant in these discussions than the likelihood itself, and we may anticipate that importance will be attached to the second derivative $d^2 \log l/d\theta^2$.

6.2.4.　M.L.E.: Kinds of Maxima

The regular case. Typical graphs of one-parameter likelihood functions are shown in Figure 6.2.7.

Figure 6.2.7(i) shows the '*regular* case' where $l(\theta)$ is a continuous differentiable function of θ and attains its maximum at the point where $dl(\theta)/d\theta = 0$. Figure 6.2.7(ii) shows the case where the maximum is attained at an extremity

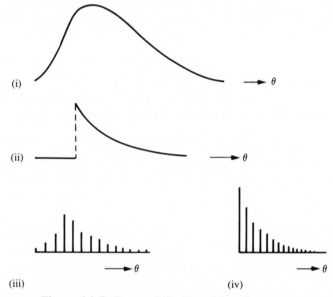

Figure 6.2.7: Types of likelihood functions.

of the parameter space Ω, and where differentiation is not applicable as a method of locating the maximum. Figure 6.2.7(iii) is the analogue of (i), and Figure 6.2.7(iv) of Figure 6.2.7(ii), in cases where the parameter space is discrete. Examples will be given below.

Whilst in all cases care must be taken to see that the proposed method of maximization is appropriate to the nature of the likelihood function in question, in practice it is the 'regular' case that is the most important and usual one. The maximum is then found by differentiating the likelihood function. If one is dealing with the common simple case of i.i.d. observations as in (6.2.5), so that the likelihood function is proportional to $\prod f(x_r, \theta)$, differentiation is facilitated by first taking logs, that is by working with the *log-likelihood*

$$\log l(\theta; x) = \sum_r \log f(x_r, \theta), \qquad (6.2.13)$$

since the equation $d \log l / d\theta = 0$ has the same roots as the '*likelihood equation*' $dl/d\theta = 0$. In principle, any logarithmic base may be used, although in practice one almost always works with base e (natural logs). Whatever the base, the notation 'log' (rather than, say, "ln") will be used throughout this chapter. (See Example 6.2.6.)

6.2.5.   Theoretical Justification for the M.L.E. Properties of the Likelihood Function

(a) *Asymptotic distribution of the derivative of the log-likelihood function*

The intuitive arguments adduced earlier in support of the method of maximum likelihood are backed by a formidable array of theoretical arguments. These show that (in 'regular' cases) the method produces estimates that for large samples are asymptotically unbiased [see § 3.3.2], consistent [see 3.3.1(c)] and efficient [see § 3.3.3(b)], and also asymptotically multivariate Normal [see II; § 13.4]; further, they are minimally sufficient [see § 3.4] when such estimates exist. In practice what this means is that the M.L.E. from a realistic, finite sample has a rather small bias (or no bias at all) and has accuracy not far from the theoretical optimum.

It turns out that it is the log-likelihood, not the likelihood itself, that is of fundamental importance. Write the likelihood (6.2.4) in any of the forms

$$l = l(\theta) = l(\theta; x) = l(\theta; x_1, x_2, \ldots, x_n),$$

according to convenience; for a given x this is a realization of the random variable

$$L = L(\theta) = l(\theta; X) = l(\theta; X_1, X_2, \ldots, X_n).$$

We first establish the property that

$$E\{\partial \log L(\theta)/\partial \theta_r\} = 0, \qquad r = 1, 2, \ldots, p. \qquad (6.2.13a)$$

[See IV, § 5.2 for partial derivatives.] This follows from the fact that

$$\frac{\partial \log L}{\partial \theta_r} = \frac{\partial \log g(X, \theta)}{\partial \theta_r}$$

where $g(x, \theta)$ is the joint p.d.f. of $X_1, X_2, \ldots, X_n$ at x, so that

$$E\{\partial \log L/\partial \theta_r\} = \int \frac{1}{g} \frac{\partial g}{\partial \theta_r} g \, dx$$

(where '$\int$' denotes an n-fold definite integral over the sample space S [see IV, § 6.4], and 'dx' denotes '$dx_1 \ldots dx_n$')

$$= \int (\partial g / \partial \theta_r) \, dx = \frac{\partial}{\partial \theta_r} \left(\int g \, dx \right)$$

(in the regular case)

$$= 0$$

since $\int g \, dx = 1$, g being a normalized density.

The second required property of $\partial \log L / \partial \theta_r$ is as follows:

The variances and covariances of the random variables $\partial \log L / \partial \theta_r$, $r = 1, 2, \ldots, p$, are given by

$$\mathrm{var} \, (\partial \log L / \partial \theta_r) = -E (\partial^2 \log L / \partial \theta_r^2)$$

and   (6.2.13b)

$$\mathrm{cov} \, (\partial \log L / \partial \theta_r, \partial \log L / \partial \theta_s) = -E (\partial^2 \log L / \partial \theta_r \, \partial \theta_s).$$

These results depend on the identity

$$E (\partial \log L / \partial \theta_r)^2 = -E (\partial^2 \log L / \partial \theta_r^2).$$

To establish this, note that

$$\partial^2 \log L / \partial \theta_r^2 = \partial^2 \log g(X, \theta) / \partial \theta_r^2$$

where $g(x, \theta)$ is the sample p.d.f. Thus

$$E (\partial^2 \log L / \partial \theta_r^2) = E \{ -g^{-2} (\partial g / \partial \theta_r)^2 + g^{-1} (\partial^2 g / \partial \theta_r^2) \}$$

$$= -E (\partial \log g / \partial \theta_r)^2 + \int (\partial^2 g / \partial \theta_r^2) \, dx$$

where, as previously, the integral is taken over the entire sample space. In the regular case

$$\int (\partial^2 g / \partial \theta_r^2) \, dx = (\partial^2 / \partial \theta_r^2) \int g \, dx$$

and this is zero since $\int g \, dx = 1$. Thus, finally,

$$E (\partial^2 \log L / \partial \theta_r^2) = -E (\partial \log L / \partial \theta_r)^2.$$

This establishes the first of the identities (6.2.11). The second may be established by similar arguments.

We then have the important result that:

The joint distribution of the random variables $\partial \log L(\theta) / \partial \theta_r$, $r = 1, 2, \ldots, p$, is approximately multivariate Normal [see II; § 13.4] with zero expectation vector and with dispersion matrix given by (6.2.13b).   (6.2.14)

For, in the case of i.i.d. observations, $\partial \log L/\partial \theta_r$ is the sum of the i.i.d. random variables $\partial \log f(X_s, \theta)/\partial \theta_r$, with positive common variance, whence by the Central Limit Theorem [see II, § 17.3], $\partial \log L/\partial \theta_r$ is asymptotically Normal and so is approximately Normal in finite samples. A more general version of the Central Limit Theorem leads to (6.2.14).

The one-parameter case

In the special case where there is only one parameter, θ, (6.2.14) reduces to the following:

The random variable $d \log L(\theta)/d\theta$ is approximately Normally distributed, with zero expectation and with variance $\tau^2(\theta)$ given by

$$\tau^2(\theta) = -E\{d^2 \log L(\theta)/d\theta^2\}. \tag{6.2.15}$$

It follows from the properties of the Normal distribution that the interval

$$(-1 \cdot 96\tau(\theta), \qquad 1 \cdot 96\tau(\theta))$$

is the central 95% probability interval [see § 4.1.3] for $d \log L/d\theta$, to the accuracy of this approximation [cf. Example 4.5.2(b)].

(b) *Confidence interval for an estimated parameter: first method*

In the following paragraphs we explain a direct and powerful approach to the problem of constructing a reasonably accurate confidence interval for a parameter estimated by maximum likelihood. It should be noted, however, that in practice a cruder but still useful approximation is usually employed. This is described in section 6.2.5(d).

First, then, the direct method. We confine ourselves to the one-parameter case, for which (6.2.14) is equivalent to the following;

with probability $0 \cdot 95$ (approximately), the observed value of $d \log l/d\theta$ will lie in the interval

$$\pm 1 \cdot 96\tau(\theta)$$

where $\{\tau(\theta)\}^2 = -E(d^2 \log l/d\theta^2)$. [See Example 4.5.2(b).]

If the function $d \log l/d\theta$ is one-to-one, to say that the function values lie in the interval $\pm 1 \cdot 96\tau(\theta)$ is equivalent to saying that θ itself lies in the interval (θ_1, θ_2), where θ_1, θ_2 are the roots of the equations

$$d \log l/d\theta = \pm 1 \cdot 96\tau(\theta).$$

This brings us to the following:

The interval (θ_1, θ_2) is a confidence interval [see § 4.2] for θ, with confidence level 0·95 approximately, where θ_1, θ_2 are the roots of the equation

$$d \log l / d\theta = \pm 1 \cdot 96 \tau(\theta), \tag{6.2.16}$$

$\tau(\theta)$ being defined in (6.2.15).

(Similarly of course for other confidence levels.) The method is illustrated in the following examples.

EXAMPLE 6.2.8.   *Confidence limits for the parameter of an exponential distri-bution.* We take the (one-parameter) exponential distribution [see II, § 10.2.3] with density of the form $f(x, \theta) = \theta e^{-\theta x}$, $\theta > 0$, where we note that $E(X) = 1/\theta$, var $(X) = 1/\theta^2$. On data $x = (x_1, \ldots, x_n)$ the likelihood is

$$l(\theta; x) = \theta^n r^{-\theta \Sigma x_r} = \theta^n e^{-n\theta \bar{x}}$$

and the log-likelihood

$$\log l = n \log \theta - n\theta\bar{x},$$

whence

$$\frac{d \log l}{d\theta} = \frac{n}{\theta} - n\bar{x}.$$

Thus the likelihood equation is

$$\frac{n}{\theta} - n\bar{x} = 0.$$

The M.L.E. is

$$\hat{\theta} = 1/\bar{x}.$$

Now look at the random variable $d \log L / d\theta$. This is

$$\frac{d \log L}{d\theta} = \frac{n}{\theta} - n\bar{X},$$

whence, since $E(\bar{X}) = E(X) = 1/\theta$,

$$E\left(\frac{d \log L}{d\theta}\right) = \frac{n}{\theta} - \frac{n}{\theta} \equiv 0,$$

in accord with (6.2.10). The variance of $d \log L / d\theta$ is

$$\tau^2(\theta) = -E\left(\frac{d^2 \log L}{d\theta^2}\right) = \frac{n}{\theta^2}$$

according to (6.2.11); since var $(\bar{x}) = 1/\theta^2$, this agrees with a direct evaluation from the expression $d \log L / d\theta = n/\theta - n\bar{x}$.

The next step is the distribution of $d \log L / d\theta$. According to (6.2.13) this is approximately Normal $(0, \sqrt{n}/\theta)$. The exact distribution is that of $n/\theta - \Sigma_1^n X_r$. Now the p.d.f. of $Y = \Sigma_1^n X_r$ at y is $\theta^n y^{n-1} e^{-\theta y}/(n-1)!$, $y > 0$ [see

§ 2.4.2]. This is a gamma distribution [see II, § 11.3] with expectation n/θ, and variance n/θ^2; with skewness $2/\sqrt{n}$ and so nearly symmetrical for moderate or large n; and with kurtosis $3\{1+(2/n)\}$ which is near the Normal value of 3. Thus for moderate values of n the distribution of $d\log L/d\theta$ is indeed approximately Normal, in accord with (6.2.12); and with expectation 0 and variance n/θ^2, so that the approximations given in (6.2.13) are exact in this case.

We now turn to the accuracy. We first obtain the exact confidence limits by finding a 95% central probability interval [see § 4.1.3] for $W = 2\theta Y = 2\theta \sum X_r = 2\theta n\bar{X}$. Since the p.d.f. of W at w is $(\frac{1}{2}w)^{n-1} e^{-w/2}/2\Gamma(n)$, the quantity $W = 2\theta \sum X_r$ is a chi-squared variable [see § 2.5.4(a)] on $\nu = 2n$ degrees of freedom, whence we can find (exactly) numbers w_1 and w_2 cutting off tail areas of $2\frac{1}{2}\%$ in the lower and the upper tail respectively. Taking the sample size n to be 40, we have $w_1 = 57\cdot153$, $w_2 = 106\cdot629$ (from the chi-squared tables). Then, with probability 95%

$$w_1 \le W \le w_2$$

i.e.

$$57\cdot153 \le 80\bar{X}\theta \le 106\cdot629$$

i.e.

$$0\cdot714/\bar{X} \le \theta \le 1\cdot333/\bar{X}.$$

In terms of the observed sample mean $1/\bar{x} = \hat{\theta}$, the interval $(0\cdot714\hat{\theta}, 1\cdot333\hat{\theta})$ has 95% probability of containing the (unknown) true value of θ. This is a *confidence interval*: (see Chapter 4): the 95% confidence limits for θ are $0\cdot714\hat{\theta}, 1\cdot333\hat{\theta}$.

These last results are exact. If however we relied on the Normal approximation (6.2.13) according to which $d\log L/d\theta$ is approximately Normal $(0, \theta/\sqrt{n})$, the argument would be that $d\log l/d\theta (=n/\theta - n\bar{x})$ is a realization of a Normal $(0, \sqrt{n}/\theta)$ variable, whence $\sqrt{n}(1-\theta\bar{x}) = \sqrt{n}(1-\theta/\hat{\theta})$ is a realization of a Normal $(0, 1)$ variable. The central 95% confidence interval (θ_1, θ_2) is obtained by solving the inequalities

$$-1\cdot96 \le \sqrt{n}(1-\theta/\hat{\theta}) \le 1\cdot96$$

so that, taking $n = 40$ as before,

$$\theta_1, \theta_2 = (1 \pm 1\cdot96/\sqrt{40})\theta = (1 \pm 0\cdot310)\theta.$$

Thus the 95% confidence interval for θ obtained from the approximation (6.2.14) (or equivalently (6.2.15), (6.2.16)) is

$$(0\cdot690\hat{\theta}, 1\cdot310\hat{\theta}),$$

which is to be compared with the exact values

$$(0\cdot714\hat{\theta}, 1\cdot333\hat{\theta}).$$

The agreement is satisfactory.

(c) *The principal practical approximation for the sampling distribution of the M.L.E.*

The approximations leading to (6.2.16) rest on the sure foundations of the Central Limit Theorems [see II, § 17.3], and can be expected to work well. It has, however, become the usual practice to push the approximation further, so as to extract explicit expressions for the (approximate) sampling distribution of the estimates themselves. These are given in the following statement, which forms what may justly be called the

Principal Practical Approximation for M.L.E.'s: (6.2.17)

(i) In the one-parameter case, where the likelihood function is $l(\theta, x)$, the sampling distribution of the M.L.E. $\hat{\theta}$ is approximately Normal with expectation θ and variance ω^2, where

$$\omega^2 = -1/E(d^2 \log L/d\theta^2).\qquad(6.2.17\text{(i)})$$

Here L denotes the random variable induced by l: that is, if $l = l(\theta; x_1, x_2, \ldots, x_n)$, $L = l(\theta; X_1, X_2, \ldots, X_n)$ where X_r is induced by x_r, $r = 1, 2, \ldots, n$.

The above expression for ω^2 may be approximated by

$$-1/(d^2 \log l/d\theta^2).\qquad(6.2.17\text{(ii)})$$

(ii) When there are p parameters $\theta_1, \theta_2, \ldots, \theta_p$, the joint sampling distribution of the M.L.E.'s $\hat{\theta}_1, \hat{\theta}_2, \ldots, \hat{\theta}_p$ is asymptotically multivariate Normal [see II, § 13.4], with expectations $\theta_1, \theta_2, \ldots, \theta_p$ and dispersion matrix $\mathbf{A}^{-1}$, where the (r, s) entry in $\mathbf{A}$ is

$$-E(\partial^2 \log L/\partial\theta_r \, \partial\theta_s),\qquad(6.2.17\text{(iii)})$$

or, approximately,

$$-\partial^2 \log l/\partial\theta_r \, \partial\theta_s$$

for $r, s = 1, 2, \ldots, p$. [For 'dispersion matrix' see II, § 13.3.1; for '$\mathbf{A}^{-1}$' see I, § 6.4.]

When, as usually happens, ω^2 in (i) or entries in $\mathbf{A}$ in (ii) depend on the (unknown) values of the parameters, numerical approximations may be obtained by replacing θ (or $\theta_1, \theta_2, \ldots, \theta_p$) by the corresponding estimate.

In the two-parameter case the dispersion matrix of the estimate, namely

$$\begin{pmatrix} \sigma_1^2 & \rho\sigma_1\sigma_2 \\ \rho\sigma_1\sigma_2 & \sigma_2^2 \end{pmatrix}$$

(where σ_1^2 is the sampling variance of $\hat{\theta}_1$, σ_2^2 that of $\hat{\theta}_2$, and ρ their sampling correlation coefficient) is approximated by

$$-\begin{pmatrix} E(\partial^2 \log L/\partial\theta_1^2) & E(\partial^2 \log L/\partial\theta_1 \, \partial\theta_2) \\ E(\partial^2 \log L/\partial\theta_1\partial\theta_2) & E(\partial^2 \log L/\partial\theta_2^2) \end{pmatrix}^{-1}\qquad(6.2.17\text{(iv)})$$

which is itself approximately equal to

$$-\begin{pmatrix} \partial^2 \log l/\partial\theta_1^2 & \partial^2 \log l/\partial\theta_1, \partial\theta_2 \\ \partial^2 \log l/\partial\theta_1\partial\theta_2 & \partial^2 \log l/\partial\theta_2^2 \end{pmatrix}^{-1} \qquad (6.2.17(v))$$

the derivatives being evaluated at $(\hat\theta_1, \hat\theta_2)$. Examples 6.4.1, 6.4.2, 6.4.3 and 6.4.4 illustrate the use of these formulae.

The arguments displayed above involve somewhat daring approximations, but the results are in most cases surprisingly accurate.

(d) *Confidence intervals and confidence regions: second method*

A direct method for constructing a good approximate confidence interval for an estimated parameter was described in section 6.2.5(b) ('first method'). We now outline a second, cruder method that rests on (6.2.17). This is the approximation almost universally used in practice.

One parameter

For the one-parameter case, the parameter θ, with estimate $\hat\theta$ and standard error ω given by (6.2.17(i)), viz.

$$-1/\omega^2 = E\{d^2 \log L/d\theta^2\}$$
$$\cong d^2 \log l/d\theta^2|_{\hat\theta}$$

has as its central 95% confidence interval the following:

$$\hat\theta \pm 2\omega$$

[cf. Example 4.5.2(b)].

As an illustration, consider the case of the exponential parameter discussed in Example 6.2.7. Here

$$d^2 \log l/d\theta^2 = -n/\theta^2$$

whence

$$\omega \approx \hat\theta/\sqrt{n}$$

and the 95% confidence interval for θ is

$$\hat\theta \pm 2\hat\theta/\sqrt{n}.$$

Several parameters

It will suffice to deal with the case of two parameters (θ_1, θ_2), estimated by the M.L.E. $(\hat\theta_1, \hat\theta_2)$. As explained in paragraph (b) of the 'Principal Practical Approximation' (6.2.17), we regard these estimates as having (approximately) a bivariate Normal sampling distribution with expectation vector (θ_1, θ_2) and dispersion matrix (6.2.17(v)). Two cases arise, one where we are interested in a (*one*-dimensional)linear combination of θ_1 and θ_2, the other where we seek a joint confidence region for θ_1 and θ_2.

Case (i): The appropriate 95% confidence interval for the linear combination $a_1\theta_1 + a_2\theta_2$ is

$$a_1\hat{\theta}_1 + a_2\hat{\theta}_2 \pm 1\cdot96\sqrt{(a_1^2 v_1 + 2a_1 a_2 c + a_2^2 v_2)},$$

approximately, where

$$\begin{pmatrix} v_1 & c \\ c & v_2 \end{pmatrix}$$

is the estimated dispersion matrix (6.2.17(v)). (See (4.9.5).)

Case (ii): Now suppose we require a joint (two-dimensional) confidence region for the pair (θ_1, θ_2). An elliptical region with confidence coefficients approximately 0·95 is given by the equation

$$\frac{(\theta_1-\hat{\theta}_1)^2}{\sigma_1^2} - \frac{2\rho}{\sigma_1\sigma_2}(\theta_1-\hat{\theta}_1)(\theta_2-\hat{\theta}_2) + \frac{(\theta_2-\hat{\theta}_2)^2}{\sigma_2^2} = 6(1-\rho^2),$$

where

$$\begin{pmatrix} \sigma_1^2 & \rho\sigma_1\sigma_2 \\ \rho\sigma_1\sigma_2 & \sigma_2^2 \end{pmatrix}$$

is the estimated dispersion matrix (6.2.17(v)). Details are explained in the discussion leading up to (4.9.6) in section 4.9.2. Explicit equations for such elliptical regions are given in Examples 6.4.3, 6.4.4.

6.2.6. Estimating a Function of θ. Invariance

In solving the likelihood equation for the M.L.E. $\hat{\theta}$ of the parameter of interest, it not infrequently turns out that it is more convenient to solve instead for some function of θ such as $\ln\theta$, $1/\theta$, etc. [see Example 4.13.2]. Call this $\phi = t(\theta)$, a one-to-one differentiable function, so that $d\phi/d\theta$ and $d\theta/d\phi$ exist and are not identically zero. Then, if $\hat{\theta}$ and $\hat{\phi}$ denote the M.L.E. of θ and of ϕ respectively,

$$\hat{\phi} = t(\hat{\theta}).$$

For example, if $\phi = 1/\theta$, $\hat{\phi} = 1/\hat{\theta}$; if $\phi = \log\theta$, $\hat{\phi} = \log\hat{\theta}$.

This follows in the regular case by writing the likelihood function of ϕ as

$$\lambda(\phi) = \lambda\{t(\theta)\} = l(\theta),$$

where [see iv, § 3.2]

$$d\lambda/d\phi = (dl/d\theta)(d\theta/d\phi).$$

Now the M.L.E. $\hat{\phi}$ of ϕ is defined such that $d\lambda/d\phi = 0$ when $\phi = \hat{\phi}$; thus

$(dl/d\theta)(d\theta/d\phi) = 0$ when $\phi = \hat{\phi}$, i.e. when $t(\theta) = \hat{\phi}$. Since by hypothesis $d\theta/d\phi \neq 0$, this latter equation is simply $dl/d\theta = 0$. It follows that $\hat{\phi} = t(\hat{\theta})$. This is often referred to as the *invariance property* of likelihood and of the M.L.E.

There is a simple relation between the M.L. approximation $v(\theta)$ for the sampling variance of $\hat{\theta}$, given in (6.2.17(ii)), and the corresponding expression $v(\phi)$ for that of $\hat{\phi}$. Since

$$d^2 \log l/d\phi^2 = (d^2 \log l/d\theta^2)(d\theta/d\phi)^2 + (d \log l/d\theta)(d^2\theta/d\phi^2),$$

and since $d \log l/d\theta = 0$ at $\theta = \hat{\theta}$, we have

$$-1/v(\phi) \cong (d^2 \log l/d\phi^2)|_{\hat{\phi}}$$
$$= (d^2 \log l/d\theta^2)|_{\hat{\theta}}(d\theta/d\phi)^2$$
$$\cong -\{1/v(\theta)\}(d\theta/d\phi)^2,$$

whence

$$v(\phi) = v(\theta)(d\phi/d\theta)^2, \qquad (6.2.18)$$

approximately, where $(d\phi/d\theta)$ is to be evaluated at the M.L. point. This, incidentally, agrees with the approximation obtained by other means in (2.7.1). For example, taking $\phi = 1/\theta$, we have

$$v(\phi) \cong \theta^{-2}v(\theta).$$

When there are two parameters involved, say θ_1 and θ_2, and ϕ is a prescribed function of θ_1 and θ_2, similar arguments lead to the approximation

$$v(\phi) = v(\theta_1)\left(\frac{\partial\phi}{\partial\theta_1}\right)^2 + 2c(\theta_1, \theta_2)\left(\frac{\partial\phi}{\partial\theta_1}\right)\left(\frac{\partial\phi}{\partial\theta_2}\right) + v(\theta_2)\left(\frac{\partial\phi}{\partial\theta_2}\right)^2 \qquad (6.2.19)$$

where $c(\theta_1, \theta_2)$ denote the covariance of $\hat{\theta}_1$ and $\hat{\theta}_2$.

6.3. EXAMPLES OF ML ESTIMATION IN ONE-PARAMETER CASES

EXAMPLE 6.3.1. *The Bernoulli (θ) distribution.* The data vector is $x = (x_1, x_2, \ldots, x_n)$, where $x_r = 0$ if the rth trial results in a 'failure' and $x_r = 1$ if a 'success'. The p.d.f. of the X_r [see II, § 5.2.1] is

$$f(x, \theta) = \theta^{x_r}(1-\theta)^{1-x_r}, \qquad x_r = 0, 1.$$

The likelihood is

$$l(\theta; x) = \theta^s(1-\theta)^{n-s}, \qquad 0 < \theta < 1,$$

where $s = \sum_1^n x_r$ denotes the total number of successes in n trials; s is a realization of a binomial (n, θ) variable. Then

$$d \log l/d\theta = s/\theta - (n-s)/(1-\theta) = (s - n\theta)/\theta(1-\theta)$$

whence the M.L.E. of θ is

$$\hat{\theta} = s/n = \text{observed proportion of successes.}$$

The approximation (6.2.16) for the sampling variance of $\hat{\theta}$ gives the value

$$-1/E\{-S/\theta^2 - (n-S)/(1-\theta)^2\}$$

(where S is binomial (n, θ))

$$= \theta(1-\theta)/n$$

which is in fact exact; The estimated value of this would of course be $\hat{\theta}(1-\hat{\theta})/n = s(n-s)/n^3$.

EXAMPLE 6.3.2.   *The Binomial (n, θ) distribution.*   This is the distribution of the total number S, of successes in n Bernoulli (θ) trials [see II, § 5.2.2]. The p.d.f. of S is

$$P(S=s) = f(s, \theta) = \binom{n}{s}\theta^s(1-\theta)^{n-s}, \qquad s = 0, 1, \ldots, n,$$

whence, removing the uninformative factor $\binom{n}{s}$, the likelihood is seen to be the same as that given in (6.3.1) of section 6.3.1—not surprisingly since s is a sufficient statistic for θ in that example. The remainder of the analysis coincides with that of Example 6.3.1. In particular the M.L.E. $\hat{\theta}$ of θ is:

$$\hat{\theta} = s/n$$

$$= \text{observed proportion of successes.}$$

To obtain the approximate confidence interval (6.2.15) for θ we note that the likelihood random variable is

$$L(\theta) = l(\theta, S) = \theta^S(1-\theta)^{n-S} \text{ where } S \text{ is Binomial } (n, \theta),$$

with

$$\log L(\theta) = S \log \theta + (n-S) \log (1-\theta),$$

$$d \log L/d\theta = S/\theta - (n-S)/(1-\theta)$$

$$= (S - n\theta)/\theta(1-\theta),$$

and

$$d^2 \log L/d\theta^2 = -S/\theta^2 - (n-S)/(1-\theta)^2.$$

Since $E(S) = n\theta$, the expression (6.2.13) for τ^2 reduces to

$$\tau^2(\theta) = n/\theta(1-\theta).$$

The procedure recommended in section 6.2.5 (b) for obtaining the 95% confidence interval for θ is to solve the inequalities

$$-1\cdot96\sqrt{\{n/\theta(1-\theta)\}} \le (s - n\theta)/\theta(1-\theta) \le 1\cdot96\sqrt{\{n/\theta(1-\theta)\}}.$$

These are precisely the same inequalities as were obtained in Example 4.7.1 by replacing the Binomial distribution by the approximating Normal.

EXAMPLE 6.3.3.   *The Poisson (θ) distribution.*   We suppose the data to be in frequency table form, that is, we are told that the observation x occurred n_x times, for $x = 0, 1, 2, \ldots, k$, each observation being a realization of the r.v. X with p.d.f. given by

$$P(X = x) = e^{-\theta}\theta^x/x!, \qquad x = 0, 1, 2, \ldots .$$

The likelihood of θ is then proportional to

$$\prod_x [(e^{-\theta}\theta^x)/x!]^{n_x}$$

Dropping the uninformative factor $1/\pi(x!)^{n_x}$ we may take the likelihood to be

$$l(\theta; x) = \prod_{x=0}^{k} (e^{-\theta}\theta^x)^{n_x}$$

$$= e^{-n\theta}\theta^{n\bar{x}} \qquad (\theta > 0)$$

where $n = \sum n_x$ denotes the total number of observations and $\bar{x} = \sum_x xn_x/n$ the sample mean. We have

$$\log l = -n\theta + n\bar{x} \log \theta$$

and

$$d \log l/d\theta = -n + n\bar{x}/\theta$$

whence the M.L.E. is the root of

$$-n + n\bar{x}/\theta = 0,$$

viz.

$$\hat{\theta} = \bar{x}.$$

Similarly

$$d \log L/d\theta = -n + n\bar{X}/\theta$$

where $\bar{X}$ is the r.v. induced by $\bar{x}$, and

$$d^2 \log L/d\theta^2 = -n\bar{X}/\theta^2$$

whence, since $E(\bar{x}) = E(X) = \theta$, (6.2.13) gives

$$\tau^2(\theta) = n/\theta.$$

The procedure of section 6.2.5(b) then leads to a requirement that we should solve the inequalities

$$-1 \cdot 96 \leq -n + n\bar{x}/\theta \leq 1 \cdot 96$$

whence, solving, the 95% confidence interval for θ, in this approximation, is

precisely the same as that obtained in Example 4.7.2 by the Normal approximation to the Poisson.

The approximate sampling expectation of $\hat{\theta}$ is given in (6.2.16) as θ, and
the approximate sampling variance is θ/n. These are both exact for this
distribution.

EXAMPLE 6.3.4.   *The geometric distribution.*   In a sequence of independent
Bernoulli trials, each with success parameter θ, let X denote the number of
trials up to and including the first success [see II, § 5.2.3], so that

$$P(X = x) = f(x, \theta) = (1 - \theta)^{x-1}\theta, \qquad x = 1, 2, \ldots, \qquad 0 < \theta < 1.$$

If we have data from n independent observations on X, with frequencies given
by:

$$\left. \begin{array}{l} \text{Magnitude } j \\ \text{Frequency } n(j) \end{array} \right\} j = 1, 2, \ldots, k,$$

where k denotes the largest magnitude observed, the likelihood of θ is

$$l(\theta) = \prod_{j=1}^{k} \{(1 - \theta)^{j-1}\theta\}^{n(j)}$$

$$= (1 - \theta)^{n(\bar{x}-1)}\theta^{n}, \qquad \bar{x} = \sum_{1}^{k} jn(j),$$

whence

$$\log l = n(\bar{x} - 1) \log (1 - \theta) + n \log \theta,$$

and the likelihood equation $d \log l/d\theta = 0$ is

$$n(\bar{x} - 1)/(1 - \theta) + n/\theta = 0.$$

The M.L.E. is then

$$\hat{\theta} = 1/\bar{x}.$$

This is a case where the approximate formula for the sampling variance is
particularly useful, since the exact evaluation of this variance is not easy. We
have

$$-d^2 \log l(\bar{X})/d\theta = n(\bar{X} - 1)/(1 - \theta)^2 + n/\theta^2.$$

Since $E(\bar{X}) = E(X) = 1/\theta$, we have

$$E(-d^2 \log l(\bar{X})/d\theta) = n/\theta^2(1 - \theta)$$

whence, by (6.2.16), the variance is approximately

$$\theta^2(1 - \theta)/n,$$

with numerical value estimated as $\hat{\theta}^2(1 - \hat{\theta})/n$.

[If we had estimated the parameter $1/\theta$ ($= \phi$, say) instead of θ (cf. § 6.2.3), the likelihood of ϕ would have been

$$l(\phi) = \left(1 - \frac{1}{\phi}\right)^{n(\bar{x}-1)} \left(\frac{1}{\phi}\right)^{n}$$

$$= (\phi - 1)^{n(\bar{x}-1)} \phi^{-n},$$

leading to the M.L.E.

$$\hat{\phi} = \bar{x}(= 1/\hat{\theta})$$

The bias of this estimate of $1/\theta$ is zero, and the M.L. approximation (6.2.16) to its sampling variance gives the exact value $\phi(\phi - 1)/n = (1 - \theta)/n\theta^2$.]

EXAMPLE 6.3.5.   *The Negative Binomial distribution.*   In a sequence of independent Bernoulli trials with success-parameter θ, let X denote the number of trials up to and including the cth success, for specified $c = 1, 2, \ldots$. Then [see II, § 5.2.4]

$$f(x, \theta) = P(X = x) = \binom{x-1}{x-c}(1-\theta)^{x-c}\theta^{c}, \qquad x = c, c+1, \ldots 0 < \theta < 1$$

The geometric distribution of Example 6.3.4 is a special case of this with $c = 1$. Calculations similar to those of that example lead to the M.L.E.

$$\hat{\theta} = c/\bar{x},$$

with sampling variance $\theta^2(1 - \theta)/nc$, approximately.

EXAMPLE 6.3.6.   *Sampling without replacement from a finite popula-tion.*   This example involves non-independent data and a discrete parameter space. The data vector is $x = (x_1, \ldots, x_n)$, where $x_r = 0$ if the rth item removed from the population is 'satisfactory', and $x_r = 1$ if it is 'defective', $r = 1, 2, \ldots, n$. The population consists of a collection of N articles (N is not a random variable!) of which an unknown number, θ are defective. The joint distribution of the X_r (the random variables whose realized values are the x_r) is most conveniently specified in terms of the p.d.f. of X_1 and, for $r = 2, 3, \ldots, n$, the *conditional* p.d.f. of X_r given $X_s = x_s$, $s = 1, 2, \ldots, r-1$. Then, for X_1, the p.d.f. is

$$f_1(x_1, \theta) = \theta^{x_1}(1 - \theta)^{1-x_1}/N, \ x_1 = 0, 1; \ \theta = 0, 1, \ldots, N;$$

for $X_r, r = 2, 3, \ldots, n$, the appropriate conditional p.d.f. at x_r is

$$f_r(x_r, \theta | x_1, x_2, \ldots, x_{r-1}) = \left(\theta - \sum_{1}^{r-1} x_s\right)^{x_r} \left\{N - \theta - \sum_{1}^{r-1}(1-x_s)\right\}^{1-x_r}/(N-r+1),$$

$$x_s = 0, 1, (s = 1, 2, \ldots, r).$$

The joint distribution of the X_r at x is

$$g(x, \theta) = f_1(x_1, \theta) \prod_{r=2}^{n} f_r(x_r, \theta | x_1, x_l, \ldots, x_{r-1})$$

which reduces to

$$g(x, \theta) = \binom{\theta}{d}\binom{N-\theta}{n-d} \bigg/ \binom{N}{n}\binom{n}{d},$$

where

$$d = \sum_1^n x_r = \text{total number of defectives in the sample,}$$

After suppressing uninformative factors we may take the likelihood as

$$l(\theta) = \theta!\,(N-\theta)!/(\theta-d)!\,(N-\theta-n+d)!, \qquad \theta = 0, 1, \ldots, N.$$

Since θ is an integer-valued variable the maximum cannot be obtained by differentiating: it is defined by the inequalities

$$l(\hat{\theta}) \geq l(\hat{\theta}-1), \qquad l(\hat{\theta}) \geq l(\hat{\theta}+1),$$

which lead to the estimate

$$\hat{\theta} = [(N+1)d/n]$$

where $[z]$ denotes the largest integer in z. Thus,

$$\hat{\theta} = Nd/n$$

nearly.

The sampling variance is discussed in the next example.

EXAMPLE 6.3.7.   *The hypergeometric distribution.*   If, in section 6.3.6, instead of recording the individual results x_r, one records only the total number d of defectives in the sample (of size n) one may use the known result that the probability of getting d defectives is [see II, § 5.3]

$$g(d, \theta) = \binom{\theta}{d}\binom{N-\theta}{n-d} \bigg/ \binom{N}{n}.$$

This gives the same likelihood function as section 6.3.6 (The relation between this example and section 6.3.6 is comparable to that between sections 6.3.2 and 6.3.1.)

The sampling variance of $\hat{\theta}$ cannot be obtained by any of the approximations given in section 6.2.5 since the parameter space is discrete. A direct evaluation from the approximate version

$$\hat{\theta} = Nd/n$$

given above shows that this estimate is unbiased, with variance

$$\frac{n\theta}{N}\left(1-\frac{\theta}{N}\right)\left(1-\frac{n-1}{N-1}\right).$$

EXAMPLE 6.3.8. *A compound example: Density of organisms by the dilution method. Use of Newton's method in solving the likelihood equation.* When estimating the density of the occurrence in water of invisible micro-organisms, one method is to take n specimens of the affected water and to incubate each in circumstances that will allow a visible colony of organisms to develop if any are present; and to repeat this with n specimens taken from a c-fold dilution— e.g. $c = 2$—of the affected water (the diluent being sterile water); and to repeat again with a c^2-fold dilution. Suppose there were $n = 5$ specimens each of size 1 cc and that on incubation all were fertile (i.e. each produced a visible colony and so contained at least one micro organism); that of five specimens from a two-fold dilution, three were fertile; and that of five specimens from a four-fold dilution just one was fertile. Thus:

Dilution	No. sterile	No. fertile
1	0	5
2	2	3
4	4	1

Denote by θ the density of organisms (number per cc) in the original water. An appropriate probability model is the Poisson family, according to which the number (R, say) of organisms in a 1 cc specimen is a Poisson (θ) variable [see II, § 5.4], whence the probability of a sterile specimen ($R = 0$) is $e^{-\theta}$, and that of a fertile one $1 - e^{-\theta}$. The number S of sterile specimens in a sample of five is then a Binomial (n, p) variable with $n = 5$, $p = e^{-\theta}$ [see II, § 5.4]. In the diluted samples θ is replaced by $\theta/2$ for the two-fold dilution, and $\theta/4$ for the four-fold. The probability of obtaining s_1, s_2, s_3 sterile specimens with undiluted source, two-fold dilution and four-fold dilution, respectively, is

$$g(\theta; s_1, s_2, s_3) = \prod_{r=1}^{3} \binom{5}{s_r} p_r^{s_r} (1-p_r)^{5-s_r}$$

where

$$p_1 = e^{-\theta}, \qquad p_2 = e^{-\theta/2}, \qquad p_3 = e^{-\theta/4},$$

whence, omitting uninformative multipliers, the likelihood of θ on our data may be taken as

$$p_1^0 (1-p_1)^5 p_2^2 (1-p_2)^3 p_3^4 (1-p_3)$$

or, in terms of $p = e^{-\theta/4}$,

$$l(p) = (1-p^4)^5 p^4 (1-p^2)^3 p^4 (1-p)$$
$$= p^8 (1-p^4)^5 (1-p^2)^3 (1-p).$$

We thus have

$$d \log l / dp = (8 - p - 7p^2 - p^3 - 35p^4)/p(1-p^4).$$

Equating this to zero gives the likelihood equation. A few trials will show that the relevant root lies between 0·60 and 0·61. We therefore take $p' = 0·605$ as the first approximation. An improved approximation p'' is then obtainable by the use of Newton's method [see III, § 5.4.1] as

$$p'' = p' - \left(\frac{d \log l(p')/dp'}{d^2 \log l(p')/dp'^2} \right).$$

Now, unless p' is a hopelessly poor approximation, the quantity $d^2 \log l(p')/dp'^2$ will be acceptably close to $d^2 \log l(\hat{p})/d\hat{p}^2 = -1/v$, where v is the expression given in (6.2.16) for the M.L. approximation to the sampling variance of $\hat{\theta}$. Thus

$$p'' \approx p' + v\{d \log l(p')/dp'\}. \tag{6.3.1}$$

Newton's method is therefore highly appropriate for the numerical solution of the likelihood equation.

In our case the expression for $d \log l/dp$ is given above, and the expression for $1/v = -d^2 \log l/dp^2$ is

$$\frac{1+14p+3p^2+140p^3}{p(1-p^4)} + \left(\frac{d \log l}{dp} \right) \left(\frac{1-5p^4}{p-p^5} \right).$$

The calculations are as follows:

$$\text{First approximation } p': \quad 0·605$$

$$d \log l/dp: \ -0·1483$$

$$v: \quad 0·0126$$

$$\text{Improved approximation } p'': \quad \underline{0·6031}$$

This is sufficiently accurate.

Thus the M.L.E. of $p = e^{-\theta/4}$ is 0·6031. To obtain 95% confidence limits, solve the equations

$$d \log l/dp = \pm 1·96w$$

(as recommended in (6.2.16)) where

$$w^2 = -d^2 \log l/dp^2|_{\hat{p}} = 1/0·0126 = (8·91)^2.$$

Thus the equations to be solved are

$$d \log l/dp = \pm 17·46$$

The roots are

$$p_1 = 0·355, \qquad p_2 = 0·760,$$

so that the 95% confidence interval for p is

$$(0·355, 0·760).$$

The corresponding results for $\theta = -4 \log_e p$ are

$$\hat{\theta} = 2 \cdot 023,$$

with 95% confidence interval

$$(1 \cdot 10, 4 \cdot 14).$$

[If instead one had used the less accurate approximation of § (6.2.5)(d) which regards $\hat{p}$ itself as Normal, with sampling variance $0 \cdot 0126 = (0 \cdot 112)^2$, the confidence interval for p would have been $\hat{p} \pm (1 \cdot 96)(0 \cdot 112)$, i.e.

$$(0 \cdot 38, 0 \cdot 82)$$

the corresponding interval for θ being

$$(0 \cdot 79, 3 \cdot 87).$$

If one had relied on the estimated sampling variance of $\hat{\theta}$ itself, namely $(4/p)$ times the estimated variance $(0 \cdot 0126)$ of p, one would have obtained as the 95% confidence interval for θ:

$$(0 \cdot 56, 3 \cdot 48).]$$

EXAMPLE 6.3.9. *Some non-regular cases*

(a) *Extremity of a uniform distribution*

If the distribution of X is uniform $(0, \theta)$, its p.d.f. is [see II, § 10.2.1]

$$f(x, \theta) = \begin{cases} \theta^{-1}, & 0 \le x \le \theta \\ 0, & \text{otherwise.} \end{cases}$$

The likelihood of θ on a data vector $x = (x_1, x_2, \ldots, x_n)$ is

$$l(\theta, x) = \prod_{r=1}^{n} f(x_r, \theta) = \begin{cases} \theta^{-n}, & 0 \le x_1, x_2, \ldots, x_n \le \theta \\ 0, & \text{otherwise} \end{cases}$$

i.e.

$$l(\theta, x) = \begin{cases} \theta^{-n}, & \theta \ge x_{max} \\ 0, & \text{otherwise.} \end{cases}$$

(Here x_{max} denotes the largest observed value). The maximizing value of θ is

$$\hat{\theta} = x_{max}.$$

Note that this is NOT a root of the equation $dl/d\theta = 0$.

(b) *Shift parameter of a shifted standard exponential*

Here the p.d.f. is

$$f(x, \theta) = \begin{cases} e^{-(x-\theta)}, & x \ge \theta \\ 0, & \text{otherwise} \end{cases}$$

and the likelihood on data $x = (x_1, x_2, \ldots, x_n)$ is

$$l(\theta, x) = \begin{cases} e^{-\Sigma(x_r - \theta)}, & \theta \le x_1, x_2, \ldots, x_n \\ 0, & \text{otherwise} \end{cases}$$

$$= \begin{cases} e^{n(\theta - \bar{x})}, & \theta \le x_{\min} \\ 0, & \text{otherwise} \end{cases}$$

(where $x_{\min}$ denotes the smallest observed values). Here again the M.L.E. is not to be found by differentiating the likelihood function. By inspection the M.L.E. is

$$\hat{\theta} = x_{\min}.$$

The approximations given in section 6.2.2 for the sampling variance in regular cases do not apply to these non-regular cases. The sampling properties of the M.L.E. must be obtained directly. Thus, in (a) above, the M.L.E. has expectation equal to $n\theta/(n+1)$, and variance $n\theta^2/(n+2)(n+1)^2$ estimated, of course, as $n\hat{\theta}^2/(n+2)(n+1)^2$.

6.4.   EXAMPLES OF M.L.E. IN MULTIPARAMETER CASES

EXAMPLE 6.4.1.   *The Normal distribution.*   If X is Normal (μ, σ), the likelihood of μ and σ on data $x = (x_1, x_2, \ldots, x_n)$ is

$$l(\mu, \sigma; x) = \sigma^{-n} \exp - \frac{1}{2\sigma^2} \sum_1^n (x_r - \mu)^2,$$

$$-\infty < \mu < \infty, \qquad \sigma > 0. \tag{6.4.11}$$

We find

$$\partial \log l/\partial \mu = \sum (x_r - \mu)/\sigma^2,$$

$$\partial \log l/\partial \sigma = -n/\sigma + \sum (x_r - \mu)^2/\sigma^3.$$

Thus the M.L.E.s are the roots of the simultaneous equations

$$\sum (x_r - \mu) = 0,$$

$$\sum (x_r - \mu)^2 = n\sigma^2,$$

whence

$$\hat{\mu} = \bar{x}, \qquad \hat{\sigma} = \{\sum (x_r - \bar{x})^2/n\}^{1/2}.$$

Next we evaluate

$$\partial^2 \log l/\partial \mu^2 = -n/\sigma^2$$

$$\partial^2 \log l/\partial \mu \partial \sigma = -2 \sum (x_r - \mu)/\sigma^3 = -2n(\bar{x} - \mu)/\sigma^3$$

$$\partial^2 \log l/\partial \sigma^2 = n/\sigma^2 - 3 \sum (x_r - \mu)^2/\sigma$$

whence

$$E(\partial^2 \log L/\partial\mu^2) = -nE(\bar{X})/\sigma^2 = -n\mu/\sigma^2$$

$$E(\partial^2 \log L/\partial\mu\partial\sigma) = -2nE(\bar{X}-\mu)/\sigma^3 = 0$$

$$E(\partial^2 \log L/\partial\sigma^2) = n/\sigma^2 - 3\sum E(X_r-\mu)^2/\sigma^4$$

$$= n/\sigma^2 - 3n/\sigma^2 = -2n/\sigma^2.$$

Thus the approximate dispersion matrix D of the M.L.E.s as given by (6.2.16) is

$$D = \begin{pmatrix} n/\sigma^2 & 0 \\ 0 & 2n/\sigma^2 \end{pmatrix}^{-1}$$

$$= \begin{pmatrix} \sigma^2/n & 0 \\ 0 & \sigma^2/2n \end{pmatrix}.$$

Thus the M.L. approximations to the sampling variances of $\hat{\mu}$ and $\hat{\sigma}$ are σ^2/n and $\sigma^2/2n$ respectively; and the corresponding result for the sampling covariance of $\hat{\mu}$ and $\hat{\sigma}$ is zero. The numerical values are, as usual, taken to be

$$v(\mu): \hat{\sigma}^2/n = \sum (x_r - \bar{x})^2/n^2$$

and

$$v(\sigma): \hat{\sigma}^2/2n = \sum (x_r - \bar{x})^2/2n^2.$$

If, instead, we take μ and $\theta = \sigma^2$ as the parameters to be estimated we find

$$\hat{\mu} = \bar{x}, \quad \hat{\theta} = \sum (x_r - \bar{x})^2/n \quad (=(\hat{\sigma})^2),$$

and

$$D = \begin{pmatrix} \theta/n & 0 \\ 0 & 2\theta^2/n \end{pmatrix} = \begin{pmatrix} \sigma^2/n & 0 \\ 0 & 2\sigma^4/n \end{pmatrix};$$

thus

$$v(\mu) = \sigma^2/n, \quad c(\mu, \sigma^2) = 0, \quad v(\sigma^2) = 2\sigma^4/n.$$

Here $\hat{\mu}$ is unbiased while $(\hat{\sigma}^2)$ has a small bias, the exact sampling expectation of $(\hat{\sigma}^2)$ being $(1-1/n)\sigma^2$.

EXAMPLE 6.4.2.   *The log-Normal distribution.*   A random variable Y is said to be log-Normally distributed if $\log Y$ ($=X$, say) is Normal. (The logarithms may conveniently be taken as natural logs.) Thus

$$Y = e^X,$$

where X is Normal, the parameters of this Normal distribution being, say, μ and σ as in section 6.6.1. The p.d.f. of Y is then

$$\frac{1}{\sigma\sqrt{2\pi}} y^{-1} \exp-\frac{1}{2\sigma^2}(\log y - \mu)^2, \quad y > 0,$$

and the likelihood of μ and σ, on data $y = (y_1, y_2, \ldots, y_n)$ is

$$l(\mu, \sigma; y) = \sigma^{-n} \exp{-\frac{1}{2\sigma^2} \sum (\log y_r - \mu)^2}, \qquad -\infty < \mu < \infty, \sigma > 0,$$

after dropping the uninformative factor $(1/\prod y_r)$. If we set

$$\log y_r = x_r, \qquad r = 1, 2, \ldots, n$$

it will be seen that the likelihood function coincides with (6.4.1). It follows therefore that the M.L. analysis of the log-Normal distribution is carried out by replacing each observation y_r by its logarithm $x_r = \log y_r$ and proceeding as for the Normal distribution.

EXAMPLE 6.4.3.   *The gamma distribution: first approximations supplied by the method of moments.*   Suppose we have a sample of n from the two-parameter gamma distribution, the p.d.f. of which at x is

$$f(x; \alpha, \beta) = \{x^{\alpha-1} e^{-x/\beta}\}/\beta^\alpha \Gamma(\alpha), \qquad x > 0 \qquad (\alpha > 0, \beta > 0),$$

where α is the shape parameter and β the scale parameter [see II, § 11.3.1]. Suppose the data given in the data-vector form. The likelihood function is

$$l(\alpha, \beta; x) = \left\{\prod_{r=1}^{n} x_r\right\}^{\alpha-1} e^{-\sum x_r/\beta} \beta^{-n\alpha} \Gamma^{-n}(\alpha)$$

and its logarithm

$$\log l = (\alpha - 1) \sum \log x_r - \beta^{-1} \sum x_r - n\alpha \log \beta - n \log \Gamma(\alpha).$$

The likelihood equations are

$$\frac{\partial \log l}{\partial \alpha} = \sum \log x_r - n \log \beta - n \Gamma'(\alpha)/\Gamma(\alpha) = 0,$$

$$\frac{\partial \log l}{\partial \beta} = \beta^{-2} \sum x_r - n\alpha/\beta = 0.$$

The second of these gives

$$\hat{\beta} = \bar{x}/\hat{\alpha}$$

where $\hat{\alpha}, \hat{\beta}$ are the M.L.E.s of α, β respectively and $\bar{x} = \sum x_r/n = $ sample mean, and the first then becomes

$$\overline{\log x} - \log \bar{x} + \log \hat{\alpha} - \psi(\hat{\alpha}) = 0$$

(where $\overline{\log x} = \sum (\log x_r)/n$ and $\psi(\alpha) = \Gamma'(\alpha)/\Gamma(\alpha)$.)

We illustrate by considering a sample known to come from a gamma distribution [see II, § 11.3] with $\alpha = \beta = 2$, each value of which was obtained by adding four consecutive entries from a table of realizations of chi-squared variables on one degree of freedom [Barnett (1965)—Bibliography F]. The sample is as follows

5·6135	2·2197	5·8971	3·8243	6·5021
0·8590	3·5452	2·4983	2·0567	0·7797
1·0184	11·3595	2·1279	2·0924	4·1648
4·9673	0·3939	4·7217	2·9399	6·8468
5·6229	3·6467	6·0812	1·9336	4·4899

$$\Sigma \log x = 27 \cdot 59314, \qquad \Sigma x = 96 \cdot 1934, \qquad \Sigma x^2 = 517 \cdot 5038$$

$$\overline{\log x} = 1 \cdot 1037, \qquad \bar{x} = 3 \cdot 8477, \qquad \overline{x^2} = 20 \cdot 702$$

$$\log \bar{x} = 1 \cdot 3475.$$

$$\log \bar{x} - \overline{\log x} = 0 \cdot 2438$$

We therefore have to solve the likelihood equation

$$h(\hat{\alpha}) = \log \hat{\alpha} - \psi(\hat{\alpha}) - 0 \cdot 2438 = 0.$$

Having solved for $\hat{\alpha}$, we then have

$$\hat{\beta} = 3 \cdot 8477 / \hat{\alpha}.$$

First approximation by the method of moments

The 'method of moments' is to equate $E(X^r)$, as a function of the parameter, to the corresponding sample moment $\overline{x^r}$ [see (2.1.7)], for $r = 1, 2, \ldots, p$, p being the number of parameters. In the two parameter gamma case this reduces to

$$E(X) = \bar{x}$$
$$E(X^2) = \overline{x^2}.$$

i.e.

$$\alpha\beta = 3 \cdot 8477$$
$$\alpha\beta^2 + \alpha^2\beta^2 = 20 \cdot 702.$$

The solutions are

$$\alpha^0 = 2 \cdot 52, \qquad \beta^0 = 1 \cdot 53.$$

These provide first approximations for the iterative solution of the likelihood equation. The iteration proceeds as follows (using the Newton–Raphson method)

Iteration No.	α	$\psi(\alpha)$	$h(\alpha)$	$\psi'(\alpha)$	$h'(\alpha)$	Correction $=-h(\alpha)/h'(\alpha)$
1	2·52	0·713	−0·32	0·49	−0·09	−0·36
2	2·160	0·521	0·0052	0·590	−0·123	0·042
3	2·202	0·5454	0·00013	0·5723	−0·1182	0·0011
4	2·203					

Thus $\hat{\alpha} = 2\cdot203$, whence $\hat{\beta} = 1\cdot747$ (These are to be compared with the actual values $\alpha = 2$, $\beta = 2$.)

The accuracy of the estimates as disclosed by the approximations in section 6.2.5(a) for the dispersion matrix V of $(\hat{\alpha}, \hat{\beta})$ is given by

$$V^{-1} = nE\begin{bmatrix} \psi'(\alpha) & 1/\beta \\ 1/\beta & \bar{X}/\beta^3 - \alpha/\beta^2 \end{bmatrix}$$

$$= n\begin{pmatrix} \psi'(\alpha) & 1/\beta \\ 1/\beta & \alpha/\beta^2 \end{pmatrix}$$

whence

$$V = \frac{1}{n}\frac{\beta^2}{\alpha\psi'(\alpha)-1}\begin{pmatrix} \alpha/\beta^2 & -1/\beta \\ -1/\beta & \psi'(\alpha) \end{pmatrix}.$$

$$= \begin{pmatrix} \sigma_1^2 & \rho\sigma_1\sigma_2 \\ \rho\sigma_1\sigma_2 & \sigma_2^2 \end{pmatrix}, \quad \text{say.}$$

With our sample the estimated values are

$$\sigma_1 = 0\cdot58, \quad \sigma_2 = 0\cdot52, \quad \rho = -0\cdot67.$$

Following section 4.9, we can construct the following (approximate) 95% confidence interval for α:

$$\alpha \in 2\cdot203 \pm (1\cdot96)(0\cdot58)$$

i.e.

$$\alpha \in (1\cdot07, 3\cdot34).$$

Alternately we can construct the following approximate 95% confidence interval for β:

$$\beta \in 1\cdot747 \pm (1\cdot96)(0\cdot52)$$

i.e.

$$\beta \in (0\cdot73, 2\cdot77)$$

but we must be careful how we interpret them when both are under consideration: in particular it is not the case that the rectangle in the (α, β) plane bounded by these limits forms a $(0\cdot95)^2$ confidence region for the pair (α, β).

We can however construct an elliptical confidence region [see § 4.9.2] with confidence coefficient $0\cdot95$, namely the region contained by the ellipse (4.9.6), viz.:

$$2\cdot923(\alpha - 2\cdot203)^2 + 4\cdot443(\alpha - 2\cdot203)(\beta - 1\cdot747)$$

$$+ 3\cdot704(\beta - 1\cdot747)^2 = 3\cdot302. \tag{6.4.2}$$

This is portrayed in Figure 6.4.1. Note that, with $\alpha = \beta = 2$, the left hand side of this expression equals $0\cdot354$, which is less than the right hand side, whence the point $(2, 2)$ lies inside the ellipse. Thus, comfortingly, the confidence region does in fact cover the correct values of (α, β).

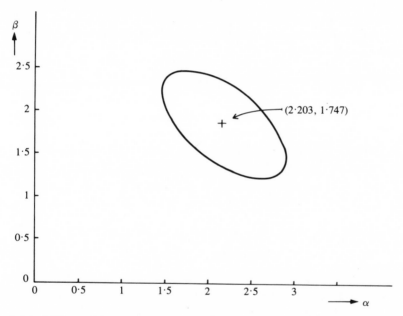

Figure 6.4.1: 95% confidence ellipse for (α, β) of Example 6.4.2. The equation of this ellipse is given in (6.4.2).

EXAMPLE 6.4.4.   In the case of the Markovian data discussed in Examples 6.2.5 and 6.2.7 the parameters α, ρ of the parameter probability matrix

$$\begin{pmatrix} \alpha & 1-n \\ 1-\beta & \beta \end{pmatrix}$$

were seen to have likelihood

$$l(\alpha, \beta) = (1 - \alpha)^{41} \alpha^3 \beta^4 (1 - \beta)^2 / (\alpha + \beta).$$

Example 6.2.7 argued that intuitively reasonable first approximations to the maximizing values $\hat{\alpha}$, $\hat{\beta}$ were the pair $(0\cdot068, 0\cdot600)$. A tabulation of values of $l(\alpha, \beta)$ in the neighbourhood of this point produces the following:

$$l(\alpha, \beta) \times 10^7$$

		α		
		0·060	0·070	0·080
	0·59	5·356	5·401	5·099
β	0·60	5·370	5·417	5·115
	0·61	5·372	5·420	5·119
	0·62	5·362	5·412	5·113

showing the maximum to be near $(0\cdot020, 0\cdot61)$.

A further tabulation, on a finer grid, in the neighbourhood of the maximum produces the following:

		α		
		0·064	0·066	0·068
	0·605	5·4405	5·4494	5·4420
β	0·606	5·4407	5·4496	5·4430
	0·607	5·4409	5·4498	5·4427
	0·608	5·4408	5·4495	5·4426

To a sufficiently close approximation, the maximum occurs at

$$\hat{\alpha} = 0\cdot066, \qquad \hat{\beta} = 0\cdot607.$$

(Compare Figure 6.2.6.)

To obtain the approximation (6.2.16) to this dispersion matrix we have to invert the matrix (6.2.17(iii)), viz.

$$-\left(\begin{matrix} \partial^2 \log l/\partial\alpha^2 & \partial^2 \log l/\partial\alpha\partial\beta \\ \partial^2 \log l/\partial\alpha\partial\beta & \partial^2 \log l/\partial\beta^2 \end{matrix} \right)\Bigg|_{\hat{\alpha},\hat{\beta}}.$$

The matrix to be inverted is

$$\left(\begin{matrix} \dfrac{41}{(1-n)^2} + \dfrac{3}{\alpha^2} - \dfrac{1}{(\alpha+\beta)^2} & -\dfrac{1}{(\alpha+\beta)^2} \\ -\dfrac{1}{(\alpha+\beta)^2} & \dfrac{2}{(1-\beta)^2} + \dfrac{4}{\beta^2} - \dfrac{1}{(\alpha+\beta)^2} \end{matrix} \right)\Bigg|_{\alpha,\beta}$$

$$= \left(\begin{matrix} 734 & -2 \\ -2 & 22 \end{matrix} \right).$$

The inverse is

$$\begin{pmatrix} 0{\cdot}00136 & 0{\cdot}00012 \\ 0{\cdot}00012 & 0{\cdot}04528 \end{pmatrix}$$

whence

$$\sigma_1 = 0{\cdot}037, \qquad \sigma_2 = 0{\cdot}213, \qquad \rho = 0{\cdot}015.$$

The approximate 95% confidence ellipse is

$$735(\alpha - 0{\cdot}068)^2 - 3{\cdot}8(\alpha - 0{\cdot}068)(\beta - 0{\cdot}607) + 22(\beta - 0{\cdot}607)^2 = 6$$

as in (4.9.6). As an example, the values $\alpha = 0{\cdot}1$, $\beta = 0{\cdot}5$, on being substituted in the left-hand side, produce a value of $1{\cdot}02$. This is less than 6 and so is inside the 95% confidence region. The values $(\alpha = 0{\cdot}2, \beta = 0{\cdot}5)$ produce a left-hand value of $13{\cdot}1$, whence the point $(0{\cdot}2, 0{\cdot}5)$ is excluded from the region: the data estimates differ significantly from the hypothesized values $(0{\cdot}2, 0{\cdot}5)$, at the $0{\cdot}5\%$ level.

6.5.   MAXIMUM LIKELIHOOD ESTIMATION OF LINEAR REGRESSION COEFFICIENTS

6.5.1.   The Meaning of 'Regression'

The reader who is not fluent in statistical jargon might well wonder at the ubiquity in statistical literature of a word which means 'the action of returning towards a place or point of departure' (the *Oxford English Dictionary*). It all seems to have started in the course of a nineteenth century study of inheritance of physical features in Man. One of the features studied was height, and it was found, not surprisingly that, on the whole, the sons of tall fathers were taller than the sons of short fathers. There was, however, less variability in height amongst the sons than amongst the fathers: there was a tendency for sons' heights to *regress* towards the average. This was demonstrated by computing the average height of sons of 56-inch fathers, the average height of sons of 58-inch fathers, ... , and plotting these averages of sons' heights against the corresponding fathers' heights. The points lay (approximately) on a straight line with positive slope, less than 45°: the regression was *linear*.

Here we had a sample from a bivariate distribution, say of the pair (X, Y) of random variables. The straight line plot was a sample version of the graph of the function

$$g(x) = E(Y|X = x).$$

The word 'regression' was adopted for this and is now universally used in probability theory for the conditional expectation of one variable, expressed as a function of the value attributed to the other by the 'condition'. If (X, Y) have a bivariate Normal distribution, for example, with $E(X) = \mu_1, E(Y) = \mu_2$, var $(X) = \sigma_1^2$, var $(Y) = \sigma_2^2$, and corr $(X, Y) = \rho$, the conditional distribution of Y, conditional on $X = x$, is Normal, with expectation

$$E(Y|X = x) = \mu_2 + \rho \frac{\sigma_2}{\sigma_1}(x - \mu_1) \tag{6.5.0}$$

and variance

$$\text{var }(Y|X = x) = \sigma_2^2(1 - \rho^2)$$

[see II, § 13.4.6). In this example, the regression of Y on X is the linear function (6.5.0).

In general, of course, the regression of one member of a bivariate pair on the other need not be linear. Nor, indeed, need we restrict ourselves to a bivariate situation. If, for example, we were concerned with the joint distribution of *three* random variables (X_1, X_2, X_3), the regression of X_1 on X_2 and X_3 would be the function

$$g(x_2, x_3) = E(X_1|X_2 = x_2, X_3 = x_3).$$

Here also there are important cases where the regression is linear: in particular this is so when (X_1, X_2, X_3) are trivariate Normal.

Statistical problems concerning regression naturally involve estimation or other inferential procedures. In a typical bivariate sample in which, say, the height x and the weight y of each individual is recorded, a study of the regression of weight on height would start with a tabulation of the data as in Table 6.5.1.

	Values of x				
	x_1	x_2	$\ldots$	x_k	
Values of y	$y(1, 1)$ $y(2, 1)$ $\vdots$ $y(n_1, 1)$	$y(1, 2)$ $y(2, 2)$ $\vdots$ $y(n_2, 2)$	$\ldots$ $\ldots$	$y(1, k)$ $y(2, k)$ $\vdots$ $y(n_k, k)$	
Number of observations	n_1	n_2	$\ldots$	n_k	n
Mean	$\bar{y}(1)$	$\bar{y}(2)$	$\ldots$	$\bar{y}(k)$	$\bar{y}$

Table 6.5.1: Data arranged in data-vector form for estimation of regression of Y on X.

(Here the x values would of course be rounded, e.g. to the nearest inch.)

Equivalently when the y values are also rounded, we might have a frequency-table version of the data, as in Table 6.5.2.

		Values of x			
		x_1	x_2	$\ldots$	x_k
Values of y	y_1 y_2 $\vdots$ y_a	$n(1, 1)$ $n(2, 1)$ $\vdots$ $n(a, 1)$	$n(1, 2)$ $n(2, 2)$ $\vdots$ $n(a, 2)$	$\ldots$	$n(1, k)$ $n(2, k)$ $\vdots$ $n(a, k)$
Number of observations		n_1	n_2	$\ldots$	n_k	n
Mean		$\bar{y}(1)$	$\bar{y}(2)$	$\ldots$	$\bar{y}_k$	$\bar{y}$

Table 6.5.2: Data arranged in frequency-table form estimation of regression of Y on X.

(A numerical example using data arranged in this form is shown in Table 6.5.3 and discussed and analyzed in Example 6.5.1.)

The designation 'regression' has however been extended in statistical practice to situations in which the x-values in Tables 6.5.2 and 6.5.3 are predetermined fixed 'levels' of a controlled variable, often called the 'treatment variable'. In these cases the y-values are regarded as the *responses* of the system to the *treatments*. A typical example would be one in which the treatments were fertilizers applied to fields, and the responses the yields of crops from those fields.

In either case, the 'regression model' is usually taken to be one based on the Normal distribution, all the y-values corresponding to a given x value (x_j say) being regarded as realizations of a Normal (μ_j, σ_j) variable, in which μ_j is taken to be a function (of specified form) of x_j. A common model is

$$\mu_j = \alpha_0 + \alpha_1 x_j + \alpha_2 x_j^2 + \ldots + \alpha_p x_j^p,$$

where p is a pre-determined integer and the *regression coefficients* $\alpha_0, \alpha_1, \ldots, \alpha_p$ are to be estimated from the data. This is 'polynomial regression'. A special case of this is 'linear regression',

$$\mu_j = \alpha + \beta x_j.$$

If, as frequently happens, it seems reasonable to take

$$\sigma_j = \sigma,$$

a constant with the same value for each value of x_j, the data are *homoscedastic*. In the heteroscedastic case we need a parametric expression for σ_j in terms of x_j: for example, σ_j may increase quadratically with x_j, and we might take

$$\sigma_j = \gamma x_j^2,$$

where γ is a parameter to be estimated.

6.5.2.   The M.L.E. Procedure for Linear Regression with Weighted Data

We consider first the case where there is just one observation (y_j) for each x-value x_j, so that $n_1 = n_2 = \ldots = n_k = 1$ in Table 6.5.1; with

$$\mu_j = \alpha + \beta x_j, \qquad j = 1, 2, \ldots, k \qquad\qquad (6.5.1)$$

the σ_j not being assumed to be equal. The likelihood equations for estimating α and β are derived from the likelihood function, expressed in terms of α, β and the σ_j (we leave the σ_j unspecified for the moment, and denote them collectively by the symbol σ), viz.

$$l(\alpha, \beta, \sigma) = \prod_{j=1}^{k} \sigma_j^{-1} \exp -\tfrac{1}{2}(y_j - \alpha - \beta x_j)^2 / \sigma_j^2$$

(compare (6.2.4)). The log-likelihood is

$$\log l = -\sum \log \sigma_j - \tfrac{1}{2}\sum (y_j - \alpha - \beta x_j)^2 / \sigma_j^2.$$

We have

$$\partial \log l / \partial \alpha = \sum (y_j - \alpha - \beta x_j) / \sigma_j^2$$

and

$$\partial \log l / \partial \beta = \sum x_j (y_j - \alpha - \beta x_j) / \sigma_j^2.$$

Equating these to zero gives the likelihood equations

$$\begin{aligned}
\alpha \sum w_j + \beta \sum w_j x_j &= \sum w_j y_j \\
\alpha \sum w_j x_j + \beta \sum w_j x_j^2 &= \sum w_j x_j y_j,
\end{aligned} \qquad\qquad (6.5.2)$$

where

$$w_j = 1/\sigma_j^2, \qquad j = 1, 2, \ldots, k.$$

The w_j are called *weights*.

The equations (6.5.2), which determine the maximum likelihood estimates $\hat{\alpha}$ and $\hat{\beta}$, coincide with the equations obtained by the method of Least Squares [see Chapter 8], in which the procedure is to minimize the 'weighted sum of squared residuals'

$$\sum (y_j - \alpha - \beta x_j)^2 / \sigma_j^2 = \sum w_j (y_j - \alpha - \beta x_j)^2 \qquad\qquad (6.5.3)$$

6.5.3.   Linear Regression with Equally Weighted Data

(i) *The estimates*

The most common estimation problem in linear regression occurs with data arranged as in Table 6.5.1 or Table 6.5.2, with

$$\sigma_j = \sigma, \qquad j = 1, 2, \ldots, k.$$

It turns out to be advantageous to express the linear formula for μ_j in a slightly different way from (6.5.1): instead of (6.5.1), we take

$$\mu_j = \alpha + \beta(x_j - \bar{x}) \tag{6.5.4}$$

where

$$\bar{x} = \sum n_j x_j / n, \qquad n = \sum n_j. \tag{6.5.5}$$

There are then three parameters to be estimated, namely α, β, σ. (The 'α' of (6.5.4) is not the same as the 'α' in (6.5.1). Calling the latter α', we have

$$\alpha - \beta\bar{x} = \alpha',$$

We shall however now work entirely in terms of the formulation (6.5.4).) The likelihood function is proportional to

$$l(\alpha, \beta, \sigma) = \prod_{j=1}^{k} f(j)$$

where $f(j)$ is given by

$$f(j) = \sigma^{-n_j} \exp -\tfrac{1}{2} \sum_{i=1}^{n_j} \{y_{ij} - \alpha - \beta(x_j - \bar{x})^2\}/\sigma^2,$$

using the notation of Table 6.5.1 with, however, $y(i, j)$ replaced by y_{ij}. The log-likelihood is

$$\log l = -n \log \sigma - \tfrac{1}{2}\sum_i \sum_j \{y_{ij} - \alpha - \beta(x_j - \bar{x})\}^2/\sigma^2,$$

$n = \sum n_j$ being the sample size. We have

$$\partial \log l/\partial\alpha = \sum_i \sum_j \{y_{ij} - \alpha - \beta(x_j - \bar{x})\}/\sigma^2$$

$$= \left\{\sum_i \sum_j y_{ij} - n\alpha - \beta \sum_j n_j(x_j - \bar{x})\right\}\Big/\sigma^2,$$

which reduces to

$$\partial \log l/\partial\alpha = \sum_i \sum_j (y_{ij} - n\alpha)/\sigma^2$$

since $\sum n_j(x_j - \bar{x}) = 0$, by the definition of $\bar{x}$; likewise

$$\partial \log l/\partial\beta = \left\{\sum_i \sum_j (x_j - \bar{x})y_{ij} - \beta \sum_j n_j(x_j - \bar{x})^2\right\}\Big/\sigma^2.$$

Equating these two quantities to zero, we obtain the likelihood equations for $\hat{\alpha}$ and $\hat{\beta}$. These reduce to

$$\hat{\alpha} = \sum_i \sum_j y_{ij}/n$$

$$= \sum_j n_j \bar{y}_j / n$$

$$= \bar{y}, \tag{6.5.6}$$

the 'grand mean'; and

$$\hat{\beta} = \sum_i \sum_j (x_j - \bar{x}) y_{ij} / \sum_j n_j (x_j - \bar{x})^2$$

$$= \sum_j n_j (x_j - \bar{x}) \bar{y}_j / \sum_j n_j (x_j - \bar{x})^2$$

$$= \left(\sum_j n_j x_j \bar{y}_j - n \bar{x} \bar{y} \right) \Big/ \left(\sum_j n_j x_j^2 - n \bar{x}^2 \right) \tag{6.5.7}$$

where $\bar{x}$ is defined in (6.5.5).

(If μ_j had been specified as in (6.5.1) instead of as in (6.5.4) we should have had two non-diagonal simultaneous equations to solve, instead of having the simple solutions (6.5.6) and (6.5.7). A more important advantage of the form (6.5.4) is that, with this form, the estimates $\hat{\alpha}$ and $\hat{\beta}$ turn out to be statistically independent of each other.)

(ii) *Sampling properties of the estimates*

In assessing the accuracy of the estimates, there is no need to use the approximate formulae developed in section 6.2.5. Since $\hat{\alpha}$ and $\hat{\beta}$ are linear functions of the y_{ij} they are realizations of Normally distributed variables. The sampling expectations, variances and covariance are found to be as follows:

$$\text{Expectations:} \quad \text{for } \hat{\alpha} \ldots \alpha$$
$$\text{for } \hat{\beta} \ldots \beta \tag{6.5.8}$$

so that the estimates are *unbiased*;

$$\text{Variances:} \quad \text{for } \hat{\alpha} \ldots \sigma^2 / n$$
$$\text{for } \hat{\beta} \ldots \sigma^2 / \sum nj(x_j - \bar{x})^2 \tag{6.5.9}$$

$$\text{Covariance:} \quad \text{of } \hat{\alpha} \text{ and } \hat{\beta} \ldots 0$$

so that $\hat{\alpha}$ and $\hat{\beta}$ are mutually independent [see II, § 13.4.2].

As regards the estimate of σ^2, the equation $\partial \log l / \partial \sigma = 0$ leads to the following:

$$\text{M.L.E. of } \sigma^2 = \sum_i \sum_j \{ y_{ij} - \hat{\alpha} - \hat{\beta}(x_j - \bar{x}) \}^2 / n$$

$$= SS / n$$

say, where SS denotes the 'sum of squares' (i.e. sum of squared *residuals*), viz.

$$SS = \sum_i \sum_j \{ y_{ij} - \hat{\alpha} - \hat{\beta}(x_j - \bar{x}) \}^2$$

$$= \sum_i \sum_j y_{ij}^2 - n \hat{\alpha}^2 - \hat{\beta}^2 \sum_{n_j} (x_j - \bar{x})^2. \tag{6.5.10}$$

In computing the value of $\sum n_j(x_j - \bar{x})^2$ it is usually preferable to use the equivalent expression $\sum n_j x_j^2 - n\bar{x}^2$.

As a matter of historical convention, it is the universal practice of statisticians to depart from the strict likelihood canon for the estimate of σ^2 and to replace the M.L. estimate by the modified version

$$\hat{\sigma}^2 = SS/(n-2). \tag{6.5.11}$$

As it happens, this is unbiased, but that in itself is not necessarily very important. What *is* important is that all the relevant tables are constructed in terms of (6.5.11). (This attitude to the divisor of SS, n or $(n-2)$, may seem somewhat casual, but as far as the accuracy of $\hat{\alpha}$ and $\hat{\beta}$ are concerned, it makes no difference at all: The use of n instead of $n-2$ would require additional computations and different entries in the relevant tables, but would lead to the same inference.)

(iii) *Confidence intervals for α and β*

As was remarked earlier, $\hat{\alpha}$ and $\hat{\beta}$ are realizations of statistically independent Normal variables, with expected values α and β respectively. Their variances will be estimated as

$$v(\alpha) = \hat{\sigma}^2/n, \qquad v(\beta) = \hat{\sigma}^2/\sum n_j(x_j - \bar{x})^2, \tag{6.5.12}$$

where $\hat{\sigma}^2$ is defined in (6.5.11), namely as

$$\hat{\sigma}^2 = SS/(n-2),$$

SS being defined in (6.5.10). It can be shown (e.g. by the methods of section 2.5.8) that

$$(n-2)\hat{\sigma}^2/\sigma^2$$

is a realization of a χ^2 variable on $n-2$ d.f. [see § 2.5.4(a)], and one which is, moreover, statistically independent of $\hat{\alpha}$ and $\hat{\beta}$. It follows that

$$\frac{(\hat{\alpha} - \alpha)/(\sigma/\sqrt{n})}{\hat{\sigma}/\sigma} = \frac{\hat{\alpha} - \alpha}{\hat{\sigma}/\sqrt{n}}$$

has Student's distribution on $n-2$ d.f. [see § 2.5.5].

Thus a central 95% confidence interval for α is

$$\hat{\alpha} \pm t_{97.5}\hat{\sigma}/\sqrt{n} \tag{6.5.13}$$

[compare Example. 4.5.2] where $t_{97.5}$ is the 97·5% point of Student's distribution in $n-2$ d.f.

Similarly a 95% confidence interval for β is

$$\hat{\beta} \pm t_{97.5}\hat{\sigma}/\sqrt{\{\sum n_j(x_j - \bar{x})^2\}}. \tag{6.5.14}$$

(iv) *Significance tests for α and β*

The arguments used in (iii) above show that the (one-sided) significance level of the difference $\hat{\alpha} - \alpha_H$ between the estimate $\hat{\alpha}$ and a hypothesized value α_H of the unknown α is the tail area lying beyond the value t' in Student's distribution on $n-2$ d.f., where

$$t' = \frac{\hat{\alpha} - \alpha_H}{\hat{\sigma}/\sqrt{n}}.$$

(6.5.15)

Similarly for $\hat{\beta} - \beta_H$. In this case we take

$$t' = \frac{\hat{\beta} - \beta_H}{\hat{\sigma}/\sqrt{\sum n_j(x_j - \bar{x})^2}}.$$

(6.5.16)

(In a two-sided test, of course, the significance level is twice the above.)

The test based on (6.5.16) is one of the most frequently used of all statistical tests, especially when $\beta_H = 0$, in which case it tests whether the y-values have any significant (linear) dependance at all on x.

(v) *Confidence interval for y(x)*

The linear model

$$\mu_j = \alpha + \beta(x_j - \bar{x})$$

of (6.5.4) is equivalent to

$$\hat{E}\{Y(x_0)\} = \hat{y}(x_0), \text{ say,}$$
$$= \hat{\alpha} + \hat{\beta}(x_0 - \bar{x}),$$

where $\hat{E}\{Y(x_0)\}$ is the estimate of the expected response at a level x_0 of the 'independent variable' x, and where $\bar{x}$ is computed from the data according to (6.5.5). Since $\hat{\alpha}$ and $\hat{\beta}$ are realizations of independent Normal variables with parameters as given in (6.5.8) and (6.5.9), it follows that $\hat{y}(x_0)$ is a realization of a Normal variable with expectation $\alpha + \beta(x_0 - \bar{x})$ and with variance estimated as

$$v\{\hat{y}(x_0)\} = \left\{ \frac{1}{n} + \frac{(x_0 - \bar{x})^2}{\sum n_j(x_j - \bar{x})^2} \right\} \hat{\alpha}^2$$

(6.5.17)

It will be seen that this has its smallest value when $x_0 = \bar{x}$, (that is at an x-value near the middle of the data values), and that it increases as x_0 moves away from $\bar{x}$, in either direction. The effect is illustrated in Figure 6.5.1, which shows the line $\hat{y}(x)$ and the curves $\hat{y}(x) \pm 2\sqrt{v\{\hat{y}(x)\}}$, which, for each value of $\bar{x}$, give an approximate 95% confidence interval for the ordinate $\alpha + \beta(x - \bar{x})$. The dangers of extrapolating to values of x for which $(x - \bar{x})$ is large will be appreciated.

The multiplier '2' is a rough-and-ready approximation. A more precise statement of the accuracy of $\hat{a} + \hat{\beta}(x_0 - \bar{x})$ as an estimate of $\alpha + \beta(x_0 - \bar{x})$ is given by the 95% confidence interval

$$\hat{y}(x_0) \pm t_{97.5}\sqrt{v\{\hat{y}(x_0)\}}, \tag{6.5.18}$$

where $t_{97.5}$ is the 97·5% point of Student's distribution on $n-2$ d.f.

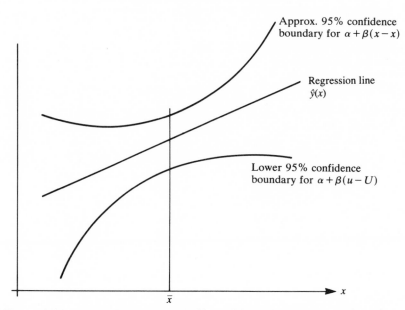

Figure 6.5.1: Regression line and 95% confidence band for $\alpha + \beta(x - \bar{x})$.

EXAMPLE 6.5.1.   *Linear regression.*   The procedures of section 6.5.3 may be illustrated by the data shown in Table 6.5.3, where the response variable is a (logarithmic) measure of the numbers of eye-facets possessed by individual insects *Drosophela melanogaster* after exposure to various temperatures. The data are recorded in the frequency table format of Table 5.8.4 (where, however, the response values are called $x_1, x_2, \ldots, x_a$ while in Table 6.5.3 they are called $y_1, y_2, \ldots, y_{25}$; while the 'treatment levels' of Table 5.8.4 correspond to the temperatures $x_1, x_2, \ldots, x_9$ of the present example).

These data were examined in Example 5.8.5 to illustrate the procedure for testing for the existence of a temperature effect. It is, of course, obvious from Table 6.5.3 that the responses are indeed temperature-dependent and this was confirmed in Example 5.8.5. We propose to analyse the data in the assumption that the response/temperature relationship is a linear regression. Before embarking on the linear regression computations however, it is as well to verify that the linear hypothesis is plausible. That this is the case is shown in Figure 6.5.2, where the array means are plotted against the temperature.

It will be seen that, apart from a kink at 23° and 25°, which might well be an acceptable random fluctuation, the means do indeed appear to accord with a linear law, the slope of the line as 'estimated' from the line drawn by eye in Figure 6.5.2 being about -0.79.

Response = log (no. of facets)		Level of treatment = Temperature (°C)								
		1	2	3	4	5	j 6	7	8	9
i	y_i	15	17	19	21	23	x_j 25	27	29	31
1	8·07	3	1	1						
2	7·07	5	2	5	1					
3	6·07	13	7	3						
4	5·07	25	9	2	1					
5	4·07	22	10	16			2			
6	3·07	12	10	12	6	1	3			
7	2·09	7	5	14	16	2	2			
8	1·09	3	4	14	21	8	9			
9	0·07		3	7	26	7	19	1		
10	−0·93		1	7	12	11	24	3	1	
11	−1·93			1	9	14	22	8	6	
12	−2·93		2	1	5	12	15	15	4	
13	−3·93				2	19	18	44	10	1
14	−4·93				1	4	4	26	6	6
15	−5·93				1	2	2	19	14	13
16	−6·93					2		11	28	9
17	−7·93					3	1	8	8	8
18	−8·93					1		2	5	5
19	−9·93								4	4
20	−10·93								10	2
21	−11·93						1		1	2
22	−12.93								0·5	1·5
23	−13·93								0·5	0·5
24	−14·93									
25	−15·93									1
Number of observations		n_1 90	n_2 54	n_3 83	n_4 100	n_5 86	n_6 122	n_7 137	n_8 98	n_9 53	$n =$ 823
Mean		4·548 $\bar{y}_1$	3·514 $\bar{y}_2$	2·492 $\bar{y}_3$	0·320 $\bar{y}_4$	−2·420 $\bar{y}_5$	−1·627 $\bar{y}_6$	−4·623 $\bar{y}_7$	−6·660 $\bar{y}_8$	−7·713 $\bar{y}_9$	−1·536 $\bar{y}$

(Reprinted with permission of Macmillan Publishing Company from *Statistical Methods for Research Workers*, 14th edition, by Sir Ronald A. Fisher, 1970. Copyright © 1970 University of Adelaide.) (See Bibliography C.)

Table 6.5.3: Influence of temperatures on number of eye-facets of *Drosophila melanogaster*.

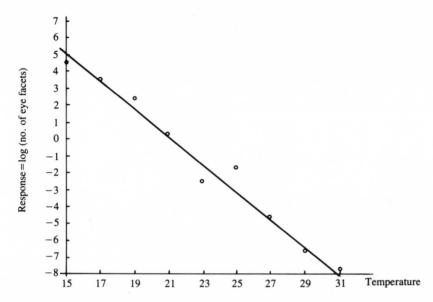

Figure 6.5.2: Average response versus temperature. Data from Table 6.5.3 (plotted as points marked ○). Estimated regression line $\hat{y}(x) = -1\cdot536 - 0\cdot7938(x - 23.277)$.

We shall therefore assume that a linear model is appropriate and follow the procedure of section 6.5.3 to estimate the parameters α and β of the model

$$\mu_j = \alpha + \beta(x_j - \bar{x})$$

of (6.5.4), leading to the estimated mean responses

$$\hat{\mu}_j = \hat{\alpha} + \hat{\beta}(x_j - \bar{x}). \tag{6.5.19}$$

In our sample, the statistics required have the following values:

$$\bar{x} = 23\cdot2770, \qquad \bar{y} = -1\cdot536$$

$$\left.\begin{array}{l} \sum n_j x_j \bar{y}_j = -44482\cdot23 \\ n\bar{x}\bar{y} = -29425\cdot11 \end{array}\right\} \text{Difference} = -15057\cdot12$$

$$\left.\begin{array}{l} \sum n_j x_j^2 = 464{,}887 \\ n\bar{x}^2 = 445{,}917 \end{array}\right\} \text{Difference} = 18{,}969.$$

It follows for (6.5.6) and (6.5.7) that

$$\hat{\alpha} = -1\cdot536, \qquad \hat{\beta} = -0\cdot7938.$$

In Table 6.5.4 we compare the values of $\hat{\mu}_j$ (see (6.5(a))) with the observed mean responses $\bar{y}_j$.

j	n_j	x_j	$\hat{\mu}_j$	$\bar{y}_j$	$\bar{y}_j - \hat{\mu}_j$
1	90	15	5·034	4·548	−0·486
2	54	17	3·446	3·514	0·068
3	83	19	1·859	2·492	0·633
4	100	21	0·271	0·320	0·049
5	86	23	−1·316	−2·420	−1·104
6	122	25	−2·904	−1·627	1·277
7	137	27	−4·491	−4·623	−0·132
8	98	29	−6·078	−6·660	−0·582
9	53	31	−7·666	−7.713	−0·047

$$\sum n_j(\bar{y}_j - \hat{\mu}_j)^2 = 395$$

Table 6.5.4: Estimated responses $\hat{\mu}_j$ compared with observed responses $\bar{y}_j$.

To estimate the standard errors of $\hat{a}$, $\hat{\beta}$ and $\hat{\mu}_j$ we need the sum of squared residuals (that is the sum of the squares of the deviations of the observed points from the fitted line given by (6.5.10). In our present notation that becomes

$$SS = \sum_r \sum_j {}' n_{ij} y_i^2 - n\hat{a}^2 - \hat{\beta}^2 \sum_j n_j(x_j - \bar{x})^2.$$

Since $\hat{a} = \bar{y}$, the first two terms on the right combine to give $\sum \sum n_{ij}(y_i - \bar{y})^2$, which is the 'total sum of squares' referred to in (5.8.18) and given numerically in Table 5.8.5 as 16,202. Thus the above becomes

$$SS = 16,202 - (0·7938)^2(18,969)$$

$$= 4249.$$

It follows that the estimate of variance given by (6.5.11) is

$$\hat{\sigma}^2 = 4249/821 = 5·175$$

and the corresponding estimates of the sampling variances of $\hat{a}$ and $\hat{\beta}$ given by (6.5.12) are, respectively

$$v_\alpha = \hat{\sigma}^2/n = 0·00629 = (0·079)^2,$$

$$v_\beta = \hat{\sigma}^2/\sum n_j(x_j - \bar{x})^2 = 0·000273 = (0·0165)^2.$$

The 95% confidence intervals for α and β given by (6.5.13) and (6.5.14) are then

$$\text{for } \alpha: \quad -1·536 \pm (1·96)(0·079)$$

$$= -1·536 \pm 0·155$$

$$= (-1·691, -1·381);$$

$$\text{for } \beta: \quad -0.794 \pm (1.96)(0.0165)$$

$$= -0.794 \pm 0.032$$

$$= (-0.826, -0.762).$$

As for the significance tests indicated in (6.5.15) and (6.5.16), we have, for example, with respect to the hypothesis that $\beta = 0$ (i.e. that the response is not temperature-dependent) the following value for Student's t:

$$t' = \hat{\beta}/\sqrt{v_\beta} = 48.$$

This huge value conclusively disposes of the hypothesis that $\beta = 0$.

The 95 % confidence band for the regression, given by (6.5.18), requires the evaluation of $v\{\hat{y}(x)\}$, given in (6.5.17), as

$$v\{\hat{y}(x)\} = \left\{ \frac{1}{823} + \frac{(x - 23.277)^2}{18,970} \right\} \times 5.1775$$

$$= 0.0063 + 0.00027(x - 23.277)^2.$$

Substituting this in (6.5.18), and taking $t_{97.5} = 1.96$, the confidence limits for $\hat{y}(x)$, for various values of x, are as shown in Table 6.5.5.

95% confidence band $\hat{y}(x) \pm 1.96\sqrt{v(x)}$ for the regression $\hat{y}(x) = \hat{\alpha} + \hat{\beta}(x - \bar{x})$, where
$$\hat{\alpha} = -1.536, \qquad \hat{\beta} = -0.7938, \qquad \bar{x} = 23.277,$$

and

$$v(x) = v\{\hat{y}(x)\} = 0.0063 + 0.00027\ (x - 23.277)^2$$

x	$\hat{y}(x)$	$\sqrt{v(x)}$	Upper confidence limit $\hat{y}(x) + 1.96\sqrt{v(x)}$	Lower confidence limit $\hat{y}(x) - 1.96\sqrt{v(x)}$
15	5.034	0.157	5.342	4.726
17	3.446	0.130	3.702	3.192
19	1.859	0.106	2.067	1.651
21	0.271	0.088	0.443	0.099
23	−1.316	0.080	−1.159	−1.473
25	−2.904	0.084	−2.739	−3.069
27	−4.491	0.100	−4.295	−4.687
29	−6.078	0.123	−5.837	−6.319
31	−7.666	0.150	−7.372	−7.960

Table 6.5.5

6.6.   ESTIMATION OF DOSAGE–MORTALITY RELATIONSHIP

6.6.1.   Quantal Responses

In section 6.5 we discussed ways of estimating the response $y(x)$ of a system to a treatment x, exemplified by the case where $y(x)$ denoted the breaking strength of a specimen of steel that had been heat-treated for x hours. Similar considerations apply to the 'life' $y(x)$ of an electric-light bulb to which a current of x volts intensity is applied. Suppose a number of lamps to be subjected for as long as necessary to x volts in a circuit which records the time $y(x)$ at which each lamp burns out; the process being repeated, with different lamps of the same kind, at several different voltages. The responses $y(x)$, for each x, will be more variable than was the case with the steel wires of section 6.5, but the procedure is essentially the same.

Suppose now, however, that equipment is no longer able to record the times at which the lamps fail. A possible procedure would then be the following: run the lamps for a fixed duration (say 2,000 hours) at voltage x, and count how many lamps have expired at the end of that time. Repeat (with different lamps) for different values of x.

For sufficiently small values of 'treatment level' all lamps will survive. For sufficiently larger voltages all (or nearly all) lamps will have failed. At intermediate values of treatment level, the inherent variability of the individual lamps will result in there being some failed lamps and some survivors, the proportion of failures increasing (on average) with treatment level.

This is an example of a 'quantal' response, typical of many kinds of biological assay, where, for example, the effectiveness of an insecticide may be measurable only in terms of the proportion of insects killed after a standard exposure to a given dosage of the treatment. The proportion of treated insects killed will be zero for sufficiently small dosages and 100% for sufficiently large dosages. At intermediate levels of dosage, individual insects behave differently, and some will die and some survive. The manner in which the proportion of deaths increases with increasing dosage is exemplified in Table 6.6.1 and in Figure 6.6.1.

Row label j	Dose of rotenone (mg/l) μ_j	$\log_{10}$ (dose) x_j	No. of insects n_j	No. killed r_j	Proportion killed p_j
1	10·2	1·07	50	44	0·88
2	7·7	0·89	49	42	0·86
3	5·1	0·71	46	24	0·52
4	3·8	0·58	48	16	0·33
5	2·6	0·41	50	6	0·12

Reprinted, with permission, from *Probit Analysis*, by D. J. Finney (1971), published by Cambridge University Press.

Table 6.6.1: Data for a test of the toxicity of rotenone to *Macrosiphoniella Sanborni*.

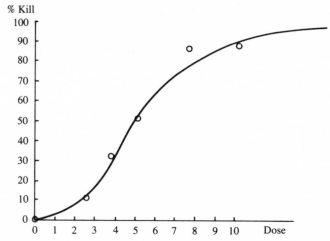

Figure 6.6.1: A sigmoid curve: Response (% kill) plotted against dose.

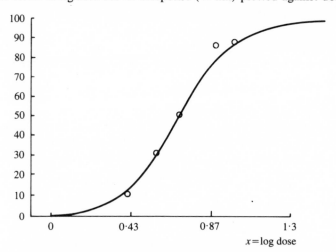

Figure 6.6.2: Response (% kill) plotted against log dose.

6.6.2.   Probability Model

The toxicity of rotenone against Macrosiphoniella sanborni is conventionally defined as that dose which would kill 50% of organisms. A rough idea of the magnitude of this may be obtained from Figure 6.6.1: as nearly as can be seen from the free-hand curve, the 50% value is about 4·8. This however, hardly counts as an estimate. To obtain the best values that the data of Table 6.6.1 can give, together with an assessment of the reliability of that value, we have to devise a *probability model* and estimate its *parameters*. The model and its analysis are described in the remaining pages of § 6.6, which closely follow the work of D. J. Finney (1971).

The model has two parts. The first part is an obvious one, namely:

(i) A given dose will kill a determinate proportion π of the entire population of organisms; in a sample, the n organisms involved react independently, so the number killed is a Binomial (n, π) random variable [see II, § 5.2.2].

The second part of the model is less obvious. This must specify a parametric formula for the dependence of the probability π on the dose μ. Here Figure 6.6.2, in which the proportion killed is plotted against the *logarithm* of the dosage, is more suggestive than Figure 6.6.1; Figure 6.6.2 resembles a typical c.d.f. (cumulative distribution function) of the Normal family. That the resemblance is in fact a close one, may be verified by plotting the data on 'probability graph paper' [see § 3.2.2(d)]. This has its probability (percentage) scale so arranged that a Normal c.d.f. plots as a straight line [see § 2.7.4]. Our data appears to adhere well to such a straight line, as will be seen in Figure 6.6.3, where the line has been 'fitted' by eye. We therefore take the following as the second part of our probability model:

(ii) In terms of the 'metameter'

$$x = \log_{10} \text{(dosage)}$$

the probability that a randomly chosen organism will die when exposed to treatment at level x is

$$\pi(x) = \int_{\infty}^{x} \{(2\pi)^{-1/2}\sigma^{-1} \exp -\tfrac{1}{2}(y-\mu)^2/\sigma^2\} \, dy$$

$$= \Phi\left(\frac{x-\mu}{\sigma}\right)$$

$$= \Phi(\alpha + \beta x) \tag{6.6.1}$$

say. (μ and σ are the 'natural' parameters of the Normal distribution, [see II, § 11.4.6] but it is more convenient in the present context to use

$$\alpha = -\mu/\sigma \quad \text{and} \quad \beta = 1/\sigma.)$$

[See Example 7.4.2 for a verification by the χ^2 test that (6.6.1) is consistent with the data.]

A remark on nomenclature: For any given value of $\pi(x)$ the corresponding value of $\alpha + \beta x$ is uniquely determined. For example, from the Normal tables, $\Phi(1\cdot96) = 0\cdot975$. Thus the value $\alpha + \beta x = 1\cdot96$ corresponds uniquely to $x = 0\cdot975$. This may be written 'inversely' [see IV, § 2.7] in the form $1\cdot96 = \Phi^{-1}(0\cdot975)$. The value $1\cdot96$ is called the Normal Equivalent Deviate (N.E.D.) of $0\cdot975$. Similarly, the N.E.D. of $0\cdot025$ is $-1\cdot96$. Early users of the method preferred to avoid negative signs, which they could do by adding 5 to the N.E.D., since N.E.D. values below -5 are never encountered in practice. This

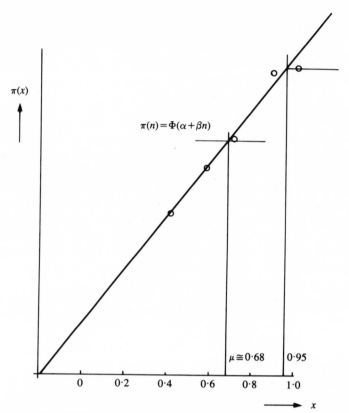

Figure 6.6.3: % kill, on probability paper scale, plotted against log dose.

modified N.E.D. is called a *probit*. Thus the following statements all mean the same thing:

$$\Phi(y) = v$$

$$y = \Phi^{-1}(v)$$

$$y \text{ is the N.E.D. of } v$$

$$y + 5 \text{ is the probit of } v.$$

(The description that follows does not use probits.)

It might be objected that the c.d.f. of some other distribution might be more suitable. The logistic for example, has been advocated. This would have

$$\pi(x) = e^{-(\alpha + \beta x)}/\{1 + e^{-(\alpha + \beta x)}\}^2$$

[see II, § 11.10]. Here, the value of $\alpha + \beta x$ defined by a given value of $\pi(x)$ is called the *logit* of $\pi(x)$ [cf. 'probit']. Yet other distributions might be considered. However, for any particular dosage–mortaility experiment, there is never enough data to discriminate between one of these and another; the Normal usually gives an adequate fit and has readily accessible tables; and the consistent use of a single family of distributions enables one to make objective comparisons between one insecticide and another. There seems little point therefore in not using the model (6.6.1).

6.6.3. The Likelihood Surface

The entomologist's desire to estimate the dosage that would produce a 50% (or 90%, or...) population kill will be satisfied if he is given the Maximum Likelihood estimates of α and β in (6.6.1) together with a statement of their accuracy. The standard measure of toxicity is in fact the '50% kill' value, that is the log-dosage x (or equivalently the actual dosage 10^x) for which $\pi(x) = 0.5$. The rough value obtained from Figure 6.6.1 may with advantage be replaced by the improved value obtained from the straight-line graph in Figure 6.6.3, fitted by eye. The estimate μ is the x-value corresponding to 50%, that is 0.68, and the estimate of σ is 0.23. Thus, from the graph, our estimate of $\pi(x)$ is $\Phi\{(x-0.68)/0.23\} = \Phi(4.4x-3.0)$.

To obtain a more sophisticated estimate, we require the likelihood function

$$l(\mu, \sigma) = \prod_j \pi_j^{r_j}(1-\pi_j)^{n_j-r_j},$$

(using the notation of Table 6.6.1). In terms of the observed proportions $p_j = r_j/n_j$ (see Table 6.6.1) the log-likelihood, which we now write as $g(\mu, \sigma)$, is

$$g(\mu, \sigma) = \log l(\mu, \sigma)$$

$$= \sum_j n_j\{p_j \log \pi_j + (1-p_j) \log (1-\pi_j)\} \qquad (6.6.2)$$

where

$$\pi_j = \pi(x_j) = \Phi\{(x_j-\mu)/\sigma\}$$

$$= \Phi(\alpha + \beta x_j) \qquad (6.6.3)$$

say, if we change the parameters, as is usual in biometry, from (μ, σ) to (α, β), with

$$\alpha = -\mu/\sigma, \qquad \beta = 1/\sigma. \qquad (6.6.4)$$

Orthodox analysis confines itself to deriving the maximum likelihood estimates $\hat{\alpha}, \hat{\beta}$ (the values of α and β which maximize (6.6.2), regarded as a function of α and β) and the usual ML approximations for the dispersion matrix of these estimates. It is however more informative to explore the likelihood surface in the neighbourhood of the maximum: the region near $\alpha = -3.0, \beta = 4.4$ will do well enough. Values of g are readily computed from specified values of α and β, using the following version of (6.6.2):

$$g(\alpha, \beta) = \sum_j \{r_j \log \Phi_j + (n_j - r_j) \log \Psi_j\} \qquad (6.6.5)$$

where r_j and n_j are given in Table 6.6.1, and where $\Phi_j = \Phi(\alpha + \beta x_j)$ and $\Psi_j = (1 - \Phi_j)$. As an illustration, Table 6.6.2 shows the material needed for the calculation when $\alpha = -3\cdot2$ and $\beta = 4\cdot4$:

x_j	$\alpha + \beta x_j$	Φ_j	r_j	Ψ_j	$n_j - r_j$
1·01	1·24	0·892	44	0·108	6
0·89	0·72	0·764	42	0·236	7
0·71	−0·08	0·469	24	0·531	22
0·58	−0·65	0·258	16	0·742	32
0·41	−1·40	0·081	6	0·919	44

Table 6.6.2: Material required for the computation of $g(\alpha, \beta)$ for $\alpha = -3\cdot2$, $\beta = 4\cdot4$.

Using an ordinary calculator, one simply enters $\Phi_j = 0\cdot892$, converts this to its logarithm, and multiplies by $r_j = 44$; repeat with $\Psi_j = 0\cdot108$ and $n_j - r_j = 6$, accumulating the results; and so on. The log-likelihood (using logs to base e) is $g(-3\cdot2, 4\cdot4) = 0\cdot12192$.

Table 6.6.3 displays a set of three-digit values of $g(\alpha, \beta)$ computed in this way for values of (α, β) near the maximum.

				α			
		−3·4	−3·2	−3·0	−2·8	−2·6	−2·4
	4·8	−121	−122	−126	−135		
	4·6	−123	−120	−122	−128		
	4·4		−122	−120	−123		
β	4·2		−126	−121	−120	−125	
	4·0			−124	−121	−121	
	3·8				−123	−120	−123
	3·6					−122	−121

Table 6.6.3: The log-likelihood $g(\alpha, \beta)$.

The table shows that the crude estimator $\alpha = -3\cdot0$, $\beta = 4\cdot4$ obtained from Figure 6.6.3 are in fact quite near the maximum. It also shows however that, for a range of values of α and β satisfying the condition

$$\alpha + \beta = 1\cdot4$$

(approximately) the likelihood is practically constant: to three-digit accuracy, $g(-3\cdot2, 4\cdot6) = g(-3\cdot0, 4\cdot4) = g(-2\cdot8, 4\cdot2) = -120$, etc. This means that the maximum is not a sharply defined peak; the surface has an almost horizontal ridge in the location indicated. The precise location of the maximum could be determined to any desired accuracy by repeating the calculations in Table 6.6.3 at finer intervals and carrying more figures, but the geometry of the surface shows that $\hat{\alpha} + \hat{\beta}$ will have a low variability, and, by similar arguments, that $\hat{\beta} - \hat{\alpha}$ will have a considerably larger variability; thus $\hat{\alpha}$ and $\hat{\beta}$ have a large negative correlation.

6.6.4   The Maximum Likelihood Estimator

The M.L. estimates are the roots of the likelihood equations:

$$\frac{\partial g}{\partial \alpha} = \sum n_j \frac{p_j - \Phi_j}{\Phi_j \Psi_j} \phi_j = 0$$

$$\frac{\partial g}{\partial \beta} = \sum n_j x_j \frac{p_j - \Phi_j}{\Phi_j \Psi_j} \phi_j = 0$$

(6.6.6)

where

$$\Phi_j = \Phi(\alpha + \beta x_j), \qquad \Psi_j = 1 - \Phi_j,$$

and

$$\phi_j = (2\pi)^{-1/2} \exp -\tfrac{1}{2} x_j^2,$$

is the value of the standard Normal density at x_j.

To solve these equations, we first need a starting approximation (α_1, β_1) for $(\hat{\alpha}, \hat{\beta})$. The values obtained from Figure 6.6.3 will do well enough. These are

$$\alpha_1 = 3 \cdot 0, \qquad \beta = 4 \cdot 4.$$

To obtain improved approximations

$$\alpha_2 = \alpha_1 + \delta \alpha_1 \qquad \beta_2 = \beta_1 + \delta \beta$$

we set

$$\partial g / \partial \alpha_2 = \partial g / \partial \beta_2 = 0$$

and expand $\partial g / \partial \alpha_2$ and $\partial g / \partial \beta_2$ about the point (α_1, β_1) by Taylor's Theorem [see IV, § 5.8] retaining only the first powers of $\delta \alpha$ and $\delta \beta$. Thus the likelihood equations become

$$\frac{\partial g}{\partial \alpha_1} + \delta \alpha \frac{\partial^2 g}{\partial \alpha_1^2} + \delta \beta \frac{\partial^2 g}{\partial \alpha_1 \partial \alpha_2} = 0$$

and

$$\frac{\partial g}{\partial \beta_1} + \delta \alpha \frac{\partial^2 g}{\partial \alpha_1 \partial \beta_1} + \delta \beta \frac{\partial y}{\partial \beta_1^2} = 0$$

(6.6.7)

approximately. Here '$\partial g / \partial \alpha_1$', is an abbreviation for '$\partial g / \partial \alpha$ evaluated at

$(\alpha_1, \beta_1)'$; etc. The expressions for $\partial g / \partial \alpha$ and $\partial g / \partial \beta$ are given in (6.6.6), from which it will be seen that (6.6.7) will involve, amongst other terms, $\partial \Phi / \partial \alpha$ and $\partial \Phi / \partial \beta$. Now

$$\frac{\partial \Phi}{\partial \alpha} = \frac{\partial}{\partial \alpha} \Phi(\alpha + \beta x) = \phi(\alpha + \beta x)$$

and

$$\frac{\partial \Phi}{\partial \beta} = \frac{\partial}{\partial \beta} \Phi(\alpha + \beta x) = x\phi(x + \beta x).$$

In the equation (6.6.7), therefore, we see from (6.6.6) that

$$\frac{\partial g}{\partial \alpha_1} = \sum_j n_j \frac{p_j - \Phi(\alpha_1 + \beta_1 x_j)}{\Phi(\alpha_1 + \beta_1 x_j)\Psi(\alpha_1 + \beta_1 x_j)} \phi(\alpha_1 + \beta_1 x_j)$$

and

$$\frac{\partial g}{\partial \beta_1} = \sum_j n_j \frac{p_j - \Phi(\alpha_1 + \beta_1 x_j)}{\Phi(\alpha_1 + \beta_1 x_j)\Psi(\alpha_1 + \beta_1 x_j)} x_j \phi(\alpha_1 + \beta_1 x_j),$$

where

$$\Psi(y) = 1 - \Phi(y)$$

for all y. In evaluating the second derivative of g we have, for example,

$$\frac{\partial^2 g}{\partial \alpha^2} = \sum_j n_j(p_j - \Phi_j) \frac{\partial}{\partial \alpha}\left(\frac{\phi_j}{\Phi_j \Psi_j}\right) - \sum_j n_j \frac{\phi_j}{\Phi_j \Psi_j} \frac{\partial \Phi_j}{\partial \alpha} \qquad (6.6.8)$$

where

$$\Phi_j = \Phi(\alpha_1 + \beta_1 x_j), \qquad \Psi_j = 1 - \Phi_j.$$

Provided α_1 and β_1 are reasonably good first approximations, $p_j - \Phi_j$ will be small in magnitude, for each j (see Table 6.6.4) and the first term in the right of (6.6.8) will be negligible in comparison with the second, in which we may note that $\partial \Phi_j / \partial \alpha = \phi_j$.

x_j	p_j	$\Phi_j = \Phi(4 \cdot 4 x_j - 3 \cdot 0)$	$\|p_j - \Phi_j\|$
1·01	0·88	0·93	0·05
0·89	0·86	0·82	0·04
0·71	0·52	0·55	0·03
0·58	0·33	0·33	0·00
0·42	0·12	0·14	0·02

Table 6.6.4: Values of $|p_j - \Phi_j|$.

It follows that, to an acceptable accuracy,

$$\frac{\partial^2 g}{\partial \alpha_1^2} = -\sum_j n_j \phi^2(\alpha_1 + \beta_1 x_j)/\Phi(\alpha_1 + \beta_1 x_j)\Psi(\alpha_1 + \beta_1 x_j)$$

$$= -\sum_j n_j w_j$$

say, where

$$w_j = \phi^2(\alpha_1 + \beta_1 x_j)/\Phi(\alpha_1 + \beta_1 x_j)\Psi(\alpha_1 + \beta_1 x_j). \qquad (6.6.9)$$

Similarly, since $\partial \Phi_j/\partial \beta = x_j \phi_j$,

$$\frac{\partial^2 g}{\partial \alpha_1 \partial \beta_1} = -\sum n_j w_j x_j.$$

Likewise

$$\frac{\partial^2 g}{\partial \beta_1^2} = -\sum n_j w_j x_j^2.$$

Thus, finally, the equations (6.6.3) for $\delta \alpha$ and $\delta \beta$ become

$$\delta \alpha \sum n_j w_j + \delta \beta \sum n_j w_j x_j = \sum n_j w_j (p_j - \Phi_j)/\phi_j$$
$$\delta \alpha \sum n_j w_j x_j + \delta \beta \sum n_j w_j x_j^2 = \sum n_j w_j x_j (p_j - \Phi_j)/\phi_j \qquad (6.6.10)$$

where the weights w_j are as defined in (6.6.9).

The procedure now becomes a straightforward iteration. Solve the simultaneous linear equations (6.6.10) for $\delta \alpha$ and $\delta \beta$, and take

$$\alpha_2 = \alpha_1 + \delta \alpha, \qquad \beta_2 = \beta_1 + \delta \beta.$$

Repeat equations (6.6.9) with α_2 and β_2 everywhere replacing α_1 and β_1. After a few iterations $\delta \alpha$ and $\delta \beta$ will become negligibly small. The procedure is illustrated in Table 6.6.5.

This table starts with a first approximation

$$\alpha_1 = -3 \cdot 0, \qquad \beta_1 = 4 \cdot 4$$

obtained from Figure 6.6.3. One then computes successively the columns of values of $\alpha_1 + \beta_1 x_j$, $\Phi_j = \Phi(\alpha_1 + \beta_1 x_j)$; $\phi_j = \phi(\alpha + \beta x_j)$; $n_j w_j = n_j \phi_j^2/\Phi_j(1 - \Phi_j)$, taking the n_j from Table 6.6.1; $p_j = r_j/n_j$; and $(p_j - \Phi_j)/\phi_j$. The values of ϕ_j and Φ_j have to be read from the Normal ordinate and integral tables: in the absence of accessible tables of ϕ_j it is not too difficult to compute the values $\phi_j = (2\pi)^{-1/2} \exp -\frac{1}{2} x_j^2$, while the Normal integral table will be found in the Appendix T3. The value of w_j may then be computed from (6.6.9) or, alternatively, obtained direct from special tables (which are, however, entered at the *probit* value of $5 + \alpha + \beta x$ and not at the 'N.E.D.' value $\alpha + \beta x$ itself. (Table IX2, Fisher and Yates (1957)—see Bibliography G.) The sums of products $\sum nw$, $\sum nwx$, $\sum nwx^2$ are then evaluated, and the equations (6.6.10) formulated. It will be seen that the adjustments to α_1 and β_1 were $\delta\alpha = 0\cdot121$ and $\delta\beta = -0\cdot231$, leading to $\alpha_2 = -2\cdot879$ and $\beta_2 = 4\cdot169$. A second cycle led to further adjustments $\delta\alpha = 0\cdot017$, $\delta\beta = 0\cdot008$, whence $\alpha_3 = -2\cdot862$, $\beta_3 = 4\cdot175$. Evidently a further iteration would leave α and β unaltered to three decimals, so we may take the estimates as

$$\hat{\alpha} = -2\cdot862, \qquad \hat{\beta} = 4\cdot175 \qquad\qquad (6.6.11)$$

and the estimated response curve as

$$\pi(x) = \Phi(4\cdot175x - 2\cdot862), \qquad\qquad (6.6.12)$$

x being the logarithm, to base 10, of the actual dosage. The 50% value $\hat{x}_{50}$ which conventionally estimates the toxicity is the value of x for which $\pi(x) = 0\cdot5$, viz. for which

$$4\cdot175x - 2\cdot862 = 0.$$

Hence

$$\hat{x}_{50} = 0\cdot685 \qquad\qquad (6.6.13)$$

and the corresponding dosage is

$$\hat{\mu}_{50} = 10^{0\cdot685} = 4\cdot84. \qquad\qquad (6.6.14)$$

6.6.5.   Reliability of the Estimates

The standard approximation to the dispersion matrix of the estimates $\hat{\alpha}$, $\hat{\beta}$ is the inverse of the matrix

$$-\begin{pmatrix} \partial^2 g/\partial\alpha_1^2 & \partial^2 g/\partial\alpha\,\partial\beta \\ \partial^2 g/\partial\alpha_1\,\partial\beta & \partial^2 g/\partial\beta^2 \end{pmatrix}$$

evaluated at $(\hat{\alpha}, \hat{\beta})$ [see § 6.2.5(c)] In our case therefore we require

$$\begin{pmatrix} \Sigma\,nw & \Sigma\,nwx \\ \Sigma\,nwx & \Sigma\,nwx^2 \end{pmatrix}^{-1}.$$

The matrix to be inverted is, to sufficient accuracy, the matrix of the coefficients of the linear simultaneous equations used in the final iteration. In our example, this was

$$\begin{pmatrix} 117\cdot48 & 83\cdot16 \\ 83\cdot16 & 80\cdot83 \end{pmatrix}$$

(see Table 6.6.5). The inverse is

$$\frac{1}{2580}\begin{pmatrix} 80\cdot83 & -83\cdot16 \\ -83\cdot16 & 117\cdot48 \end{pmatrix} = \begin{pmatrix} 0\cdot0313 & -0\cdot0322 \\ -0\cdot0322 & 0\cdot0455 \end{pmatrix}.$$

First approximations								
α	β	x	$\alpha + \beta x$	Φ	ϕ	nw	p	$(p - \Phi)/\phi$
$-3{\cdot}0$	$4{\cdot}4$	$1{\cdot}009$	$1{\cdot}44$	$0{\cdot}925$	$0{\cdot}141$	$14{\cdot}50$	$0{\cdot}880$	$-0{\cdot}319$
		$0{\cdot}886$	$0{\cdot}90$	$0{\cdot}816$	$0{\cdot}266$	$23{\cdot}68$	$0{\cdot}857$	$0{\cdot}154$
		$0{\cdot}708$	$0{\cdot}12$	$0{\cdot}548$	$0{\cdot}396$	$29{\cdot}12$	$0{\cdot}522$	$-0{\cdot}066$
		$0{\cdot}580$	$-0{\cdot}45$	$0{\cdot}327$	$0{\cdot}361$	$28{\cdot}32$	$0{\cdot}333$	$0{\cdot}017$
		$0{\cdot}415$	$-1{\cdot}17$	$0{\cdot}121$	$0{\cdot}201$	$19{\cdot}40$	$0{\cdot}120$	$-0{\cdot}005$

$$\sum nw = 114{\cdot}12 \qquad \sum nw(p - \Phi)/\phi = -2{\cdot}42$$

$$\sum nwx = 80{\cdot}05 \qquad \sum nwx(p - \Phi)/\phi = -2{\cdot}64$$

$$\sum nwx^2 = 60{\cdot}29$$

$$114{\cdot}12\delta\alpha + 80{\cdot}05\delta\beta = -2{\cdot}42$$

$$80{\cdot}05\delta\alpha + 60{\cdot}201\delta\beta = -2{\cdot}64$$

$$\delta\alpha = 0{\cdot}121, \qquad \delta\beta = -0{\cdot}231$$

Second approximations								
α	β	x	$\alpha + \beta x$	Φ	ϕ	nw	p	$(p - \Phi)/\phi$
$-2{\cdot}879$	$4{\cdot}169$	$1{\cdot}009$	$1{\cdot}330$	$0{\cdot}908$	$0{\cdot}164$	$16{\cdot}30$	$0{\cdot}880$	$-0{\cdot}171$
		$0{\cdot}886$	$0{\cdot}815$	$0{\cdot}792$	$0{\cdot}286$	$24{\cdot}45$	$0{\cdot}857$	$0{\cdot}227$
		$0{\cdot}708$	$0{\cdot}073$	$0{\cdot}528$	$0{\cdot}398$	$21{\cdot}96$	$0{\cdot}522$	$-0{\cdot}015$
		$0{\cdot}580$	$-0{\cdot}461$	$0{\cdot}323$	$0{\cdot}359$	$28{\cdot}27$	$0{\cdot}333$	$0{\cdot}028$
		$0{\cdot}415$	$-1{\cdot}149$	$0{\cdot}125$	$0{\cdot}206$	$19{\cdot}30$	$0{\cdot}120$	$-0{\cdot}024$

$$\sum nw = 117{\cdot}48 \qquad \sum nw(p - \Phi)/\phi = 2{\cdot}654$$

$$\sum nwx = 83{\cdot}16 \qquad \sum nwx(p - \Phi)/\phi = 2{\cdot}062$$

$$\sum nwx^2 = 80{\cdot}83$$

$$117{\cdot}48\delta\alpha + 83{\cdot}16\delta\beta = 2{\cdot}654$$

$$83{\cdot}16\delta\alpha + 80{\cdot}83\delta\beta = 2{\cdot}062$$

$$\delta\alpha = 0{\cdot}017, \qquad \delta\beta = 0{\cdot}008$$

Table 6.6.5: Computation of the coefficients of (6.6.10), and iteration.

If we write this as

$$\begin{pmatrix} v(\alpha) & c(\alpha, \beta) \\ c(\alpha, \beta) & v(\beta) \end{pmatrix},$$

where $v(\alpha)$ and $v(\beta)$ are the sampling variances of $\hat{\alpha}$ and $\hat{\beta}$ respectively, and $c(\alpha, \beta)$ the covariance, we have

$$v(\alpha) = 0 \cdot 0313 = (0 \cdot 177)^2, \qquad v(\beta) = 0 \cdot 0455 = (0 \cdot 213)^2$$

and

$$c(\alpha, \beta) = -0 \cdot 0322. \tag{6.6.15}$$

The correlation coefficient of $\hat{\alpha}$ and $\hat{\beta}$ is thus

$$-0 \cdot 0322/(0 \cdot 177)(0 \cdot 213) = -0 \cdot 85.$$

The correlation is large and negative, in agreement with the conclusions drawn in section 6.6.3 from the geometry of the log-likelihood surface.

Confidence band for $\pi(x)$

The M.L. estimate of the response function $\pi(x)$ is

$$\hat{\pi}(x) = \Phi(\hat{\alpha} + \hat{\beta}x),$$

and the sampling variance of this, $v\{\hat{\pi}(x)\}$ say, for a given x, is obtainable (to the usual degree of approximation) by applying (6.2.19), viz.

$$v\{\hat{\pi}(x)\} = \{\phi(\hat{\alpha} + \hat{\beta}x)\}^2\{v(\alpha) + 2xc(\alpha, \beta) + x^2 v(\beta)\}.$$

A rough representation of the 95% confidence band for $\pi(x)$ would therefore be the zone lying between the curves

$$\hat{\pi}(x) \pm 2\sqrt{v\{\hat{\pi}(x)\}}. \tag{6.6.16}$$

Some representative values of these limits are given in Table 6.6.6, and illustrated graphically in Fig. 6.6.4.

x	$\hat{\pi}(x)$	$\hat{\pi}(x) + 2\sqrt{v\{\hat{\pi}(x)\}}$	$\hat{\pi}(x) - 2\sqrt{v\{\hat{\pi}(x)\}}$
0·2	0·021	0·047	—
0·4	0·117	0·161	0·073
0·6	0·360	0·430	0·290
0·8	0·684	0·717	0·618
1·0	0·905	0·943	0·867
1·2	0·985	0·995	0·975

Table 6.6.6: Approximate 95% confidence band for $\pi(x)$.

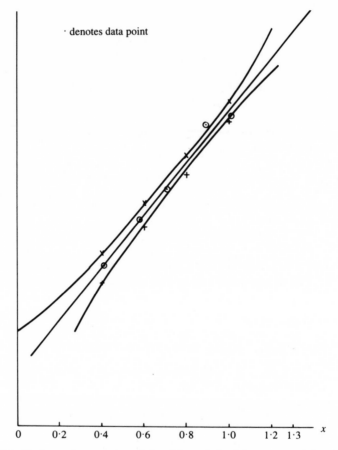

Figure 6.6.4: Estimated response $\pi(x) = \Phi(\alpha + \beta x)$ with upper and lower 95% confidence limit. $\odot$ denotes data points.

6.6.6.   Dose Required for a Specified Response

(i) *Crude argument*

When the response $\pi(x)$ at a treatment level x is given by

$$\pi(x) = \Phi(\alpha + \beta x),$$

the treatment level x_{50} required to achieve a 50% response is given by

$$0\cdot 50 = \Phi(\alpha + \beta x_{50}),$$

whence

$$\alpha + \beta x_{50} = \Phi^{-1}(0\cdot 50)$$

$$= 0,$$

$$x_{50} = -\alpha/\beta.$$

The M.L.E. of x_{50} is therefore

$$\hat{x}_{50} = -\hat{\alpha}/\hat{\beta}$$

$$= 0 \cdot 685. \tag{6.6.17}$$

This is the log-dosage. The actual dosage is

$$\hat{\mu}_{50} = 10^{0 \cdot 685} = 4 \cdot 84 \tag{6.6.18}$$

(cf. (6.6.13), (6.6.14)).

To the degree of approximation to which we are working, the sampling variance of $\hat{x}_{50}$ is

$$v(\hat{x}_{50}) = \left[\frac{1}{\beta^2} v(\alpha) - \frac{2\alpha}{\beta^3} c(\alpha, \beta) + \frac{\alpha^2}{\beta^4} v(\beta) \right] (\hat{\alpha}, \hat{\beta})$$

$$= (0 \cdot 022)^2$$

[see (6.2.19)], whence a sufficiently accurate 95% confidence interval for x_{50} is

$$\hat{x}_{50} \pm 2 \sqrt{v(\hat{x}_{50})} = 0 \cdot 685 \pm 0 \cdot 044$$

$$= (0 \cdot 64, 0 \cdot 73). \tag{6.6.19}$$

The corresponding interval for the natural dosage $u = 10^x$ is $(4 \cdot 36, 5 \cdot 37)$. (Similar considerations apply to x_{10}, x_{25}, etc.)

(ii) *More precise argument*

An alternative, more elaborate (and more accurate) procedure for the assessment of the accuracy of $\hat{x}_{50}$ is to use Fieller's Theorem [see Example 4.5.8], according to which the 95% confidence limits for x_{50} are the roots λ_1, λ_2 of the quadratic

$$(\lambda b + a)^2 = t_{97 \cdot 5}^2 v$$

[see (4.5.14)], where

$$a = \hat{\alpha} = -2 \cdot 862, \qquad b = \hat{\beta} = 4 \cdot 175,$$

$$v = \lambda^2 v(\beta) + 2\lambda c(\alpha, \beta) + v(\alpha)$$

[see (4.5.12)]

$$= 0 \cdot 313 \lambda^2 - 0 \cdot 0644 \lambda + 0 \cdot 0455$$

on using the values given in (6.6.15), and $t_{97 \cdot 5}$ denotes the 97·5% point of Student's distribution on the appropriate number of degrees of freedom. Bearing in mind the approximations employed in deriving the values (6.6.15), one may take $t = 2 \cdot 0$ (the precise Normal distribution value being (1·96) so that the quadratic becomes

$$(4 \cdot 175\lambda - 2 \cdot 862)^2 = 4(0 \cdot 0313\lambda^2 - 0 \cdot 0644\lambda + 0 \cdot 0455),$$

or

$$17 \cdot 30\lambda^2 - 23 \cdot 64\lambda + 8 \cdot 01 = 0,$$

whence

$$\lambda_1, \lambda_2 = 0 \cdot 641, 0 \cdot 729.$$

These are indistinguishable (in this example) from the values obtained by the simpler approach of (6.6.19).

We conclude that the M.L.E. and the 95% confidence interval for the 50% value of the log-dosage x are, respectively,

$$0 \cdot 685; \qquad (0 \cdot 641, 0 \cdot 729).$$

The corresponding figures for the natural dosage 10^x are

$$4 \cdot 84; \qquad (4 \cdot 38, 5 \cdot 36).$$

For further details, see Finney (1971).

6.7. MAXIMUM-LIKELIHOOD ESTIMATES FROM GROUPED, CENSORED OR TRUNCATED DATA

6.7.1. Discrete Data

Consider the frequency table:

Magnitude of observed quantity	0	1	2	...	k
Frequency	$n(0)$	$n(1)$	$n(2)$		$n(k)$

where observations are made on a population with discrete p.d.f. $f(y, \theta)$ at y, $y = 0, 1, \ldots$. In our table k is the largest observed value. The data might be available in a less complete way, e.g. truncated, grouped, or censored.

(i) *Truncation*

One common form is:

Magnitude	0	1	2	3	...	k
Frequency	Not available			$n(3)$	...	$n(k)$

Here the data has been *truncated*: the frequencies $n(0)$, $n(1)$ and $n(2)$ are simply not available (e.g. through a failure of the recording equipment). The available observations must then be regarded as coming from the conditional population, with p.d.f. at y given by

$$f_c(y, \theta) = f(y, \theta)/\{f(0, \theta) + f(1, \theta) + f(2, \theta)\} \qquad y = 3, 4, \ldots.$$

For example, if the original population is Poisson (θ), the truncated distribution (truncated as above) will have p.d.f.

$$f_c(y, \theta) = \{e^{-\theta}\theta^y/y!\}/\{e^{-\theta}(1 + \theta + \tfrac{1}{2}\theta^2)\}$$

$$= \theta^y/(1 + \theta + \tfrac{1}{2}\theta^2)y!, \qquad y = 2, 3, \ldots$$

and the likelihood will be proportional to

$$\prod_{y=3}^{k} \{f_c(y, \theta)\}^{n(y)}.$$

Dropping uninformative factors, we take the likelihood function as

$$l(\theta) = \theta^{\Sigma yn(y)}/(1 + \theta + \tfrac{1}{2}\theta^2)^{\Sigma n(y)}$$

$$= \theta^{n\bar{y}}/(1 + \theta + \tfrac{1}{2}\theta^2)^n$$

where, as usual, $\bar{y}$ denotes the sample mean and n the sample size. The M.L.E. $\hat{\theta}$ is the relevant root of the likelihood equation $d \log l/d\theta = 0$, which in this example reduces to the quadratic equation

$$(\tfrac{1}{2}\bar{y} - 1)\theta^2 + (\bar{y} - 1)\theta + \bar{y} = 0.$$

(ii) *Grouping, censoring*

Another common form is as follows:

Magnitude	0	1	...	z	$z+1$	$z+2$	$z+3$	$z+4$	$\geq z+5$
Frequency	$n(0)$	$n(1)$	...	$n'(z)$		$n'(z+2)$			$n'(z+5)$

The frequencies $n(z)$ and $n(z+1)$ are not separately available: they have been *grouped*, and only their sum $n'(z) = n(z) + n(z+1)$ is available. Similarly the frequencies $n(z+2)$, $n(z+3)$, $n(z+4)$ have been grouped. As for the frequencies $n(z+5)$, $n(z+6), \ldots$, none of these is separately available: all that is known is the number $n'(z+5)$ of observations for which the observed magnitude was equal to or greater than $z+5$. This is *censoring*. (We may similarly have censoring at the lower magnitudes; or censoring at both extremities.)

Here the likelihood is

$$l(\theta) = \prod_{y=0}^{z-1} \{f(y, \theta)\}^{n(y)}\{f(z, \theta) + f(z+1, \theta)\}^{n'(z)}\{f(z+2, \theta) + f(z+3, \theta)$$

$$+ f(z+4, \theta)\}^{n'(z+2)}\{G(z+5, \theta)\}^{n'(z+5)},$$

where $G(u, \theta) = g(u, \theta) + g(u+1, \theta) + \ldots$, the upper tail c.d.f.

In practice, data may be grouped but not censored, or censored but not grouped, or both grouped and censored; with or without truncation. In all such cases it may be taken for granted that the likelihood equation will have to be solved by numerical methods.

EXAMPLE 6.7.1.   *Poisson data with grouped cells.*   As an example of grouped discrete data consider the frequency table given in section 3.2.2. If we assume the underlying distribution to be Poisson with parameter θ, the likelihood function for θ is

$$l(\theta) = \left[\prod_{r=0}^{11} \{e^{-\theta}\theta^r/r!\}^{f_r} \right] \left\{ e^{-\theta}\left(\frac{\theta^{12}}{12!}+\frac{\theta^{13}}{13!}+\frac{\theta^{14}}{14!}\right) \right\}^2,$$

the values of $f_0, f_1, \ldots, f_{11}$ being tabulated.

In this example, the effect of the grouping (two observations in the class '12, 13 or 14') is clearly going to be small, and the maximizing value of θ may be expected to be very close to what would be obtained of these two observations both corresponded to the value '13'. This would give as our first approximation $\hat{\theta}_1$ the mean value of the doctored data, namely 3·871. The correct value of the maximum likelihood estimate is that which maximizes the function

$$\log l = \sum_{r=0}^{11} f_r \log'(e^{-\theta}\theta^r/r!) + 2 \log \left\{ e^{-\theta}\left(\frac{\theta^{12}}{12!}+\frac{\theta^{13}}{13!}+\frac{\theta^{14}}{14!}\right) \right\},$$

viz. the approximate root of the equation obtained on differentiating this with respect to θ and equating to zero. The equation to be solved is

$$-2608 + \frac{10070}{\theta} + \frac{24}{\theta}\frac{1+\theta/12+\theta^2/156}{1+\theta/13+\theta^2/182} = 0.$$

The relevant root is $\theta^* = 3\cdot867$, very close to the approximation $\hat{\theta}$, as anticipated.

6.7.2.   Continuous data: Grouping. Sheppard's Correction

Observations on a continuous random variable may be presented either 'as observed' or in the form of a grouped frequency table. Suppose the p.d.f. at x of the population in question is $f(x, \theta)$. If the observations $x_1, x_2, \ldots, x_n$ are recorded to a sufficiently high degree of accuracy they may properly be regarded as realizations of a continuous variable. Their joint sampling density at $(x_1, x_2, \ldots, x_n)$, if they are independent, will then be $\prod_1^n f(x_r, \theta)$ and this gives the likelihood of θ. In practice the degree of accuracy is always finite; an observed value of x_r purporting to be, say, 1·02, in reality corresponds to the inequality '1·015 $< x_r <$ 1·025', whence the likelihood of θ on that datum is not $f(1\cdot02, \theta)$ but $\int_{1\cdot015}^{1\cdot025} f(x, \theta) \, dx = F(1\cdot025, \theta) - F(1\cdot015, \theta)$, where $F(x)$ is the c.d.f. of the population studied.

Suppose the data from a continuous distribution to be recorded in a frequency table, with $n(r)$ as the frequency in the cell $(x_r - \frac{1}{2}h, x_r + \frac{1}{2}h)$, $r = 1, 2, \ldots, k$. Denote the p.d.f. of X at x by $f(x, \theta)$, and its c.d.f. by $F(x, \theta)$.

If one treated the data as if it had arisen from a discrete distribution, with $n(r)$ as the frequency of the observation x_r, the likelihood would be computed as

$$l_1(\theta) = \prod_{r=1}^{k} \{f(x_r, \theta)\}^{n(r)},$$

whence

$$\log l_1(\theta) = \sum n(r) \log f(x_r, \theta),$$

and the M.L.E. would be obtained as the root (θ_1, say) of the equation $\partial \log l_1(\theta)/\partial\theta = 0$.

If however one takes account of the actual nature of the data, the (correct) likelihood function is found to be

$$l(\theta) = \prod_{r=1}^{k} \{F(x_r + \tfrac{1}{2}h, \theta) - F(x_r - \tfrac{1}{2}h, \theta)\}^{n(r)},$$

which may be maximized by an appropriate computer algorithm. If however h is small enough, say $h < s/2$, we may write

$$l(\theta) = \prod \{hf(x_r, \theta) + \tfrac{1}{12}h^3 f''(x_r, \theta) + \ldots\}^{n(r)},$$

neglecting h^4 and higher powers in the Taylor expansion [see IV, § 3.6]. Here primes denote differentiation with respect to θ. Thus

$$\log l(\theta) = n \log h + \log l_1(\theta) + \frac{1}{24} h^2 \sum n(r) \frac{f''(x_r, \theta)}{f(x_r, \theta)}$$

to the same order of accuracy, whence

$$\frac{\partial \log l}{\partial \theta} = \frac{\partial \log l_1}{\partial \theta} + \frac{1}{24} h^2 \sum n(r) \frac{\partial}{\partial \theta} \left(\frac{f''(x_r, \theta)}{f(x_r, \theta)} \right).$$

If we apply this to the Normal (μ, σ) distribution, taking firstly, θ to be the expectation μ, we have

$$\partial \log l_1/\partial \mu = n\sigma^{-2}(\bar{x} - \mu).$$

Thus the grouped estimate of μ is

$$\mu_1^* = \bar{x}.$$

The corrected likelihood equation reduces to

$$n\sigma^{-2}(\bar{x} - \mu) - h^2 n(\bar{x} - \mu)/12\sigma^4 = 0$$

whence the corrected M.L.E. of μ is

$$\hat{\mu} = \bar{x}.$$

Thus, to this accuracy, $\hat{\mu} = \mu_1^*$.

Now consider the grouped estimate of σ. We have

$$\partial \log l_1/\partial \sigma = -n/\sigma + \sum n(r)(x_r - \mu)^2/\sigma^3.$$

Equating this to zero and solving simultaneously with the corresponding equation for μ gives

$$n(s^2 - \sigma^2)/\sigma^3 = 0$$

where

$$s^2 = \sum n(r)(x_r - \bar{x})^2/n.$$

Thus the grouped estimate is

$$(\sigma^2)_1^* = s^2.$$

The corrected likelihood equation reduces to

$$\sigma^2 = s^2 + \tfrac{2}{3}h^2 - \tfrac{1}{4}h^2 q^4/s^4$$

where q^4 is the grouped fourth sample moment $\sum n(r)(x_r - \bar{x})^4/n$. If we further approximate by replacing q^4 and s^4 by the corresponding population moments we have $q^4/s^4 \cong 3$, whence (to the order of accuracy employed—i.e. neglecting terms of order h^4 and above)

$$\hat{\sigma}^2 \cong s^2 - h^2/12.$$

Thus the corrected M.L.E. is obtained from the grouped estimate by deducting the quantity $h^2/12$, where h is the grouping interval. (This is known as Sheppard's adjustment.)

E.H.L.

6.8.   FURTHER READING

Maximum likelihood procedures usually reduce in the end to problems depending on the numerical inversion of matrices and the numerical solution of equations. In many cases, these numerical solutions have been tabulated. In the case of the estimation of the density of organisms by the dilution method, for example (cf. Example 6.3.8), values of the estimate and its accuracy are tabulated in Table VIII2 of Fisher and Yates (1957) [see Bibliography G]. A table for the estimation of the parameters of the Logistic distribution [see II, § 11.10] is given in Berkson (1960), etc.

The general principles, with many examples, are well documented in the references cited in section 3.6, to which should be added Fisher (1959). A useful short specialized study is to be found in Shenton and Bowman (1977).

The uses of likelihood and the method of maximum likelihood were championed and largely developed by Fisher. A one-volume reprint of his collected papers, many of which appear in the reprint with a brief contemporary (i.e. 1950) comment written by the author provides a superb historical sketch of the growth of twentieth-century statistics in general and likelihood methods in particular [Fisher (1950)—see Bibliography D]. A formal rigorous mathematical treatment of the basic properties is given in Cramér (1946) [see Bibliography C]. A modern text which uses likelihood methods almost exclusively and with clarity is Kalbfleisch (1979) [see Bibliography C]. For an authoritative treatment of the dosage-mortality problem see Finney (1971).

Finney, D. J. (1971). *Probit Analysis*, Cambridge University Press.

Berkson J (1960). Nomogram for Fitting the Logistic Function by Maximum Likelihood, *New Statistical Table No. XXIX* (*Biometrika*), Cambridge University Press.

Shenton, L. R. and Bowman, K. L. (1977), *Maximum Likelihood Estimation in Small Samples*, Griffin.

CHAPTER 7

Chi-squared. Tests of Goodness of Fit and of Independence and Homogeneity

7.1. VALIDITY OF THE MODEL

7.1.1. Introduction

In attempting to give a bird's eye view of Statistics in Chapter I, we included (in § 1.1) the concepts of formulating a probability model, obtaining estimates of its parameters, assessing the reliability of those estimates, and *verifying the validity of the model.*

Quite simple procedures are usually capable of ensuring that the model is not grossly unsuitable, but the question whether an alternative model might be better is not so easily answered. In example 6.5.1, the data displayed in Table 6.5.3 are analysed in terms of a homoscedastic Normal linear regression model. Each column is assumed to be the frequency table of a sample from a Normal distribution. The nine Normal distributions involved are assumed to have equal variances, and their expected values are assumed to be a linear function of the temperature. Plots of the cumulative frequencies of some of the columns on 'probability graph paper' [see § 3.2.2(d)] are shown in Figure 7.1.1. It is at least plausible to claim that the points lie on or near to a set of parallel lines, indicating that the assumption of Normal distributions with equal variances is not unreasonable. More refined methods may then be used to provide an objective verification of the assumption of Normality (such as the tests using third and fourth moments described in section 5.9 and Example 5.9.1 or the chi-squared test described in Example 7.4.1 or the test of equality of variances [see § 5.10.1).

As to the linearity assumption, here again a primitive first-stage test is available, viz. to plot column means against temperature, and note whether they do indeed appear to lie on or near a straight line (see Figure 6.5.2). This graphical procedure, however, is incapable of detecting a small systematic departure from linearity. For this, a more refined procedure must be used. This is described below in section 7.1.2. It is an example of a *goodness of fit* test.

355

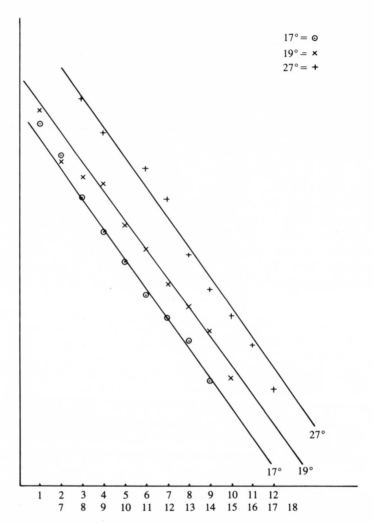

Figure 7.1.1: Graphical test of homoscedesticity by inspecting plot of empirical c.d.f.'s on probability paper for parallelism. Plots of c.d.f.'s of columns of Table 6.5.3.

Likewise, in section 6.6, a procedure is described for assessing the toxicity of an insecticide from dosage-mortality data, the analysis being carried out in terms of a response curve of a particular ('probit') kind. That the model is reasonably plausible was established by graphical evidence in Figure 6.6.3, but this kind of evidence is inherently incapable of detecting small departures from the assumed model, and an objective goodness of fit test is required. Such a test is described below in Example 7.4.2.

7.1.2.   Adequacy of a Linear Regression Model

In section 6.5.3 and in Example 6.5.1 we were concerned with the fitting of a linear regression model to data for which, conceivably, a curvilinear regression of some kind might have been more appropriate. The estimation procedures adopted took it for granted that the linear model indicated by Figure 6.5.2 was adequate. With sparser data it might have been difficult to go further: large samples are obviously required to detect small deviations. In the example actually considered the arrays are well-populated, and an objective test is available. This depends on the fact that, provided the basic assumption of homoscedasticity is valid, the 'within-arrays' sum of squares w of Table 5.8.5 provides one estimate of the variance σ^2, namely $w/(n-k) = 4\cdot708$, while the expression (6.5.11) provides another, independent, estimate. The first is valid whether or not the regression is linear, while the validity of the second depends on the existence of linearity. If therefore the two estimates are incompatible the assumption of linearity is discredited. The procedure is to break down the 'between arrays' sum of squares b ($=12{,}370$) of Table 5.8.5 into a part attributable to the linear regression and a part representing deviations from the regression. The latter is immediately available from Table 6.5.4; it is

$$\sum n_j(\bar{y}_j - \hat{\mu}_j)^2 = 395.$$

It follows that the former equals $12{,}370 - 395 = 11{,}975$. We then have the following analysis of variance table:

Source of variation	Sum of squares	d.f.	Mean square
Linear regression	11,975	1	
Deviations from the linear regression	395	7	56·4
Total (= 'between columns' sum of squares b)	12,370	8	

If we combine this with Table 5.8.5 we finally obtain the following analysis of the data:

Source of variation	Sum of squares	d.f.	Mean square
Within arrays (= random fluctuations)	3,832	814	4·708
Linear regression	11,975	1	
Deviations from the Linear regression	395	7	56·4
Total $=\sum\limits_{i,j} n_{ij}(y_i - \bar{y})^2$	16,202	822	

If the linear hypothesis were valid, the two 'mean square' entries in this table would be independent estimates of the same quantity (σ^2), and their ratio $56\cdot4/4\cdot708$ ($\simeq 12$) a realization of the F distribution on 7 and 814 degrees of freedom. The expected value of F is very nearly 1, and a value as large as 12 is very improbable indeed. The significance level is considerably less than $0\cdot1\%$ (available tables do not allow of a more precise statement) and the linear hypothesis must therefore be regarded as conclusively discredited. For a satisfactory description of the data, therefore, one would need a more elaborate model than the linear one, possibly something like

$$\mu_j = \alpha_0 + \alpha_1(x_j - \bar{x}) + \alpha_2(x_j - \bar{x})^2 + \alpha_3(x_j - \bar{x})^3.$$

7.2. KARL PEARSON'S DISTANCE STATISTIC: THE χ^2 TEST

The discussion in section 7.1 suggests that it may in suitable cases be possible to devise an objective test of the question whether or not the probability model used in an analysis is consistent with the data. One might expect that such tests, when they exist, would necessarily have to be tailored specifically to the nature of the data structures involved. Most goodness-of-fit tests carried out in practice, however, are not specific in that way; they are based on a quite general idea, invented by Karl Pearson, [Pearson (1900)] of using a statistic that provides a measure of the discrepancy between the data and the model. The data-vector, or, equivalently, the frequency table, is regarded as specifying the coordinates of a point in multidimensional space, and the discrepancy is evaluated as a generalized distance [see § (7.2.2)] between that point and the corresponding point representing the vector of expected frequencies provided by the model. The main characteristic feature of the procedure is that the sampling distribution of the distance function is in most cases excellently approximated by a χ^2 distribution [see § 2.5.4(a)]. This holds whatever the distribution prescribed by the probability model, only the number of degrees of freedom being dependent on a more detailed examination of the model and of the data structure.

The test that emerges may be exemplified by a model that prescribes a univariate probability distribution, with data in the form of a frequency table, as in the following examples.

EXAMPLE 7.2.1. *Weldon's dice data.* These are data on the scores obtained in a now classical experiment with dice. In repeated throws of 12 dice, a score exceeding 4 on any one die was regarded as a 'success', and the number of successes among the 12 dice was recorded on 26,306 occasions. These frequencies are recorded in Table 7.2.1 (from Fisher (1970) Table 10—Bibliography C) and compared with those to be expected on the assumption that each of the dice was a true one. In the latter case the probability of a success with any one die would be $\theta = \frac{1}{3}$, and the probability of r successes among 12 dice ($r = 0, 1, \ldots, 12$) would be given by the Binomial distribution [see II, § 5.2.2]:

$$p_r = \binom{12}{r}\left(\frac{1}{3}\right)^r\left(\frac{2}{3}\right)^{12-r}$$

$$= \{12!/r!(12-r)!\}2^{12-r}/3^{12}, \qquad r = 0, 1, \ldots, 12,$$

and the expected frequency of the outcome 'r successes' would be 26,036 p_r [see II, Table 6.4.2]. The discrepancies appear to be quite large. It could well be that the dice are not truly unbiased. Since the individual scores are not identified the best we can do to allow for bias is to treat the dice as if they were all equally biased, and to estimate that bias. If we denote by θ this common success-probability we find from the data the Maximum Likelihood Estimate of θ to be

$$\theta^* = \bar{r} = \sum rf_r/12\sum f_r = 0\cdot33770$$

[see Example 6.3.2]. If we recalculate the expected frequencies on the assumption that the success-parameter is the same for all dice, with the value θ^*, we obtain the expected frequencies

$$e_r = 26,306\, p_r^*$$

where

$$p_r^* = \binom{12}{r}(\theta^*)^r(1-\theta^*)^{12-r}, \qquad r = 0, 1, \ldots, 12.$$

Number of dice, out of 12 scoring more than 4 $= r$	Observed frequency $= f_r$	Expected frequencies (with $\theta = \frac{1}{3}$) $= e_r$	Discrepancy $f_r - e_r$
0	185	202·75	−17·75
1	1149	1216·50	−67·50
2	3265	3345·39	−80·37
3	5475	5575·61	−100·61
4	6114	6272·56	−158·56
5	5194	5018·05	170·05
6	3067	2927·20	139·80
7	1331	1254·51	76·49
8	403	392·04	10·96
9	105	87·12	17·88
10	14	13·07	0·83
11	4	1·19	2·81
12	0	0·05	0·05

Table 7.2.1: Observed and expected frequencies of r successes

The values are given in Table 7.2.2. This model appears to give a better fit, most (but not quite all) of the discrepancies being smaller in absolute magnitude than in Table 7.2.1.

r	Observed frequency f_r	Expected frequency with biased dice $(\theta^* = 0\cdot33770)$ $e_r = 26{,}306\, p_r^*$	Discrepancy $f_r - e_r$
0	185	187·38	−2·38
1	1149	1146·51	2·49
2	3265	3215·24	49·76
3	5475	5464·70	10·30
4	6114	6269·35	−155·35
5	5194	5114·65	79·35
6	3067	3042·54	24·46
7	1331	1329·73	1·27
8	403	423·76	−20·76
9	105	96·03	8·97
10	14	14·69	−0·69
11	4	1·36	2·64
12	0	0·06	−0·06

Reprinted with permission of Macmillan Publishing Company from *Statistical Methods for Research Workers* by Sir Ronald A. Fisher, 14th Edition (1970).

Table 7.2.2: Observed and expected frequencies of *r* successes with equally biased dice

What is evidently needed, however, is a method of combining the individual discrepancies into a single statistic which may be taken as a measure of the distance between the vector of observed frequencies and the vector of expected frequencies. This is precisely what is provided by *Karl Pearson's Goodness of Fit Statistic* x^2, defined as

$$x^2 = \sum_r \frac{(e_r - f_r)^2}{e_r} \tag{7.2.1}$$

where f_r is the observed frequency in the cell labelled 'r', and e_r is the corresponding expected frequency, and the sum is carried out over all observations.

It was shown in section 2.9.4 that the joint sampling distribution of the f_r is Multinomial (k), where k denotes the number of cells ($k = 13$ in Table 7.2.2), with index $n = \sum f_r$ and probability parameters (in this case) $p_0, p_1, \ldots, p_{12}$ where p_r denotes the probability of an observation falling in the cell 'r', $r = 0, 1, \ldots, 12$. Thus

$$e_r = np_r$$

The sampling distribution of x^2 may be shown [see Cramér (1946), §§ 30.1, 30.3—Bibliography C] to be approximately that of the sum of squares of $k - s$ standard Normals (where s is the number of parameters fitted). This approximation is derived from the fact that, just as the Binomial may under appropriate circumstances be approximated by the Normal, so may the Multinomial be approximated by the Multivariate Normal. The approximations are remarkably robust, but they do require that the e_r should not be too small. (How small? See § 7.3.) When the expected frequency in a cell falls below a critical value (see § 7.3) that cell must be amalgamated with one or more of its immediate neighbours so as to bring the amalgamated expected frequency above the forbidden level; and the observed frequencies must be correspondingly amalgamated.

Evidently the statistic will have a small value when the agreement between observations and expectation is good, and a large value when the agreement is poor. The question whether a given value is small enough to be attributable to chance fluctuations, indicating that the agreement is acceptable, has to be discussed in terms of the sampling distribution of the statistic x^2 of (7.2.1).

Sampling distribution of Karl Pearson's statistic

The sampling distribution of x^2 is (approximately) *the chi squared distribution*, with degrees of freedom ν given by the following rule:

$$\nu = a - b - 1 \tag{7.2.2}$$

where

> a is the effective number of entries in the 'observed frequencies' column (counting each group of amalgamated frequencies as a single entry),

and

> b is the number of parameters of the probability model that have had to be estimated from the data.

The test is, for this reason, called (Karl Pearson's) χ^2 test of goodness of fit.

It follows from the definition, and this is illustrated in the examples that follow, that, for $\nu > 1$, x^2 lumps together all discrepancies between observation and expectation, whatever their nature. A departure from the hypothesis being tested will always tend to increase the value of x^2, and the significance level is therefore always calculated by the single-tail (one-sided) procedure [see § 5.2.3] as

$$S.L. = P(\chi^2_\nu \geq x^2).$$

The case $\nu = 1$ is a special one to which the above does not apply. It is discussed separately below, in section 7.4.1.

Nomenclature

Readers are warned that Pearson's statistic, which we have called x^2, is often—indeed usually—called Pearson's χ^2, or, simply, χ^2. We thus have the confusing situation that a statistic that is called χ^2 has a sampling distribution that is *approximately* the χ^2 distribution. In this book we attempt to avoid this ambiguity by reserving the symbol χ^2 to denote the random variable having the chi squared distribution, or the name of that distribution.

EXAMPLE 7.2.2.   *Pearson's statistic for the dice data of Example* 7.2.1.   As an example we present the computation of x^2 for the dice data of Example 7.2.1, firstly on the hypothesis of true dice and secondly on the hypothesis of equally biased dice with $\theta^* = 0{\cdot}33770$.

In the case of the true dice hypothesis, the expected frequencies $e_{11} = 1{\cdot}19$ and $e_{12} = 0{\cdot}05$ (see Table 7.2.2) may be too small to allow the χ^2 distribution to be used as a sufficiently accurate approximation to the sampling distribution of x^2. (Advice on acceptably small expected frequencies is given below.) The sum

$$e_{10} + e_{11} + e_{12} = 13{\cdot}07 + 1{\cdot}19 + 0{\cdot}05 = 14{\cdot}31$$

is however large enough, and that is why the observed frequencies $f_{10} + f_{11} + f_{12} = 18$ have been amalgamated in Table 7.2.3. The amalgamated total of 18 has been compared with the correspondingly amalgamated expectations $e_{10} + e_{11} + e_{12} = 14{\cdot}31$ to make the contribution $(18 - 14{\cdot}31)^2 / 14{\cdot}31 = 0{\cdot}952$. This contribution to the value of x^2 is shown as the last entry in the penultimate column of Table 7.2.3. Thus although the original frequency table contained 13 entries in the frequency column (namely $f_0, f_1, \ldots, f_{12}$), amalgamation has reduced the effective number to 11, namely $f_0, f_1, \ldots, f_9$ and $(f_{10} + f_{11} + f_{12})$. The value of x^2 is 35·491.

In the 'true dice' situation under discussion the value of θ is specified by the hypothesis $(\theta = \frac{1}{3})$ and no parameter has had to be estimated from the data. The number of degrees of freedom is thus $\nu = '11 - 1 = 10$.

The significance level of the x^2 statistic with respect to the 'true dice' hypothesis is thus, approximately,

$$P(\chi^2_{10} \geq 34{\cdot}491).$$

This is less than $0{\cdot}005 = 0{\cdot}5\%$: the evidence against the hypothesis is therefore very strong indeed.

In the case of the 'equally biased dice' hypothesis, reference to Table 7.2.2 again indicates that $e_{11} = 1{\cdot}36$ and $e_{12} = 0{\cdot}06$ are too small for safety in using the χ^2 approximation, and amalgamation of the last three frequencies is again advised. The value of x^2 turns out to be 8·179, and this time the number of degrees of freedom is

$$\nu = 11 - 1 - 1 = 9$$

No. of successes out of 12	Observed frequency	$(e_r - f_r)^2 / e_r$	
r	f_r	True dice $\theta = \frac{1}{3}$	Biased dice $\theta^* = 0.33770$
0	185	1·554	0·030
1	1149	3·745	0·005
2	3265	1·931	0·770
3	5475	1·815	0·019
4	6114	4·008	3·849
5	5194	6·169	1·231
6	3067	6·677	0·197
7	1331	4·664	0·001
8	403	0·306	1·017
9	105	3·670	0·838
10⎫ 11⎬amalgamated 12⎭	14⎫ 4⎬ 0⎭	$\left\{0·952\right\}$	$\left\{0·222\right\}$
		$x^2 = 35·491$ $\nu = 10$	$x^2 = 8·179$ $\nu = 9$

Table 7.2.3

since one parameter (θ) has been estimated from the data. The significance level of the statistics with respect to the hypothesis of equally biased dice is thus

$$P(\chi_9^2 \geq 8·179).$$

This is about 0·5, a very large probability. The data must therefore be regarded as consistent with the hypothesis.

EXAMPLE 7.2.3.   *Testing the fit of a Poisson distribution.*   The data in Table 3.2.1 give the frequencies with which 0, 1, 2, . . . particles were emitted from a certain radio-active source in intervals of duration 7·5 seconds. The probability model, dictated by the physics of radio-activity, attributes to R, the number of particles emitted in a randomly chosen $7\frac{1}{2}$-second interval, the Poisson probability distribution

$$P(R = r) = \pi_r(\theta) = e^{-\theta}\theta^r / r!, \qquad r = 0, 1, \ldots$$

[see II, § 20.1], where θ is proportional to the intensity of radio-activity of the specimen. The value of θ may be estimated from the data, the maximum-likelihood estimate being $\theta^* = 3·867$ [see Example 6.7.1]. The estimated expected frequency of the observation r ($r = 0, 1, 2, \ldots$) would then be $n\pi_r(\theta^*)$ where n ($= 2608$) is the sample size. The numerical values are given in Table 7.2.4, the frequencies f_{12}, f_{13} and f_{14} having been amalgamated as $f_{12} + f_{13} + f_{14} = 2$ in the recording of the experimental results.

No. of particles emitted r	Observed frequency f_r	Estimated expected frequency e_r	$(f_r - e_r)^2/e_r$
0	57	54·56	0·109
1	203	210·99	0·303
2	383	407·95	1·526
3	525	525·85	0·001
4	532	508·37	1·099
5	408	393·17	0·560
6	273	253·40	1·516
7	139	139·98	0·007
8	45	67·66	7·589
9	27	29·07	0·148
10	10	11·24	0·137
11	4	3·95	0·001
12–14	2	1·76	0·001
Pearson's statistic			15·355

Table 7.2.4

It is often stated in texts that the smallest expected frequency that may safely be used without invalidating the χ^2 approximation is about 5; if this advice is adopted the entries in the last two rows of the table ought to be amalgamated to give

r		f_r	e_r	$(f_r - e_r)^2/e_r$
$\left.\begin{array}{l}11\\12\text{–}14\end{array}\right\}$ amalgamated		{6}	{5·71}	0·015

leading to

$$x^2 = 15\cdot368$$

with degrees of freedom

$$\nu = 12 - 1 - 1$$
$$= 10,$$

there being now 12 effective frequencies, namely $f_0, f_1, \ldots, f_{10}$ and $(f_{11} + f_{12} + f_{13} + f_{14})$.

The significance level of this value of x^2 with respect to the hypothesis that the underlying distribution is a Poisson one is

$$P(\chi^2_{10} \geq 15\cdot368) \approx 0\cdot11.$$

This is a large probability, and the data must be regarded as upholding the Poisson hypothesis.

According to authoritive investigations by W. G. Cochran, however, the recommendation that expected frequencies should not be less than five is unduly restrictive, and, for a distribution of the kind we are dealing with, a minimum of one is safe enough [see § 7.3]. In our example therefore no amalgamation would be required beyond that already present in Table 7.2.4. The value of x^2 is then 15·355, with $\nu = 13 - 1 - 1 = 11$; and since

$$P(\chi^2_{11} \geq 15 \cdot 355) > 0 \cdot 1$$

the data are clearly consistent with the Poisson hypothesis; the same result as before.

7.3.   AMALGAMATING LOW FREQUENCY CELLS: COCHRAN'S CRITERIA

The attractiveness, to the user, of Pearson's statistic depends on its sampling distribution having the widely-tabulated χ^2 form, to a sufficiently close approximation. Naturally certain precautions must be observed for this to be the case. Of these the most important are the following

(i) when expected frequencies are based on estimated values of parameters, the estimates must be obtained by an efficient procedure such as maximum likelihood; and

(ii) the expected frequencies must not be too small.

Just how small the expected frequencies may be without invalidating the χ^2 approximation is not an entirely simple question to answer. As was mentioned above, many texts quote 5 (or even 10) as the smallest safe value, advocating the amalgamation of adjoining cells when necessary to achieve the recommended minimum. Whilst such conservative practices undoubtedly reduce the danger of invalidating the χ^2 approximation, they also, unfortunately, have the effect of reducing the sensitivity of the test, a consequence which could be an important one if the resulting number of effective cells were small. In the interests of sensitivity one should avoid cell-amalgamation except when really necessary.

The advice of the eminent statistician W. G. Cochran, based on long experience, is as follows [Cochran (1952), (1954)]:

Amalgamating cells in a frequency table when carrying out a χ^2 test:
   With unimodal distributions, where expected frequencies will be small only in the tails, arrange matters so that the minimum expectation in each tail is at least one.

7.4.   THE χ^2 TEST FOR CONTINUOUS DISTRIBUTIONS

The principles involved in working with data from a continuous distribution, presented as a frequency table, are identical with those applied in the discrete case. Some additional computational difficulties however may arise if the grouping interval is coarse.

The estimation of parameters from a frequency table of (rounded) continuous data has been discussed in section 6.7.2. Similar considerations arise in the computation of expected frequencies. If the data are grouped in 'cells' $x_r \pm \frac{1}{2}h$, $r = 1, 2, \ldots, k$, and the hypothesis to be tested is that the p.d.f. of the underlying random variable at x is $f(x; \theta)$, with c.d.f. $F(x; \theta)$, the expected frequency (on this hypothesis) in the rth cell is estimated as

$$n\{F(x_r + \tfrac{1}{2}h; \theta^*) - F(x_r - \tfrac{1}{2}h; \theta^*)\},$$

θ^* being the Maximum Likelihood Estimate of θ (see § 6.7.2) and n the sample size. Unless h is large it is usually sufficiently accurate to replace this by the simpler expression $nhf(x_r; \theta^*)$.

As regards the smallest safe frequencies it is recommended that one should adhere to:

Cochran's criteria for χ^2 tests with continuous data:

> Amalgamate cells at the tail(s), if necessary, so that the smallest expected frequency is not less than 1. To maximize the sensitivity of the test, the size of the grouping interval should be small enough to keep the expectations down to 12 per cell for a sample of size $n = 200$, 20 per cell for $n = 400$, and 30 per cell for $n = 1000$.

[Cochran (1952), (1954)].

EXAMPLE 7.4.1. χ^2 *test for Normality.*   A sample of rainfall data (reproduced in Table 7.4.1) was discussed in Example 5.9.1 and tested for skewness and kurtosis. The sample estimates of the coefficients of skewness and of kurtosis did not differ significantly from the values appropriate to a Normal distribution. We now submit the same data to a χ^2 test, tentatively adopting the working hypothesis that the data do in fact come from a Normal (μ, σ) distribution.

The sample mean was found in Example 5.9.1 to be 28·62 inches; this is the Maximum Likelihood Estimate of the expected value μ of the presumed Normal distribution. The sample second moment about the mean was found to be $m_2 = 23 \cdot 013$ in². Applying Sheppard's correction for grouping (see § 6.7.2) we find the estimate of σ^2 to be

$$23 \cdot 013 - 1/12 = 22 \cdot 930 = (4 \cdot 788)^2,$$

whence our estimate of σ is 4·788. The expected frequencies are therefore those appropriate to a Normal (28·62, 4·788) distribution. The expected frequency in the cell centred on x_r inches will therefore be

$$90\left\{\Phi\left(\frac{x_r + \frac{1}{2} - 23 \cdot 013}{4 \cdot 788}\right) - \Phi\left(\frac{x_r - \frac{1}{2} - 23 \cdot 013}{4 \cdot 788}\right)\right\} \approx 90\phi \frac{(x_r - 23 \cdot 013)}{4 \cdot 788} \qquad (7.4.1)$$

Rainfall	Observed frequency	Rainfall	Observed frequency
16	1 ⎫	29	8 ⎫ (17)
17	0 ⎪ (4)	30	9 ⎭
18	0 ⎪		
19	3 ⎭	31	6 ⎫ (13)
		32	7 ⎭
20	2 ⎫		
21	3 ⎬ (5)	33	4 ⎫ (8)
22	0 ⎭	34	4 ⎭
23	3 ⎫ (5)	35	4 ⎫ (7)
24	2 ⎭	36	3 ⎭
25	12 ⎫ (16)	37	3 ⎫
26	4 ⎭	38	0 ⎬ (4)
		39	1 ⎭
27	7 ⎫ (11)		
28	4 ⎭	Total	90

Table 7.4.1

to an acceptable accuracy. To illustrate how the calculation proceeds with coarse and not necessarily constant cell widths we amalgamate cells as indicated in Table 7.4.1 and evaluate the expected frequencies in the manner shown in Table 7.4.2.

Boundary value x	$\dfrac{x-28\cdot62}{4\cdot788}=u$	$\Phi(u)$	Difference	$\times 90$
<16·5				0·513
16·5	−2·531	0·0057	0·0217	1·952
19·5	−1·905	0·0284	0·0722	6·498
22·5	−1·278	0·1006	0·0943	8·487
24·5	−0·860	0·1949	0·1340	12·060
26·5	−0·443	0·3289	0·1611	14·499
28·5	−0·025	0·4900	0·1628	14·652
30·5	+0·393	1−0·3472	0·1382	12·438
32·5	0·810	1−0·2090	0·0995	8·955
34·5	1·228	1−0·1095	0·0596	5·364
36·5	1·646	1−0·0499	0·0375	3·375
39·5	2·272	1−0·0114	0·0114	1·026
				89·8

Table 7.4.2

Table 7.4.3 shows the observed frequencies (in the amalgamated cells), the corresponding expected frequencies, the further amalgamations required to avoid very small expected frequencies in the tails, and finally the value of x^2.

x	Observed frequency n_r	Expected frequency e_r	$(n_r - e_r)^2/e_r$
< 19.5	4	2·463	0·959
19·5–22·5	5	6·498	0·345
22·5–24·5	5	8·487	1·433
24·5–26·5	16	12·060	1·287
26·5–28·5	11	14·499	0·844
28·5–30·5	17	14·652	0·376
30·5–32·5	13	12·438	0·025
32·5–34·5	8	8·955	0·102
34·5–36·5	7	5·364	0·499
36·5–39·5	4	3·375	0·116
> 39.5	0	1·026	1·026
			7·487

Table 7.4.3

The value of x^2 is 7·487. The number of degrees of freedom is

$$11 - 1 - 2 = 8.$$

The significance level is $P(\chi_8^2 \geq 7 \cdot 487) > 0 \cdot 40$. This very high probability shows that the Normal hypothesis is consistent with the data.

EXAMPLE 7.4.2. χ^2 *test for the model used in the dosage-mortality determination of section* 6.6.   In section 6.6 we were concerned with the estimation of the toxicity of an insecticide by a method which assumed that the probability $\pi(x)$ of an insect's succumbing after exposure to a dose measured in terms of

$$\log (\text{dose}) = x$$

was given by

$$\pi(x) = \Phi(\alpha + \beta x),$$

where Φ denoted the standard Normal integral. The Maximum Likelihood Estimates of α and β were found to be

$$\alpha^* = -2 \cdot 862, \qquad \beta^* = 4 \cdot 175$$

[see (6.6.10)]. The numbers of insects involved, and the number killed at the various dosages (given in Table 6.6.1) together with the estimated probabilities $\pi^*(x) = \Phi(\alpha^* + \beta^* x)$ are as follows:

x_j	No. of insects n_j	No. killed r_j	Probability of kill $\pi_j = \pi(x_j)$	Expected number killed $e_j = n_j \pi(x_j)$
1·01	50	44	0·9123	45·62
0·89	49	42	0·8040	39·40
0·71	46	24	0·5488	25·25
0·58	48	16	0·3300	15·84
0·41	50	6	0·1251	6·26

Each row records the number of observed successes (r_j) in a Binomial (n_j, π_j) distribution, for which the expected number of successes is $n_j \pi_j$; implicitly it also records the number of failures ($n_j - r_j$), with expected frequency $n_j(1 - \pi_j)$. The contribution of that row to the total value of Pearson's x^2 is

$$x_j^2 = \frac{(r_j - n_j\pi_j)^2}{n_j\pi_j} + \frac{(n_j - r_j - n_j + n_j\pi_j)^2}{n_j(1 - \pi_j)}$$

$$= \frac{(r_j - n_j\pi_j)^2}{n_j}\left\{\frac{1}{\pi_j} + \frac{1}{1 - \pi_j}\right\}$$

$$= (r_j - n\pi_j)^2 / n_j\pi_j(1 - \pi_j),$$

the sampling distribution of which is (approximately) χ^2. If π_j were known, without the involvement of estimated parameters, the number of degrees of freedom would be 1 (2 cells, 0 parameters; $2 + 0 - 1 = 1$). Thus, for the whole set of data,

$$x^2 = \sum_1^5 x_j^2 = \sum_1^5 (r_j - n_j\pi_j)^2 / n_j\pi_j(1 - \pi_j),$$

the sampling distribution of which is χ^2. The five contributions $x_1^2, x_2^2, \ldots, x_5^2$ make up in all 5 d.f., but now we can allow for the two parameters that were estimated (α and β), so that finally, the number of d.f. is 3. Thus the significance level of the data with respect to the hypothesis that

$$\pi(x) = \Phi(\alpha + \beta x)$$

is

$$P(\chi_3^2 \geq x^2) = P(\chi_3^2 \geq 1\cdot68)$$

$$> 0\cdot50.$$

This very large value indicates that the Φ hypothesis is consistent with the data.

7.5. CROSS-CLASSIFIED FREQUENCY TABLES (CONTINGENCY TABLES). TESTS OF INDEPENDENCE AND OF HOMOGENEITY

7.5.1. 2×2 Tables; the Special Case of a Single Degree of Freedom

A typical '2×2' table [see also § 5.4.2] is shown in Table 7.5.1, which gives data (due to Greenwood and Yule) on the numbers of people in a given community who were attacked by cholera, and the numbers not attacked, cross-classified into those who had been inoculated against the disease and those who had not.

	Not attacked	Attacked	Total
Inoculated	1625	5	1630
Not inoculated	1022	11	1033
Total	2647	16	2663

Table 7.5.1: Effect of inoculation on cholera infection. (Reprinted with permission of Macmillan Publishing Company from *Statistical Methods for Research Workers*, 14th edition, by Sir Ronald A. Fisher. Copyright © 1970 University of Adelaide.)

The four entries in the body of the table, namely 1625, 5, 1022, 11, are a set of frequencies, and what we have is a frequency table arranged as a square instead of in the more usual columnar form. This frequency table is, in principle, amenable to the χ^2 test of agreement with a proposed hypothesis.

There are however certain characteristic features of 2×2 tables which merit special attention, related to

(i) the need in certain cases for a 'continuity correction' to alleviate the strain placed upon the continuous χ^2 approximation by the discreteness of the data and its exact sampling distribution, and

(ii) the fact that, for 2×2 tables, a straightforward one-sided χ^2 test becomes a two-sided test of the discrepancy between observation and expectation.

These features are elaborated below.

The null hypothesis

The question raised by Table 7.4.1 is: does inoculation significantly affect the chance of being infected? We adopt, tentatively, the 'null hypothesis' that the inoculation has no such effect and that the advantage that appears to derive from inoculation is the result of chance fluctuations. We must therefore compare the entries in the table with the corresponding entries that would be expected if the hypothesis were true.

The expected frequencies

The hypothesis implies that, of the 2663 people at risk, the expected proportion infected after inoculation will be the same as the expected proportion infected among the non-inoculated subjects, the common value of this proportion being the proportion infected in the whole sample, namely $p = 16/2663$. These expected proportions are set out in Table 7.4.2. Clearly, we are identifying the idea of the *independence* of the propositions

   A: a randomly chosen member of the inoculated set will be infected

and

   B: a randomly chosen member of the non-inoculated set will be infected,

   with the idea of *homogeneity* among the expected proportions:

	Not attacked	Attacked	(Total numbers)
Inoculated	$1-p$	p	(1630)
Not inoculated	$1-p$	p	(1033)
(Total numbers)	(2647)	(16)	(2663)
	$1-p=\dfrac{2647}{2663}$	$p=\dfrac{16}{2663}$	

Table 7.5.2: Expected proportions infected on the hypothesis that inoculation has no effect.

The expected frequency in any cell, on the null hypothesis, can now be filled in by multiplying the proportion (p or $1-p$) by the marginal total of the relevant row (1630 for the inoculated category, 1033 for the others). This leads to the following table for the expected frequencies:

$1630(1-p)=1620\cdot21$	$5p=9\cdot79$	1630
$1033(1-p)=1026\cdot79$	$11p=6\cdot21$	1033
2647	16	

Table 7.5.3: Expected frequencies corresponding to Table 7.4.2.

Only one entry need be evaluated by multiplying the expected proportion by the marginal frequency, since the others may be filled in by subtraction.

The value of χ^2, (without the continuity correction)

Tabulating the observed and expected frequencies together we have the following figures, entered in each cell as

                    Observed frequency
                    (expected frequency)
                    [difference]

1625	5
(1620·21)	(9·79)
[4·79]	[−4·79]

1022	11
(1026·79)	(6·21)
[−4·79]	[4·79]

Table   7.5.4:   Observed   frequencies, expected frequencies
and discrepancies.

Thus the value of Pearson's statistic, in a straightforward calculation, is

$$x^2 = \frac{4 \cdot 79^2}{1620 \cdot 21} + \frac{4 \cdot 79^2}{9 \cdot 79} + \frac{4 \cdot 79^2}{1026 \cdot 79} + \frac{4 \cdot 79^2}{6 \cdot 21} = 6 \cdot 07. \tag{7.5.1}$$

The number of degrees of freedom is $\nu = 1$

The computation of the expected frequencies has implicitly involved the estimation of *two* parameters since, for example, the probability p_{11} that an inoculated person will not be attacked is equal (on the null hypothesis) to the product $p_1.p_{.1}$, where $p_1.$ is the probability that a randomly chosen person will have been inoculated (estimated as $p_1^* = 1630/2663$) and $p_{.1}$ is the probability that a randomly chosen person will not be attacked (estimated as $p_{.1}^* = 2647/2663$). The estimated expected frequency in the 'inoculated, not attacked' cell will then be $2663 p_1^*.p_{.1}^* = 1630 \times 2647/2663$. Since the expectations must add up in accordance with the observed marginal totals, this single calculation suffices to determine the expected entries in all four cells, and no further parameters are required. [See § 5.4.2.]

The number of degrees of freedom is then

$$\nu = \binom{\text{number of}}{\text{cells}} - 1 - \binom{\text{number of}}{\text{parameters}}$$

$$= 4 - 1 - 2$$

$$= 1.$$

The continuity correction

In the exact test of the independence hypothesis described in Examples 5.4.1 and 5.4.4, the significance level is the sum of the probabilities (on the null hypothesis) of getting the 2×2 table actually observed and of all more extreme tables which have the same marginal totals. To see what is meant by 'more extreme' tables consider the following. In our example the frequencies that are to be expected on the null hypothesis [see Table 7.4.3] are

$$1620 \cdot 21 \qquad 9 \cdot 79$$
$$1026 \cdot 79 \qquad 6 \cdot 21,$$

while the observed frequencies are

$$1625 \qquad 5$$
$$1022 \qquad 11.$$

5 attacked after inoculation is a small frequency (i.e. smaller than expectation), and the more extreme cases referred to would be those presented by tables with 4, 3, 2, 1 and 0 attacked after inoculation. Preserving the marginal totals, these tables form the set

$$\begin{array}{ccccc}
1625 \;\; 5 & 1626 \;\; 4 & 1627 \;\; 3 & 1629 \;\; 1 & 1630 \;\; 0 \\
1022 \;\; 11 & 1021 \;\; 12 & 1020 \;\; 13 & 1018 \;\; 15 & 1017 \;\; 16
\end{array} \qquad (7.5.2)$$

As there is only one degree of freedom we need pay attention to only one entry in each table, say the one in the upper right-hand corner, representing the 'attacked after inoculation' category. The significance level, in relation to the independence hypothesis, is then

$$p(5) + p(4) + \ldots + p(0)$$

where $p(r)$ is the probability on that hypothesis of obtaining r as the number attacked after inoculation, $r = 0, 1, \ldots, 5$. The exact test of Example 5.4.4 obtains the probabilities $p(r)$ and adds them up. In the χ^2 approximation the sampling distribution is continuous and the sum must be replaced by an integral. We may denote the separate probabilities $p(r)$ by the ordinates shown in Figure 7.5.1, or, equivalently, by the rectangles shown in Figure 7.5.2. These rectangles have unit width and so the sum of the $p(r)$ is numerically equal to the total area of the rectangles, that is by the integral between 5·5 and −0·5 of the step-function that bounds them. The rectangle corresponding to $p(0)$ has a very small area and so it does not matter in practice whether the integral runs from 5·5 to −0·5 or from 5·5 to 0: the important point is that its right hand end point is 5·5, not 5. It was pointed out by Frank Yates [see Yates (1934)] that the effect of this could be achieved by replacing the observed table

$$\begin{array}{cc}
1625 & 5 \\
1022 & 11
\end{array}$$

by a modified table

$$\begin{array}{cc}
1624 \cdot 5 & 5 \cdot 5 \\
1022 \cdot 5 & 10 \cdot 5
\end{array} \qquad (7.5.3)$$

in which the '5' had been increased to 5·5, and all the other entries correspondingly changed in such a way as to preserve the marginal totals. In this modification the expected frequencies remain unchanged, and the procedure may be

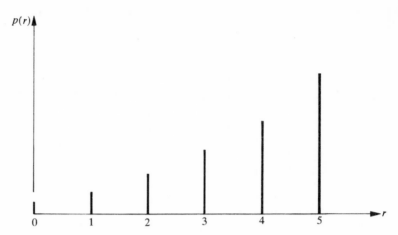

Figure 7.5.1: Ordinates representing $p(0)$, $p(1)$, ..., $p(5)$. The sum of these ordinates is the significance level of the data in relation to the independence hypothesis.

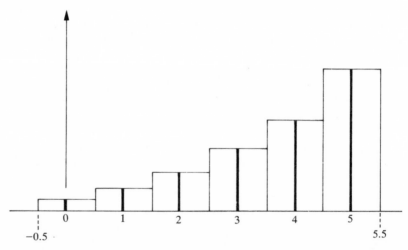

Figure 7.5.2: Rectangles of unit width, with heights equal to the ordinates $p(0)$, $p(1)$, ..., $p(5)$ of Figure 7.5.1. The sum of the ordinates is numerically equal to the sum of the areas of the rectangles, that is the integral of the bounding step-function between -0.5 and 5.5.

summarized as follows: it is known as

Yates's continuity correction for a χ^2 test in 2×2 tables:

Before calculating χ^2, reduce by 0.5 the absolute value of the difference between each of the observed and expected frequencies.

In our case this results in our reducing the absolute value (4·79) of these discrepancies as obtained in Table 7.5.4 to the modified value 4·29, the table thus being modified to the following:

$$\begin{array}{cc} 4 \cdot 29 & -4 \cdot 29 \\ -4 \cdot 29 & 4 \cdot 29. \end{array}$$

The modified value of x^2 is

$$(4 \cdot 29)^2 \left(\frac{1}{1620 \cdot 1} + \frac{1}{9 \cdot 79} + \frac{1}{1026 \cdot 79} + \frac{1}{6 \cdot 21} \right) = 4 \cdot 873. \qquad (7.5.4)$$

Significance level

Because the discrepancies between the expected frequencies and the continuity-corrected observed frequencies are squared in computing Pearson's statistic x^2, the value $x^2 = 4 \cdot 873$ obtained in (7.5.4) would equally have arisen had it been calculated from the table

$$\begin{array}{cc} 1615 \cdot 92 & 11 \cdot 08 \\ 1031 \cdot 08 & 1 \cdot 92 \end{array}$$

which has the same marginal totals and, therefore, the same expected frequencies, but in which each discrepancy has had its sign reversed as compared with the actual continuity-corrected frequencies

$$\begin{array}{cc} 1624 \cdot 5 & 5 \cdot 5 \\ 1022 \cdot 5 & 10 \cdot 5 \end{array}$$

of (7.4.3). If, therefore, we took as our significance level the probability

$$P(\chi_1^2 \geq 4 \cdot 873)$$

we should be taking into account not just the sum P_1 of the probabilities of the observed table and all more extreme tables (7.4.2), but also the sum P_2 of the probabilities of tables having

$$15 \cdot 08, \qquad 16 \cdot 08, \qquad 17 \cdot 08, \qquad 18 \cdot 08, \qquad 19 \cdot 08$$

in their 'attacked after inoculation' cell. This would distort the significance level, since what we want is the sum P_1 only.

To separate this out we note that χ_1^2 is equivalent to the square of a standard Normal variable U [see § 2.5.4(a)], whence

$$P(\chi_1^2 \geq x^2) = P(U^2 \geq x^2)$$
$$= P(U \geq |x|) + P(-U \leq -|x|)$$
$$= P(U \geq |x|) + P(U \leq -|x|)$$
$$= 2P(U \geq |x|)$$

since U has the same distribution as $-U$. Here the term $P(U \geq |x|)$ corresponds to one set of discrepancies and $P(U \le -|x|)$ to the set with reversed signs. Thus our significance level is

$$SL = P(U \ge |x|)$$

$$= \tfrac{1}{2}P(\chi_1^2 \ge x^2).   \qquad (7.5.5)$$

In our example, x^2 was found in (7.5.2) to be $4 \cdot 873 = (2 \cdot 207)^2$, Thus

$$SL = P(U \ge 2 \cdot 207)$$

$$= 0 \cdot 014.$$

(The exact value found in Example 5.4.4 was $0 \cdot 015$.) The conclusion is that the independence hypothesis is discredited: the inoculation does have some preventive effect.

Cochran's criteria for the validity of the χ^2 method of testing 2×2 tables:

Cochran's advice for 2×2 tables is as follows [Cochran (1952), (1954)]:

If the sum of the four frequencies is less than 20, use Fisher's exact test. [See § 5.4.2.] Likewise, if the sum is between 20 and 40, and the smallest expectation is less than 5, use Fisher's exact test. If the sum is 40 or above, the χ^2 test may safely be used, provided the *continuity correction* is employed.

The following example shows that even this advice may be on the conservative side. In the example, the sum of the four frequencies is only 30, but, all the same, the χ^2 procedure (with continuity correction) gives an acceptable approximation to the exact result obtained in Example 5.4.5.

EXAMPLE 7.5.1.   *Criminality and twinning.*   The data refer to a set of 30 male criminals, each of whom had a twin brother. The 30 subjects were cross classified (a) according to the nature of the twinning (identical or fraternal) and (b) according to the criminality or otherwise of the brother. The figures were as shown in Table 7.5.5.

	Brother convicted	Brother not convicted	
Identical twin	10	3	13
Fraternal twin	2	15	17
	12	18	30

Table 7.5.5

A straightforward computation of the expected frequencies, assuming no relation between the nature of the twinning and the criminality of the twin, leads to the following expected frequencies:

5·2	7·8	13
6·8	10·2	17
12	18	30

Pearson's statistic, calculated directly from these figures, is

$$(4\cdot8)^2\left(\frac{1}{5\cdot2}+\frac{1}{7\cdot8}+\frac{1}{6\cdot8}+\frac{1}{10\cdot2}\right),$$

Application of the *continuity correction*, which reduces the absolute value of the discrepancy between the observed and expected frequency in a cell by 0·5, leads to the corrected value of the statistic as

$$x^2 = (4\cdot3)^2\left(\frac{1}{5\cdot2}+\frac{1}{7\cdot8}+\frac{1}{6\cdot8}+\frac{1}{10\cdot2}\right) = 10\cdot46.$$

Clearly, a one-sided test is called for, and the corresponding significance level is therefore

$$\tfrac{1}{2}P(\chi_1^2 \geq 10\cdot46) = 0\cdot0006.$$

The result compares well with the value 0·0005 obtained by the exact test. It is highly significant and decisively discredits the null hypothesis.

7.5.2. k × m tables

In section 7.4.1 we discussed contingency tables based on a double dichotomy, each individual in the sample being classified as (a) belonging to category A or not, and also (b) belonging to category B or not. When the first dichotomy is replaced by a subdivision into k categories, and the second by m $(k, m \geq 2)$ we obtain a so-called $k \times m$ contingency table:

	B_1	B_2	$\ldots$	B_j	$\ldots$	B_m	Total
A_1	n_{11}	n_{12}	$\ldots$	n_{1j}	$\ldots$	n_{1m}	r_1
A_2	n_{21}	n_{22}	$\ldots$	n_{2j}	$\ldots$	n_{2m}	r_2
$\vdots$							$\vdots$
A_i	n_{i1}	n_{i2}	$\ldots$	n_{ij}	$\ldots$	n_{im}	r_i
$\vdots$							$\vdots$
A_k	n_{k1}	n_{k2}	$\ldots$	n_{kj}	$\ldots$	n_{km}	r_k
Total	c_1	c_2	$\ldots$	c_j	$\ldots$	c_m	t

Here for example the data might be a sample of school-children classified according to hair colour (coded as $A_1, A_2, \ldots, A_k$) and also according to quality of teeth (coded as $B_1, B_2, \ldots, B_m$), there being n_{ij} children with hair of category A_i and teeth of category B_j. On the null hypothesis of independence (i.e. that hair colour does not affect tooth quality and is not affected by it) the expected frequency corresponding to the observed frequency n_{ij} is $e_{ij} = r_i c_j / t$, and Pearson's statistic is

$$x^2 = \sum_i \sum_j \frac{(n_{ij} - e_{ij})^2}{e_{ij}}.$$

The sampling distribution of this, when the null hypothesis is correct, is approximately that of χ^2 on ν degrees of freedom, where

$$\nu = (k-1)(m-1),$$

and the significance level of x^2 is

$$P\{\chi^2 \geq x^2\}.$$

$2 \times k$ tables

In the case of $2 \times k$ (or $k \times 2$) contingency tables, independence is equivalent to *homogeneity*, as in the (2×2) table discussed in section 7.4.1. Thus, with data as follows:

		Tooth quality				
		1	2	...	k	Total
Hair colour	Fair	n_{11}	n_{12}	...	n_{1k}	r_1
	Dark	n_{21}	n_{22}	...	n_{2k}	r_2
Total		c_1	c_2	...	c_k	t

independence of tooth quality and hair colour is equivalent to the homogeneity of expected proportions, as exhibited in the following array

	1	2	...	k
Fair	p	p	...	p
Dark	$1-p$	$1-p$	...	$1-p$

where $p = r_1/t$. The corresponding expected frequencies will be

r_1c_1/t	r_1c_2/t	$\ldots$	r_1c_k/t
r_2c_1/t	r_2c_2/t	$\ldots$	r_2c_k/t

Thus, to test the hypothesis that the proportion of 'successes' is the same in each of k populations, one may carry out a χ^2 test based on the statistic

$$x^2 = \sum_j \frac{(n_{1j} - r_1c_j/t)^2}{r_1c_j/t} + \sum_j \frac{(n_{2j} - r_2c_j/t)^2}{r_2c_j/t},$$

taking this to be a realization of χ^2 on $k-1$ d.f.

7.6.   INDEX OF DISPERSION

7.6.1.   Index of Dispersion for Binomial Samples

In the special case where the c_r are all equal (with common value c, say) the table may be written in the form

x_1	x_2	$\ldots$	x_k	$k\bar{x}$
$c - x_1$	$c - x_2$	$\ldots$	$c - x_k$	$k(c - \bar{x})$
c	c	$\ldots$		kc

The value of p is $\bar{x}/c$, and the expected frequencies are

$\bar{x}$	$\bar{x}$	$\ldots$	$\bar{x}$
$c - \bar{x}$	$c - \bar{x}$	$\ldots$	$c - \bar{x}$

whence

$$x^2 = \sum_j \frac{(x_j - \bar{x})^2}{\bar{x}} + \sum_j \frac{\{(c - x_j) - (c - \bar{x})\}^2}{c - \bar{x}}$$

$$= \sum_j (x_j - \bar{x})^2 / cpq, \qquad (p = \bar{x}/c, \, q = 1 - p = 1 - \bar{x}/c)$$

$$= \sum_j (x_j - \bar{x})^2 / \bar{x}q.$$

This form of x^2 is called the *index of dispersion* for Binomial samples. It may be used to test whether the observed frequencies $x_1, x_2, \ldots, x_k$ of 'successes' in k samples of equal size c are consistent with the hypothesis that they all come from the same Binomial (c, p) distribution, since, if they do, x^2 is a realization of χ^2 on $k-1$ d.f.

Binomial data

In a sample of 100 children from each of five geographical regions of Britain the numbers of left-handed children were as follows:

Region	1	2	3	4	5	Mean
No of left-handed children	x_1	x_2	x_3	x_4	x_5	$\bar{x}$

The question is whether the regions are homogeneous in regard to the occurrence of left-handedness. On the hypothesis that they are, the index of dispersion

$$x^2 = \sum_{j=1}^{5} (x_j - \bar{x})^2 / \bar{x}(1 - \bar{x}/100)$$

will be a realization of χ_4^2, and the significance level of x^2 is

$$P(\chi_4^2 \geq x^2).$$

7.6.2.   Index of Dispersion for Poisson Samples

There is also a χ^2-distributed index of dispersion for Poisson samples. Although this is not an obvious special case of a contingency table, its resemblance to the index for Binomial samples discussed above provides some justification for dealing with it at this point.

The distribution of dust particles in the atmosphere may be assumed under suitable conditions to be Poissonian. In one form of equipment a standard volume of air is drawn through a tube and the dust particles captured on a filter paper. The number of particles is determined from their total weight. If from (say) ten such readings taken in various London boroughs the individual numbers of particles are $x_1, x_2, \ldots, x_{10}$, with mean value $\bar{x}$, it may be of interest to question whether the mean density of particles is constant. With such a small number of readings it is not practicable to form a frequency table from which to carry out a χ^2 test of agreement with the supposed Poisson distribution (as in Example 7.2.3); nevertheless a test can be carried out. The appropriate index of dispersion is the statistic

$$x^2 = \sum_{j} (x_j - \bar{x})^2 / \bar{x}.$$

On the null hypothesis that all the x_j come from the same Poisson distribution, the sampling distribution of x^2 is χ^2 on 9 degrees of freedom, and the significance level of x^2 with respect to this hypothesis is

$$P(\chi_9^2 \geq x^2).$$

E.H.L.

7.7. FURTHER READING AND REFERENCES

Mention has already been made in the text of Pearson's original χ^2 paper, of Cochran's studies of the effect of small frequencies, and of Yates's continuity correction. References for these are given below. A proof that the asymptotic sampling distribution of Pearson's statistic is the χ^2 distribution with the appropriate number of degrees of freedom will be found in Cramér (1946), Chapter 30 [Bibliography C]. Many practical examples are given in Fisher (1970) [Bibliography C]. Most general texts give examples of the uses of the χ^2 statistic—see for example Hald (1952), or Mood, Graybill and Boes (1974), both listed in Bibliography C.

Cochran, W. G. (1952). The χ^2 Test of Goodness of Fit, *Annals of Maths. Statistics*, **23**, 315.

Cochran, W. G. (1954). Some Methods of Strengthening the Common χ^2 Tests, *Biometrics* (1954) 417.

Yates, F. (1934). Contingency Tables Involving Small Numbers and the χ^2 Test, *Supplement to Journal Royal Statist. Soc.* **i**, 217.

Pearson, K. (1900). On a Criterion that a System of Deviations from the Probable in the Case of a Correlated Systems of Variables is Such that it can be Reasonably Supposed to have Arisen in Random Sampling, *Phil. Mag.* **50**, 157.

CHAPTER 8

Least Squares Estimation and the Analysis of Variance

8.1. LEAST SQUARES ESTIMATION FOR GENERAL MODELS

The method of least squares has been briefly touched on in section 3.5.2. We now move on to a more detailed study of this very important estimation procedure.

We consider n observations y_i, $i = 1, 2, \ldots, n$, regarded as realized values of n random variables Y_i, $i = 1, 2, \ldots, n$, whose expected values $E(Y_i)$ depend on p unknown real parameters θ_j, $j = 1, 2, \ldots, p$ $(p \leq n)$ through n known functions f_i, $i = 1, 2, \ldots, n$, as follows

$$E(Y_i) = f_i(\theta_1, \theta_2, \ldots, \theta_p) = f_i(\boldsymbol{\theta})$$

where $\boldsymbol{\theta} = (\theta_1, \theta_2, \ldots, \theta_p)'$. Thus, $Y_i = f_i(\boldsymbol{\theta}) + \varepsilon_i$, where ε_i is the random 'error' in the ith observation; the ε_i are assumed to be independent and equally serious, with zero expectations. We denote their (unknown) realized values by $e_1, e_2, \ldots, e_n$ so that $y_i = f_i(\boldsymbol{\theta}) + e_i$.

Suppose that $\mathbf{u} = (u_1, u_2, \ldots, u_p)'$, a vector function of $y_1, y_2, \ldots, y_n$, is proposed as an estimate [see Definition 3.1.1] of $\boldsymbol{\theta}$. Consider the sum of the squared differences between the observations and their estimated expectations based on $\mathbf{u}$, viz.

$$S(\mathbf{u}) = \sum_{i=1}^{n} [y_i - f_i(\mathbf{u})]^2. \tag{8.1.1}$$

The closer the estimated expectations are to the actual observations, the smaller S will be; hence the value of S can be used as a measure of how well the observations are fitted by the model when $\boldsymbol{\theta}$ is estimated by $\mathbf{u}$. The *Principle of Least Squares*, due to Gauss, says that $\boldsymbol{\theta}$ should be estimated by $\tilde{\boldsymbol{\theta}} = (\tilde{\theta}_1, \tilde{\theta}_2, \ldots, \tilde{\theta}_p)'$ where $\tilde{\boldsymbol{\theta}}$ is chosen to minimize this sum of squared deviations, i.e. the minimum value of $S(\mathbf{u})$, for variations in $\mathbf{u}$, is at $\mathbf{u} = \tilde{\boldsymbol{\theta}}$.

Under general conditions, the least squares estimate (L.S.E.) $\tilde{\boldsymbol{\theta}}$ may be found by solving the 'normal equations'

$$\left[\frac{\partial S(\mathbf{u})}{\partial u_j} \right]_{\mathbf{u} = \tilde{\boldsymbol{\theta}}} = 0, \qquad j = 1, 2, \ldots, p.$$

These reduce to

$$\sum_{i=1}^{n} \{y_i - f_i(\tilde{\boldsymbol{\theta}})\} \left[\frac{\partial f_i(\mathbf{u})}{\partial u_j}\right]_{\mathbf{u}=\tilde{\boldsymbol{\theta}}} = 0, \qquad j = 1, 2, \ldots, p. \qquad (8.1.2)$$

The properties of estimators [see Definition 3.1.1] produced by this method depend on the distribution of $Y_1, Y_2, \ldots, Y_n$; for example, when the ε_i are independent, identically distributed Normal variables, the L.S.E. $\tilde{\boldsymbol{\theta}}$ coincides with the maximum likelihood estimate [see Chapter 6]. Least Squares estimators are not generally unbiased [see § 3.3.2]; however, in an important special case discussed below—the linear model—least squares estimators *are* unbiased, and also have an optimum property which is discussed later in section 8.2.3.

8.2. LEAST SQUARES ESTIMATION FOR FULL-RANK LINEAR MODELS. NORMAL EQUATIONS. THE GAUSS–MARKOV THEOREM

8.2.1. Examples of Least Squares Estimation

EXAMPLE 8.2.1. *Repeated measurements of a single entity.*   A single numerical quantity, θ_1, is measured, subject to error, n times. The observations have the form $y_i = \theta_1 + e_i$, where e_i is the error in the ith observation. In this case the sum of squares (8.1.1) reduces to

$$S(\mathbf{u}) = S(u_1) = \sum_{i=1}^{n} (y_i - u_1)^2,$$

and the normal equation (8.1.2)—there is only one in this case—to

$$\sum_{i=1}^{n} (y_i - \tilde{\theta}_1) = 0,$$

so that the L.S.E. of θ_1 is $\tilde{\theta}_1 = \sum y_i / n = \bar{y}$.

EXAMPLE 8.2.2. *Nine observations on five gravity constants.*   A more interesting example involving the determination of physical constants uses the following data:

$$g_2 = 981 \cdot 1880 \qquad g_4 - g_1 = -0 \cdot 0030 \qquad g_4 - g_5 = 0 \cdot 1390$$

$$g_3 = 981 \cdot 2000 \qquad g_4 - g_3 = 0 \cdot 0647 \qquad g_1 = 981 \cdot 2670$$

$$g_3 - g_2 = 0 \cdot 0140 \qquad g_4 - g_5 = 0 \cdot 1431 \qquad g_1 = 981 \cdot 2690$$

where $g_1, g_2, \ldots, g_5$ are the values of the acceleration due to gravity at five selected sites. As a system of mathematical equations to be solved, these are clearly inconsistent, this being the result of experimental error; in fact, the above 'equations' are just approximations and we can describe the situation accurately by writing $g_2 + e_1 = 981 \cdot 1880$, $g_4 - g_1 + e_2 = -0 \cdot 0030$, etc. In the notation of section 8.1, (the parameters being $g_1, g_2, \ldots, g_5$), we have $f_1(g_1, g_2, \ldots, g_5) = g_2$, $f_2(g_1, g_2, \ldots, g_5) = g_4 - g_1, \ldots$; and the sum of squares to be minimized is

$$S(\mathbf{u}) = (981 \cdot 1880 - u_2)^2 - (-0 \cdot 0030 - u_4 + u_1)^2 + \ldots + (981 \cdot 2690 - u_1)^2.$$

Differentiation with respect to $u_1, u_2, \ldots, u_5$, respectively, and substitution of $u_i = \tilde{g}_i$, $i = 1, 2, \ldots, 5$ produces the normal equations:

$$3\tilde{g}_1 - \tilde{g}_4 - 1962 \cdot 5390 = 0$$
$$2\tilde{g}_2 - \tilde{g}_3 - 981 \cdot 1740 = 0$$
$$-\tilde{g}_2 + 3\tilde{g}_3 - \tilde{g}_4 - 981 \cdot 1493 = 0$$
$$-\tilde{g}_1 - \tilde{g}_3 + 4\tilde{g}_4 - 2\tilde{g}_5 - 0 \cdot 3438 = 0$$
$$-2\tilde{g}_4 + 2\tilde{g}_5 + 0 \cdot 2821 = 0.$$

This is a full-rank (regular) set of five non-homogeneous linear equations [see I; 5.7] in five unknowns, whose (unique) solution is

$$\tilde{g}_1 = 981 \cdot 2681, \qquad \tilde{g}_2 = 981 \cdot 1873, \qquad \tilde{g}_3 = 981 \cdot 2006,$$
$$\tilde{g}_4 = 981 \cdot 2652, \qquad \tilde{g}_5 = 981 \cdot 1241$$

(continued in Example 8.2.4).

8.2.2. Matrix Formulation for Linear Models

Examples 8.2.1 and 8.2.2 are concerned with linear models; in both cases, each observation is expressed as a linear function of parameters, plus an error. In terms of section 8.1, the function f_i is a linear function of the parameter vector $\boldsymbol{\theta} = (\theta_1, \theta_2, \ldots, \theta_p)'$ and Y_i may be written as

$$Y_i = a_{i1}\theta_1 + a_{i2}\theta_2 + \ldots + a_{ip}\theta_p + \varepsilon_i, \qquad i = 1, \ldots, n,$$

or

$$Y_i = \sum_{j=1}^{p} a_{ij}\theta_j + \varepsilon_i,$$

where the a_{ij} are known constants. Writing $\mathbf{Y} = (Y_1, Y_2, \ldots, Y_n)'$, $\boldsymbol{\varepsilon} = (\varepsilon_1, \varepsilon_2, \ldots, \varepsilon_n)'$ and denoting the $(n \times p)$ matrix [see IV, § 6.2] of constants $\{a_{ij}\}$ by $\mathbf{A}$ (this is known as the *design* matrix) we can write the model concisely as $\mathbf{Y} = \mathbf{A}\boldsymbol{\theta} + \boldsymbol{\varepsilon}$. In terms of realized values, $\mathbf{y} = \mathbf{A}\boldsymbol{\theta} + \mathbf{e}$, where $\mathbf{y} = (y_1, y_2, \ldots, y_n)'$ is the observed value of $\mathbf{Y}$ and $\mathbf{e} = (e_1, e_2, \ldots, e_n)'$ is the vector of errors. The sum of squares to be minimized is

$$S(\mathbf{u}) = \sum_{i=1}^{n} \left\{ y_i - \sum_{j=1}^{p} a_{ij}u_j \right\}^2 = (\mathbf{y} - \mathbf{A}\mathbf{u})'(\mathbf{y} - \mathbf{A}\mathbf{u}).$$

The normal equations (8.1.1) for the estimates $\tilde{\theta}_j$ are linear in this case, and may be written as

$$\sum_{i=1}^{n}\left\{y_i - \sum_{j=1}^{p} a_{ij}\tilde{\theta}_j\right\}a_{ij}=0, \qquad j=1,2,\ldots,p,$$

or, in matrix form

$$\mathbf{A'y}-\mathbf{A'A\tilde{\theta}}=\mathbf{0}.$$

For future reference, we write these as

$$\mathbf{A'A\tilde{\theta}}=\mathbf{A'y} \qquad\qquad (8.2.1)'$$

We assume that the p columns of A are linearly independent; this is the full-rank [see I, § 5.6] case, the rank $r(\mathbf{A})$ of $\mathbf{A}$ being p. It follows that the symmetric matrix $\mathbf{A'A}$ has rank $r(\mathbf{A'A})=p$, and since $\mathbf{A'A}$ is $(p\times p)$, it is non-singular [see I, Definition 6.4.2] and has an inverse $(\mathbf{A'A})^{-1}$ [see I, § 6.4]. The unique solution of the normal equations is thus

$$\tilde{\boldsymbol{\theta}}=(\mathbf{A'A})^{-1}\mathbf{A'y}, \qquad\qquad (8.2.2)$$

a linear function of the observations, $y_1, y_2, \ldots, y_n$, and this is the least squares estimate (L.S.E.) of $\boldsymbol{\theta}$. The case $r(\mathbf{A})<p$ will be discussed in Chapter 10.

EXAMPLE 8.2.3.   *Continuation of Example* 8.2.1.   For the measurement model of Example 8.2.1, the matrix $\mathbf{A}$ is an $(n \times 1)$ vector of units, $\mathbf{A'A}=n$ and $\mathbf{A'y}=\sum_{i=1}^{n} y_i$.

EXAMPLE 8.2.4.   *Continuation of Example* 8.2.2.   For example 8.2.2 the parameter vector $\boldsymbol{\theta}=(g_1, g_2, \ldots, g_5)'$ and we have

$$
\begin{bmatrix} y_1 \\ y_2 \\ y_3 \\ y_4 \\ y_5 \\ y_6 \\ y_7 \\ y_8 \\ y_9 \end{bmatrix}
=
\begin{bmatrix} 981{\cdot}1880 \\ -0{\cdot}0030 \\ 0{\cdot}1390 \\ 981{\cdot}2000 \\ 0{\cdot}0647 \\ 981{\cdot}2670 \\ 0{\cdot}0140 \\ 0{\cdot}1431 \\ 981{\cdot}2690 \end{bmatrix}
=
\begin{bmatrix} 0 & 1 & 0 & 0 & 0 \\ -1 & 0 & 0 & 1 & 0 \\ 0 & 0 & 0 & 1 & -1 \\ 0 & 0 & 1 & 0 & 0 \\ 0 & 0 & -1 & 1 & 0 \\ 1 & 0 & 0 & 0 & 0 \\ 0 & -1 & 1 & 0 & 0 \\ 0 & 0 & 0 & 1 & -1 \\ 1 & 0 & 0 & 0 & 0 \end{bmatrix}
\begin{bmatrix} g_1 \\ g_2 \\ g_3 \\ g_4 \\ g_5 \end{bmatrix}
+
\begin{bmatrix} e_1 \\ e_2 \\ e_3 \\ e_4 \\ e_5 \\ e_6 \\ e_7 \\ e_8 \\ e_9 \end{bmatrix}
$$

Here

$$
\mathbf{A'A}=
\begin{bmatrix}
3 & 0 & 0 & -1 & 0 \\
0 & 2 & -1 & 0 & 0 \\
0 & -1 & 3 & -1 & 0 \\
-1 & 0 & -1 & 4 & -2 \\
0 & 0 & 0 & -2 & 2
\end{bmatrix}
$$

while $\tilde{\boldsymbol{\theta}} = (\mathbf{A}'\mathbf{A})^{-1}\mathbf{A}'\mathbf{y}$ is given by

$$\begin{bmatrix} 0\cdot4211 & 0\cdot0526 & 0\cdot1053 & 0\cdot2632 & 0\cdot2632 \\ 0\cdot0526 & 0\cdot6316 & 0\cdot2632 & 0\cdot1579 & 0\cdot1579 \\ 0\cdot1053 & 0\cdot2632 & 0\cdot5263 & 0\cdot3158 & 0\cdot3158 \\ 0\cdot2632 & 0\cdot1579 & 0\cdot3158 & 0\cdot7895 & 0\cdot7895 \\ 0\cdot2632 & 0\cdot1579 & 0\cdot3158 & 0\cdot7895 & 1\cdot2895 \end{bmatrix} \begin{bmatrix} 1962\cdot5390 \\ 981\cdot1740 \\ 981\cdot1493 \\ 0\cdot3438 \\ -0\cdot2821 \end{bmatrix} = \begin{bmatrix} 981\cdot2681 \\ 981\cdot1873 \\ 981\cdot2006 \\ 981\cdot2652 \\ 981\cdot1241 \end{bmatrix}$$

(continued in Example 8.2.8).

EXAMPLE 8.2.5.   *Regression.*   Frequently the relationship between a variable Y and several 'independent' variables $x_1, x_2, \ldots, x_s$ is approximated by a 'linear model' of the form

$$E(Y) = \beta_0 + \beta_1 h_1(x_1, x_2, \ldots, x_s) + \beta_2 h_2(x_1, x_2, \ldots, x_s) + \ldots$$
$$+ \beta_q h_q(x_1, x_2, \ldots, x_s)$$

[cf. § 3.5.5] where $h_1, h_2, \ldots, h_q$ are specified functions of $x_1, x_2, \ldots, x_s$, and $\beta_0, \beta_1, \ldots, \beta_q$ are unknown parameters, (replacing the $\theta_1, \theta_2, \ldots, \theta_p$ of the general formulation above). The observations $y_1, y_2, \ldots, y_n$ are taken at various combinations of known values of the variables $x_1, x_2, \ldots, x_s$. A common case is *multiple linear regression,* [cf. 6.5.1] in which $h_j(x_1, x_2, \ldots, x_s) = x_j$ and $q = s$, so that the model becomes

$$E(Y) = \beta_0 + \beta_1 x_1 + \ldots + \beta_s x_s.$$

[The case where there is only one *regressor variable* x_1 is discussed in Example 6.5.1]. The terminology here can be confusing; this is a *linear model* because it is a linear function of the parameters $\beta_0, \beta_1, \ldots, \beta_s$; the *regression* is called linear because the dependence of $E(Y)$ on each independent variable, x_r, is linear. In contrast, the following is linear in the parameters $\beta_0, \beta_1, \ldots, \beta_s$ and so is a linear model:

$$E(Y) = \beta_0 + \beta_1 x + \beta_2 x^2 + \ldots + \beta_s x^s,$$

but it expresses non-linear dependence of $E(Y)$ on the single independent regressor variable x, and so is not a linear regression: it is *polynomial* regression. Non-linear dependence of $E(Y)$ on several regressor variables can be postulated in a linear model such as

$$E(Y) = \beta_0 + \beta_1 x_1 + \beta_{11} x_1^2 + \beta_2 x_2 + \beta_{22} x_2^2 + \beta_{12} x_1 x_2 + \beta_{123} x_1 x_2 x_3,$$

in which the parameters are re-labelled for greater clarity.

To write these models in matrix form, suppose that the observed values of the variables $x_1, x_2, \ldots, x_s$ for the ith observation on Y are $x_{i1}, \ldots, x_{is}$. To illustrate, in the case of multiple linear regression we have

$$y_i = \beta_0 + \beta_1 x_{i1} + \beta_2 x_{i2} + \ldots + \beta_s x_{is} + e_i,$$

so that

$$\mathbf{y} = \mathbf{A}\boldsymbol{\beta} + \mathbf{e},$$

where

$$\mathbf{A} = \begin{bmatrix} 1 & x_{11} & x_{12} & \cdots & x_{1s} \\ 1 & x_{21} & x_{22} & \cdots & x_{2s} \\ \vdots & \vdots & \vdots & & \vdots \\ 1 & x_{n1} & x_{n2} & \cdots & x_{ns} \end{bmatrix}, \quad \boldsymbol{\beta} = \begin{bmatrix} \beta_0 \\ \beta_1 \\ \vdots \\ \beta_s \end{bmatrix}.$$

Thus

$$\mathbf{A}'\mathbf{A} = \begin{bmatrix} n & \sum x_{i1} & \cdots & \sum x_{is} \\ \sum x_{i1} & \sum x_{i1}^2 & \cdots & \sum x_{i1}x_{is} \\ \vdots & \vdots & & \vdots \\ \sum x_{is} & \sum x_{i1}x_{is} & \cdots & \sum x_{is}^2 \end{bmatrix}, \quad \mathbf{A}'\mathbf{y} = \begin{bmatrix} \sum y_i \\ \sum y_i x_{i1} \\ \vdots \\ \sum y_i x_{is} \end{bmatrix}.$$

The observed values of any particular regressor variable, say x_j, need not be all different, but we suppose that the values of these variables as a whole are such that $\mathbf{A}$ has full rank ($s+1$ in this example); this would not be the case if, for instance, all n observations were taken at the same set of values of $x_1, \ldots, x_s$. To obtain the least squares estimates $\tilde{\beta}_0, \ldots, \tilde{\beta}_s$ we require $(\mathbf{A}'\mathbf{A})^{-1}$, which will normally require the numerical inversion of $\mathbf{A}'\mathbf{A}$ by computer; explicit formulas can be given in the special case of linear regression on one variable, for which, in the above notation,

$$E(Y) = \beta_0 + \beta_1 x_1.$$

It is convenient in this case to drop the suffix on x and write simply $E(Y) = \beta_0 + \beta_1 x$. Denoting the observed values of x by $x_1, x_2, \ldots, x_n$ we have

$$(\mathbf{A}'\mathbf{A})^{-1} = \frac{1}{n \sum (x_i - \bar{x})^2} \begin{bmatrix} \sum x_i^2 & -n\bar{x} \\ -n\bar{x} & n \end{bmatrix},$$

where $\bar{x} = (\sum x_i)/n$. Using $\tilde{\boldsymbol{\beta}} = (\mathbf{A}'\mathbf{A})^{-1}\mathbf{A}'\mathbf{y}$ gives the least squares estimates

$$\tilde{\beta}_0 = \bar{y} - \tilde{\beta}_1 \bar{x}, \qquad \tilde{\beta}_1 = \frac{\sum (x_i - \bar{x})(y_i - \bar{y})}{\sum (x_i - \bar{x})^2},$$

where $\bar{y} = (\sum y_i)/n$.

For the polynomial regression model, we have

$$y_i = \beta_0 + \beta_1 x_i + \beta_2 x_i^2 + \ldots + \beta_s x_i^s + e_i,$$

where x_i is the value of x when the ith observation is made. Writing $\mathbf{y} = \mathbf{A}\boldsymbol{\beta} + \mathbf{e}$,

$$\mathbf{A} = \begin{bmatrix} 1 & x_1 & x_1^2 & \cdots & x_1^s \\ 1 & x_2 & x_2^2 & \cdots & x_2^s \\ \vdots & \vdots & \vdots & & \vdots \\ 1 & x_n & x_n^2 & \cdots & x_n^s \end{bmatrix}, \quad \mathbf{A}'\mathbf{A} = \begin{bmatrix} n & \sum x_i & \sum x_i^2 & \cdots & \sum x_i^s \\ \sum x_i & \sum x_i^2 & \sum x_i^3 & \cdots & \sum x_i^{s+1} \\ \vdots & \vdots & \vdots & & \vdots \\ \sum x_i^s & \sum x_i^{s+1} & \sum x_i^{s+2} & \cdots & \sum x_i^{2s} \end{bmatrix}$$

In this case also, numerical inversion of $\mathbf{A}'\mathbf{A}$ will usually be necessary.

Finally, as an example of a nonlinear model, consider

$$E(Y) = \beta_0 + \beta_1 x_1 + \beta_1^2 x_2 + \beta_0 \beta_1 x_3.$$

Although this exhibits linear dependence of $E(Y)$ on x_1, x_2 and x_3, it is not linear in the parameters and we cannot express it in the matrix form $E(\mathbf{Y}) = \mathbf{A}\boldsymbol{\beta}$, so the formulas used above do not apply.

EXAMPLE 8.2.6. *The one-way classification.* Suppose that data is divided into I groups, with J_i observations in group i, $i = 1, 2, \ldots, I$, these being denoted by $y_{i1}, y_{i2}, \ldots, y_{iJ_i}$, for $i = 1, 2, \ldots, I$. We can display the data as

Group	Observations	Mean of group
1	$y_{11}, y_{12}, \ldots, y_{1J_1}$	$y_1.$
2	$y_{21}, y_{22}, \ldots, y_{2J_2}$	$y_2.$
$\vdots$		$\vdots$
I	$y_{I1}, y_{I2}, \ldots, y_{IJ_I}$	$y_I.$

Assuming that the observations in group i are a random sample from a distribution with expectation μ_i, the model is

$$y_{ij} = \mu_i + e_{ij}, \qquad i = 1, 2, \ldots, I; \qquad j = 1, 2, \ldots, J_i,$$

where e_{ij} is a realization of a random variable with zero mean. We can present this in the form $\mathbf{y} = \mathbf{A}\boldsymbol{\theta} + \mathbf{e}$ as follows

$$
\begin{bmatrix} y_1 \\ \vdots \\ y_n \end{bmatrix}
=
\begin{bmatrix} y_{11} \\ \vdots \\ y_{1J_1} \\ y_{21} \\ \vdots \\ y_{2J_2} \\ \vdots \\ y_{I1} \\ \vdots \\ y_{IJ_I} \end{bmatrix}
=
\begin{bmatrix}
1 & 0 & 0 & 0 \\
\vdots & \vdots & \vdots & \vdots \\
1 & 0 & 0 & 0 \\
0 & 1 & 0 & 0 \\
\vdots & \vdots & \vdots & \vdots \\
0 & 1 & 0 & 0 \\
\vdots & \vdots & \vdots & \vdots \\
0 & 0 & 0 & 1 \\
\vdots & \vdots & \vdots & \vdots \\
0 & 0 & 0 & 1
\end{bmatrix}
\begin{bmatrix} \mu_1 \\ \vdots \\ \mu_I \end{bmatrix}
+
\begin{bmatrix} e_{11} \\ \vdots \\ e_{1J_1} \\ e_{21} \\ \vdots \\ e_{2J_2} \\ \vdots \\ e_{I1} \\ \vdots \\ e_{IJ_I} \end{bmatrix}
\tag{8.2.3}
$$

In this case $\mathbf{A}'\mathbf{A}$ is a diagonal matrix [see IV, § 6.7] with diagonal elements $J_1, J_2, \ldots, J_I$; we write

$$\mathbf{A}'\mathbf{A} = \text{diag}\{J_1, J_2, \ldots, J_I\},$$

so

$$(\mathbf{A}'\mathbf{A})^{-1} = \text{diag}\{1/J_1, 1/J_2, \ldots, 1/J_I\},$$

and

$$\mathbf{A}'\mathbf{y} = \left(\sum_{j=1}^{J_1} y_{1j}, \sum_{j=1}^{J_2} y_{2j}, \ldots, \sum_{j=1}^{J_I} y_{Ij} \right)'.$$

Thus $\tilde{\mu}_i = \sum_{j=1}^{J_i} y_{ij}/J_i = y_{i+}/J_i = y_{i\cdot}$, the average of the observations in group i [here, and elsewhere, we indicate summation over a particular subscript by a $+$ sign and averaging by a dot].

We can also exhibit the one-way classification model in regression form as

$$E(Y) = \mu_1 x_1 + \mu_2 x_2 + \ldots + \mu_I x_I,$$

where the regressor variables $x_1, x_2, \ldots, x_I$ are such that, for $i = 1, \ldots, I$, we have $x_i = 1$ for the observations in group i and $x_i = 0$ otherwise; the values of x_i are thus given in column i of the design matrix, $\mathbf{A}$. In such a case, the regressor variables are qualitative, serving only to indicate the data structure, in contrast to the illustrations given in Example 8.2.5 where the x_i were the values of quantitative variables.

EXAMPLE 8.2.7.   *The one-way classification with a concomitant variable.*   In addition to classifying data into I groups, we may also have available supplementary observations on a related quantitative variable z known as a '*concomitant variable*'. In such a case a possible model is

$$y_{ij} = \mu_i + \phi z_{ij} + e_{ij}.$$

In regression form this is just

$$E(Y) = \mu_1 x_1 + \ldots + \mu_I x_I + \phi z,$$

and the parameter vector for this model is $\boldsymbol{\theta}' = (\mu_1, \ldots, \mu_I, \phi)$, while the design matrix is given by augmenting $\mathbf{A}$ of the previous example (Equation 8.2.3) by an extra column equal to $(z_{11}, \ldots, z_{1J_1}, z_{21}, \ldots, z_{2J_2}, \ldots, z_{I1}, \ldots, z_{IJ_I})'$.

This idea extends in an obvious way to models with several concomitant variables. The design matrix is then composed of several columns of 0 or 1 elements indicating the data structure, followed by several columns giving the values of the supplementary variables.

8.2.3.   Properties of Least Squares Estimates

How good are least squares estimates? The method can be appraised by considering the statistical properties of the least squares estimator [see Definition 3.1.1] $\tilde{\boldsymbol{\Theta}} = (\tilde{\Theta}_1, \tilde{\Theta}_2, \ldots, \tilde{\Theta}_p)' = (\mathbf{A}'\mathbf{A})^{-1}\mathbf{A}'\mathbf{Y}$, whose realized value is $\tilde{\boldsymbol{\theta}} = (\mathbf{A}'\mathbf{A})^{-1}\mathbf{A}'\mathbf{y}$. Clearly, the distribution of $\tilde{\boldsymbol{\Theta}}$ depends on the distribution of $\mathbf{Y}$. This section is concerned only with first and second moments [see II, § 9.11] of $\tilde{\boldsymbol{\Theta}}$ (other distributional properties will be considered later in section 8.3.1) and here we shall assume only that the random errors of observation have zero expectations and equal variances, and are uncorrelated [see II, § 9.6.3], the earlier assumption in section 8.1 of independent errors not being necessary for the immediately following results. Our assumptions are thus

$$E(\boldsymbol{\varepsilon}) = \mathbf{0}, \qquad \mathscr{D}(\boldsymbol{\varepsilon}) = \sigma^2 \mathbf{I}_n,$$

where $\mathscr{D}(\cdot)$ is the dispersion matrix [see II, Definition 9.6.3]. These results imply that

$$E(\mathbf{Y}) = \mathbf{A}\boldsymbol{\theta}, \qquad \mathscr{D}(\mathbf{Y}) = \sigma^2 \mathbf{I}_n.$$

The L.S.E. $\tilde{\boldsymbol{\Theta}}$ then has the following properties:

Property 1.   *Unbiasedness*

$$E(\tilde{\boldsymbol{\Theta}}) = \boldsymbol{\theta};$$

for $j = 1, \ldots, p$, θ_j is estimated without bias by $\tilde{\Theta}_j$.

Property 2.   *Dispersion matrix*

The dispersion matrix of $\tilde{\boldsymbol{\Theta}}$ is given by

$$\mathscr{D}(\tilde{\boldsymbol{\Theta}}) = \sigma^2 (\mathbf{A}'\mathbf{A})^{-1}.$$

Thus the variance of $\tilde{\Theta}_i$ is given by the ith diagonal element of this matrix, while the covariance cov $(\tilde{\Theta}_i, \tilde{\Theta}_j)$ is equal to the (i, j) element.

Property 3.   *Minimal variance (Gauss–Markov theorem)*

The L.S. estimator $\tilde{\Theta}_j$ of θ_j is a linear function of the observation variables $Y_1, Y_2, \ldots, Y_n$ with the property that its variance is not greater than the variance of any other linear function of $Y_1, Y_2, \ldots, Y_n$ which is also unbiased for θ_j, i.e. if $E(\boldsymbol{\ell}'\mathbf{Y}) = \theta_j$, where $\boldsymbol{\ell} = (\ell_1, \ell_2, \ldots, \ell_n)'$ is a given constant vector, then

$$\text{var } \tilde{\Theta}_j \leq \text{var } \boldsymbol{\ell}'\mathbf{Y}, \quad \text{for all } \boldsymbol{\ell}.$$

In other words, amongst all unbiased linear estimates of θ_j, the L.S.E. $\tilde{\Theta}_j$ is best in the sense of minimum variance; it is the minimum variance unbiased linear estimator (MVULE) of θ_j [see § 3.3.2].

More generally, if we consider the estimation of a known linear combination of the parameters, say $\boldsymbol{\lambda}'\boldsymbol{\theta} = \lambda_1\theta_1 + \lambda_2\theta_2 + \ldots + \lambda_p\theta_p$, where $\boldsymbol{\lambda} = (\lambda_1, \lambda_2, \ldots, \lambda_p)'$ is known, we can say that $\boldsymbol{\lambda}'\tilde{\boldsymbol{\Theta}} = \lambda_1\tilde{\Theta}_1 + \lambda_2\tilde{\Theta}_2 + \ldots + \lambda_p\tilde{\Theta}_p$ is unbiased for $\boldsymbol{\lambda}'\boldsymbol{\theta}$ and has variance no larger than any other unbiased estimator of $\boldsymbol{\lambda}'\boldsymbol{\theta}$ which is also linear in the observation variables $Y_1, Y_2, \ldots, Y_n$, i.e. $\boldsymbol{\lambda}'\tilde{\boldsymbol{\Theta}}$ is the MVULE of $\boldsymbol{\lambda}'\boldsymbol{\theta}$; its variance is

$$\boldsymbol{\lambda}'\mathscr{D}(\tilde{\boldsymbol{\Theta}})\boldsymbol{\lambda} = \sigma^2 \boldsymbol{\lambda}'(\mathbf{A}'\mathbf{A})^{-1}\boldsymbol{\lambda}$$

[see IV, § 6.2]. In particular the MVULE of $E(Y_i) = \sum_{j=1}^{p} a_{ij}\theta_j$ is $\sum_{j=1}^{p} a_{ij}\tilde{\Theta}_j$.

The property expressed by the Gauss–Markov theorem does not necessarily mean that the Principle of Least Squares always leads to good estimators; an estimator may be best in a restricted class and still be mediocre. For instance, it is possible to construct nonlinear functions of $Y_1, \ldots, Y_n$ which are unbiased for $\theta_1, \ldots, \theta_p$ and which outperform the L.S. estimators when the errors $\varepsilon_1, \ldots, \varepsilon_n$ are not Normally distributed (see below). On the other hand, if $\varepsilon_1, \ldots, \varepsilon_n$ are Normally distributed, it can be shown that the L.S. estimators are then the same as the maximum likelihood estimators [see § 3.5.5], so that in this case, in addition to the properties listed above, they have various properties which make them attractive when compared with other estimators, linear or nonlinear. The Gauss–Markov theorem requires only the weak assumptions noted above for its validity, and may be regarded as a justification for the use of Least Squares in circumstances when we are unable or unwilling to make more specific assumptions about the distribution of the errors. When the distributional form of the errors is known, other methods (e.g. maximum likelihood) would usually be preferred.

In the case of the measurement model

$$Y_i = \theta_1 + \varepsilon_i,$$

with uncorrelated errors and $E(\varepsilon_i) = 0$, var $\varepsilon_i = \sigma^2$, the MVULE of θ_1 is $\tilde{\Theta}_1 = \bar{Y}$, and var $\tilde{\Theta}_1 = \sigma^2(\mathbf{A}'\mathbf{A})^{-1} = \sigma^2/n$. For various symmetric non-Normal error distributions, the median of $Y_1, \ldots, Y_n$ (a nonlinear function of $Y_1, \ldots, Y_n$) has smaller variance than $\bar{Y}$.

EXAMPLE 8.2.8.   *The gravity example (Examples 8.2.2 and 8.2.4) considered further.*   In the gravity example we found that

$$(\mathbf{A}'\mathbf{A})^{-1} = \begin{bmatrix} 0{\cdot}4211 & 0{\cdot}0526 & 0{\cdot}1053 & 0{\cdot}2632 & 0{\cdot}2632 \\ 0{\cdot}0526 & 0{\cdot}6316 & 0{\cdot}2632 & 0{\cdot}1579 & 0{\cdot}1579 \\ 0{\cdot}1053 & 0{\cdot}2632 & 0{\cdot}5263 & 0{\cdot}3158 & 0{\cdot}3158 \\ 0{\cdot}2632 & 0{\cdot}1579 & 0{\cdot}3158 & 0{\cdot}7895 & 0{\cdot}7895 \\ 0{\cdot}2632 & 0{\cdot}1579 & 0{\cdot}3158 & 0{\cdot}7895 & 1{\cdot}2895 \end{bmatrix}.$$

We note that since the diagonal elements of $(\mathbf{A}'\mathbf{A})^{-1}$ are unequal the variances of the L.S. estimators $\tilde{G}_1, \ldots, \tilde{G}_5$ are not the same, i.e. $g_1, \ldots, g_5$ are estimated with differing precisions, resulting from the asymmetrical way in which these parameters appear in the model equations. In particular, the most precise estimate is $\tilde{g}_1$ (for var $\tilde{G}_1 = 0{\cdot}4211\sigma^2$) and the least precise estimate is $\tilde{g}_5$ (for var $\tilde{G}_5 = 1{\cdot}2895\sigma^2$). We can also note some other consequences of the structure of the model equations, such as

(i) cov $(\tilde{G}_4, \tilde{G}_j) = $ cov $(\tilde{G}_5, \tilde{G}_j)$, $j = 1, \ldots, 4$
(ii) cov $(\tilde{G}_2, \tilde{G}_3) = $ cov $(\tilde{G}_1, \tilde{G}_4)$.

As well as individual estimates of the acceleration due to gravity at the sites considered, we might be interested in the average value of g at the five sites, i.e. $\bar{g} = (g_1 + g_2 + g_3 + g_4 + g_5)/5$. Under the conditions noted the MVULE of this linear function is $(\tilde{G}_1 + \tilde{G}_2 + \tilde{G}_3 + \tilde{G}_4 + \tilde{G}_5)/5$, which is realized as $981 \cdot 2091$. The variance is $\sigma^2 \boldsymbol{\lambda}'(\mathbf{A}'\mathbf{A})^{-1}\boldsymbol{\lambda}$, where $\boldsymbol{\lambda}' = (1/5, 1/5, 1/5, 1/5, 1/5)$; this is just $\sigma^2 \times$ the arithmetic mean of the elements of $(\mathbf{A}'\mathbf{A})^{-1}$, and is equal to $0 \cdot 3611\sigma^2$. (Continued in Example 8.2.12).

EXAMPLE 8.2.9. *Linear regression on one variable.* For linear regression on one variable, for which $E(Y) = \beta_0 + \beta_1 x$, the MVUL estimators of β_0 and β_1 are

$$\tilde{B}_0 = \bar{Y} - \tilde{B}_1 \bar{x}, \qquad \tilde{B}_1 = \frac{\sum (x_i - \bar{x})(Y_i - \bar{Y})}{\sum (x_i - \bar{x})^2}.$$

Note that, since the regressor, x, is not a random variable, its fixed values $x_1, \ldots, x_n$ enter these formulae in exactly the same way as in the previous expressions for $\tilde{\beta}_0, \tilde{\beta}_1$ (given in Example 8.2.5) from which the above are derived.

By property 1 above, $E(\tilde{B}_0) = \beta_0$, $E(\tilde{B}_1) = \beta_1$. Furthermore, using property 2, above, and $(\mathbf{A}'\mathbf{A})^{-1}$ as given in Example 8.2.5,

$$\text{var } \tilde{B}_0 = \frac{\sigma^2 \sum x_i^2}{n \sum (x_i - \bar{x})^2}, \qquad \text{var } \tilde{B}_1 = \frac{\sigma^2}{\sum (x_i - \bar{x})^2}.$$

Note also that $\tilde{B}_0$ and $\tilde{B}_1$ are correlated; their covariance is given by the off-diagonal element of $\sigma^2(\mathbf{A}'\mathbf{A})^{-1}$, which is $\sigma^2 \bar{x}/\sum (x_i - \bar{x})^2$.

EXAMPLE 8.2.10. *The one-way layout.* In the one-way layout (Example 8.2.6) the MVUL estimators of the group means $\mu_1, \ldots, \mu_I$ are $Y_1., \ldots, Y_I.$ respectively, with variances $\sigma^2/J_1, \ldots, \sigma^2/J_I$, respectively. The off-diagonal elements of $(\mathbf{A}'\mathbf{A})^{-1}$ are zero, so that any two of these estimators are uncorrelated. To obtain the MVULE of a linear combination of $\mu_1, \ldots, \mu_I$ we construct the same linear combination of $Y_1., \ldots, Y_I.$. For example, the MVULE of $\mu_r - \mu_s$ is $Y_r. - Y_s.$, with variance $\sigma^2\{1/J_r + 1/J_s\}$. For $J_1\mu_1 + J_2\mu_2 + \ldots + J_I\mu_I$ the MVULE is $J_1 Y_1. + J_2 Y_2. + \ldots + J_I Y_I.$, the variance of which is $\sum J_i^2 \text{ var } Y_i. = \sigma^2 \sum J_i = n\sigma^2$, where n is the total number of observations. The variance can also be found directly on observing that $\sum J_i Y_i. = \sum Y_{i+} = Y_{++}$, the sum of all the observations.

8.2.4. Residuals

In practice the error variance σ^2 is unknown. Since σ^2 is involved as a common factor in the variances and covariances of $\tilde{\Theta}_1, \ldots, \tilde{\Theta}_p$ and in the variances of linear functions of these, we need to estimate σ^2 in order to estimate these quantities. The estimated standard deviations of $\tilde{\Theta}_1, \ldots, \tilde{\Theta}_p$ will then give some idea of the accuracy of the estimates of $\theta_1, \ldots, \theta_p$, and similarly for linear functions of the parameters.

To estimate σ^2 note that if we knew all the θ_i's, then we could determine the actual errors from

$$\mathbf{e} = \mathbf{y} - \mathbf{A}\boldsymbol{\theta}.$$

Estimation of σ^2 would then be based on

$$\mathbf{e}'\mathbf{e} = (\mathbf{y} - \mathbf{A}\boldsymbol{\theta})'(\mathbf{y} - \mathbf{A}\boldsymbol{\theta}),$$

which is just $\sum e_i^2$. This is a realization of the random variable $\boldsymbol{\varepsilon}'\boldsymbol{\varepsilon} = \sum \varepsilon_i^2$, having expectation $n\sigma^2$; hence the estimate of σ^2 would be $(\sum e_i^2)/n$. As the θ_i's are in fact unknown, we modify this procedure and use their L.S. estimates instead. Thus the MVULE of $\mathbf{e}$ is $\tilde{\mathbf{e}} = \mathbf{y} - \mathbf{A}\tilde{\boldsymbol{\theta}}$, and we base the estimate of σ^2 on $\tilde{\mathbf{e}}'\tilde{\mathbf{e}} = \sum \tilde{e}_i^2$, where $\tilde{e}_i$, the *ith residual*, is given by

$$\tilde{e}_i = y_i - \sum_{j=1}^{p} a_{ij}\tilde{\theta}_j.$$

Note that $\sum \tilde{e}_i^2$ is, in fact, the minimum value of

$$S(\mathbf{u}) = (\mathbf{y} - \mathbf{A}\mathbf{u})'(\mathbf{y} - \mathbf{A}\mathbf{u})$$

which, by definition, occurs when $\mathbf{u} = \tilde{\boldsymbol{\theta}}$; $\sum \tilde{e}_i^2$ is the sum of squared residuals (S.S.R.), which we denote by r; thus $r = \sum \tilde{e}_i^2 = S(\tilde{\boldsymbol{\theta}})$. It turns out that the random variable R of which r is a realization, no longer has expectation $n\sigma^2$; instead its expectation is reduced to $(n-p)\sigma^2$, so that our estimate of σ^2 is

$$\tilde{\sigma}^2 = \frac{1}{n-p} \sum_{i=1}^{n} \tilde{e}_i^2 = \frac{r}{n-p} \tag{8.2.4}$$

Thus an unbiased estimator of $\mathscr{D}(\tilde{\boldsymbol{\Theta}})$ is $(\mathbf{A}'\mathbf{A})^{-1}R/(n-p)$.

We estimate the standard deviation of $\tilde{\Theta}_j$ by

$$\tilde{\sigma} \times \{(j, j) \text{ element of } (A'A)^{-1}\}^{1/2},$$

and the standard deviation of the linear function $\boldsymbol{\lambda}'\tilde{\boldsymbol{\Theta}} = \lambda_1\tilde{\Theta}_1 + \ldots + \lambda_p\tilde{\Theta}_p$ by

$$\tilde{\sigma}\{\boldsymbol{\lambda}'(\mathbf{A}'\mathbf{A})^{-1}\boldsymbol{\lambda}\}^{1/2}.$$

It will be seen that (8.2.4) breaks down when $n = p$, i.e. when the number of observations is the same as the number of parameters. In this case, although we have enough observations to estimate $\theta_1, \ldots, \theta_p$, having done this there is no further information in the data on which to base an estimate of σ^2, (in fact $r \equiv 0$, as the observations can be fitted perfectly). In practice, therefore, since estimation of σ^2 is essential for judging the accuracy of the estimates of $\theta_1, \ldots, \theta_p$, we must ensure that $n > p$.

It will be shown later (Equation 8.3.7) that it is unnecessary to find the individual residuals in order to evaluate r; instead we may use the formula

$$r = \mathbf{y}'\mathbf{y} - \tilde{\boldsymbol{\theta}}'\mathbf{A}'\mathbf{y} \tag{8.2.5}$$

i.e. from $\sum y_i^2$, the sum of squares of all the observations, we subtract the product of the L.S.E. of θ with the right hand side of the normal equations as presented in Equation (8.2.1).

EXAMPLE 8.2.11. *Residuals in Example 8.2.1 and Example 8.2.3.* In the measurement model $y_i = \theta_1 + e_i$, $i = 1, 2, \ldots, n$, the ith residual, $\tilde{e}_i$, is given by

$$\tilde{e}_i = y_i - \tilde{\theta}_1 = y_i - \bar{y}.$$

The sum of squared residuals is thus $r = \sum (y_i - \bar{y})^2$; alternatively, we can write this as $r = \sum y_i^2 - n\bar{y}^2$, this formula being given directly on using Equation 8.2.5.
    For the estimate of σ^2 we have

$$\tilde{\sigma}^2 = \sum (y_i - \bar{y})^2/(n-1).$$

EXAMPLE 8.2.12. *Residuals in the gravity example.* For the gravity example of Examples 8.2.2, 8.2.4 and 8.2.8, the residuals are $(981 \cdot 1880 - \tilde{g}_2)$, $(-0 \cdot 0030 - \tilde{g}_4 + \tilde{g}_1), \ldots, (981 \cdot 2690 - \tilde{g}_1)$ i.e. $0 \cdot 0007$, $-0 \cdot 0001$, $-0 \cdot 0021$, $-0 \cdot 0006$, $0 \cdot 0001$, $-0 \cdot 0011$, $0 \cdot 0007$, $0 \cdot 0020$, $0 \cdot 0009$, whence $r = S(\tilde{\theta}) = \sum \tilde{e}_i^2 = 0 \cdot 00001179$. In this case $n = 9$, $p = 5$ and so $\tilde{\sigma}^2 = r/4 = 0 \cdot 00000295$. The estimated variances of $\tilde{G}_1, \ldots, \tilde{G}_5$ are (using the matrix $(\mathbf{A}'\mathbf{A})^{-1}$ given in Example 8.2.8) $0 \cdot 4211\tilde{\sigma}^2$, $0 \cdot 6316\tilde{\sigma}^2$, $0 \cdot 5263\tilde{\sigma}^2$, $0 \cdot 7895\tilde{\sigma}^2$, $1 \cdot 2895\tilde{\sigma}^2$, respectively; using the estimated value of $\tilde{\sigma}$ we obtain the estimated standard deviations $0 \cdot 0011, 0 \cdot 0014, 0 \cdot 0012, 0 \cdot 0015, 0 \cdot 0017$ respectively. The variance associated with the estimate of $\bar{g} = (g_1 + \ldots + g_5)/5$ was found in Example 8.2.8 to be $0 \cdot 3611\sigma^2$; the estimated standard deviation is therefore $(0 \cdot 3611)^{1/2}\tilde{\sigma} = 0 \cdot 0010$ (continued in Example 8.3.1).

EXAMPLE 8.2.13. *Residuals in simple linear regression.* In the regression model $y_i = \beta_0 + \beta_1 x_i + e_i$, (cf. Example 8.2.9) we have

$$r = \sum_{i=1}^{n} (y_i - \tilde{\beta}_0 - \tilde{\beta}_1 x_i)^2$$

where $\tilde{\beta}_0 = \bar{y} - \tilde{\beta}_1 \bar{x}$, $\tilde{\beta}_1 = \sum (x_i - \bar{x})(y_i - \bar{y})/\sum (x_i - \bar{x})^2$.
    Substituting for $\tilde{\beta}_0$ and expanding gives

$$r = \sum (y_i - \bar{y})^2 - 2\tilde{\beta}_1 \sum (x_i - \bar{x})(y_i - \bar{y}) + \tilde{\beta}_1^2 \sum (x_i - \bar{x})^2.$$

The central term is $-2\tilde{\beta}_1^2 \sum (x_i - \bar{x})^2$; whence

$$r = \sum y_i^2 - n\bar{y}^2 - \tilde{\beta}_1^2 \sum (x_i - \bar{x})^2.$$

It is a simple matter to show that this is equal to the alternative expression found by using Equation 8.2.5, viz.

$$r = \sum y_i^2 - \tilde{\beta}_0 \sum y_i - \tilde{\beta}_1 \sum x_i y_i.$$

EXAMPLE 8.2.14. *Residuals in a one-way layout.* In the one-way layout with model

$$y_{ij} = \mu_i + e_{ij}, \qquad i = 1, 2, \ldots, I; \qquad j = 1, 2, \ldots, J_i$$

(cf. Example 8.2.6) we have, for the (i, j) residual

$$\tilde{e}_{ij} = y_{ij} - \tilde{\mu}_i = y_{ij} - y_{i.},$$

so that the sum of squared residuals is

$$r = \sum_{i=1}^{I} \sum_{j=1}^{J_i} (y_{ij} - y_{i.})^2.$$

This is the sum of squares of observations from their group means. The number of parameters is I, so that $\tilde{\sigma}^2 = r/(n-I)$, where $n = \sum_{i=1}^{I} J_i$.

An alternative expression for r, found from Equation 8.2.5 is

$$r = \sum_i \sum_j y_{ij}^2 - \sum_i \tilde{\mu}_i y_{i+} = \sum_i \sum_j y_{ij}^2 - \sum_i J_i y_{i.}^2.$$

8.2.5. Orthogonal Designs and Orthogonal Polynomials

The model is termed *orthogonal* when the design matrix $\mathbf{A}$ is such that $\mathbf{A'A}$ is a diagonal matrix [see I, § 6.7], say $\text{diag}(A_{11}, \ldots, A_{pp})$. In this case $(\mathbf{A'A})^{-1} = \text{diag}(A_{11}^{-1}, \ldots, A_{pp}^{-1})$, and the vector solution $\tilde{\boldsymbol{\theta}} = (\mathbf{A'A})^{-1}\mathbf{A'y}$ may be written in explicit scalar form as

$$\tilde{\theta}_j = \frac{1}{A_{jj}} \sum_{i=1}^{n} y_i a_{ij}, \qquad j = 1, \ldots, p$$

a result which is obvious on inspecting the normal equations, which now take the simple form

$$\sum_{i=1}^{n} y_i a_{ij} - A_{jj} \tilde{\theta}_j = 0, \qquad j = 1, \ldots, p.$$

The one-way layout of Example 8.2.6 is an orthogonal model. The ease with which the normal equations can be solved is a considerable computational convenience of orthogonal models. We illustrate this point further by considering the case of *orthogonal polynomials*.

Suppose that it is required to fit a polynomial regression model, of the form

$$y = \beta_0 + \beta_1 x + \beta_2 x^2 + \ldots + \beta_s x^s + e,$$

to data (y_i, x_i), $i = 1, 2, \ldots, n$.

The matrices $\mathbf{A}$ and $\mathbf{A'A}$ are given in Example 8.2.5 and it is clear that the design is not orthogonal. The model can be re-written in various equivalent ways, of the form

$$y = \alpha_0 P_0(x) + \alpha_1 P_1(x) + \alpha_2 P_2(x) + \ldots + \alpha_s P_s(x) + e, \qquad (8.2.6)$$

where, for $r = 0, 1, \ldots, s$, $P_r(x)$ is a polynomial of degree r in x, $(P_0(x) \equiv 1)$, the coefficients of which (together with $\beta_0, \ldots, \beta_s$) determine $\alpha_0, \ldots, \alpha_s$. For this new model, the design matrix is

$$\mathbf{A} = \begin{bmatrix} P_0(x_1) & P_1(x_1) & \cdots & P_s(x_1) \\ P_0(x_2) & P_1(x_2) & \cdots & P_s(x_2) \\ \vdots & \vdots & & \vdots \\ P_0(x_n) & P_1(x_n) & \cdots & P_s(x_n) \end{bmatrix},$$

so that $\mathbf{A}'\mathbf{A}$ has (j, k) element A_{jk} given by

$$A_{jk} = \sum_{i=1}^{n} P_j(x_i)P_k(x_i).$$

Thus $\mathbf{A}'\mathbf{A}$ will be a diagonal matrix provided that

$$\sum_{i=1}^{n} P_j(x_i)P_k(x_i) = 0, \qquad j = 0, 1, \ldots, s; \qquad k = 0, 1, \ldots, s; \quad j \neq k.$$

Polynomials satisfying this condition are termed *orthogonal* and are uniquely determined by $x_1, x_2, \ldots, x_n$ apart from multiplicative constants. The jth diagonal element of $\mathbf{A}'\mathbf{A}$ is $A_{jj} = \sum_{i=1}^{n} P_j^2(x_i)$, whence the least squares estimate of α_j is

$$\tilde{\alpha}_j = \sum_{i=1}^{n} y_i P_j(x_i) \bigg/ \sum_{i=1}^{n} P_j^2(x_i); \qquad j = 0, 1, \ldots, s,$$

while the variance associated with this estimate is $\sigma^2 / \sum_{i=1}^{n} P_j^2(x_i)$. In particular, since $P_0(x) \equiv 1$, we have $\sum y_i P_0(x_i) = \sum y_i$ and $A_{11} = n$, so $\tilde{\alpha}_0 = \bar{y}$ and the L.S.E. of α_0 has variance σ^2/n. We note that, since $\mathbf{A}'\mathbf{A}$ is a diagonal matrix, the L.S. estimators of the parameters are always uncorrelated in orthogonal designs.

From (8.2.5), we see that the sum of squared residuals may be written as

$$r = \sum y_i^2 - \tilde{\alpha}_0 y_i - \tilde{\alpha}_1 \sum y_i P_1(x_i) - \ldots - \tilde{\alpha}_s \sum y_i P_s(x_i),$$

or alternatively, since $\sum y_i P_j(x_i) = \tilde{\alpha}_j \sum P_j^2(x_i)$, as

$$r = \sum y_i^2 - \tilde{\alpha}_0^2 n - \tilde{\alpha}_1^2 \sum P_1^2(x_i) - \ldots - \tilde{\alpha}_s^2 \sum P_s^2(x_i).$$

We mentioned above that orthogonal polynomials are only determined up to multiplicative constants. It does not matter which constants are chosen, for if we consider

$$y = \alpha_0 P_0(x) + \alpha_1' P_1'(x) + \ldots + \alpha_s' P_s'(x) + e$$

where $P_j'(x) = c_j P_j(x)$, for $j = 1, 2, \ldots, s$, (the c_j's being constants), then it is easy to see that $\tilde{\alpha}_j' = \tilde{\alpha}_j/c_j$, so that $\sum_1^s \alpha_j' P_j'(x) = \sum_1^s \alpha_j P_j(x)$; also the residual is the same in both cases.

EXAMPLE 8.2.15.   *Orthogonal polynomials in simple linear regression.*   The earlier discussion (Example 8.2.5) of linear regression on one variable $(s = 1)$

used the model

$$y_i = \beta_0 + \beta_1 x_i + e_i, \qquad i = 1, 2, \ldots, n.$$

This model is not orthogonal; $\mathbf{A'A}$, $(\mathbf{A'A})^{-1}$, $\tilde{\beta}_0$, $\tilde{\beta}_1$ were given in Example 8.2.5. An alternative model which *is* orthogonal is

$$y_i = \beta_0' + \beta_1(x_i - \bar{x}) + e_i, \qquad i = 1, 2, \ldots, n.$$

Here the x variable is measured about the mean $\bar{x}$ of its observed values $x_1, x_2, \ldots, x_n$, and β_0' satisfies $\beta_0 = \beta_0' - \beta_1 \bar{x}$. For this new model we have

$$\mathbf{A'A} = \begin{pmatrix} n & \sum (x_i - \bar{x}) \\ \sum (x_i - \bar{x}) & \sum (x_i - \bar{x})^2 \end{pmatrix} = \begin{pmatrix} n & 0 \\ 0 & \sum (x_i - \bar{x})^2 \end{pmatrix},$$

$$\mathbf{A'y} = \begin{pmatrix} \sum y_i \\ \sum y_i(x_i - \bar{x}) \end{pmatrix},$$

whence $\tilde{\beta}_0' = \bar{y}$, $\tilde{\beta}_1 = \sum y_i(x_i - \bar{x})/\sum (x_i - \bar{x})^2$. Thus $\tilde{\beta}_1$ is the same for both models, while $\tilde{\beta}_0 = \tilde{\beta}_0' - \tilde{\beta}_1 \bar{x}$. Further we have,

$$\operatorname{var} \tilde{B}_0' = \sigma^2/n, \qquad \operatorname{var} \tilde{B}_1 = \sigma^2/\sum (x_i - \bar{x})^2,$$

and $\tilde{B}_0'$, $\tilde{B}_1$ are uncorrelated (whereas $\tilde{B}_0$, $\tilde{B}_1$ were not).

In the case of multiple linear regression, we can use the same technique to produce a partial orthogonality, by re-writing the model

$$y_i = \beta_0 + \beta_1 x_{i1} + \beta_2 x_{i2} + \ldots + \beta_s x_{is} + e_i$$

in the alternative 'mean-corrected' form

$$y_i = \beta_0' + \beta_1(x_{i1} - x_{\cdot 1}) + \beta_2(x_{i2} - x_{\cdot 2}) + \ldots + \beta_s(x_{is} - x_{\cdot s}) + e_i,$$

where $x_{\cdot j}$ is the mean of the observations $x_{1j}, x_{2j}, \ldots, x_{nj}$ on x_j. In this case β_0' satisfies $\beta_0 = \beta_0' - \sum_1^s \beta_j x_{\cdot j}$. For the second model the first row of $\mathbf{A'A}$ is $[n, \sum (x_{i1} - x_{\cdot 1}), \ldots, \sum (x_{is} - x_{\cdot s})]$, i.e. $(n, 0, 0, \ldots, 0)$; it follows that the first row of $(\mathbf{A'A})^{-1}$ is $(n^{-1}, 0, 0, \ldots, 0)$, whence the L.S. estimator of β_0' is just $\bar{y} = \sum y_i/n$. Furthermore, each of the L.S. estimators of $\beta_1, \beta_2, \ldots, \beta_s$ is uncorrelated with the L.S. estimator of β_0'.

Returning to the orthogonal polynomial model of (8.2.6), it turns out that when the x-values are unequally spaced, the polynomials must be specially constructed for each set of x-values, which negates the advantage of the orthogonal model in simplifying the solution of the normal equations. However, when the values of x are equally spaced, and all different, we can use polynomials with a standardized argument, for which tables are available.

We consider the standardized variable $u = (x - \bar{x})/h$, where $\bar{x} = \sum x_i/n$ and $h = x_{i+1} - x_i$, $i = 1, 2, \ldots, n-1$. Thus, the values of u corresponding to

$x_1, x_2, \ldots, x_n$, viz. $u_1, u_2, \ldots, u_n$ are (with obvious re-arrangement)

$$0, \pm 1, \pm 3, \ldots, \pm(n-1)/2, \quad \text{if } n \text{ is odd};$$

$$\pm\tfrac{1}{2}, \pm\tfrac{3}{2}, \ldots, \pm(n-1)/2, \quad \text{if } n \text{ is even};$$

i.e. $u_i = i - (n+1)/2$, $i = 1, 2, \ldots, n$.

The model actually fitted is then of the form

$$y = \gamma_0 + \gamma_1 Q_1(u) + \ldots + \gamma_s Q_s(u) + e,$$

where the Q_i's are orthogonal polynomials chosen for tabular and computational convenience so that each takes integral values at $u_1, u_2, \ldots, u_n$. Biometrika Tables (Table 47) [Pearson and Hartley (1966)—see Bibliography G] gives the functional form of $Q_j(u)$ for $j = 1, 2, \ldots, 6$ and for $3 \le n \le 52$, together with the numerical values of $Q_j(u_i)$ and $\sum_{i=1}^{n} Q_j^2(u_i)$. We illustrate the method with a numerical example.

EXAMPLE 8.2.16. *Fitting a cubic with orthogonal polynomials.* The following monthly rainfall record was recorded:

Month	Jan	Feb	Mar	Apr	May	Jun	Jul	Aug	Sep	Oct	Nov	Dec
Rainfall	2·52	2·68	2·92	2·85	3·26	3·44	3·01	2·51	2·54	2·17	2·07	2·35

We code the months $1, 2, \ldots, 12$ and take these as the values of x. The rainfall recorded is the corresponding value of Y. We shall fit various polynomial models to this data, in the variable $u = x - 13/2$, using the following table:

u	$-\tfrac{11}{2}$	$-\tfrac{9}{2}$	$-\tfrac{7}{2}$	$-\tfrac{5}{2}$	$-\tfrac{3}{2}$	$-\tfrac{1}{2}$	$\tfrac{1}{2}$	$\tfrac{3}{2}$	$\tfrac{5}{2}$	$\tfrac{7}{2}$	$\tfrac{9}{2}$	$\tfrac{11}{2}$
Rainfall, y	2·52	2·68	2·92	2·85	3·26	3·44	3·01	2·51	2·54	2·17	2·07	2·35
$Q_1(u)$	−11	−9	−7	−5	−3	−1	1	3	5	7	9	11
$Q_2(u)$	55	25	1	−17	−29	−35	−35	−29	−17	1	25	55
$Q_3(u)$	−33	3	21	25	19	7	−7	−19	−25	−21	−3	33

The last three rows are obtained from the tables referred to above, from which we also find that

$$\sum Q_1^2(u_i) = 572, \qquad \sum Q_2^2(u_i) = 12{,}012 \quad \text{and} \quad \sum Q_3^2(u_i) = 5148.$$

We begin by fitting the model

$$y = \gamma_0 + \gamma_1 Q_1(u) + e_{(1)}.$$

From the data we calculate

$$\sum y_i = 32\cdot32 \quad \text{and} \quad \sum y_i Q_1(u_i) = -16\cdot84.$$

The least squares estimates of γ_0 and γ_1 are then

$$\tilde{\gamma}_0 = \bar{y} = 2\cdot693333$$

$$\tilde{\gamma}_1 = \sum y_i Q_1(u_i) / \sum Q_1^2(u_i) = -0\cdot029441.$$

From the tables, with $n = 12$, we find that $Q_1(u) = 2u$; the fitted polynomial is thus

$$y = 2 \cdot 6933 - 0 \cdot 0589 u.$$

On substitution of $u = x - 13/2$ this becomes

$$y = 3 \cdot 0761 - 0 \cdot 0589 x.$$

This is the result we would get by applying least squares to the non-orthogonal model

$$y = \beta_0 + \beta_1 x + e_{(1)}.$$

We now fit the quadratic model

$$y = \gamma_0 + \gamma_1 Q_1(u) + \gamma_2 Q_2(u) + e_{(2)}.$$

The estimates of γ_0 and γ_1 are unchanged; the only calculation required is $\sum y_i Q_2(u_i) = -93 \cdot 02$, giving

$$\tilde{\gamma}_2 = \sum y_i Q_2(u_i) / \sum Q_2^2(u_i) = -0 \cdot 007744.$$

From the tables $(n = 12)$ we have $Q_2(u) = 3u^2 - 143/4$, so that the fitted polynomial is

$$y = 2 \cdot 6933 - 0 \cdot 0589 u - 0 \cdot 0077 (3u^2 - 35 \cdot 75),$$

or, in terms of x,

$$y = 2 \cdot 3714 + 0 \cdot 2431 x - 0 \cdot 0232 x^2.$$

This is the result we would get by applying least squares to the natural quadratic model

$$y = \beta_0 + \beta_1 x + \beta_2 x^2 + e_{(2)}.$$

Note that, in contrast to the orthogonal model, the estimates of β_0 and β_1 are not the same as in the first-degree model.

To fit the cubic model

$$y = \gamma_0 + \gamma_1 Q_1(u) + \gamma_2 Q_2(u) + \gamma_3 Q_3(u) + e_{(3)},$$

we require only $\sum y_i Q_3(u_i)$, which is equal to $36 \cdot 98$, giving

$$\tilde{\gamma}_3 = \sum y_i Q_3(u_i) / \sum Q_3^2(u_i) = 0 \cdot 007183.$$

The tables show that $Q_3(u) = (2/3)[u^3 - 21 \cdot 25 u]$, so that the fitted cubic is

$$y = 2 \cdot 6933 - 0 \cdot 0589 u - 0 \cdot 0077 (3u^2 - 35 \cdot 75) + 0 \cdot 0048 (u^3 - 21 \cdot 25 u),$$

or

$$y = 1 \cdot 7177 + 0 \cdot 2348 x - 0 \cdot 1166 x^2 + 0 \cdot 0048 x^3,$$

and this is the result we would obtain on applying least squares to the model

$$y = \beta_0 + \beta_1 x + \beta_2 x^2 + \beta_3 x^3 + e_{(3)}.$$

The question which of these models best fits the data will be discussed later in section 8.3.5; the procedure involves fitting several models of increasing degree. The use of orthogonal polynomials in this context is thus particularly advantageous as the additional calculations at each stage are slight, whereas the use of non-orthogonal models necessitates a matrix inversion at each stage (such computational ease is, of course, less important when the calculations are carried out using a computer).

8.2.6.   Modifications for Observations of Unequal Accuracy; Weighted Least Squares

So far we have supposed that the observations $y_1, y_2, \ldots, y_n$ are equally reliable, i.e. that the random errors $\varepsilon_1, \varepsilon_2, \ldots, \varepsilon_n$ have equal variances. This is not always the case: $\mathcal{D}(\varepsilon)$ may be diagonal but with unequal diagonal elements $\sigma_1^2, \sigma_2^2, \ldots, \sigma_n^2$. We suppose that these variances are known up to a constant factor σ^2, which is unknown, i.e. $\sigma_i^2 = v_i \sigma^2$, where v_i, $i = 1, 2, \ldots, n$ is known. We have

$$Y_i = \sum_{j=1}^{p} a_{ij} \theta_j + \varepsilon_i, \qquad i = 1, 2, \ldots, n.$$

Re-scaling by the transformation $Z_i = Y_i v_i^{-1/2}$ we obtain the new model

$$Z_i = \sum_{j=1}^{p} b_{ij} \theta_j + \delta_i, \qquad i = 1, 2, \ldots, n$$

where $b_{ij} = a_{ij} v_i^{-1/2}$, $\delta_i = \varepsilon_i v_i^{-1/2}$.
 Now for $i = 1, \ldots, n$,

$$E(\varepsilon_i) = 0 \quad \text{implies } E(\delta_i) = 0,$$

and

$$\text{var } \varepsilon_i = v_i \sigma^2 \quad \text{implies var } \delta_i = \sigma^2.$$

Clearly, if the ε_i's are uncorrelated, then so are the δ_i's. Thus we can apply the methods already discussed to this new model, the least squares estimates of $\theta_1, \theta_2, \ldots, \theta_p$ being found by minimizing

$$S(\mathbf{u}) = \sum_{i=1}^{n} \left[z_i - \sum_{j=1}^{n} b_{ij} u_j \right]^2$$

where z_i is the realized value of Z_i. In terms of the original model we have (substituting $z_i = y_i v_i^{-1/2}$, $b_{ij} = a_{ij} v_i^{-1/2}$)

$$S(\mathbf{u}) = \sum_{i=1}^{n} v_i^{-1} \left[y_i - \sum_{j=1}^{n} a_{ij} u_j \right]^2.$$

Thus, instead of minimizing the sum of squared error estimates as we did previously (in 8.2.2), we now weight each term in that sum by the inverse of the variance of the corresponding random variable. This gives more weight to the observations for which the errors are likely to be small, an intuitively attractive procedure, referred to as '*weighted least squares*'.

Expressed in matrix terms, the new model is $Z = B\theta + \delta$, with the design matrix $B = W^{-1}A$, where $W = \text{diag}(v_1^{1/2}, v_2^{1/2}, \ldots, v_n^{1/2})$ and $Z = (Z_1, Z_2, \ldots, Z_n)'$ given by $Z = W^{-1}Y$.

The least squares estimate of θ is thus

$$\tilde{\theta} = (B'B)^{-1}B'z,$$

where z is the realized value of Z.

Substituting $B = W^{-1}A$ and noting that W is symmetric, we have

$$\tilde{\theta} = (A'W^{-1}W^{-1}A)^{-1}A'W^{-1}W^{-1}y.$$

Now $WW = \text{diag}(v_1, v_2, \ldots, v_n) = V$, say, so that

$$\tilde{\theta} = (A'V^{-1}A)^{-1}A'V^{-1}y, \tag{8.2.7}$$

and this is the value of u which minimizes the weighted sum of squares $S(u) = (y - Au)'V^{-1}(y - Au)$.

The dispersion matrix of $\tilde{\Theta}$ is

$$\mathcal{D}(\tilde{\Theta}) = \sigma^2(B'B)^{-1} = \sigma^2(A'V^{-1}A)^{-1}.$$

EXAMPLE 8.2.17.   *Repeated observations.*   Suppose that Y_i is the average of m_i observations, all with the same expectation $\sum_{j=1}^{p} a_{ij}\theta_j$, and common variance σ^2, $i = 1, 2, \ldots, n$. If all observations are independent, then Y_i has variance σ^2/m_i and the above results apply with $V = \text{diag}(m_1^{-1}, m_2^{-1}, \ldots, m_n^{-1})$.

EXAMPLE 8.2.18.   *A linear law passing through the origin.*   Consider the regression model

$$E(Y_i) = \theta x_i, \qquad i = 1, 2, \ldots, n.$$

Here, $A = (x_1, x_2, \ldots, x_n)'$. Using (8.2.7) we find

$$\tilde{\theta} = \sum_{i=1}^{n} \frac{x_i y_i}{v_i} \bigg/ \sum_{i=1}^{n} \frac{x_i^2}{v_i},$$

while the variance of $\tilde{\Theta}$ is $\sigma^2/\sum(x_i^2/v_i)$.

As an illustration, suppose that Y denotes rate of heat loss from a standard type of house (e.g. three bedroom semi-detached) and x represents the difference between internal and external temperature. In such a case, the variance of Y will usually increase as x increases. A simple form of dependence is $v_i = cx_i$, i.e. the variance of Y_i is proportional to the temperature difference.

In this case we readily find that

$$\tilde{\theta} = \frac{\sum y_i}{\sum x_i} = \frac{\bar{y}}{\bar{x}},$$

and var $\tilde{\Theta} = {}_c\sigma^2/\sum x_i = c\sigma^2/(n\bar{x})$.

Alternatively, if the standard deviation of Y_i is proportional to x_i, i.e. $v_i = cx_i^2$, then

$$\tilde{\theta} = \frac{1}{n}\sum_{i=1}^{n}\frac{y_i}{x_i} \quad \text{and} \quad \text{var } \tilde{\Theta} = c\sigma^2/n.$$

8.2.7. Modifications for Non-independent Observations

The method used in section 8.2.6 is a special case of a more general technique, which allows us to deal with observations, that are not only unequally accurate but non-independent, provided that the dispersion matrix of the errors is known up to a constant factor, i.e. we suppose that $\mathcal{D}(\varepsilon) = \sigma^2 V$, where V is a known positive definite [see I, Theorem 7.9.1] symmetric matrix, now not necessarily diagonal.

For such a V, it can be shown that there is a (not necessarily unique) nonsingular symmetric matrix W such that $W'W = V$, for example by the Choleski factorization [see III, § 4.2]. The original model is

$$Y = A\theta + \varepsilon, \quad \text{with } E(\varepsilon) = 0, \qquad \mathcal{D}(\varepsilon) = \sigma^2 V.$$

Pre-multiplying both sides by W^{-1} we obtain the new model

$$Z = B\theta + \delta,$$

where $Z = W^{-1}Y$, $B = W^{-1}A$, $\delta = W^{-1}\varepsilon$.

Now

$$E(\delta) = W^{-1}E(\varepsilon) = 0,$$

and

$$\mathcal{D}(\delta) = W^{-1}\mathcal{D}(\varepsilon)(W^{-1})'$$

$$= \sigma^2 W^{-1}VW^{-1}, \quad \text{since } W \text{ is symmetric}$$

$$= \sigma^2 W^{-1}WWW^{-1} = \sigma^2 I_n.$$

The new model is thus in standard form, and $\tilde{\theta} = (B'B)^{-1}B'z$ gives the minimum value of $S(u) = (z - Bu)'(z - Bu)$. In terms of the original model,

$$S(u) = (y - Au)V^{-1}(y - Au).$$

Although W is no longer diagonal, it is symmetric, so the algebra given in subsection 8.2.6 still applies and

$$\tilde{\theta} = (A'V^{-1}A)^{-1}A'V^{-1}y,$$
$$\mathcal{D}(\tilde{\Theta}) = \sigma^2(A'V^{-1}A)^{-1}.$$

The sum of squared residuals is

$$r = S(\tilde{\mathbf{\theta}}) = (\mathbf{y} - \mathbf{A}\tilde{\mathbf{\theta}})' \mathbf{V}^{-1} (\mathbf{y} - \mathbf{A}\tilde{\mathbf{\theta}}),$$

and the error variance σ^2 is estimated by $\tilde{\sigma}^2 = r/(n-p)$.

EXAMPLE 8.2.19.   *Least squares analysis of order statistics.*   Suppose that $y_1, y_2, \ldots, y_n$ is a sample of independent observations from a distribution depending only on two parameters μ and σ, which we assume to be the expected value and standard deviation (the following also applies more generally when μ and σ are arbitrary location and scale parameters). We consider the estimation of μ and σ from the *ordered* sample $y_{(1)}, y_{(2)}, \ldots, y_{(n)}$, where $y_{(1)} < y_{(2)} < \ldots < y_{(n)}$ [see § 14.3].

Let $s_j = (y_j - \mu)/\sigma$, $j = 1, 2, \ldots, n$. These are observations from a standardized distribution which does not depend on any parameters; ordering them in ascending order as $s_{(1)}, s_{(2)}, \ldots, s_{(n)}$ (so that $s_{(j)} = (y_{(j)} - \mu)/\sigma$) we therefore have for the corresponding induced random variables

$$E(S_{(j)}) = m_j, \qquad \text{var}(S_{(j)}) = v_{jj}, \qquad \text{cov}(S_{(j)}, S_{(k)}) = v_{jk}.$$

These quantities may be found explicitly as their values depend on the form of the original distribution but not on μ and σ. In terms of the corresponding random variables $Y_{(1)}, Y_{(2)}, \ldots, Y_{(n)}$ we have

$$E(Y_{(j)}) = \mu + \sigma m_j, \qquad \text{var}(Y_{(j)}) = \sigma^2 v_{jj}, \qquad \text{cov}(Y_{(j)}, Y_{(k)}) = \sigma^2 v_{jk},$$

so that we may now apply the above results with

$$\mathbf{\theta} = \begin{pmatrix} \mu \\ \sigma \end{pmatrix}, \qquad \mathbf{A}' = \begin{pmatrix} 1 & 1 & \cdots & 1 \\ m_1 & m_2 & \cdots & m_n \end{pmatrix}, \qquad \mathbf{V} = \{v_{jk}\}.$$

8.3.   ANALYSIS OF VARIANCE AND TESTS OF HYPOTHESES FOR FULL-RANK DESIGNS

8.3.1.   Normal Theory

The methods and results given so far have not depended on the shape of the distribution of the errors. In most applications, however, we require not only estimates of parameters and their standard errors, but confidence regions [see § 4.9] and methods of testing hypotheses [see Chapter 2] of interest.

For these purposes, we assume now that the error distribution is Normal, i.e. ε_i is $N(0, \sigma)$ for $i = 1, 2, \ldots, n$. The assumption that ε_i and ε_j are uncorrelated now implies their independence [see II, Theorem 13.4.1]. The vector of errors, $\mathbf{\varepsilon}$, is thus multivariate Normal [see II, § 13.4] MVN $(0, \sigma^2 I_n)$. In terms of the observations themselves,

$$Y_i \quad \text{is independent of } Y_j, \qquad i \neq j,$$

and

$$Y_i \quad \text{is} \quad N\left(\sum_{j=1}^{p} a_{ij}\theta_j, \sigma\right), \qquad i = 1, \ldots, n.$$

The vector of observations $\mathbf{Y}$ is multivariate Normal MVN $(\mathbf{A\theta}, \sigma^2\mathbf{I}_n)$. It follows that the least-squares estimator $\tilde{\mathbf{\Theta}} = (\mathbf{A'A})^{-1}\mathbf{A'Y}$ is MVN with expectation $\mathbf{\theta}$ and with dispersion matrix $(\mathbf{A'A})^{-1}\mathbf{A'}\mathcal{D}(\mathbf{Y})\mathbf{A}(\mathbf{A'A})^{-1} = (\mathbf{A'A})^{-1}\sigma^2$. The sum of squared residuals is

$$\sum_{i=1}^{n}\left\{y_i - \sum_{j=1}^{n}a_{ij}\tilde{\theta}_j\right\}^2 = (\mathbf{y} - \mathbf{A\tilde{\theta}})'(\mathbf{y} - \mathbf{A\tilde{\theta}})$$

$$= \mathbf{y'y} - \tilde{\mathbf{\theta}}'\mathbf{A'y} - \mathbf{y'A\tilde{\theta}} + \tilde{\mathbf{\theta}}'\mathbf{A'A\tilde{\theta}}.$$

Since $\tilde{\mathbf{\theta}} = (\mathbf{A'A})^{-1}\mathbf{A'y}$, this reduces to $\mathbf{y'Cy}$ where

$$\mathbf{C} = \mathbf{I} - \mathbf{A}(\mathbf{A'A})^{-1}\mathbf{A'} = \mathbf{C'} = \mathbf{C}^2.$$

Since $\mathbf{C} = \mathbf{C}^2$, it follows from Theorem 2.5.1 that R/σ^2 is a χ^2 variable, the degrees of freedom being $n - p$. The linear combination $\mathbf{\lambda'}\tilde{\mathbf{\Theta}}$ is Normal with expectation $\mathbf{\lambda'\theta}$ and variance $\mathbf{\lambda'}(\mathbf{A'A})^{-1}\mathbf{\lambda}\sigma^2$.

8.3.2. Confidence Regions

A confidence interval for an individual parameter, say θ_j, can be obtained in the usual way, using Student's distribution [see Example 4.5.3]. We have the following property:

$$(\tilde{\Theta}_j - \theta_j)/\{A^{jj}R/(n-p)\}^{1/2}$$

has Student's distribution with $n - p$ d.f. [see § 2.5.5] where $(\mathbf{A'A})^{-1} = \{A^{ij}\}$. A central $100(1 - \alpha)\%$ confidence interval for θ_j is thus given by

$$(\tilde{\theta}_j \pm t^{(\alpha)}_{n-p}\tilde{\sigma}\sqrt{A^{jj}}),$$

where $t^{(\alpha)}_{n-p} = t_{n-p}(1 - \tfrac{1}{2}\alpha)$ is the upper $100(1 - \tfrac{1}{2}\alpha)\%$ point of Student's distribution on $n - p$ d.f. For an orthogonal design this is just

$$(\tilde{\theta}_j \pm t^{(\alpha)}_{n-p}\tilde{\sigma}A_{jj}^{-1/2}),$$

where A_{jj} is the jth diagonal element of $\mathbf{A'A}$. Similarly a central $100(1 - \alpha)\%$ confidence interval for $\mathbf{\lambda'\theta}$ is given by

$$(\mathbf{\lambda'}\tilde{\mathbf{\theta}} \pm t^{(\alpha)}_{n-p}\tilde{\sigma}\{\mathbf{\lambda'}(\mathbf{A'A})^{-1}\mathbf{\lambda}\}^{1/2}).$$

When two or more parameters are considered together we must remember that the individual intervals cannot be 'multiplied' to construct a confidence box, because they are not independent (even if the $\tilde{\Theta}_j$ are) owing to the fact that each uses the same residuals [cf. § 4.9]. A valid confidence interval for several parameters $\theta_{j_1}, \theta_{j_2}, \ldots, \theta_{j_q}$, $(q \leq p)$, may be obtained as follows. Let

$\boldsymbol{\phi} = (\theta_{j_1}, \theta_{j_2}, \ldots, \theta_{j_q})'$. The least squares estimator of $\boldsymbol{\phi}$ is $\tilde{\boldsymbol{\Phi}} = (\tilde{\Theta}_{j_1}, \ldots, \tilde{\Theta}_{j_q})'$. The dispersion matrix of $\tilde{\boldsymbol{\Phi}}$, which we denote by $\sigma^2 \mathbf{Q}$, is found by removing from $\sigma^2 (\mathbf{A}'\mathbf{A})^{-1}$ all rows and columns relating to parameters not included in $\boldsymbol{\phi}$, i.e. we retain rows (cols) $j_1, j_2, \ldots, j_q$. It can be shown that

$$(\tilde{\boldsymbol{\Phi}} - \boldsymbol{\phi})'\mathbf{Q}^{-1}(\tilde{\boldsymbol{\Phi}} - \boldsymbol{\phi})/\{qR/(n-p)\}$$

has an $F_{q,n-p}$ distribution [see § 2.5.6]. A $100(1-\alpha)\%$ confidence region for $\boldsymbol{\phi}$ is thus [cf. § 4.9.2]

$$\{\boldsymbol{\phi} : (\tilde{\boldsymbol{\Phi}} - \boldsymbol{\phi})'\mathbf{Q}^{-1}(\tilde{\boldsymbol{\Phi}} - \boldsymbol{\phi}) \le q\tilde{\sigma}^2 F_{q,n-p}(1-\alpha)\}$$

where $F_{q,n-p}(1-\alpha)$ is the upper $100(1-\alpha)\%$ point.

In general, the set of points thus defined is a q-dimensional ellipsoid centred at $\tilde{\boldsymbol{\Phi}}$; for orthogonal designs it is a q-dimensional sphere.

EXAMPLE 8.3.1. *Confidence ellipsoid for the gravity constants of Example 8.2.2.* For the gravity example, we found

$$\tilde{g}_1 = 981{\cdot}2681, \quad \text{with standard error } \tilde{\sigma}\sqrt{A^{11}} = 0{\cdot}0011,$$

$$\tilde{g}_5 = 981{\cdot}1241, \quad \text{with standard error } \tilde{\sigma}\sqrt{A^{55}} = 0{\cdot}0017,$$

$$\tilde{\tilde{g}} = 981{\cdot}2091, \quad \text{with standard error } 0{\cdot}0010.$$

For this example $n = 9$, $p = 5$; so, assuming a Normal error distribution, 95% confidence intervals for the corresponding parameters are found using $t_4(0{\cdot}975) = 2{\cdot}776$; we obtain

$$g_1 : (981{\cdot}2681 \pm 0{\cdot}0031) = (981{\cdot}2650, 981{\cdot}2712)$$

$$g_5 : (981{\cdot}1241 \pm 0{\cdot}0047) = (981{\cdot}1194, 981{\cdot}1288)$$

$$\bar{g} : (981{\cdot}2091 \pm 0{\cdot}0028) = (981{\cdot}2063, 981{\cdot}2119).$$

It is tempting to use the rectangle defined by

$$\{(g_1, g_5) : 981{\cdot}2650 \le g_1 \le 981{\cdot}2712, 981{\cdot}1194 \le g_5 \le 981{\cdot}1288\}$$

as a confidence set for (g_1, g_5); such a procedure is only valid if the individual intervals are independent (when its confidence coefficient is $(0{\cdot}95)^2$); as they are not, we must proceed as follows [cf. § 4.9.2]. The dispersion matrix of $\tilde{\boldsymbol{\Phi}} = (\tilde{G}_1, \tilde{G}_5)$ is

$$\sigma^2 \begin{pmatrix} 0{\cdot}4211 & 0{\cdot}2632 \\ 0{\cdot}2632 & 1{\cdot}2895 \end{pmatrix}.$$

A $100(1-\alpha)\%$ confidence region for (g_1, g_5) is then

$$\left\{ (g_1, g_5) : (\tilde{g}_1 - g_1, \tilde{g}_5 - g_5) \begin{pmatrix} 0{\cdot}4211 & 0{\cdot}2632 \\ 0{\cdot}2632 & 1{\cdot}2895 \end{pmatrix}^{-1} \begin{pmatrix} \tilde{g}_1 - g_1 \\ \tilde{g}_5 - g_5 \end{pmatrix} \le 2\tilde{\sigma}^2 F_{2,4}(1-\alpha) \right\},$$

which defines an elliptical region in the g_1, g_5 plane, centred at $(\tilde{g}_1, \tilde{g}_5) =$

(981·2681, 981·1241), the extent of which depends on α (larger ellipses corresponding to smaller values of α).

EXAMPLE 8.3.2. *Confidence interval for a regression.* On the assumption that the second degree polynomial model fitted in Example 8.2.16 is correct, we can obtain a confidence interval for the true average rainfall in a particular month, say August, as follows.

We have

$$y = \gamma_0 + \gamma_1 Q_1(u) + \gamma_2 Q_2(u) + e_{(2)}.$$

Now, August corresponds to $u = 3/2$, so the expected rainfall in August is

$$\mu_8 = \gamma_0 + \gamma_1 Q_1(\tfrac{3}{2}) + \gamma_2 Q_2(\tfrac{3}{2}).$$

The L.S. estimate of μ_8 is $\tilde{\mu}_8 = \tilde{\gamma}_0 + \tilde{\gamma}_1 Q_1(\tfrac{3}{2}) + \tilde{\gamma}_2 Q_2(\tfrac{3}{2})$; evaluating this expression we find $\tilde{\mu}_8 = 2 \cdot 83$.

Now, due to the orthogonality, the L.S. estimators of $\gamma_0, \gamma_1, \gamma_2$ are uncorrelated, whence the variance of the L.S. estimator of μ_8 is

$$\sigma^2 \{1/n + Q_1^2(\tfrac{3}{2})/\textstyle\sum Q_1^2(u_i) + Q_2^2(\tfrac{3}{2})/\textstyle\sum Q_2^2(u_i)\},$$

which is equal to $0 \cdot 1691 \sigma^2$. In this case we have $n = 12$, $p = 3$, so a 95% confidence interval for μ_8 is

$$\tilde{\mu}_8 \pm t_9(0 \cdot 975)(0 \cdot 1691)^{1/2} \tilde{\sigma}$$

where $\tilde{\sigma} = r/(n - p)$, r being the SSR for the quadratic model. To find r we use

$$r = \textstyle\sum y_i^2 - \tilde{\gamma}_0 \sum y_i - \tilde{\gamma}_1 \sum y_i Q_1(u_i) - \tilde{\gamma}_2 \sum y_i Q_2(u_i),$$

which gives $r = 0 \cdot 706$, whence $\tilde{\sigma}^2 = r/9 = 0 \cdot 078$. The computed interval for μ_8 is thus $(2 \cdot 83 \pm 0 \cdot 26)$, i.e. $(2 \cdot 57, 3 \cdot 09)$.

8.3.3. Testing Hypotheses

In addition to estimating the parameters in the model $E(\mathbf{Y}) = \mathbf{A}\boldsymbol{\theta}$, and computing standard errors or confidence intervals, we often wish to test various hypotheses [see Chapter 5] concerning these parameters. A typical hypothesis H imposes constraints on the parameter vector $\boldsymbol{\theta}$, e.g. by specifying the values of one or more components of $\boldsymbol{\theta}$, or, more generally, by specifying the values of one or more linear combinations of these components. Now $\tilde{\boldsymbol{\theta}}$ will not generally satisfy the conditions of H, so we must apply least squares again to estimate $\boldsymbol{\theta}$ under the restrictions implied by H. This involves minimization of $S(\mathbf{u}) = (\mathbf{y} - \mathbf{A}\mathbf{u})'(\mathbf{y} - \mathbf{A}\mathbf{u})$ subject to H, leading to the L.S. estimate $\boldsymbol{\theta}^*$, which does satisfy H and which is generally not equal to $\tilde{\boldsymbol{\theta}}$.

As in the unconstrained case, the minimized sum of squares, viz. $S(\boldsymbol{\theta}^*)$, is a measure of how well the model (in this case the model subject to H) fits the data when $\boldsymbol{\theta}$ is estimated by $\boldsymbol{\theta}^*$, and $S(\boldsymbol{\theta}^*)$ is equal to the sum of squared residuals for this restricted model, which we denote by r_H. Obviously, the constrained minimum $S(\boldsymbol{\theta}^*)$ will satisfy $S(\boldsymbol{\theta}^*) \geq S(\tilde{\boldsymbol{\theta}})$; in other words, $r_H \geq r$. The difference $r_H - r$ is a measure of the extra fit achieved when the parameters are not restricted by H. If this difference is small—i.e. if the restricted model accounts for the variability in the data nearly as well as the unrestricted model—then H would appear reasonable; on the other hand, a large difference—indicating a much better fit when H is ignored—would tend to discredit H. Of course, a yardstick is required to judge 'smallness' or 'largeness', and the assessment essentially involves (after making standardizing adjustments) a comparison of $r_H - r$ with r. The tests will be developed on the assumption of Normal errors; as pointed out earlier the L.S.E. $\tilde{\boldsymbol{\theta}}$ is actually the M.L.E. in this case—correspondingly, the tests we describe are likelihood-ratio tests [see § 5.5]. The following sections will exemplify the procedure in detail.

8.3.4. The Basic Identities

For the original (i.e. unrestricted) model, we have the following important identity in the variable $\mathbf{u}$:

$$(\mathbf{y} - \mathbf{Au})'(\mathbf{y} - \mathbf{Au}) \equiv (\tilde{\boldsymbol{\theta}} - \mathbf{u})'\mathbf{A}'\mathbf{A}(\tilde{\boldsymbol{\theta}} - \mathbf{u}) + (\mathbf{y} - \mathbf{A}\tilde{\boldsymbol{\theta}})'(\mathbf{y} - \mathbf{A}\tilde{\boldsymbol{\theta}}) \qquad (8.3.1)$$

or, briefly,

$$S(\mathbf{u}) \equiv (\tilde{\boldsymbol{\theta}} - \mathbf{u})'\mathbf{A}'\mathbf{A}(\tilde{\boldsymbol{\theta}} - \mathbf{u}) + S(\tilde{\boldsymbol{\theta}}). \qquad (8.3.2)$$

This may be shown by writing $\mathbf{y} - \mathbf{Au}$ in the form $(\mathbf{y} - \mathbf{A}\tilde{\boldsymbol{\theta}}) + \mathbf{A}(\tilde{\boldsymbol{\theta}} - \mathbf{u})$, whence (8.3.1) follows on noting that $(\mathbf{y} - \mathbf{A}\tilde{\boldsymbol{\theta}})'\mathbf{A}(\tilde{\boldsymbol{\theta}} - \mathbf{u}) = 0$ since, by the normal equations, $\mathbf{A}'\mathbf{y} - \mathbf{A}'\mathbf{A}\tilde{\boldsymbol{\theta}} = 0$.

Substituting the particular value $\mathbf{u} = \boldsymbol{\theta}$ in (8.3.2) we have, since $S(\tilde{\boldsymbol{\theta}}) = r$, the equation

$$S(\boldsymbol{\theta}) = (\tilde{\boldsymbol{\theta}} - \boldsymbol{\theta})'\mathbf{A}'\mathbf{A}(\tilde{\boldsymbol{\theta}} - \boldsymbol{\theta}) + r. \qquad (8.3.3)$$

Remembering that $S(\boldsymbol{\theta}) = \mathbf{e}'\mathbf{e}$, we may write the corresponding equation in terms of the induced random variables as

$$\boldsymbol{\varepsilon}'\boldsymbol{\varepsilon} = (\tilde{\boldsymbol{\Theta}} - \boldsymbol{\theta})'\mathbf{A}'\mathbf{A}(\tilde{\boldsymbol{\Theta}} - \boldsymbol{\theta}) + R. \qquad (8.3.4)$$

Under the distributional assumptions of section 8.3.1, $\boldsymbol{\varepsilon}'\boldsymbol{\varepsilon}$ is the sum of the squares of n independent Normal variables, each with zero mean and variance σ^2; whence $\boldsymbol{\varepsilon}'\boldsymbol{\varepsilon}$ is $\sigma^2\chi_n^2$ [see § 2.5.4(a)]. Since $\tilde{\boldsymbol{\Theta}}$ is MVN $(\boldsymbol{\theta}, \sigma^2(\mathbf{A}'\mathbf{A})^{-1})$, the distribution of the first term on the right of (8.3.4) may be found using a standard theorem [see II, Theorem 13.4.5], and is thus $\sigma^2\chi_p^2$. Moreover, it can be shown that the two terms on the right are independent; it then follows from Theorem 2.5.4 that R is a $\sigma^2\chi_{n-p}^2$ variable. Thus, under these stronger assumptions we can easily confirm that $E(R) = (n - p)\sigma^2$, as stated in section 8.2.4.

Equation 8.3.3 above cannot be used to calculate the value of r since it involves sums of squares about unknown expected values; however, substituting $\mathbf{u} = \mathbf{0}$ in Equation 8.3.2 we obtain

$$S(\mathbf{0}) = \tilde{\mathbf{\theta}}\mathbf{A}'\mathbf{A}\tilde{\mathbf{\theta}} + S(\tilde{\mathbf{\theta}}).$$

Noting that $S(\mathbf{0}) = \mathbf{y}'\mathbf{y}$ this may be written as

$$\mathbf{y}'\mathbf{y} = \tilde{\mathbf{\theta}}\mathbf{A}'\mathbf{A}\tilde{\mathbf{\theta}} + r. \tag{8.3.5}$$

This is known as an analysis of variance (anova) identity [cf. § 5.8.7]. It shows how the *total sum of squares* $\mathbf{y}'\mathbf{y} = \sum y_i^2$ can be split up into two parts: $\tilde{\mathbf{\theta}}\mathbf{A}'\mathbf{A}\tilde{\mathbf{\theta}}$, the *sum of squares 'due to' the fitted model* (SSM), and r, the *residual sum of squares* (SSR), representing the residual variation in the data remaining after the model has been fitted, i.e.

$$\text{Total SS} = \text{SSM} + \text{SSR}.$$

In order to test various hypotheses, we shall develop in later sections a number of other anova identities by partitioning SSM into components associated with different sources of variation.

In (8.3.5) there are no unknowns, and we can calculate r from it as

$$r = \mathbf{y}'\mathbf{y} - \tilde{\mathbf{\theta}}'\mathbf{A}'\mathbf{A}\tilde{\mathbf{\theta}}, \tag{8.3.6}$$

thus avoiding the squaring and summing of separate individual residuals. From the normal equations $\mathbf{A}'\mathbf{y} = \mathbf{A}'\mathbf{A}\tilde{\mathbf{\theta}}$, we see that SSM can also be written as $\tilde{\mathbf{\theta}}'\mathbf{A}'\mathbf{y}$, i.e. the product of $\tilde{\mathbf{\theta}}$ and the right hand side of the normal equations. Hence r is also given by

$$r = \mathbf{y}'\mathbf{y} - \tilde{\mathbf{\theta}}'\mathbf{A}'\mathbf{y}, \tag{8.3.7}$$

a formula which was used earlier for the calculation of r in Examples 8.2.11 to 8.2.14.

We noted above that the total SS $\mathbf{y}'\mathbf{y}$ is equal to $S(\mathbf{0})$, which is the 'residual SS' under $H : \mathbf{\theta} = \mathbf{0}$, i.e. the 'residual' when no parameters are fitted to the data; $\text{SSM} = \tilde{\mathbf{\theta}}'\mathbf{A}'\mathbf{A}\tilde{\mathbf{\theta}} = S(\mathbf{0}) - S(\tilde{\mathbf{\theta}})$ is the amount by which this residual is reduced as a result of fitting the model $E(\mathbf{Y}) = \mathbf{A}\mathbf{\theta}$. To emphasize this we shall denote SSM by $r(\mathbf{\theta})$ and refer to this quantity as the reduction in total SS due to fitting $\mathbf{\theta}$.

We now let $\mathbf{\theta}^*$ denote the L.S. estimate of $\mathbf{\theta}$ subject to the hypothesis H. Substituting $\mathbf{u} = \mathbf{\theta}^*$ in the basic identity (8.3.2), we have

$$S(\mathbf{\theta}^*) = (\tilde{\mathbf{\theta}} - \mathbf{\theta}^*)\mathbf{A}'\mathbf{A}(\tilde{\mathbf{\theta}} - \mathbf{\theta}^*) + r.$$

Now $S(\mathbf{\theta}^*) = (\mathbf{y} - \mathbf{A}\mathbf{\theta}^*)'(\mathbf{y} - \mathbf{A}\mathbf{\theta}^*) = r_H$, the residual SS for the model restricted by H, so that the extra fit (i.e. extra reduction $r_H - r$ in the total SS) when the unrestricted model is fitted is given by

$$r_H - r = (\tilde{\mathbf{\theta}} - \mathbf{\theta}^*)'\mathbf{A}'\mathbf{A}(\tilde{\mathbf{\theta}} - \mathbf{\theta}^*). \tag{8.3.8}$$

Alternatively, we can think of this increase in residual as the loss of fit when $\mathbf{\theta}$ is constrained by H; it is often called the sum of squares due to the hypothesis, H. Although (8.3.8) will be useful for finding the extra reduction in some of the simpler cases which follow, the general procedure is to calculate both residuals and subtract.

8.3.5. Tests of Hypotheses in Orthogonal Designs

We consider an orthogonal model in which $\mathbf{A}'\mathbf{A}$ is diag $(A_{11}, \ldots, A_{pp})$ with the L.S.E. $\tilde{\boldsymbol{\theta}}$ having jth component $\tilde{\theta}_j = \sum_i y_i a_{ij} / A_{jj}$, $j = 1, \ldots, p$. The diagonal structure of $\mathbf{A}'\mathbf{A}$ allows us to express the sum of squares due to the fitted model in simple explicit form as

$$\text{SSM} = \tilde{\boldsymbol{\theta}}'\mathbf{A}'\mathbf{A}\tilde{\boldsymbol{\theta}} = \sum_{i=1}^{p} A_{ii}\tilde{\theta}_i^2, \tag{8.3.9}$$

and (8.3.6) becomes

$$r = \sum_{i=1}^{n} y_i^2 - \sum_{i=1}^{p} A_{ii}\tilde{\theta}_i^2. \tag{8.3.10}$$

The most elementary hypothesis of interest is that a particular parameter in the model, say θ_t, is zero, i.e. the parameter θ_t is not needed to model the data. Such a hypothesis may have various interpretations depending on the context; for example, it may represent (a) the absence of an effect in a one-way layout where the groups differ in respect of treatment received, or (b) it may denote merely that a term of degree t is not required in a polynomial regression model.

For orthogonal models, the fact that each of the p normal equations involves only one parameter makes it extremely easy to re-estimate $\boldsymbol{\theta}$ under $H: \theta_t = 0$. We simply ignore the tth equation (which involves θ_t only) since this is no longer applicable, and the rest are unchanged. Thus, the L.S. estimates of all the components of $\boldsymbol{\theta}$ except θ_t are the same whether H is true or not, i.e. $\tilde{\boldsymbol{\theta}}$ and $\boldsymbol{\theta}^*$ differ only in the tth component, which in $\boldsymbol{\theta}^*$ is zero by hypothesis. Hence the extra reduction when θ_t is included in the model is, from (8.3.8), $r_H - r = A_{tt}\tilde{\theta}_t^2$; we shall denote this by $r(\theta_t)$, or more briefly, by r_t. As mentioned in section 8.3.3, the test of H is carried out by comparing r_t with an appropriate standard. The exact procedure to be followed is suggested by noting that

(i) r_t is a realization of the random variable $R_t = A_{tt}\tilde{\Theta}_t^2$;

(ii) $\tilde{\Theta}_t$ is Normal $(\theta_t, \sigma A_{tt}^{-1/2})$, so that if H is true $(\theta_t = 0)$, R_t is a $\sigma^2\chi_1^2$ variable, with expectation σ^2;

(iii) r is a realization of R, which is a $\sigma^2\chi_{n-p}^2$ variable, with expectation $(n-p)\sigma^2$ [see § 2.5.4].

We therefore compare the realized values of R_t and $R/(n-p)$, since both quantities have expectation σ^2 when H is true. In addition, it can be shown that R_t and R are independent, so that $F = (n-p)R_t/R$ has the F-distribution [see § 2.5.6] with degrees of freedom 1 and $(n-p)$ when H is true. The appropriate critical region can be found by observing that when H is false we have $E(R_t) > \sigma^2$, a property which follows on writing R_t as

$$R_t = A_{tt}(\tilde{\Theta}_t - \theta_t + \theta_t)^2 = A_{tt}(\tilde{\Theta}_t - \theta_t)^2 + 2A_{tt}(\tilde{\Theta}_t - \theta_t)\theta_t + A_{tt}\theta_t^2,$$

whence $E(R_t) = \sigma^2 + A_{tt}\theta_t^2$.

Thus, although small values of F are possible when H is untrue, we do not regard such events as evidence against H, the critical region being simply of the form $f \geq c$, a constant, where $f = (n-p)r_t/r$. A test of size α therefore rejects the hypothesis that $\theta_t = 0$ if $f \geq F_{1,n-p}(1-\alpha)$. Noting that $f^{1/2} = \tilde{\theta}_t/(\tilde{\sigma}A_{tt}^{-1/2})$, where $\tilde{\sigma}^2 = r/(n-p)$, [see § 5.8.2] we see that this is equivalent to the t-test which rejects the hypothesis that $\theta_t = 0$ when $|\tilde{\theta}_t| > \tilde{\sigma}A_{tt}^{-1/2}t_{n-p}(1-\tfrac{1}{2}\alpha)$.

We can use a similar approach to test whether several components of $\boldsymbol{\theta}$ are simultaneously zero. Let $\boldsymbol{\phi}$ denote a subset of $(\theta_1, \ldots, \theta_p)$, comprising m members and let H specify $\boldsymbol{\phi} = \mathbf{0}$. To estimate $\boldsymbol{\theta}$ under H, observe that the normal equations for the restricted model are just those for the full model modified only by the removal of the m equations corresponding to the parameters in $\boldsymbol{\phi}$. Thus $\tilde{\boldsymbol{\theta}}$ and $\boldsymbol{\theta}^*$ differ only in the places corresponding to the parameters in $\boldsymbol{\phi}$, $\boldsymbol{\theta}^*$ having zeros in these positions. Hence, using (8.3.8), the extra fit when $\boldsymbol{\theta}$ is not constrained by H is

$$r_H - r = \sum_{\phi} A_{ii}\tilde{\theta}_i^2, \tag{8.3.11}$$

where $\sum_{\phi}$ denotes summation over values of i corresponding to the parameters in $\boldsymbol{\phi}$. We denote this extra reduction when the parameters of $\boldsymbol{\phi}$ are included in the fit by $r(\boldsymbol{\phi})$. Note that we have

$$r(\boldsymbol{\phi}) = \sum_{\phi} r_i; \tag{8.3.12}$$

in particular,

$$\text{SSM} = r(\boldsymbol{\theta}) = \sum_{i=1}^{p} r_i. \tag{8.3.13}$$

This important formula shows how the reduction in TSS due to fitting the model $E(\mathbf{Y}) = \mathbf{A}\boldsymbol{\theta}$ can be split up into p components, the jth component being the contribution of θ_j to this reduction. Note that the contribution to $r(\boldsymbol{\phi})$ of any parameter included in $\boldsymbol{\phi}$ is always the same, irrespective of which other parameters belong to $\boldsymbol{\phi}$; we shall see that this property does not hold in non-orthogonal designs.

Now $r(\boldsymbol{\phi})$ is a realization of $R(\boldsymbol{\phi}) = \sum_{\phi} A_{ii}\tilde{\Theta}_i^2$, so that if H is true $R(\boldsymbol{\phi})$ is the sum of the squares of m independent Normal variables, each with zero mean and variance σ^2, whence $R(\boldsymbol{\phi})$ is a $\sigma^2\chi_m^2$ variable under H. We therefore compare $r(\boldsymbol{\phi})/m$ with $r/(n-p)$, i.e. if H is true, the test is again the comparison of two estimates of σ^2. In fact, $R(\boldsymbol{\phi})$ is independent of R so that $F = (n-p)R(\boldsymbol{\phi})/mR$ is distributed as $F_{m,n-p}$ when H is true. The expectation of $R(\boldsymbol{\phi})$ is increased when H is false, and a test of size α rejects $\boldsymbol{\phi} = \mathbf{0}$ if $f \geq F_{m,n-p}(1-\alpha)$, where $f = (n-p)r(\boldsymbol{\phi})/mr$.

These procedures are usually conveniently summarized in analysis of variance (anova) tables. The basic anova $\mathbf{y}'\mathbf{y} = \text{SSM} + r$ is displayed in the following form:

Source	Sum of squares	Degrees of freedom	Mean square
Due to model	$\text{SSM} = \tilde{\boldsymbol{\theta}}'\mathbf{A}'\mathbf{A}\tilde{\boldsymbol{\theta}} = r(\boldsymbol{\theta})$ $= \sum_{i=1}^{p} A_{ii}\tilde{\theta}_i^2$	p	
Residual	$r = \mathbf{y}'\mathbf{y} - \text{SSM}$	$n - p$	$r/(n-p) = \tilde{\sigma}^2$
Total	$\mathbf{y}'\mathbf{y}$	n	

We can refine this table so as to facilitate the testing of hypotheses by partitioning SSM, as follows

Source	SS	DF	MS	f
Due to θ_1	$r_1 = A_{11}\tilde{\theta}_1^2$	1	r_1	$f_1 = r_1/\tilde{\sigma}^2$
Due to θ_2	$r_2 = A_{22}\tilde{\theta}_2^2$	1	r_2	$f_2 = r_2/\tilde{\sigma}^2$
$\vdots$	$\vdots$	$\vdots$	$\vdots$	$\vdots$
Due to θ_p	$r_p = A_{pp}\tilde{\theta}_p^2$	1	r_p	$f_p = r_p/\tilde{\sigma}^2$
Residual	$r = \mathbf{y}'\mathbf{y} - \sum_{i=1}^{p} r_i$	$n - p$	$\tilde{\sigma}^2 = r/(n-p)$	
Total	$\mathbf{y}'\mathbf{y}$	n		

In this table, 'due to θ_j' refers to the contribution $A_{jj}\tilde{\theta}_j^2$ of θ_j to SSM; it is the amount by which r would be increased if θ_j were omitted from the model. Thus, to test $H: \theta_1 = 0$, we have $r_H - r = r_1$, and we refer f_1 to $F_{1,n-p}$; to test $H: \theta_1 = \theta_2 = 0$, we have $r_H - r = r_1 + r_2$, and we refer $(r_1 + r_2)/2\tilde{\sigma}^2$ to $F_{2,n-p}$, etc.

It is sometimes of interest to test whether a parameter has a specified non-zero value, and this only requires a slight modification of the above procedure. Suppose we want to test $H: \theta_t = c_t$, where c_t is given. As before, $\tilde{\boldsymbol{\theta}}$ and $\boldsymbol{\theta}^*$ differ only in the tth component, which is now c_t in $\boldsymbol{\theta}^*$. Consequently, the extra reduction in $S(\boldsymbol{0}) = \mathbf{y}'\mathbf{y}$ due to fitting θ_t freely instead of fixing $\theta_t = c_t$ is $A_{tt}(\tilde{\theta}_t - c_t)^2$; the rest of the argument proceeds as before but using this as r_t. Thus, when H is true, $R_t = A_{tt}(\tilde{\Theta}_t - c_t)^2$ is distributed as $\sigma^2\chi_1^2$, and the test consists of referring $f = A_{tt}(\tilde{\theta}_t - c_t)^2/\tilde{\sigma}^2$ to $F_{1,n-p}$.

Similarly, if we require to test $H: \boldsymbol{\phi} = \mathbf{c}$, where $\boldsymbol{\phi}$ is a set of m θ_j's, and $\mathbf{c}$ is a given m-tuple, we calculate the increase in residual as

$$r_H - r = \sum_{\phi} A_{ii}(\tilde{\theta}_i - c_i)^2. \tag{8.3.14}$$

When H is true, this is a realization of a random variable distributed as $\sigma^2 \chi_m^2$, and we refer $(r_H - r)/m\tilde{\sigma}^2$ to the $F_{m,n-p}$ distribution.

We illustrate the foregoing techniques in the following examples.

EXAMPLE 8.3.3.   *One-way layout: to test whether all means have a specified common value.*   For the one-way layout with model

$$y_{ij} = \mu_i + e_{ij}, \qquad i = 1, 2, \ldots, I; \qquad j = 1, 2, \ldots, J_i$$

the reduction due to fitting μ_i is $r_i = A_{ii}\tilde{\mu}_i^2$. We have $A_{ii} = J_i$ (the number of observations in group i) and $\tilde{\mu}_i = \bar{y}_i.$ (the average of the observations in group i), so that $r_i = J_i \bar{y}_i^2.$. The anova is thus

Source	SS	DF
Due to μ_1	$r_1 = J_1 y_1^2.$	1
Due to μ_2	$r_2 = J_2 y_2^2.$	1
$\vdots$	$\vdots$	$\vdots$
Due to μ_I	$r_I = J_I y_I^2.$	1
Residual	$r = \mathbf{y'y} = \sum_1^I r_i$	$n - I$
Total	$\mathbf{y'y}$	$n = \sum_1^I J_i$

In fact, this table is of limited interest as there is not usually much point in the group context in testing whether individual groups (or collections of groups) have zero means. Instead let us consider testing the hypothesis $H_0: \mu_1 = \mu_2 = \ldots = \mu_I = \mu_0$, where μ_0 is given. In the above notation we have $\Phi = (\mu_1, \mu_2, \ldots, \mu_I)'$ and $\mathbf{c} = (\mu_0, \mu_0, \ldots, \mu_0)'$. using (8.3.14), we have

$$r_{H_0} - r = \sum_{i=1}^{I} J_i(y_i. - \mu_0)^2.$$

When H_0 is true, this is an observation from $\sigma^2 \chi_I^2$, and we can assess the strength of the evidence against H_0 by referring $(n - I)(r_{H_0} - r)/Ir$ to the $F_{I,n-I}$ distribution.

EXAMPLE 8.3.4.   *Test for individual coefficients of a polynomial regression, using orthogonal polynomials.*   For data (y_i, x_i), $i = 1, 2, \ldots, n$, an orthogonal polynomial model is fitted, of the form

$$y = \alpha_0 + \alpha_1 P_1(x) + \alpha_2 P_2(x) + \ldots + \alpha_s P_s(x) + e.$$

The least squares estimate of α_j is

$$\tilde{\alpha}_j = \sum_i y_i P_j(x_i) / \sum_i P_j^2(x_i)$$

and the sum of squares (reduction) due to α_j is

$$r(\alpha_j) = A_{jj}\tilde{\alpha}_j^2 = \tilde{\alpha}_j^2 \sum_i P_j^2(x_i) = \left[\sum_i y_i P_j(x_i)\right]^2 \Big/ \left[\sum_i P_j^2(x_i)\right],$$

giving the analysis of variance

Source	SS	DF	MS
Due to α_0	$r(\alpha_0) = n\bar{y}^2$	1	r_0
Due to α_1	$r(\alpha_1) = \tilde{\alpha}_1^2 \sum P_1^2(x_i)$	1	r_1
$\vdots$	$\vdots$	$\vdots$	$\vdots$
Due to α_s	$r(\alpha_s) = \tilde{\alpha}_s^2 \sum P_s^2(x_i)$	1	r_s
Residual	$r = \mathbf{y'y} - \sum_{i=0}^{s} r(\alpha_i)$	$n-s-1$	$\tilde{\sigma}^2 = r/(n-s-1)$
Total	$\mathbf{y'y} = \sum_{i=1}^{n} y_i^2$	n	

Now, if (on the basis of previous experience or theoretical analysis) the model is known to be correct, we can test whether any α_i is zero, as indicated above, by referring $r(\alpha_i)/\tilde{\sigma}^2$ to $F_{1,n-s-1}$. For example, suppose it is desired to test whether the data can be adequately represented by a polynomial of lower degree, say $(s-1)$. Since $P_s(x)$ is the only component of the model containing an x^s term, this means testing $H: \alpha_s = 0$, and we can test this hypothesis by referring $r(\alpha_s)/\tilde{\sigma}^2$ to $F_{1,n-s-1}$. More commonly, however, the experimenter merely wishes to fit a polynomial to given data as an approximation to the exact form, which is unknown. In particular, the degree of the appropriate approximating polynomial is not known in advance. Of course, an exact fit to the data can be achieved using a polynomial of degree $(n-1)$; what we want, however, is a satisfactory approximation by a polynomial of low degree. An approach to this problem is to fit successively polynomials of increasing degree, using the corresponding analysis of variance and F-test to assess the contribution of the latest term added to the model. The procedure is terminated when two consecutive terms are judged to be zero; this is necessary because if a polynomial of odd-degree fits the data, the terms of even degree (apart from the constant term) are quite likely to make little contribution, and vice versa. We illustrate the method using the rainfall example considered earlier (Example 8.2.15). A plot of the data suggests that a quadratic or, possibly, cubic approximation will suffice.

We begin by fitting the first-degree model

$$y = \gamma_0 + \gamma_1 Q_1(u) + e_{(1)}.$$

We have $\tilde{\gamma}_0 = 2.693333$, $\tilde{\gamma}_1 = -0.029441$, and the anova is:

Source	SS	DF	MS
Due to γ_0	$r(\gamma_0) = n\bar{y}^2 = 87.049$	1	
Due to γ_1	$r(\gamma_1) = \tilde{\gamma}_1^2 \sum Q_1^2(u_i) = 0.496$	1	0.496
Residual	$r_1 = \mathbf{y'y} - r(\gamma_0) - r(\gamma_1) = 1.426$	10	0.1426
Total	$\mathbf{y'y} = 88.971$	12	

We see that γ_1 does not greatly reduce the residual from the value it would have if only γ_0 were fitted, viz. 1.922. For a more precise assessment, we compare the reduction of 0.496 with the appropriately scaled quantity, viz. the estimated error variance 0.1426, and this gives an F-ratio of 3.48. The probability of an observation from $F_{1,10}$ exceeding this value is well over 0.05 ($F_{1,10}(0.95) = 4.96$, $F_{1,10}(0.9) = 3.285$) so the test gives no evidence for rejecting the hypothesis $\gamma_1 = 0$.

A plot of the data quickly shows that it cannot be described satisfactorily by $y = \gamma_0 + e_{(0)}$, demonstrating the need for the stopping rule mentioned above. We now fit the quadratic model

$$y = \gamma_0 + \gamma_1 Q_1(u) + \gamma_2 Q_2(u) + e_{(2)}.$$

The estimates of γ_0 and γ_1 are the same as in the first-degree model, and $\tilde{\gamma}_2 = -0.007744$. The anova is

Source	SS	DF	MS
Due to γ_0	$r(\gamma_0) = 87.049$	1	
Due to γ_1	$r(\gamma_1) = 0.496$	1	
Due to γ_2	$r(\gamma_2) = \tilde{\gamma}_2^2 \sum Q_2^2(u_i) = 0.720$	1	0.720
Residual	$r_2 = \mathbf{y'y} - \sum r(\gamma_i) = 0.706$	9	0.078
Total	$\mathbf{y'y} = 88.971$	12	

The F-ratio in this case is 9.23 and, since $F_{1,9}(0.975) = 7.21$ and $F_{1,9}(0.99) = 10.56$, there is evidence for rejecting the hypothesis that $\gamma_2 = 0$. We proceed to fit the cubic model

$$y = \gamma_0 + \gamma_1 Q_1(u) + \gamma_2 Q_2(u) + \gamma_3 Q_3(u) + e_{(3)},$$

for which $\tilde{\gamma}_3 = 0.007183$, and the anova is:

Source	SS	DF	MS
Due to γ_0	$r(\gamma_0) = 87.049$	1	
Due to γ_1	$r(\gamma_1) = 0.496$	1	
Due to γ_2	$r(\gamma_2) = 0.720$	1	
Due to γ_3	$r(\gamma_3) = \tilde{\gamma}_3^2 \sum Q_3^2(u_i) = 0.266$	1	0.266
Residual	$r_3 = \mathbf{y'y} - \sum r(\gamma_i) = 0.440$	8	0.055
Total	$\mathbf{y'y} = 88.971$	12	

This time the F-ratio is 4.84. Since $F_{1,8}(0.95) = 5.32$ and $F_{1,8}(0.9) = 3.46$ we conclude that there is insufficient evidence to reject the hypothesis that $\gamma_3 = 0$.

For the quartic model,

$$y = \gamma_0 + \gamma_1 Q_1(u) + \gamma_2 Q_2(u) + \gamma_3 Q_3(u) + \gamma_4 Q_4(u) + e_{(4)}$$

we find that $\tilde{\gamma}_4 = 0.005527$, and the anova is

Source	SS	DF	MS
Due to γ_0	$r(\gamma_0) = 87.049$	1	
Due to γ_1	$r(\gamma_1) = 0.496$	1	
Due to γ_2	$r(\gamma_2) = 0.720$	1	
Due to γ_3	$r(\gamma_3) = 0.266$	1	
Due to γ_4	$r(\gamma_4) = \tilde{\gamma}_4^2 \sum Q_4^2(u_i) = 0.245$	1	0.245
Residual	$r_4 = \mathbf{y'y} - \sum r(\gamma_i) = 0.195$	7	0.028
Total	$\mathbf{y'y} = 88.971$	12	

The F-ratio is 8.66; from tables we have $F_{1,7}(0.975) = 8.07$ and $F_{1,7}(0.99) = 12.25$. Do we reject '$\gamma_4 = 0$'? Before doing this, the following points should be considered:

(a) The F-test for, say, $\gamma_j = 0$, is based on the assumption that the jth-degree model is correct; in particular that the errors are independent and $e_{(j)}$ is an observation from $N(0, \sigma)$. If, in fact, a model of higher degree is required, the residual r_j for the jth degree model will not be an appropriate estimate of σ^2 (as the test assumes) since it will be biased by the contributions of the non-zero higher degree terms which it includes. Gross errors resulting from this can usually be avoided by plotting the data and observing the requirement for two consecutive zero coefficients. In this example the first test suggests that γ_1 may be zero; however, a different conclusion is possible if the test is carried out in conjunction with a residual from a later fit, e.g. if r_2 is appropriate for the estimation of σ^2, the relevant F-ratio is 6·36.

(b) The relative importance of the various parameters can be judged directly by comparing their reductions. Thus, γ_3 and γ_4 achieve about the same reduction in the total sum of squares, and this is about half of the contribution due to γ_1. This implies a similar appraisal of $H_3: \gamma_3 = 0$ and $H_4: \gamma_4 = 0$, neither of which should be rejected if $H_1: \gamma_1 = 0$ is not rejected.

Bearing these remarks in mind, it is clear that a flexible approach is required in the interpretation of the F-tests, and that instead of an automatic rejection of $H_4: \gamma_4 = 0$ on the grounds that 8·66 is beyond the upper $2\frac{1}{2}\%$ point of $F_{1,7}$, it is more appropriate to decide that the evidence against $\gamma_4 = 0$ is weak. We therefore conclude that the quadratic model $y = 2\cdot3714 + 0\cdot2431x - 0\cdot0232x^2$ is a suitable approximation to the data.

8.3.6. Hypothesis Testing in Non-orthogonal Designs

The testing of a hypothesis H in a non-orthogonal design follows the same basic procedure outlined in section 8.3.3; i.e. the extra reduction (fit) obtained when θ is not restricted by H, viz. $r(\theta) - r(\theta|H)$ is compared (after standardiz-ation) with the error variance as estimated by $\tilde{\sigma}^2 = r/(n-k)$, where $r = y'y - r(\theta)$ is the residual for the full model. As a result of the more complex structure of the normal equations, the detailed implementation of the pro-cedure is more complicated than in the orthogonal case, for which we recall that, if H places no restrictions on θ_t, the L.S. estimate of θ_t is the same whether H is true or not. This very convenient feature is, in general, absent in non-orthogonal designs (although it could hold for exceptional components of θ occurring in isolation in their 'own' normal equations).

To illustrate the changes, consider the question of testing whether the parameter θ_t is required i.e. testing the hypothesis $H: \theta_t = 0$. The extra reduc-tion when θ_t is fitted is, by (8.3.8),

$$r_H - r = (\tilde{\theta} - \theta^*)'\mathbf{A}'\mathbf{A}(\tilde{\theta} - \theta^*),$$

where θ^* is the least squares estimate of θ under H, and r_H is the corresponding residual. In the orthogonal case this expression involved only θ_t; in the general case, however, in which the estimates of the other parameters are not unaffected by H, it depends on these as well. Let us denote this reduction by $r(\theta_t|\phi)$, where ϕ denotes the other parameters fitted. For an orthogonal design we have $r(\theta_t|\phi) = A_{tt}\tilde{\theta}_t^2$, which is independent of ϕ, whatever ϕ may be, so that we can speak unambiguously of the reduction, $r(\theta_t)$ in TSS due to fitting a given parameter θ_t, irrespective of what other parameters (if any) are fitted. This implies a unique partition (i.e. analysis of variance) of the SSM, consisting, as we saw earlier, of the separate reductions due to each parameter fitted, in the form

$$\text{SSM} = r(\boldsymbol{\theta}) = \tilde{\boldsymbol{\theta}}'\mathbf{A}'\mathbf{A}\tilde{\boldsymbol{\theta}} = \sum_{i=1}^{p} A_{ii}\tilde{\theta}_i^2 = \sum_{i=1}^{p} r(\theta_i).$$

In the non-orthogonal case $r(\theta_t|\boldsymbol{\phi})$ depends on which parameters are included in $\boldsymbol{\phi}$, and SSM may be partitioned in various different ways, of the form

$$\text{SSM} = r(\theta_{t_1}) + r(\theta_{t_2}|\theta_{t_1}) + r(\theta_{t_3}|\theta_{t_1}, \theta_{t_2}) + \ldots$$
$$+ r(\theta_{t_p}|\theta_{t_1}, \theta_{t_2}, \ldots, \theta_{t_{p-1}}).$$

In this expression $r(\theta_{t_1})$ is the reduction in TSS due to fitting θ_{t_1} alone; $r(\theta_{t_2}|\theta_{t_1})$ is the extra reduction obtained by fitting θ_{t_2} as well. When both parameters are fitted, the reduction is $r(\theta_{t_1}, \theta_{t_2})$; subtracting $r(\theta_{t_1})$ gives $r(\theta_{t_2}|\theta_{t_1})$, which generally depends on both parameters, but in the orthogonal case only on θ_{t_2}. Similarly $r(\theta_{t_3}|\theta_{t_1}, \theta_{t_2})$ is the extra reduction achieved when θ_{t_3} is fitted in addition to θ_{t_1} and θ_{t_2}; it is equal to $r(\theta_{t_1}, \theta_{t_2}, \theta_{t_3}) - r(\theta_{t_1}, \theta_{t_2})$; and so on. The final term is the extra reduction when θ_{t_p} is fitted in addition to all the other parameters. It can be shown that, if $\theta_{t_p} = 0$, this reduction is distributed, independently of the residual, as $\sigma^2\chi_1^2$; hence the final term is appropriate for testing the hypothesis that $\theta_{t_p} = 0$ (in the context of the full model $E(\mathbf{Y}) = \mathbf{A}\boldsymbol{\theta}$) by an F-test.

EXAMPLE 8.3.5.   *Test for coefficients of a polynomial regression in a non-orthogonal design.*   We reconsider the question of obtaining an adequate polynomial representation of the rainfall data, this time using natural polynomial models. We remind the reader that the following methods of analysis apply without modification to fitting polynomials when the x-values are not equally spaced. We begin by fitting the first-degree polynomial

$$y = \beta_0 + \beta_1 x + e_{(1)}.$$

Denoting the design matrix for this model by $\mathbf{A}_1$, the normal equations $\mathbf{A}_1'\mathbf{A}_1\tilde{\boldsymbol{\beta}} = \mathbf{A}_1'\mathbf{y}$ are

$$\begin{pmatrix} n & \sum x_i \\ \sum x_i & \sum x_i^2 \end{pmatrix} \begin{pmatrix} \tilde{\beta}_0 \\ \tilde{\beta}_1 \end{pmatrix} = \begin{pmatrix} \sum y_i \\ \sum x_i y_i \end{pmatrix}.$$

For the rainfall data these are

$$\begin{pmatrix} 12 & 78 \\ 78 & 650 \end{pmatrix} \begin{pmatrix} \tilde{\beta}_0 \\ \tilde{\beta}_1 \end{pmatrix} = \begin{pmatrix} 32{\cdot}32 \\ 201{\cdot}66 \end{pmatrix},$$

and the solutions are $\tilde{\beta}_0 = 3{\cdot}0761$, $\tilde{\beta}_1 = -0{\cdot}0589$. The reduction due to fitting this model is, by (8.3.6), $r(\beta_0, \beta_1) = \tilde{\boldsymbol{\beta}}'\mathbf{A}_1'\mathbf{A}_1\tilde{\boldsymbol{\beta}}$. To calculate this, we use the more convenient alternative formula,

$$r(\beta_0, \beta_1) = \tilde{\boldsymbol{\beta}}'\mathbf{A}_1'\mathbf{y} = \tilde{\beta}_0 \sum y_i + \tilde{\beta}_1 \sum x_i y_i,$$

obtaining $r(\beta_0, \beta_1) = 87{\cdot}545$.

The hypothesis of interest is $H_1: \beta_1 = 0$, under which the model is $y = \beta_0 + e_{(0)}$. The normal equation for this model is found from above by deleting the last row and column of the matrix and the last row of the vectors, whence $\beta_0^* = \bar{y} = 2 \cdot 693333$. Instead of using (8.3.8) it is easier to find the extra reduction as follows. The reduction for this model is $r(\beta_0) = \beta_0^* \sum y_i = n\bar{y}^2 = 87 \cdot 049$, whence the extra reduction if we fit β_1 as well is $r(\beta_1|\beta_0) = r(\beta_0, \beta_1) - r(\beta_0) = 0 \cdot 496$. The anova for testing $H_1: \beta_1 = 0$ is

$$\mathbf{y}'\mathbf{y} = r(\beta_0) + r(\beta_1|\beta_0) + r_1,$$

and the table is

Source	SS	DF	MS
Due to β_0	$r(\beta_0) = 87 \cdot 049$	1	
Due to $\beta_1|\beta_0$	$r(\beta_1|\beta_0) = 0 \cdot 496$	1	$0 \cdot 496$
Residual	$r_1 = \mathbf{y}'\mathbf{y} - r(\beta_0, \beta_1) = 1 \cdot 426$	10	$0 \cdot 1426$
Total	$\mathbf{y}'\mathbf{y} = 88 \cdot 971$	12	

The argument is now the same as before; we conclude that, in the first-degree model there is insufficient evidence to reject $H_1: \beta_1 = 0$, and it is clear that at least a quadratic fit is required. We proceed to fit $y = \beta_0 + \beta_1 x + \beta_2 x^2 + e_{(2)}$. Denoting the design matrix for this model by $\mathbf{A}_2$, the normal equations $\mathbf{A}_2'\mathbf{A}_2\tilde{\boldsymbol{\beta}} = \mathbf{A}_2'\mathbf{y}$ may be written explicitly as

$$\begin{pmatrix} n & \sum x_i & \sum x_i^2 \\ \sum x_i & \sum x_i^2 & \sum x_i^3 \\ \sum x_i^2 & \sum x_i^3 & \sum x_i^4 \end{pmatrix} \begin{pmatrix} \tilde{\beta}_0 \\ \tilde{\beta}_1 \\ \tilde{\beta}_2 \end{pmatrix} = \begin{pmatrix} \sum y_i \\ \sum x_i y_i \\ \sum x_i^2 y_i \end{pmatrix}.$$

Substituting the values for the rainfall data and inverting the matrix we obtain

$$\tilde{\beta}_0 = 2 \cdot 3714, \qquad \tilde{\beta}_1 = 0 \cdot 2431, \qquad \tilde{\beta}_2 = -0 \cdot 0232.$$

The reduction in the total sum of squares due to fitting this quadratic model is

$$r(\beta_0, \beta_1, \beta_2) = \tilde{\boldsymbol{\beta}}'\mathbf{A}_2'\mathbf{y} = \tilde{\beta}_0 \sum y_i + \tilde{\beta}_1 \sum x_i y_i + \tilde{\beta}_2 \sum x_i^2 y_i.$$

Evaluating this expression we find $r(\beta_0, \beta_1, \beta_2) = 88 \cdot 265$. Under the hypothesis $H_2: \beta_2 = 0$, the model is the first-degree model analysed previously, for which we found $r(\beta_0, \beta_1) = 87 \cdot 545$. The additional fit produced by β_2 is thus $r(\beta_2|\beta_0, \beta_1) = r(\beta_0, \beta_1, \beta_2) - r(\beta_0, \beta_1) = 0 \cdot 720$. The anova is

$$\mathbf{y}'\mathbf{y} = r(\beta_0, \beta_1) + r(\beta_2|\beta_0, \beta_1) + r_2,$$

and the table is

Source	SS		DF	MS
Due to β_0, β_1	$r(\beta_0, \beta_1) = 87 \cdot 545$		2	
Due to $\beta_2 \vert \beta_0, \beta_1$	$r(\beta_2 \vert \beta_0, \beta_1) = 0 \cdot 720$		1	$0 \cdot 720$
Residual	$r_2 = \mathbf{y}'\mathbf{y} - r(\beta_0, \beta_1, \beta_2) = 0 \cdot 706$		9	$0 \cdot 078$
Total	$\mathbf{y}'\mathbf{y} = 88 \cdot 971$		12	

The interpretation is as before: we conclude that $\beta_2 \neq 0$.

We now fit the cubic model $y = \beta_0 + \beta_1 x + \beta_2 x^2 + \beta_3 x^3 + e_{(3)}$. On substituting the appropriate values in the normal equations

$$\begin{pmatrix} n & \sum x_i & \sum x_i^2 & \sum x_i^3 \\ \sum x_i & \sum x_i^2 & \sum x_i^3 & \sum x_i^4 \\ \sum x_i^2 & \sum x_i^3 & \sum x_i^4 & \sum x_i^5 \\ \sum x_i^3 & \sum x_i^4 & \sum x_i^5 & \sum x_i^6 \end{pmatrix} \begin{pmatrix} \tilde{\beta}_0 \\ \tilde{\beta}_1 \\ \tilde{\beta}_2 \\ \tilde{\beta}_3 \end{pmatrix} = \begin{pmatrix} \sum y_i \\ \sum x_i y_i \\ \sum x_i^2 y_i \\ \sum x_i^3 y_i \end{pmatrix}$$

and inverting the matrix, we find

$$\tilde{\beta}_0 = 1 \cdot 7177, \qquad \tilde{\beta}_1 = 0 \cdot 2348, \qquad \tilde{\beta}_2 = -0 \cdot 1166, \qquad \tilde{\beta}_3 = 0 \cdot 0048.$$

The reduction due to fitting this model is

$$r(\beta_0, \beta_1, \beta_2, \beta_3) = \tilde{\beta}_0 \sum y_i + \tilde{\beta}_1 \sum x_i y_i + \tilde{\beta}_2 \sum x_i^2 y_i + \tilde{\beta}_3 \sum x_i^3 y_i,$$

which we evaluate as $88 \cdot 531$. We now want to test $H_3: \beta_3 = 0$. Under H_3, the model is the quadratic previously fitted, for which the reduction was $r(\beta_0, \beta_1, \beta_2) = 88 \cdot 265$. The extra reduction given by β_3 is

$$r(\beta_3 \vert \beta_0, \beta_1, \beta_2) = r(\beta_0, \beta_1, \beta_2, \beta_3) - r(\beta_0, \beta_1, \beta_2) = 0 \cdot 266.$$

The anova is

$$\mathbf{y}'\mathbf{y} = r(\beta_0, \beta_1, \beta_2) + r(\beta_3 \vert \beta_0, \beta_1, \beta_2) + r_3,$$

in tabular form

Source	SS		DF	MS
Due to $\beta_0, \beta_1, \beta_2$	$r(\beta_0, \beta_1, \beta_2) = 88 \cdot 265$		3	
Due to $\beta_3 \vert \beta_0, \beta_1, \beta_2$	$r(\beta_3 \vert \beta_0, \beta_1, \beta_2) = 0 \cdot 266$		1	$0 \cdot 266$
Residual	$r_3 = \mathbf{y}'\mathbf{y} - r(\beta_0, \beta_1, \beta_2, \beta_3) = 0 \cdot 440$		8	$0 \cdot 055$
Total	$\mathbf{y}'\mathbf{y} = 88 \cdot 971$		12	

The conclusion, as earlier, is that the evidence is too weak to reject H_3. The quartic model $\mathbf{y} = \beta_0 + \beta_1 x + \beta_2 x^2 + \beta_3 x^3 + \beta_4 x + e_{(4)}$ may be fitted similarly. The procedure involves inverting a 5×5 matrix, and the anova is

$$\mathbf{y'y} = r(\beta_0, \beta_1, \beta_2, \beta_3) + r(\beta_4 | \beta_0, \beta_1, \beta_2, \beta_3) + r_4;$$

the table is

Source	SS	DF	MS
Due to $\beta_0, \beta_1, \beta_2, \beta_3$	$r(\beta_0, \beta_1, \beta_2, \beta_3) = 88 \cdot 531$	4	
Due to $\beta_4 \| \beta_0, \beta_1, \beta_2, \beta_3$	$r(\beta_4 \| \beta_0, \beta_1, \beta_2, \beta_3) = 0 \cdot 245$	1	$0 \cdot 245$
Residual	$r_4 = \mathbf{y'y} - r(\beta_0, \beta_1, \beta_2, \beta_3, \beta_4) = 0 \cdot 195$	7	$0 \cdot 028$
Total	$\mathbf{y'y} = 88 \cdot 971$	12	

The interpretation of the F-test is the same as in the previous discussion using the orthogonal model and we conclude that the quadratic model $y = 2 \cdot 3714 + 0 \cdot 2431 x - 0 \cdot 0232 x^2$ provides an adequate fit to the data.

To reiterate, in an orthogonal model, there is essentially one analysis of variance, expressed in (8.3.9); the reduction by means of which we assess the contribution of a given parameter to the model or any sub-model is always the same. For non-orthogonal models the anova depends on the hypothesis under consideration. Thus, the basic anova for the first degree model

$$\mathbf{y'y} = r(\beta_0, \beta_1) + r_1$$

is refined, in the example above, to

$$\mathbf{y'y} = r(\beta_0) + r(\beta_1 | \beta_0) + r_1$$

since we wish to test $\beta_1 = 0$, $r(\beta_1 | \beta_0)$ being the contribution of β_1, additional to that of β_0, in this first-degree model. Although it is of no particular interest in the example discussed above, we can assess the contribution of β_0 additional to β_1 by using

$$\mathbf{y'y} = r(\beta_1) + r(\beta_0 | \beta_1) + r_1$$

this division being appropriate for testing $\beta_0 = 0$ in the first degree model. The first term in this expression is not equal to $r(\beta_1 | \beta_0)$; it is found by fitting the model $y = \beta_1 x + e$, for which the normal equation is

$$\tilde{\beta}_1 \sum x_i = \sum x_i y_i,$$

and for which $r(\beta_1) = \tilde{\beta}_1 \sum x_i y_i$.

It should be noted that the additional reduction $r(\beta_1 | \beta_0)$ is only relevant to testing the hypothesis $\beta_1 = 0$ in the first-degree model. In the example above, the test of $\beta_2 = 0$ in the quadratic model involved the anova

$$\mathbf{y'y} = r(\beta_0, \beta_1) + r(\beta_2 | \beta_0, \beta_1) + r_2.$$

This can be split further into

$$\mathbf{y'y} = r(\beta_0) + r(\beta_1|\beta_0) + r(\beta_2|\beta_0, \beta_1) + r_2,$$

but the second term is not appropriate for testing whether $\beta_1 = 0$ in this model. When this hypothesis is of interest we use the partition

$$\mathbf{y'y} = r(\beta_0, \beta_2) + r(\beta_1|\beta_0, \beta_2) + r_2,$$

where $r(\beta_1|\beta_0, \beta_2) = r(\beta_0, \beta_1, \beta_2) - r(\beta_0, \beta_2)$. We obtain $r(\beta_0, \beta_2)$ by fitting the quadratic model under the hypothesis $\beta_1 = 0$; i.e. the model $y = \beta_0 + \beta_2 x^2 + e$. The normal equations for this model are

$$\begin{pmatrix} n & \sum x_i^2 \\ \sum x_i^2 & \sum x_i^4 \end{pmatrix} \begin{pmatrix} \beta_0^* \\ \beta_2^* \end{pmatrix} = \begin{pmatrix} \sum y_i \\ \sum x_i^2 y_i \end{pmatrix}.$$

For the rainfall data these are

$$\begin{pmatrix} 12 & 650 \\ 650 & 60710 \end{pmatrix} \begin{pmatrix} \beta_0^* \\ \beta_2^* \end{pmatrix} = \begin{pmatrix} 32 \cdot 32 \\ 1610 \cdot 2 \end{pmatrix},$$

from which we find $\beta_0^* = 2 \cdot 991691$, $\beta_2^* = -0 \cdot 005508$. The reduction due to fitting this model is

$$r(\beta_0, \beta_2) = \beta_0^* \sum y_i + \beta_2^* \sum x_i^2 y_i = 87 \cdot 822,$$

and the anova table is

Source	SS	DF	MS
Due to β_0, β_2	$r(\beta_0, \beta_2) = 87 \cdot 822$	2	
Due to $\beta_1|\beta_0, \beta_2$	$r(\beta_1|\beta_0, \beta_2) = 0 \cdot 443$	1	0·443
Residual	$r_2 = 0 \cdot 706$	9	0·078
Total	$\mathbf{y'y} = 88 \cdot 971$	12	

The F-ratio is $5 \cdot 68$, and we have $F_{1,9}(0 \cdot 95) = 5 \cdot 12$, $F_{1,9}(0 \cdot 975) = 9 \cdot 21$. We can see from this and the earlier tables that

(a) The quadratic model without the linear term is better than the first-degree model.
(b) In the quadratic model, the quadratic term (additional reduction of $0 \cdot 720$) is more important than the linear term (additional reduction of $0 \cdot 443$).
(c) The additional contribution of the linear term is less (though not much less) in the quadratic model than in the first-degree model; formal

F-tests, however, attach greater significance to the former. This is due to the use of different residuals for the comparison. In practice, the contribution of such lower-order terms in a polynomial regression model would not usually be of interest until an overall decision had been taken regarding the degree of polynomial required for an acceptable approximation to the data.

We can similarly test whether a group of parameters is redundant. Let $\boldsymbol{\phi}$ be a subset of $\boldsymbol{\theta}$, and let $\boldsymbol{\psi}$ denote the parameters not included in $\boldsymbol{\phi}$. Under the hypothesis $H: \boldsymbol{\phi} = \mathbf{0}$, the model $\mathbf{y} = \mathbf{A}\boldsymbol{\theta} + \mathbf{e}$ becomes $\mathbf{y} = \mathbf{W}\boldsymbol{\psi} + \mathbf{e}$, where the matrix $\mathbf{W}$ is obtained from $\mathbf{A}$ by deleting the columns corresponding to the parameters in $\boldsymbol{\phi}$. The normal equations for the hypothesized model are thus

$$\mathbf{W'W}\boldsymbol{\psi}^* = \mathbf{W'y},$$

giving the least squares estimate of $\boldsymbol{\psi}$ under H, as $\boldsymbol{\psi}^* = (\mathbf{W'W})^{-1}\mathbf{W'y}$. This is not the same as $\tilde{\boldsymbol{\psi}}$, the least squares estimate of $\boldsymbol{\psi}$ in the full model, which consists of the appropriate components of $\tilde{\boldsymbol{\theta}} = (\mathbf{A'A})^{-1}\mathbf{A'y}$.

The reduction for the restricted model is $r(\boldsymbol{\psi}) = \boldsymbol{\psi}^{*\prime}\mathbf{W'W}\boldsymbol{\psi}^* = \boldsymbol{\psi}^{*\prime}\mathbf{W'y}$; thus the extra reduction when the full model is fitted is $r(\boldsymbol{\theta}) - r(\boldsymbol{\psi}) = \tilde{\boldsymbol{\theta}}'\mathbf{A'y} - \boldsymbol{\psi}^{*\prime}\mathbf{W'y}$. It can be shown that when $\mathbf{H}: \boldsymbol{\phi} = \mathbf{0}$ is true, this extra reduction is distributed independently of the full-model residual as $\sigma^2 \chi_m^2$, where m is the number of parameters in $\boldsymbol{\phi}$.

The anova table is

Source	SS	DF	MS
Due to $\boldsymbol{\psi}$	$r(\boldsymbol{\psi})$	$p - m$	
Due to $\boldsymbol{\phi}\vert\boldsymbol{\psi}$	$r(\boldsymbol{\phi}\vert\boldsymbol{\psi}) = r(\boldsymbol{\theta}) - r(\boldsymbol{\psi})$	m	$r(\boldsymbol{\phi}\vert\boldsymbol{\psi})/m$
Residual	$r = \mathbf{y'y} - r(\boldsymbol{\theta})$	$n - p$	$\tilde{\sigma}^2 = r/(n-p)$
Total	$\mathbf{y'y}$	n	

Under $H: \boldsymbol{\phi} = \mathbf{0}$, the two mean squares are independent estimates of σ^2, based on m and $(n-p)$ degrees of freedom, respectively, so we can test H by referring $(n-p)r(\boldsymbol{\phi}\vert\boldsymbol{\psi})/mr$ to $F_{m,n-p}$.

EXAMPLE 8.3.6. We illustrate the above technique by testing the hypothesis $H: \beta_1 = \beta_2 = 0$, in the quadratic model fitted to the rainfall data, which was $y = 2 \cdot 3714 + 0 \cdot 2431x - 0 \cdot 0232x^2$.

In the above notation, $\boldsymbol{\phi} = (\beta_1, \beta_2)'$, $\boldsymbol{\psi} = \beta_0$. We found earlier that $r(\boldsymbol{\theta}) = r(\beta_0, \beta_1, \beta_2) = 88 \cdot 265$. Under H, the model is $y = \beta_0 + e_0$, which we also considered earlier, finding $r(\beta_0) = 87 \cdot 049$.

The anova is thus

Source	SS	DF	MS
Due to β_0	$r(\beta_0) = 87 \cdot 049$	1	
Due to $\beta_1, \beta_2 \vert \beta_0$	$r(\beta_1, \beta_2 \vert \beta_0) = 1 \cdot 216$	2	0·608
Residual	$r_2 = \mathbf{y}'\mathbf{y} - r(\beta_0, \beta_1, \beta_2) = 0 \cdot 706$	9	0·078
Total	$\mathbf{y}'\mathbf{y} = 88 \cdot 971$	12	

The mean square ratio of 7·79 is fairly strong evidence against H ($F_{2,9}(0 \cdot 975) = 5 \cdot 71$, $F_{2,9}(0 \cdot 99) = 8 \cdot 025$).

As well as deciding whether certain parameters may be omitted from the model, we may want to see if the data is consistent with the hypothesis that various parameters have specified values.

EXAMPLE 8.3.7. *Test of consistency of gravity data with specified parameter values.* For the gravity data of Examples 8.2.2, 8.2.4, 8.2.8 and 8.2.12, the question of testing whether various parameters are required does not arise; other types of hypotheses which may be of interest are illustrated by H_1: $g_1 = 981$, H_2: $g_1 = g_2 = g_3 = g_4 = g_5 = 981 \cdot 2$ and H_3: $g_1 = g_4 = 981 \cdot 26$, $g_2 = g_3 = 981 \cdot 2$. We shall now test these on the usual assumption that the errors are independent observations from $N(0, \sigma^2)$.

Since H_1 involves only one parameter, it can be tested by a simple t-test. Thus, if H_1 is true, $(\tilde{g}_1 - 981)$ divided by its estimated standard deviation is an observation from Student's distribution with 4 degrees of freedom (since our estimate of σ^2 is based on a residual having 4 degrees of freedom).

Now

$$\frac{\tilde{g}_1 - 981}{\tilde{\sigma}\sqrt{0 \cdot 4211}} = \frac{0 \cdot 2681}{0 \cdot 0011} = 244, \quad \text{approximately.}$$

For a two-tailed test at the 5% level we compare this with $t_4(0 \cdot 975) = 2 \cdot 78$. Clearly, the hypothesis is untenable (at any level!).

An equivalent procedure is to check whether the hypothesized value of 981 falls in the central 95% confidence interval for g_1, which was found in Example 8.3.1 to be (981·2650; 981·2712). The hypothesized value is well outside the indicated range. An advantage of this approach is that we can see at a glance which values of g_1—viz. below 981·2650 and above 981·2712—are inconsistent with the data and would lead to rejection of H_1 by a t-test, at the 5% level.

We now develop the F-test for H_1; the same approach will be used for H_2 and H_3, for which t-tests are not available. Under H_1, the model equations of Example 8.2.2 become

$$981 \cdot 1880 = g_2 + e_1, \qquad -0 \cdot 0030 = g_4 - 981 + e_2, \qquad 0 \cdot 1390 = g_4 - g_5 + e_3,$$

$$981 \cdot 2000 = g_3 + e_4, \qquad 0 \cdot 0647 = g_4 - g_3 + e_5, \qquad 981 \cdot 2670 = 981 + e_6,$$

$$0 \cdot 0140 = g_3 - g_2 + e_7, \qquad 0 \cdot 1431 = g_4 - g_5 + e_8, \qquad 981 \cdot 2690 = 981 + e_9.$$

These can be written in matrix form as

$$
\begin{bmatrix}
981 \cdot 1880 \\
\cdot 980 \cdot 9970 \\
0 \cdot 1390 \\
981 \cdot 2000 \\
0 \cdot 0647 \\
0 \cdot 2670 \\
0 \cdot 0140 \\
0 \cdot 1431 \\
0 \cdot 2690
\end{bmatrix}
=
\begin{bmatrix}
1 & 0 & 0 & 0 \\
0 & 0 & 1 & 0 \\
0 & 0 & 1 & -1 \\
0 & 1 & 0 & 0 \\
0 & -1 & 1 & 0 \\
0 & 0 & 0 & 0 \\
-1 & 1 & 0 & 0 \\
0 & 0 & 1 & -1 \\
0 & 0 & 0 & 0
\end{bmatrix}
\begin{bmatrix}
g_2 \\
g_3 \\
g_4 \\
g_5
\end{bmatrix}
+
\begin{bmatrix}
e_1 \\
e_2 \\
e_3 \\
e_4 \\
e_5 \\
e_6 \\
e_7 \\
e_8 \\
e_9
\end{bmatrix}
$$

or,

$$\mathbf{z} = \mathbf{B\phi} + \mathbf{e}.$$

We note that the matrix $\mathbf{B}$ has two zero rows, these corresponding to equations 6 and 9, in which the error is known exactly. We have

$$
\mathbf{B'B} =
\begin{bmatrix}
2 & -1 & 0 & 0 \\
-1 & 3 & -1 & 0 \\
0 & -1 & 4 & -2 \\
0 & 0 & -2 & 2
\end{bmatrix},
\qquad
\mathbf{B'z} =
\begin{bmatrix}
981 \cdot 1740 \\
981 \cdot 1493 \\
981 \cdot 3438 \\
-0 \cdot 2821
\end{bmatrix}.
$$

The least squares estimates of $g_2, \dots, g_5$ under H_1, denoted by $g_2^*, \dots, g_5^*$ are obtained by solving the normal equations $\mathbf{B'B\phi}^* = \mathbf{B'z}$; we find that

$$\mathbf{\phi}^* = (981 \cdot 1538,\ 981 \cdot 1336,\ 981 \cdot 0976,\ 980 \cdot 9566)'.$$

We observe that these estimates of $g_2, \dots, g_5$ are not the same as those found in Example 8.2.2 for the unrestricted model, as a result of the non-orthogonality of the model. The residual sum of squares under H_1 is $r_{H_1} = \sum (e_i^*)^2$, where e_i^* is the least squares estimate under H_1 of e_i (of course, $e_6^* = e_6$ and $e_9^* = e_9$). We obtain the e_i^* by replacing g_2, g_3, g_4, g_5 in the equations for the restricted model by their least squares estimates, whence

$$e_1^* = 0.0342, \qquad e_2^* = -0.1006, \qquad e_3^* = -0.0020,$$

$$e_4^* = 0.0664, \qquad e_5^* = 0.1007, \qquad e_6^* = 0.2670,$$

$$e_7^* = 0.0342, \qquad e_8^* = 0.0021, \qquad e_9^* = 0.2690.$$

These give $r_{H_1} = 0.1706675$ which can also be found using the formula $r_{H_1} = \mathbf{z}'\mathbf{z} - \boldsymbol{\phi}^{*\prime}\mathbf{B}'\mathbf{z}$.

The full model is $\mathbf{y} = \mathbf{A}\boldsymbol{\theta} + \mathbf{e}$, where $\boldsymbol{\theta} = (g_1, g_2, \ldots, g_5)'$ and $\mathbf{y}$ and $\mathbf{A}$ are given in Example 8.2.4. The residual for the full model is $r = \sum \tilde{e}_i^2$, where $\tilde{e}_i$ is the unrestricted least squares estimate of e_i. The $\tilde{e}_i$ were found earlier, in Example 8.2.12, to be

$$\tilde{e}_1 = 0.00071, \qquad \tilde{e}_2 = -0.00011, \qquad \tilde{e}_3 = -0.00205,$$

$$\tilde{e}_4 = -0.00058, \qquad \tilde{e}_5 = 0.00011, \qquad \tilde{e}_6 = -0.00106,$$

$$\tilde{e}_7 = 0.00071, \qquad \tilde{e}_8 = 0.00205, \qquad \tilde{e}_9 = 0.00094.$$

Apart from the third and eighth, these are all much smaller than in the restricted model, casting doubt on the hypothesis. For the formal test, we square and sum, obtaining $r = 0.00001178$, so that $r_{H_1} - r = 0.17065572$. It can be shown that if the hypothesis is true (i.e. $g_1 = 981$), then $R_{H_1} - R$ is distributed as $\sigma^2 \chi_1^2$ independently of R, which is $\sigma^2 \chi_4^2$. The appropriate F-ratio is therefore $4(r_{H_1} - r)/r$, which is approximately 58×10^3, than which few higher F-ratios can have been observed!

We now consider H_2: $g_1 = g_2 = g_3 = g_4 = g_5 = 981.2$. In appraising such a hypothesis, the individual confidence intervals for the parameters may be misleading, owing to correlation between the estimates. Thus, using the separate intervals, the hypothesized value of 981.2 would appear reasonable for g_3 but not for the others; in general, the overall judgement could go either way, depending on the structure of $(\mathbf{A}'\mathbf{A})^{-1}$—although in this instance we can expect rejection of H_2, as the hypothesized value is so far outside three of the intervals. An overall confidence region based test can be constructed by determining, at the chosen level, the five-dimensional confidence ellipsoid for $(g_1, g_2, g_3, g_4, g_5)$, rejecting H_2 if the point $(981.2, 981.2, 981.2, 981.2, 981.2)$ is not contained therein. Instead, we shall construct the F-test in the usual way.

Under H_2, there are no parameters to estimate and the errors are known exactly, viz.

$$e_1^+ = -0.0120, \qquad e_2^+ = -0.0030, \qquad e_3^+ = 0.1390,$$

$$e_4^+ = 0, \qquad e_5^+ = 0.0647, \qquad e_6^+ = 0.0670,$$

$$e_7^+ = 0.0140, \qquad e_8^+ = 0.1431, \qquad e_9^+ = 0.0690,$$

whence, $r_{H_2} = \sum (e_i^+)^2 = 0.0535837$. If H_2 is true, R_{H_2} is distributed as $\sigma^2 \chi_9^2$, and it can be shown that $R_{H_2} - R$ is distributed independently of R as $\sigma^2 \chi_5^2$; the test statistic is thus

$$\frac{(r_{H_2} - r)/5}{r/4} = 3638.$$

As anticipated, the hypothesis is massively rejected.

Finally, we have H_3: $g_1 = g_4 = 981 \cdot 26$, $g_2 = g_3 = 981 \cdot 2$, under which the model equations are

$$-0 \cdot 0120 = e_1, \qquad -0 \cdot 0030 = e_2, \qquad 0 \cdot 1390 = 981 \cdot 26 - g_5 + e_3, \qquad 0 = e_4,$$

$$0 \cdot 0647 = 0 \cdot 06 + e_5, \qquad 0 \cdot 0070 = e_6, \qquad 0 \cdot 0140 = e_7,$$

$$0 \cdot 1431 = 981 \cdot 26 - g_5 + e_8, \qquad 0 \cdot 0090 = e_9.$$

In matrix form, these are

$$
\begin{bmatrix}
-0 \cdot 0120 \\
-0 \cdot 0030 \\
-981 \cdot 121 \\
0 \\
0 \cdot 0047 \\
0 \cdot 0070 \\
0 \cdot 0140 \\
-981 \cdot 1169 \\
0 \cdot 0090
\end{bmatrix}
=
\begin{bmatrix}
0 \\
0 \\
-1 \\
0 \\
0 \\
0 \\
0 \\
-1 \\
0
\end{bmatrix}
(g_5) +
\begin{bmatrix}
e_1 \\
e_2 \\
e_3 \\
e_4 \\
e_5 \\
e_6 \\
e_7 \\
e_8 \\
e_9
\end{bmatrix}
$$

or,

$$\mathbf{w} = \mathbf{C} g_5 + \mathbf{e}.$$

Now $\mathbf{C}'\mathbf{C} = 2$ and $\mathbf{C}'\mathbf{w} = 1962 \cdot 2379$, so the L.S.E. of g_5 under H_3 is $981 \cdot 11895$. All the errors are known except e_3 and e_8, which we estimate by $\hat{e}_3 = -0 \cdot 00205$ and $\hat{e}_8 = 0 \cdot 00205$. Squaring and summing, $r_{H_3} = 0 \cdot 0005095$, whence $r_{H_3} - r = 0 \cdot 00049772$. It can be shown that when H_3 is true, $R_{H_3} - R$ is distributed independently of R as $\sigma^2 \chi_4^2$. The test statistic is therefore $(r_{H_3} - r)/r = 42 \cdot 25$, and the hypothesis can be confidently rejected $(F_{4,4}(0 \cdot 995) = 23 \cdot 15, F_{4,4}(0 \cdot 999) = 53.44)$.

8.3.7. Group Orthogonality

In some cases, although the design is not fully orthogonal in the sense considered so far, a partial orthogonality is present in that the parameters fall into orthogonal groups. By this we mean that $\boldsymbol{\theta}' = (\theta_1, \theta_2, \ldots, \theta_p)$ can (after appropriate re-ordering, if necessary) be exhibited as

$$\boldsymbol{\theta}' = (\boldsymbol{\phi}_1', \boldsymbol{\phi}_2', \ldots, \boldsymbol{\phi}_s')$$

where, for $i = 1, 2, \ldots, s$, $\boldsymbol{\phi}_i$ comprises certain components of $\boldsymbol{\theta}$, say p_i in number, with $\sum_{i=1}^{s} p_i = p$, and where the design matrix $\mathbf{A}$ (correspondingly re-arranged, if required) is such that $\mathbf{A'A}$ is in block diagonal form, i.e.

$$\mathbf{A'A} = \begin{bmatrix} \mathbf{A}_1 & & 0 \\ & \mathbf{A}_2 & \\ & & \ddots \\ 0 & & \mathbf{A}_s \end{bmatrix}$$

where, apart from the diagonally-placed square submatrices $\mathbf{A}_1, \ldots, \mathbf{A}_s$ (of orders $p_1 \times p_1, \ldots, p_s \times p_s$, respectively) all entries are zero. It follows that the inverse of $\mathbf{A'A}$ is given by

$$(\mathbf{A'A})^{-1} = \begin{bmatrix} \mathbf{A}_1^{-1} & & 0 \\ & \mathbf{A}_2^{-1} & \\ & & \ddots \\ 0 & & \mathbf{A}_s^{-1} \end{bmatrix}$$

This shows that the L.S. estimators of elements of $\boldsymbol{\theta}$ in different $\boldsymbol{\phi}$'s are uncorrelated; the correlations between elements of $\boldsymbol{\theta}$ in the same group, say $\boldsymbol{\phi}_i$, are found by dividing the off-diagonal elements of $\mathbf{A}_i^{-1}$ by the square roots of the appropriate diagonal elements.

It follows that the model sum of squares $r(\boldsymbol{\theta}) = \tilde{\boldsymbol{\theta}}'(\mathbf{A'A})^{-1}\tilde{\boldsymbol{\theta}}$ may be partitioned in the form

$$r(\boldsymbol{\theta}) = \sum_{i=1}^{s} r(\boldsymbol{\phi}_i),$$

where $r(\boldsymbol{\phi}_i) = \tilde{\boldsymbol{\phi}}_i \mathbf{A}_i^{-1} \tilde{\boldsymbol{\phi}}_i$ is the reduction due to fitting the parameters represented by $\boldsymbol{\phi}_i$. Moreover, this is the reduction due to $\boldsymbol{\phi}_i$, irrespective of whether any of the other parameters are fitted or not; this is clear from the form of $\mathbf{A'A}$, which shows that the L.S. estimate of any $\boldsymbol{\phi}_j$ is not affected by any hypothesis involving the other parameters. In particular, under $H: \boldsymbol{\phi}_i = \mathbf{0}$, the L.S. estimates of the other $\boldsymbol{\phi}$'s are unchanged and the extra reduction when H is not imposed is $r(\boldsymbol{\phi}_i)$. Hence we can test H by considering the ratio

$$\frac{r(\boldsymbol{\phi}_i)/p_i}{r/(n-p)}$$

which will be a realization of $F_{p_i, n-p}$ if H is true.

On the other hand, to test whether just some, rather than all, of the parameters in $\boldsymbol{\phi}_i$ are necessary, would involve the usual complications of non-orthogonality, viz. changed L.S. estimates of the remaining parameters of $\boldsymbol{\phi}_i$ together with the reduction due to the parameters under test being dependent on which other elements of $\boldsymbol{\phi}_i$ were fitted. The same difficulties would arise if we wanted to examine the need for a group of parameters which included θ_i's from different $\boldsymbol{\phi}_j$'s.

We shall meet several examples of this partial orthogonality later when we consider models for classified data (see Chapter 10).

8.3.8. The General Linear Hypothesis

The hypotheses discussed so far are all examples of the general linear hypothesis, which also includes important cases not yet illustrated. A general linear hypothesis, H, imposes linear constraints on the parameter vector $\boldsymbol{\theta}$, i.e. it specifies the values of certain known linear combinations of $\theta_1, \theta_2, \ldots, \theta_p$, by means of equations such as

$$h_{11}\theta_1 + h_{12}\theta_2 + \ldots + h_{1p}\theta_p = c_1$$
$$h_{21}\theta_2 + h_{22}\theta_2 + \ldots + h_{2p}\theta_p = c_2$$
$$\ldots$$
$$h_{m1}\theta_1 + h_{m2}\theta_2 + \ldots + h_{mp}\theta_p = c_m,$$

where the coefficients h_{ij} and the constants c_i are known. In matrix form these equations may be written as

$$\mathbf{H}\boldsymbol{\theta} = \mathbf{c},$$

where $\mathbf{H}$ is the $m \times p$ matrix with (i, j) element h_{ij}, and $\mathbf{c}' = (c_1, c_2, \ldots, c_m)$.

For example,

(a) $H_a: \theta_1 = 0, \theta_2 = 0, \ldots, \qquad \theta_p = 0$
(b) $H_b: \theta_p = 0$
(c) $H_c: \theta_1 = c_1, \theta_2 = c_2, \ldots, \qquad \theta_p = c_p,$

are all cases already dealt with.

An example of an important general linear hypothesis not yet encountered is

(d) $H_d: \theta_1 - \theta_2 = 0, \theta_2 - \theta_3 = 0, \ldots, \theta_{p-1} - \theta_p = 0$

which may be stated more concisely as $H_d: \theta_1 = \theta_2 = \ldots = \theta_p$.

Note the distinction between H_d, in which the common value is unspecified, and the special case of H_c in which it is, viz. $c_1 = c_2 = \ldots = c_p$.

To test H, we apply the same general procedure as in the particular cases already discussed: using the full-model residual as datum, we assess the importance of the extra reduction in the total sum of squares when the model is not constrained by H.

For the full model $E(\mathbf{Y}) = \mathbf{A}\boldsymbol{\theta}$, we have, as usual,

$$\mathbf{y}'\mathbf{y} = r(\boldsymbol{\theta}) + r,$$

where $r(\boldsymbol{\theta})$ is the reduction due to $\boldsymbol{\theta}$, on p degrees of freedom, and r is the residual, on $n - p$ degrees of freedom.

For the restricted model, $E(\mathbf{Y}) = \mathbf{A}\boldsymbol{\theta}$ subject to H, we denote the reduction in $\mathbf{y}'\mathbf{y}$ by $r(\boldsymbol{\theta}|H)$, so that

$$\mathbf{y}'\mathbf{y} = r(\boldsymbol{\theta}|H) + r_H,$$

where r_H is the residual under H.

We have seen that, assuming the errors are independent observations from $N(0, \sigma)$, R is $\sigma^2 \chi^2_{n-p}$. It can further be shown that, if H is true, the extra reduction $R(\theta) - R(\theta|H) = R_H - R$ is distributed independently of R as $\sigma^2 \chi^2_h$, where $h \leq p$ is the number of mathematically independent equations specified by H; h is known as the order of H and is equal to the rank of the matrix H.

In the examples above, we note that H_a and H_c both impose p independent conditions on θ (in each case the matrix of the hypothesis is the unit matrix $\mathbf{I}_p$) so each is of order p; H_b is of order 1; H_d comprises $(p-1)$ independent conditions on θ, so its order is $(p-1)$.

As specified, these hypotheses are all of full rank, there being no redundant conditions. For (a) and (c), $h = m = p$; for (b) $h = m = 1$; for (d) $h = m = p - 1$. When $m > h$, some of the equations are superfluous; this is illustrated by augmenting H_d by constraints such as $\theta_1 = \theta_3$, $2\theta_2 - \theta_4 - \theta_5 = 0$, which impose no additional restrictions but increase m to $(p+1)$.

The basis of the above distributional result is essentially that, when H is true, we can use the equations defining H to eliminate h of the parameters in θ. This is done by solving for these parameters (if a choice is possible, it does not matter which h are chosen) in terms of the remaining $(p-h)$; denote the former by ϕ and the latter by ψ. The solutions for ϕ are then substituted back into the original model. By this means, the restricted model is thus expressed, via a new design matrix $\mathbf{D}$, in terms of $(p-h)$ parameters which are themselves not subject to any restrictions, i.e. under H, the model may be written as $E(\mathbf{Y}) = \mathbf{D}\psi$, where ψ has $(p-h)$ components. The standard theory applies to this model, so that the L.S.E. of ψ under H is (assuming $\mathbf{D}'\mathbf{D}$ is nonsingular) $\psi^* = (\mathbf{D}'\mathbf{D})^{-1}\mathbf{D}'\mathbf{y}$, and the residual is $r_H = \mathbf{y}'\mathbf{y} - \psi^{*\prime}\mathbf{D}'\mathbf{y}$. With the usual assumptions about the errors, R_H is distributed as $\sigma^2 \chi^2_{n-(p-h)}$. Since $R_H = (R_H - R) + R$, it only remains to establish the independence of $(R_H - R)$ and R to deduce that $(R_H - R)$ is $\sigma^2 \chi^2_h$. The hypothesis H can be tested by the usual statistic

$$f = \frac{(r_H - r)/h}{r/(n-p)}.$$

When H is true, f is a realization of a random variable with the $F_{h,n-p}$ distribution; the upper tail of the distribution is used as the expectation is increased when H is false.

We can summarize these results in an anova table for testing H, as follows

Source	SS	DF	MS
Reduction due to θ	$r(\theta) = \tilde{\theta}'\mathbf{A}'\mathbf{y}$	p	
Reduction due to $\theta\|H$	$r(\theta\|H) = \psi^{*\prime}\mathbf{D}'\mathbf{y}$	$p-h$	
Extra reduction	$r(\theta) - r(\theta\|H)$	h	$\{r(\theta) - r(\theta\|H)\}/h$
Residual	$r = \mathbf{y}'\mathbf{y} - r(\theta)$	$n-p$	$\tilde{\sigma}^2 = r/(n-p)$
Total	$\mathbf{y}'\mathbf{y}$	n	

In this table, $r(\boldsymbol{\theta})$, the reduction due to fitting the model $E(\mathbf{Y}) = \mathbf{A}\boldsymbol{\theta}$, is divided into two parts: first, the amount $r(\boldsymbol{\theta}|H)$ by which the model reduces $\mathbf{y'y}$ when H is true, and secondly—the remaining part of $r(\boldsymbol{\theta})$—the extra reduction achieved when $\boldsymbol{\theta}$ is not bound by the constraints of H.

All the previous examples are versions of this table; for instance, if H specifies $\boldsymbol{\phi} = \mathbf{0}$, where $\boldsymbol{\phi}$ consists of m parameters from $\boldsymbol{\theta}$, we have $h = m$ and $r(\boldsymbol{\theta}|H) = r(\boldsymbol{\psi})$, where $\boldsymbol{\psi}$ denotes the remaining parameters of $\boldsymbol{\theta}$. Thus $r(\boldsymbol{\theta}) - r(\boldsymbol{\theta}|H) = r(\boldsymbol{\theta}) - r(\boldsymbol{\psi}) = r(\boldsymbol{\phi}|\boldsymbol{\psi})$, and we have the anova $r(\boldsymbol{\theta}) = r(\boldsymbol{\psi}) + r(\boldsymbol{\phi}|\boldsymbol{\psi})$ used in the previous discussions of this hypothesis.

We now present some further examples in this format and illustrate some of the more general hypotheses which we can test using the above theory.

EXAMPLE 8.3.8.   *Test for equality of means in a one-way layout.*   In Example 8.3.3 we showed how to test the hypothesis H_0, that the groups have a common, known mean, μ_0. We can now also consider the more general hypothesis H, that the means have a common, unknown value. As the design is orthogonal, the extra reduction can easily be found using (8.3.8), as in the previous discussion. To illustrate the general method we shall find r_H explicitly, and for purposes of comparison we shall first treat H_0 this way.

We recall from Example 8.2.14 that the residual for the full model $y_{ij} = \mu_i + e_{ij}$ is given by

$$r = \sum_{i=1}^{I} \sum_{j=1}^{J_i} (y_{ij} - y_{i.})^2 = \sum_{i=1}^{I} \sum_{j=1}^{J_i} y_{ij}^2 - \sum_{i=1}^{I} J_i y_{i.}^2.$$

where $y_{i.}$ is the average of the J_i observations in group i; r is a realization of R which is distributed as $\sigma^2 \chi_{n-I}^2$.

Under H_0, the model is $y_{ij} = \mu_0 + e_{ij}$. There are no unknown parameters left so

$$r_{H_0} = \sum_{i=1}^{I} \sum_{j=1}^{J_i} e_{ij}^2 = \sum_{i=1}^{I} \sum_{j=1}^{J_i} (y_{ij} - \mu_0)^2$$

which equals

$$\sum_{i=1}^{I} \sum_{j=1}^{J_i} y_{ij}^2 - 2\mu_0 \sum_{i=1}^{I} J_i y_{i.} + n\mu_0^2 \quad \text{where } n = \sum_{i=1}^{I} J_i,$$

whence

$$r_{H_0} - r = \sum_{i=1}^{I} J_i y_{i.}^2 - 2\mu_0 \sum_{i=1}^{I} J_i y_{i.} + n\mu_0^2 = \sum_{i=1}^{I} J_i (y_{i.} - \mu_0)^2.$$

The full model contains I parameters, each of which is equal to μ_0 under H_0, which is therefore of order I. The test statistic is therefore

$$\frac{(r_{H_0} - r)/I}{r/(n-I)} = \frac{\sum_{i=1}^{I} J_i (y_{i.} - \mu_0)^2 / I}{\sum_{i=1}^{I} \sum_{j=1}^{J_i} (y_{ij} - y_{i.})^2 / (n-I)}.$$

When H_0 is true, this is an observation from $F_{I, n-I}$. However, in most practical situations we are more interested in testing whether the μ_i are all equal without specifying their common value, which would actually be unknown. Thus we want to test $H: \mu_1 = \mu_2 = \ldots = \mu_I = \mu$, where μ is unknown; as noted above, this is of order $(I-1)$.

Under H, the model is $y_{ij} = \mu + e_{ij}$, so treating μ simply as an algebraic variable for the moment, we need to minimize with respect to μ the sum of squares

$$S(\mu) = \sum_{i=1}^{I} \sum_{j=1}^{J_i} (y_{ij} - \mu)^2.$$

Differentiating $S(\mu)$ with respect to μ and replacing μ by $\tilde{\mu}$, we obtain the normal equation

$$\sum_{i=1}^{I} \sum_{j=1}^{J_i} (y_{ij} - \tilde{\mu}) = 0,$$

the solution of which is $\tilde{\mu} = y_{..}$, the overall mean of all the observations. The residual for this model is then

$$r_H = S(\tilde{\mu}) = \sum_{i=1}^{I} \sum_{j=1}^{J_i} (y_{ij} - y_{..})^2,$$

and the extra reduction is

$$r_H - r = \sum_{i=1}^{I} \sum_{j=1}^{J_i} (y_{ij} - y_{..})^2 - \sum_{i=1}^{I} \sum_{j=1}^{J_i} (y_{ij} - y_{i.})^2,$$

which simplifies to give

$$r_H - r = \sum_{i=1}^{I} J_i (y_{i.} - y_{..})^2.$$

When H is true, $R_H - R$ is distributed as $\sigma^2 \chi_{I-1}^2$. The test statistic is

$$\frac{(r_H - r)/(I-1)}{r/(n-I)} = \frac{\sum_{i=1}^{I} J_i (y_{i.} - y_{..})^2/(I-1)}{\sum_{i=1}^{I} \sum_{j=1}^{J_i} (y_{ij} - y_{i.})^2/(n-I)},$$

this being an observation from $F_{I-1,n-I}$ if H is true.

EXAMPLE 8.3.9. We return to the gravity data, for which the following hypotheses could be of interest:

$$H_1: g_1 = g_2 = g_3 = g_4 = g_5; \qquad H_2: \bar{g} = 981 \cdot 2; \qquad H_3: g_1 = g_4.$$

We note that H_1 assumes a common value for the parameters but this value (g, say) is unspecified; H_1 is thus more general than the second hypothesis considered in Example 8.3.7, which stated that the common value is $981 \cdot 2$, and we can expect a better fit in the present case. To test H_1 we require the residual sum of squares for the model restricted by H_1, which is specified by

$$981 \cdot 1880 = g + e_1, \qquad -0 \cdot 0030 = e_2, \qquad 0 \cdot 1390 = e_3,$$
$$981 \cdot 2000 = g + e_4, \qquad 0 \cdot 0647 = e_5, \qquad 981 \cdot 2670 = g + e_6,$$
$$0 \cdot 0140 = e_7, \qquad 0 \cdot 1431 = e_8, \qquad 981 \cdot 2690 = g + e_9$$

or, in matrix form

$$
\begin{bmatrix}
981 \cdot 1880 \\
-0 \cdot 0030 \\
0 \cdot 1390 \\
981 \cdot 2000 \\
0.0647 \\
981 \cdot 2670 \\
0 \cdot 0140 \\
0 \cdot 1431 \\
981 \cdot 2690
\end{bmatrix}
=
\begin{bmatrix}
1 \\
0 \\
0 \\
1 \\
0 \\
1 \\
0 \\
0 \\
1
\end{bmatrix}
(g) +
\begin{bmatrix}
e_1 \\
\vdots \\
e_9
\end{bmatrix},
$$

whence $\tilde{g} = (981 \cdot 1880 + 981 \cdot 2000 + 981 \cdot 2670 + 981 \cdot 2690)/4 = 981 \cdot 2310$.

The residuals are thus $-0 \cdot 0430, -0 \cdot 0030, 0 \cdot 1390, -0 \cdot 0310, 0 \cdot 0647, 0 \cdot 0360,$ $0 \cdot 0140, 0 \cdot 1431, 0 \cdot 0380$, whence $r_{H_1} = 0 \cdot 0497397$. This is only slightly smaller than the residual under the hypothesis of a common value of $981 \cdot 2$, which was $0 \cdot 0535837$, so we can expect another large test ratio. In computing the latter we must remember that as the common value is unspecified the order of H_1 is 4. We have $r_{H_1} - r = 0 \cdot 04972792$ and the required test value is $\{(r_{H_1} - r)/4\}/\{r/4\} = 4221$. Clearly, we cannot entertain the hypothesis of a common value.

We turn to H_2: $\bar{g} = 981 \cdot 2$; again we note that this is more general than the hypothesis of a common value of $981 \cdot 2$ considered earlier in Example 8.3.7. There are two ways of finding the residual when the model is subject to H_2. First, we can use the constraint to eliminate one parameter, say g_2, by solving the equation $g_1 + g_2 + g_3 + g_4 + g_5 = 5 \times 981 \cdot 2$ for g_2 and writing the model equations in terms of g_1, g_3, g_4, g_5 only. Thus $g_2 = 4906 - g_1 - g_3 - g_4 - g_5$, whence the model equations are unchanged except for equations 1 and 7, which become

$$981 \cdot 1880 = 4906 - g_1 - g_3 - g_4 - g_5 + e_1,$$

and

$$0 \cdot 0140 = g_1 + 2g_3 + g_4 + g_5 - 4906 + e_7.$$

In matrix form we have

$$
\begin{bmatrix}
-3924\cdot812 \\
-0\cdot0030 \\
0\cdot1390 \\
981\cdot2000 \\
0\cdot0647 \\
981\cdot2670 \\
4906\cdot0140 \\
0\cdot1431 \\
981\cdot2690
\end{bmatrix}
=
\begin{bmatrix}
-1 & -1 & -1 & -1 \\
-1 & 0 & 1 & 0 \\
0 & 0 & 1 & -1 \\
0 & 1 & 0 & 0 \\
0 & -1 & 1 & 0 \\
1 & 0 & 0 & 0 \\
1 & 2 & 1 & 1 \\
0 & 0 & 1 & -1 \\
1 & 0 & 0 & 0
\end{bmatrix}
\begin{bmatrix}
g_1 \\
g_3 \\
g_4 \\
g_5
\end{bmatrix}
+
\begin{bmatrix}
e_1 \\
\vdots \\
e_9
\end{bmatrix}
$$

Since in this formulation g_1, g_3, g_4, g_5 are not subject to any restrictions we can now apply the standard theory to obtain r_{H_2}. Denoting the above by $z = D\psi$, we have

$$
D'D =
\begin{bmatrix}
5 & 3 & 1 & 2 \\
3 & 7 & 2 & 3 \\
1 & 2 & 6 & 0 \\
2 & 3 & 0 & 4
\end{bmatrix},
\qquad
D'z =
\begin{bmatrix}
10793\cdot365 \\
14717\cdot975 \\
8831\cdot169 \\
8830\cdot544
\end{bmatrix}.
$$

The L.S.E. of ψ is $\psi^* = (g_1^*, g_3^*, g_4^*, g_5^*) = (D'D)^{-1}D'z$ whence

$$g_1^* = 981\cdot2625, \; g_3^* = 981\cdot1929, \; g_4^* = 981\cdot2536, \; g_5^* = 981\cdot1101.$$

These are the L.S.E.'s of g_1, g_3, g_4, g_5 under H_2. The L.S.E. under H_2 of the linear function $4906 - g_1 - g_3 - g_4 - g_5$, i.e. g_2, is $g_2^* = 4906 - g_1^* - g_3^* - g_4^* - g_5^* = 981\cdot1809$.

The residual r_{H_2} can now be found either from $r_{H_2} = z'z - \psi^*D'z$ or by calculating the individual residuals followed by squaring and summing; we find that $r_{H_2} = 0\cdot00023854$.

The alternative procedure is to retain all five parameters and minimize

$$S = (981\cdot1880 - u_2)^2 + (-0\cdot0030 - u_4 + u_1)^2 + \ldots + (981\cdot2690 - u_1)^2$$

subject to the condition $u_1 + u_2 + u_3 + u_4 + u_5 = 4906$.

Introducing a Lagrange multiplier λ, we seek the unconditional minimum of $T = S + \lambda(u_1 + u_2 + u_3 + u_4 + u_5 - 4906)$. Differentiating T with respect to $u_1, \ldots, u_5$ respectively and substituting $u_i = g_i^*$, $i = 1, \ldots, 5$ we have

$$3g_1^* - g_4^* - 1962\cdot5390 + \lambda = 0$$

$$2g_2^* - g_3^* - 981\cdot1740 + \lambda = 0$$

$$-g_2^* + 3g_3^* - g_4^* - 981\cdot1493 + \lambda = 0$$

$$-g_1^* - g_3^* + 4g_4^* - 2g_5^* - 0\cdot3438 + \lambda = 0$$

$$-2g_4^* + 2g_5^* + 0\cdot2821 + \lambda = 0.$$

The restriction means that these estimates must satisfy $g_1^* + g_2^* + g_3^* + g_4^* + g_5^* = 4906$. The six equations are to be solved for $g_1^*, \ldots, g_5^*$ (at this stage λ is a nuisance parameter to be eliminated). We obtain the same L.S. estimates of $g_1, \ldots, g_5$ under H_2 as previously found by the elimination method. We can now obtain the residuals under H_2 by substituting these estimates in the original model equations.

For the test we require $r_{H_2} - r = 0\cdot00022675$; noting that H_2 is of order 1, the test statistic is just $(r_{H_2} - r)/(r/4) = 76\cdot93$. The notion that the average value of g at the five sites is $981\cdot2$ is extremely implausible.

For our final illustration using the gravity data, consider H_3: $g_1 = g_4$, under which the equations are

$$981\cdot1880 = g_2 + e_1, \qquad -0\cdot0030 = e_2, \qquad 0\cdot1390 = g_1 - g_5 + e_3,$$

$$981\cdot2000 = g_3 + e_4, \qquad 0\cdot0647 = g_1 - g_3 + e_5, \qquad 981\cdot2670 = g_1 + e_6,$$

$$0\cdot0140 = g_3 - g_2 + e_7, \qquad 0\cdot1431 = g_1 - g_5 + e_8, \qquad 981\cdot2690 = g_1 + e_9$$

or,

$$
\begin{bmatrix}
981\cdot1880 \\
-0\cdot0030 \\
0\cdot1390 \\
981\cdot2000 \\
0\cdot0647 \\
981\cdot2670 \\
0\cdot0140 \\
0\cdot1431 \\
981\cdot2690
\end{bmatrix}
=
\begin{bmatrix}
0 & 1 & 0 & 0 \\
0 & 0 & 0 & 0 \\
1 & 0 & 0 & -1 \\
0 & 0 & 1 & 0 \\
1 & 0 & -1 & 0 \\
1 & 0 & 0 & 0 \\
0 & -1 & 1 & 0 \\
1 & 0 & 0 & -1 \\
1 & 0 & 0 & 0
\end{bmatrix}
\begin{bmatrix}
g_1 \\
g_2 \\
g_3 \\
g_5
\end{bmatrix}
+
\begin{bmatrix}
e_1 \\
\vdots \\
e_9
\end{bmatrix}
$$

Denoting these by $\mathbf{y} = \mathbf{D\psi}$, we have

$$
\mathbf{D'D} =
\begin{bmatrix}
5 & 0 & -1 & -2 \\
0 & 2 & -1 & 0 \\
-1 & -1 & 3 & 0 \\
-2 & 0 & 0 & 2
\end{bmatrix},
\qquad
\mathbf{D'y} =
\begin{bmatrix}
1962.8828 \\
981.1740 \\
981.1493 \\
-0.2821
\end{bmatrix}.
$$

We find the L.S.E.'s of g_1, g_2, g_3, g_5 under H_3 to be

$$g_1^* = 981\cdot26739, \qquad g_2^* = 981\cdot18774, \qquad g_3^* = 981\cdot20148,$$

$g_5^* = 981\cdot12634$; the residuals are $e_1^* = 0\cdot00026$,

$$e_2^* = -0\cdot0030, \qquad e_3^* = -0\cdot00205, \qquad e_4^* = -0\cdot00148, \qquad e_5^* = -0\cdot00121,$$

$$e_6^* = -0\cdot00039, \qquad e_7^* = 0\cdot00026, \qquad e_8^* = 0\cdot00205, \qquad e_9^* = 0\cdot00161,$$

whence $r_{H_3} = 0.00001503$ and $r_{H_3} - r = 0.00000324$. The order of H_3 is 1, so we calculate $(r_{H_3} - r)/(r/4) = 1.099$. Now $F_{1,4}(0.95) = 7.71$, so the evidence against $g_1 = g_4$ is weak.

D.W.

8.4. FURTHER READING

A guide to further reading on the subjects of least squares and the analysis of variance will be found at the end of Chapter 10.

CHAPTER 9

The Design of Comparative Experiments

9.1. HISTORICAL INTRODUCTION

In the physical sciences in the nineteenth century, the nature of the entities studied and the sophisticated level of laboratory technique available reduced experimental error to a level at which competently executed repetitions of an experiment were able to reproduce the original observations closely enough for all practical purposes. In the biological sciences however the situation was different. 'Most biological material is inherently variable, and the sweet simplicity and reproducibility of physical and chemical experiments is consequently lacking. Statistical problems therefore began to obtrude.' (Frank Yates (1937).)

The statistical problems in question were posed by the difficulties of interpreting data from non-reproducible experiments. These in turn led to the concept of so arranging the experiments that the effects of the problem-causing variability were minimized.

One of the disciplines in which a systematic attack along these lines was first developed was in scientific agriculture, a subject which in its modern form came into existence with the establishment in 1843 of the now famous agricultural research establishment at Rothamsted in England. Here, field trials were initiated to assess the effects of various fertilizers on crop yields, and to compare the yields of different varieties. The scientists concerned had to contend not only with the wide variability of their experimental materials—the fertilities of their plots, the quality of the seed, the level of rainfall, etc.,—but also with the fact that each of their experiments necessarily required something like a year for its completion.

In this environment it was inevitable that research should be carried out to develop deliberately planned designs in the intended experimental programme with a view to reducing the consequences of the built-in variability to as low a level as possible, and to making possible an objective assessment of the accuracy of the resulting conclusions.

A typical experiment would be concerned with a comparison between the yield (of wheat, say) achieved when no fertilizer was applied (the 'control'), and that resulting from a given quantity of nitrogenous fertilizer; or, more realistically, with multiple comparisons between the control yield and the yields from various levels of each of several nitrogenous, phosphoric and potassium-bearing fertilizers.

By the end of the century, some general principles of design had been established and much data accumulated. The next step forward in the interpretation of these records and in the invention of error-reducing designs for improving on them in future work came with the arrival at Rothamsted of R. A. Fisher. Fisher immediately began work on the problem of how best to apply modern statistical principles to agricultural research work, and in so doing was led to parallel and interacting research of his own in the principles of statistical inference. The problems and techniques were not peculiar to agricultural research, and there came into being a new science dealing with the design and analysis of comparative experiments in areas like agriculture, biochemistry, materials science, chemical engineering, etc., etc., where experimental variability is high. This has now become a substantial professional discipline in its own right, usually going under the abbreviated title of 'experimental Design and Analysis'.

The aim of this chapter is to describe in broad outline the basic concepts of experimental design, including randomization, blocking, balance, interactions, confounding. These are in the main qualitative concepts and their description here is on the whole also qualitative. The analysis of experimental data consists in the estimation by least squares of the parameters of a suitable (linear) model which corresponds to the design, and the testing of hypotheses (such as that a particular parameter is zero, or that two parameters are equal) by standard analysis-of-variance techniques.

In the language of Chapter 8 the models used in most designs are 'singular'— that is, for reasons of symmetry, they employ more parameters than the model can support, with compensating side conditions to eliminate the redundancy.

The appropriate modification of Least Squares theory is described in Chapter 10.

9.2.   THE TEA-TASTING LADY

Fisher has given a famous exposition of the principles of experimental design in an example (possibly apocryphal) of an experiment designed to test the validity of a certain lady's claim that she could tell by tasting whether or not, in a given cup of tea, the tea had been poured into the cup before the milk. The design he presents [Fisher (1950)] exhibits the following features:

Replication (repetition): this is an obviously necessary ingredient; no experimenter would be willing to draw any conclusion from the subject's correct or incorrect identification of milk/tea order in a single cup.

Balance: the lady is to be offered equal numbers of tea-first and milk-first cups, so as not to bias her judgment.

Randomization: this essential design ingredient refers to the order in which the cups are to be presented to the taster. The randomization of this order is in fact the only feature which makes possible the application of statistical principles to the analysis of the results.

Sensitivity: Fisher points out that unless the number of cups involved exceeds a certain minimum, no conclusions may validly be drawn: that is, an experiment may be completely insensitive if it is too small. Beyond that minimum size however, the greater the number of replications the greater (within limits) is the sensitivity of the experiment.

Homogeneity: the above argument cannot be taken too far; when the number of cups exceeds a certain level the alleged difference in flavour due to the priority of tea or milk will be masked by temperature differences, ageing effects, saturation of the lady's taste buds, and so on. These are departures from homogeneity which will complicate or invalidate the analysis.

The analysis proceeds as follows. First, a clear statement of the purpose of the experiment, which is to provide evidence for or against the lady's claimed discriminatory powers. This is formulated in terms of the *null hypothesis* that she has in fact no discriminatory powers at all (in relation to milk-tea pouring order).

Suppose she is presented, in random order, with $2n$ cups of tea, n being milk-first and n tea-first. On the null hypothesis the probability of her making r correct identifications is

$$p_r = \binom{n}{n-r}\binom{n}{r} \Big/ \binom{2n}{n}$$

$$= \binom{n}{r}^2 \Big/ \binom{2n}{n}, \qquad r = 0, 1, \ldots, n;$$

[see II, § 5.3] where, e.g.,

for $n = 2$: $p_0 = 1/6$, $\quad p_1 = 2/3$, $\quad p_2 = 1/6$;

for $n = 3$: $p_0 = 1/20$, $\quad p_1 = 9/20$, $\quad p_2 = 9/20$, $\quad p_3 = 1/20$;

for $n = 4$: $p_0 = 1/70$, $\quad p_1 = 16/70$, $\quad p_2 = 36/70$,

$\qquad p_3 = 16/70$, $p_4 = 1/70$.

Thus, if $n = 2$ (i.e. two cups of each kind) there is a one in six chance of completely correct identification even if she has no discriminatory power. This is a high probability, and so even a completely correct set of identifications provide no evidence against the null hypothesis. (This is the case of zero sensitivity mentioned earlier.) With $n = 4$ the correct identification of all four tea-first cups (and therefore also of all four milk-first cups) would be unlikely if she had no discriminatory power, the chance of achieving this being only one in seventy.

It might however be objected that she ought to be allowed an occasional error, the requirement that she has to classify every single cup correctly being unduly severe. Now, with $n = 4$, the correct identification of three or more of the four tea-first cups has a probability on the null hypothesis of $17/70$, nearly one in four. That is far too high a value to discredit the null hypothesis. If, therefore, we wished to allow the subject the opportunity of validating her claim even if she makes one error, a larger number of cups than four of each kind will be required: for example, with $n = 7$, the significance level [see § 5.2.2] of the result that $n - 1$ correct choices were made out of n is 0.015, a level that most workers would accept as being low enough to discredit the null hypothesis [see Table 5.2.1] and thus provide evidence in support of the lady's claim.

9.3.   A MORE ELABORATE EXAMPLE: DARWIN'S EXPERIMENT

Whilst the tea-tasting lady provides a valuable introduction to the principles involved in experimental design, it is atypical in the following respects: (i) it seeks merely to provide evidence for or against the existence of an alleged effect, where a more general design would be concerned with estimating the magnitude of the effect; (ii) more general designs employ the device of 'blocking' to allow the experiment to be enlarged (i.e. to include more replications) without losing the benefits of the relatively homogeneous experimental material and environment available to a small experiment.

Both of these additional features are exhibited in the discussion (again by Fisher [see Fisher (1951)]) of a celebrated experiment carried out by Charles Darwin to establish whether or not cross-fertilized plants were superior (in size to self-fertilized ones of the same species, and, if so, by how much. The experiment, with Fisher's gloss on it, has been described in Example 5.7.1. The classical principles incorporated by Darwin included the following features:

   (i) '*Blocking*': direct comparison was made only between the two members of a pair of plants, one selfed and one crossed, which were grown in the same pot and which therefore had as nearly as possible the same environment. This device reduces the effect of the unavoidable difference between one environment and another, the fertility of the soil in one pot and that in another, etc.

  (ii) '*Replication*': more than one pot was used (of course!) In this particular experiment there were in fact 15 replicates (repetitions of the basic experimental module). By averaging over all 15 replicates Darwin was able to minimise the ratio of experimental error to the size of the effect under investigation.

 (iii) '*Balance*': the two kinds of plants were used in equal numbers, so as not to give unfair emphasis to either.

The additional ingredient supplied by Fisher, in his discussion of the experiment, was

    (iv) '*Randomization*': this feature was in actuality not employed by Darwin, but was specified by Fisher as a prerequisite that ought to have been included to make possible a valid analysis of the experiment. By 'randomization' in this context is meant, for each of the 15 pots, the carrying out of an act of physical randomization (such as tossing a coin) to determine which of the two seeds, selfed or crossed, should occupy the northern (say) and which the southern position in the pot. In his exposition Fisher argued that, on the null hupothesis that there was no systematic difference between crossed and selfed seeds, the observed differences in growth rates could be attributed only to environmental differences. The set of differences actually observed could then be compared with those that might have been observed with the other permutations, (equiprobable, on the null hypothesis) of seed-to station allocations that could have resulted from the randomization procedure. Indeed, from this point of view, it is only the imposition of randomization that allows one to use probability considerations at all.

An analysis of the experiment based on the '*randomization model*', i.e. on the concept that the data forms a sample from a population of equiprobable potential data sets induced by randomization, has been presented in Example 5.7.1. It will be seen that, even in such a simple experiment as this, the computations are non-trivial. In a more elaborate experiment the computational effort involved in a full-blown randomization analysis are apt to be extensive, and, possibly because the subject grew up in the precomputer age, alternative models were sought. One of the leading contenders takes the view that the 15 pairs of seeds actually used form a random sample from a (bivariate) population of pairs of plants, with a certain value for the expected difference $\delta = E(X - Y)$ between the height of a crossed and a selfed plant. The obvious estimate of δ being the mean $\bar{d}$ of the observed differences $d_r = x_r - y_r$, one relies on the Central Limit Theorem to justify (or at least to motivate) the procedure of regarding the observations d_r as realizations of independent Normally distributed random variables. Practically identical conclusions regarding the significance of the observed differences are obtained whether one uses the randomization model or one of the versions of the Normal distribution model.

This 'Normal distribution' point of view seems at first to invalidate Fisher's argument that requires a particular kind of randomization as an essential ingredient in the design of an experiment. In fact, however, it is generally agreed that randomization *is* an essential ingredient, if not for the Fisherian reasons adduced above, then for other equally cogent ones, such as that without it experimental errors could not be regarded as mutually independent, that the various treatments to be compared would not all be fairly represented, and so on.

9.4. COMPLETE RANDOMIZED BLOCK DESIGNS

In Darwin's experiment there were only two 'treatments' to be compared, and in each replication these were represented by two seedlings, easily accomodated in a relatively small 'block' consisting of an ordinary gardener's pot in which an acceptable degree of homogeneity in fertility and in ambient environment could easily be achieved. In agricultural (and other) experiments however the treatments are commonly larger in number and also require more space,— field-sized plots in the case of crop-growing trials; alternatively, time might be a limiting factor; an experiment might require several hours perhaps in chemical engineering, where the treatments to be compared may be the presence of one of several solvents, and the 'block' may be the working day. These factors all tend in the direction of 'over-saturating' the blocks: it might well be impossible to fit the treatments into the available block size (e.g. not enough hours in the day), or, if the block can be enlarged to contain all the treatments, it might then be so large as to have lost its homogeneity. There is therefore a limit beyond which a direct generalization of Darwin's design becomes impractical.

When however all 'treatments' can be accommodated in one block (e.g. each of three varieties of wheat in one field) without too great a sacrifice of homogeneity the principle of 'complete blocking' is evidently a sound one. Each of the three varieties of wheat would be allocated to one of three plots in the field, the allocation of variety to plot within each block being, of course, randomized. This gives us the (complete) *randomized blocks* design. Thus for example with four treatments *A, B, C, D*, and five blocks, the 'maps' of the five fields might be as illustrated in Figure 9.4.1. Comparisons between the yields of the treatments would be made only *within* blocks, thus eliminating as far as possible the differences in fertility etc. *between* blocks.

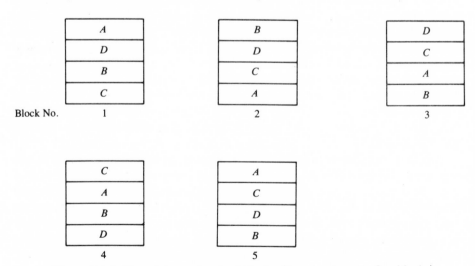

Figure 9.4.1: Complete randomized blocks (four treatments, five blocks).

9.5.   TREATMENTS AT ONE LEVEL AND AT SEVERAL LEVELS

In the five-block design exhibited in Figure 9.4.1, the entities to be compared (the 'treatments') may be four varieties of wheat; or they may be four different manurial treatments (A = control, with no fertilizer; B, C, D = conventional amounts of nitrogenous (N), phosphoric (P) and potassium (K) fertilizers, respectively); or they may include quantitatively different 'levels' of two fertilizers, each at two levels or intensities: (A = light application of N and light application of P; B = light application of N with heavy application of P; C = heavy N with light P; D = heavy N with heavy P). There are two 'treatments' (N, P) each at two levels, each level of N being combined with each level of P. This is an example of a 'fully crossed' experiment, capable of supplying information not only about the separate effects of nitrogen and phosphate, but also about the way their joint presence affects the yields.

This latter concept, called *interaction*, is evidently not relevant in the first illustration, where we are comparing four varieties of wheat. Here each of the four 'treatments' (varieties) is present at one level only. The *differential effect* of the treatments, for example the differential yield of variety A as compared with variety B, may be assessed as the difference $a - b$ between the yield a of variety A and the yield b of variety B in a given block. Any contribution that the block characteristics may make to the yield in that block is assumed to be the same for A as for B, and to be additive in each case, so that the difference $a - b$ in any one block is uninfluenced by the block characteristics. A valid measure of the *differential effect* of A with respect to B may therefore be obtained by averaging the individual block estimates $a - b$ over all blocks.

9.6.   THE NEED FOR DEVICES TO REDUCE BLOCK SIZE

To appreciate the number of treatments that can be required in even a modest inquiry, consider the case where one is seeking to measure the effects of two different fertilizers on, say, the yield of wheat. What we should like is to know the function

$$z = f(x, y)$$

giving the yield (z) per acre when x units of phosphate and y units of nitrate are applied to a field of ordinary fertility under ordinary climatic conditions. Quite apart from the problems posed by the presence of large experimental 'error' we have the problem of constructing, as accurately as possible, the form of the '*response surface*' $z = f(x, y)$ from a finite number of observations $(z_1; x_1, y_1), (z_2; x_2, y_2), \ldots, (z_k; x_k, y_k)$. Even if only four levels of each fertilizer were used (zero, light, medium and heavy applications, say) the number of treatment combinations of phosphate at level r with nitrate at level s, (r, s, = 1, 2, 3, 4) would amount to 16. With more than two treatments, each at several levels, the number of treatment combinations may well be too large to be accommodated in a single relatively homogeneous block.

This difficulty is not confined to experiments where one is assessing the effects of different combinations of various levels of several treatment intensities; it can arise in any comparative experiment. Designs have therefore been developed which attempt to obviate this difficulty by various devices for reducing the size of the blocks without sacrificing too much information.

9.7. BALANCED INCOMPLETE BLOCKS FOR COMPARING SINGLE-LEVEL TREATMENTS

In the complete randomized blocks design exhibited in Figure 9.4.1 each block contained all four treatments *A*, *B*, *C*, *D*. Suppose instead that for practical reasons only three treatments could be accommodated in a block. Such a block would be *incomplete* (i.e. would not contain all treatments). For reasons of *balance*, a certain symmetry between blocks would be necessary. In the *balanced incomplete blocks* design the 'balance' requirements are

each treatment must occur once, and only once, in each block,

the number of blocks containing a given treatment must be the same for all treatments,

and

the number of blocks containing a given pair of treatments must be the same for all pairs.

EXAMPLE 9.7.1.   *Four treatments in six blocks.*   An example, with four treatments and six blocks, would be:

$$
\left.
\begin{array}{l}
\text{Block 1: } A, B \\
\text{Block 2: } B, C \\
\text{Block 3: } C, D \\
\text{Block 4: } B, D \\
\text{Block 5: } A, C \\
\text{Block 6: } A, D
\end{array}
\right\}
\quad
\begin{array}{l}
\text{the allocation of treatments to units} \\
\text{in each block being randomized.}
\end{array}
$$

An alternative method of exhibiting this is:

	Block	1	2	3	4	5	6
	A	X				X	X
Treatment	B	X	X		X		
	C		X	X		X	
	D			X	X		X

Here the differential effect of A and B may be estimated as $a - b$ from Block 1, that of B and C as $b - c$ from block 2, and so on. This is a rather simple example in the sense that each block contains only one pair of treatments. A more interesting example, with four treatments, three in each of four blocks, is

	Block			
	1	2	3	4
A	X	X	X	
B	X	X		X
C	X		X	X
D		X	X	X

(Treatments)

Here the differential effect of A and B may be estimated from $a - b$ in Block 1 and also from $a - b$ in Block 2; the average of these gives the effective estimate. Similarly for A and C from Blocks 1 and 3, etc.

It will be clear that the construction of balanced incomplete block designs depends heavily on combinatorial analysis. Tables are available showing the designs that exist for various numbers of treatments, block sizes and numbers of blocks (see Bibliography).

9.8.   THE COMPLETE FACTORIAL DESIGN

When we are concerned with treatments which may be applied in a combined way to the experimental units, possibly at more than one level of intensity, the 'complete' design calls for blocks each of which contains every combination of treatment levels. Such a design is called 'completely crossed', or a 'complete factorial' design. For example, with two fertilizers A and B, a 'completely crossed 2×3' experiment with A present at two levels A_0 and A_1, and B at three levels B_0, B_1 and B_2, all $2 \times 3 = 6$ combinations of A_r with B_s would be represented:

	Level of factor B		
	B_0	B_1	B_2
A_0	X	X	X
A_1	X	X	X

(Level of factor A)

With a third factor C, at two levels C_0 and C_1, all $2 \times 3 \times 2 = 12$ combinations of A_r, B_s and C_t would be represented:

		C_0			C_1	
	B_0	B_1	B_2	B_0	B_1	B_2
A_0	X	X	X	X	X	X
A_1	X	X	X	X	X	X

9.8.1.   Main Effects and Interactions: Two Factors at Two Levels

Consider first a 2^2 design, that is one with two factors, A and B, each at two levels, a 'lower' level (A_0, B_0) and an 'upper level' (A_1, B_1). (The 'lower' level of A might in practice represent the complete absence of the treatment A, and similarly for B.) The most obvious procedure that suggests itself for the analysis is the following. Code the treatment levels and the corresponding yields in a given block as follows:

Level of A	Code	Level of B	Code	Yield
A_0	x_0	B_0	y_0	$z(x_0, y_0)$
A_1	x_1	B_0	y_0	$z(x_1, y_0)$
A_0	x_0	B_1	y_1	$z(x_0, y_1)$
A_1	x_1	B_1	y_1	$z(x_1, y_1)$

Then (neglecting for the moment the existence of experimental error) use the data to fit some convenient functional form $z = f(x, y)$, e.g.

$$z = \beta_0 + \beta_{10}x + \beta_{01}y + \beta_{11}xy;$$

or, more realistically, do not ignore the errors, and fit the function $f(x, y)$ by some smoothing procedure such as Least Squares, possibly following this up by producing the contours of the surface.

Procedures of this kind are currently in common use, particularly when—as in chemical engineering, for example—one is interested in locating the maximum of a response surface in circumstances where new experiments with different combinations of factor levels may be carried out relatively quickly, so that 'hill-climbing' techniques may be used to approach the maximum ('evolutionary operation') and where the contours in the neighbourhood of the maximum may be well approximated by similar confocal ellipses. During the greater part of the history of our subject however this approach did not seem to be a fruitful one. The primary object of the experiment was seen (effectively) as a general exploration of the response surface rather than the

location of its maxima. The method which was evolved, which readily general-
izes to any number of factors at any number of levels, depends on the concepts
of *main effects* and *interactions*. In the 2×2 design, (two factors, each at two
levels) the *effect* (or *main effect*) *of A* is proportional to the difference between

> (i) the average of all yields from treatment combinations involving the
> *higher* level of *A*

and

> (ii) the average of all yields from treatment combinations involving the
> *lower* level of *A*.

The *main effect of B* is defined similarly.
The *interaction* between *A* and *B* is proportional to the difference between

> (i) the effect of increasing *A* while holding *B* fixed at its higher level

and

> (ii) the effect of increasing *A* while holding *B* fixed at its lower level.

(In the last definition the letters *A* and *B* may of course be interchanged.)
Thus the interaction is a measure of the non-additivity of the effects of *A* and
B on the yield: an increase in carbon content may have one effect on the
strength of steel containing no tungsten, and quite a different effect if the mix
contains an appreciable amount of tungsten. [cf. § 11.3.5]
    The notation $z = f(x, y)$ is not very appropriate to this way of looking at
things. A useful and widely used notation is the following:

Level of A	Level of B	Code
A_0	B_0	(1)
A_1	B_0	a
A_0	B_1	b
A_1	B_1	ab

Thus '*a*' indicates that *A* is at its higher level, '*b*' that *B* is at its higher level,
'*ab*' that both are at the higher level; while '(1)' containing neither *a* nor *b*,
indicates that both factors are at the lower level. There will be no ambiguity
if we use the same code to designate not only the treatment level combinations
but also, when necessary, the corresponding yield. Thus the main effect of *A*
is proportional to

$$(a + ab) - ((1) + b).$$

Note that this expression can be obtained by expanding the algebraic expression

$$(a - 1)(b + 1),$$

provided that in the expansion '1' is interpreted as '(1)', the code for the yield obtained when both treatments are at their lower level.

Similarly the main effect of B is proportional to

$$(b+ab)-((1)+a)=(a+1)(b-1)$$

(with the same convention regarding '(1)'), and the interaction of A and B is proportional to

$$(ab-b)-(a-(1))$$

or equivalently to

$$(ab-a)-(b-(1)).$$

Subject to the convention mentioned earlier about '1' and '(1)', both of these may be regarded as the expansion of

$$(a-1)(b-1).$$

In a block containing one representative of each of the four treatment combinations (1), a, b, ab, in the absence of variability the four yields could be expressed in terms of four equivalent but more intuitively informative statistics consisting of (i) the three measures of treatment effects, namely the two main effects and the interaction, and (ii) the 'block effect', measured by the overall average for the block. On the usual additivity assumption this latter influences all yields within the block to an equal (additive) extent. Thus the treatment effects, estimated from within the block, are not influenced by the block effect.

In reality of course the variability of the responses can*not* be ignored. Suppose there are k blocks in all, each containing (in randomized order) all four treatment combinations. Then the main effects and the interaction may be estimated by the average, over all blocks, of the appropriate single-block statistics. The k individual block means will have been used to estimate the 'block effects'. From the original $4k$ observations, therefore, $3+k$ quantities will have been estimated, leaving $4k-(3+k)=3(k-1)$ observations available for the estimation of the variability upon which may be based tests for the significance of the estimated effects.

9.8.2. Main Effects and Interactions: Three factors at Two Levels

For a 3^2 design, that is one with three factors A, B and C, each at two levels, the coding would be as follows:

Level of A	Level of B	Level of C	Code
A_0	B_0	C_0	(1)
A_1	B_0	C_0	a
A_0	B_1	C_0	b

Level of A	Level of B	Level of C	Code
A_1	B_1	C_0	ab
A_0	B_0	C_1	c
A_1	B_0	C_1	ac
A_0	B_1	C_1	bc
A_1	B_1	C_1	abc

The main effect of A, as before, is proportional to the difference between

(i) the average of all yields from treatment combinations involving the *higher* level of A (the sum of all such yields being

$$a + ab + ac + abc)$$

and

(ii) the average of all yields from treatment combinations involving the *lower* level of A, (the sum of all such yields being

$$(1) + b + c + bc).$$

Thus the *main effect of A* is proportional to

$$(a + ab + ac + abc) - ((1) + a + b + c + bc)$$

and this, subject to the convention that '1' denotes '(1)', is the expansion of the algebraic expression

$$(a - 1)(b + 1)(c + 1).$$

Similarly the main effects of B and C respectively are those combinations of yields obtained by expanding

$$\text{(for } B\text{): } (a + 1)(b - 1)(c + 1),$$

$$\text{(for } C\text{): } (a + 1)(b + 1)(c - 1).$$

This time there are four interactions, namely the three second-order interactions, A with B, A with C, and B with C, plus the triple interaction of A, B and C. Their definitions are similar to those used in the 2^2 experiment, the AB interaction for example being the difference between

(i) the effect of increasing A while holding B fixed at its higher level, averaged over all C levels

and

(ii) the corresponding effect obtained while holding B fixed at its lower level.

Here the effect (i) is proportional to

$$(ab + abc) - (b + bc)$$

and the effect (ii) to

$$(a+ac)-((1)+c).$$

Thus the interaction of A with B (the 'AB interaction') is that combination of yields indicated by

$$(ab+abc)-(b+bc)-(a+ac)+((1)+c).$$

This expression may be obtained by formally expanding

$$(a-1)(b-1)(c+1).$$

Similarly the AC interaction is obtainable from

$$(a-1)(b+1)(c-1),$$

and the BC interaction from

$$(a+1)(b-1)(c-1).$$

Finally, the triple interaction ABC represents the difference between what might be called the partial AB interaction at the higher level of C namely

$$(abc-bc)-(ac-c),$$

and the corresponding combination at the lower value of C, namely

$$(ab-b)-(a-(1)),$$

whence the triple interaction is obtainable from a formal expansion of the expression

$$(a-1)(b-1)(c-1).$$

In the above calculations the yields referred to are all obtained from a single block. In practice there would be several blocks—k say. The treatment effects would be estimated by averaging, over all blocks, the separate estimates from the individual blocks. Thus the main effect of A would be estimated by computing the statistic $(a-1)(b+1)(c+1)$ from each block separately, and averaging over all blocks. Similarly for the other main effects and interactions. The $8k$ individual observations ($2^3 = 8$ per block, k blocks) would thus be utilized to estimate the seven treatment effects (three main effects, 2 second-order interactions, one triple interaction) and the k block means, leaving effectively $8k-7-k=7(k-1)$ observations for the estimation of variability.

9.8.3. Main Effects and Interactions in a 2^t Design:

In general, in a complete blocked experiment in which each of t factors is present at two levels in each block, (making a total of 2^t observations per block) an analysis of the yields within any one block would enable one to evaluate

the average yield for that block,

the t main effects, one for each factor,

the $\binom{t}{2} = \frac{1}{2}t(t-1)$ second order interactions,

the $\binom{t}{3} = \frac{1}{6}t(t-1)(t-2)$ third order interactions,

. . .

. . .

the $\binom{t}{t-1} = t$ interaction of order $t-1$,

and

the (unique) tth order interaction.

These add up to a total of

$$\binom{t}{1} + \binom{t}{2} + \ldots + \binom{t}{t} = 2^t - 1$$

estimates of treatment effects. In practice the estimates would be averaged over the k blocks. In addition the k separate block means would be evaluated as measures of the block effects. After evaluating these statistics, $2^t - 1 + k$ in number, from the $2^t k$ original observations, there would be effectively

$$2^t k - (2^t - 1 + k) = (2^t - 1)(k - 1)$$

observations available for the evaluation of the variability.

9.9. INCOMPLETE BLOCKS: CONFOUNDING

When, for reasons of economy or practicality or convenience not all treatment combinations are represented in each relatively homogeneous 'block' of a factorial experiment, various strategies are available for the design of the (necessarily) incomplete blocks. One of the chief criteria in the comparison of the available incomplete blocks designs is based on the concept of *confounding*.

Suppose that in a 2^3 experiment it is possible to accommodate only four treatment combinations per block. Since there are in all eight treatment combinations to be considered, it will not be possible for all blocks to be (randomized) replicates of one another. Consider the consequences of the following distribution of treatments:

Block 1: (1), a, b, ab
Block 2: c, ac, bc, abc

The yields may be represented as follows:

> Block 1: (1), a, b, ab
> Block 2: $c + \lambda$, $ac + \lambda$, $bc + \lambda$, $abc + \lambda$

where λ represents the differential block effect due to, for example, the additional fertility of Block 2 as compared with Block 1. Block 2 differs from Block 1 in that all yields in the former are subjected to the combined (but additive) consequences of

(i) having factor C at its higher level

and

(ii) being increased relative to the yields on Block 1 by the addition of the 'Block effect' λ.

Factor C is said to be *confounded* with blocks, and this distorts some (but not all) of the estimates of treatment effects. To carry out the calculations it is helpful to use the following table (Table 9.9.1) in which the column entries give the signs to be attached to the corresponding row symbols in the expression to be used, in an experiment with complete blocks, for the treatment effect designated at the head of the column.

	A	B	AB	C	AC	BC	ABC
(1)	−	−	+	−	+	+	−
†a	+	−	−	−	−	+	+
†b	−	+	−	−	+	−	+
ab	+	+	+	−	−	−	−
†*c	−	−	+	+	−	−	+
*ac	+	−	−	+	+	−	−
*bc	−	+	−	+	−	+	−
†*abc	+	+	+	+	+	+	+

Table 9.9.1

Here the main effects are denoted by A, B, C, respectively, and the interaction of A with B by AB, etc. The treatment combinations that occur in Block 2, for which the yields are subject to the increment λ, are marked with an asterisk. In a complete factorial design, the interaction AB is computed from the formula:

$$(1) - a - b + ab + c - ac - bc + abc.$$

In the confounded design under consideration the (asterisked) yields c, ac, bc, abc are replaced by $c + \lambda$, $ac + \lambda$, $bc + \lambda$, $abc + \lambda$ respectively. Thus their use in the conventional formula for the interaction AB leads to two positive and two negative appearances of the intrusive λ so that the resulting contribution of λ is zero. The *interaction AB* is therefore correctly estimated. It is easy to see from the table of signs that the same applies to all main effects and interactions *except the main effect of C*, for which the formula would yield an answer too large by an amount 4λ. The effect of confounding treatment C with blocks in this way is to render it impossible to estimate the main effect of C. The art of confounding—and therefore to a large extent the art of designing comparative experiments—is to distribute the treatment combinations among the incomplete blocks in such a way that the treatment effects that cannot be estimated (or that can be estimated with lesser accuracy than the others) are those which the experimenter is most willing to sacrifice. Usually the main effects are of greater interest than the highest order interactions. In the 2^3 experiment referred to above, with blocks capable of accommodating only four treatments each, the following arrangement confounds the triple interaction with blocks, but preserves the main effects and the second order interactions:

Block 1: (1), *ab*, *ac*, *bc*
Block 2: *a*, *b*, *c*, *abc*.

(The treatment combinations allocated to Block 2 by this scheme are marked with a dagger (†) in the Table. If each daggered yield is regarded as suffering an increase λ due to the block effect, it will be seen that, in the formula for estimating the interaction *BC*, for example, two of the Block 2 yields occur with positive sign, and two with negative sign, so that the net contribution of the block effect λ is zero. Similarly for the other second-order interactions and for the main effects.

9.10. PARTIAL CONFOUNDING

In an ordinarily replicated experiment, with say four replications, we should have four replications of each of the Block 1 and Block 2 patterns described above. The estimates of the three main effects and of the three second-order interactions would have the precision appropriate to this number of replications, and there would be no information at all about the triple interaction.

It is possible however to modify the replicates so that different interactions are confounded in these modified replicates. For example:

Modified replication	I		II		III		IV	
	(1)	*a*	(1)	*a*	(1)	*a*	(1)	*b*
	ab	*b*	*ab*	*b*	*b*	*c*	*a*	*ab*
	ac	*c*	*c*	*ac*	*ac*	*ab*	*bc*	*c*
	bc	*abc*	*abc*	*bc*	*abc*	*bc*	*abc*	*ac*
Confounding	*ABC*		*AB*		*AC*		*BC*	

Here, for example, the triple interaction cannot be estimated from the first replication, but can from each of the others. The net result is that, as compared with an experiment of the same total size in which however the 'replicates' were all true replicates of one another, this modified design estimates the main effects with the same precision, estimates the second order interactions with lower precision but in addition estimates the triple interaction (with the same precision, as it happens, as the second order interactions.)

This device is known as partial confounding.

9.11. FACTORS AT THREE OR MORE LEVELS

Completely crossed experimental designs are not of course restricted to factors at two levels. The concepts of main effects, interactions, and confounding may be extended to designs in which the factors have three or more levels. Here however we restrict ourselves to some simple special cases.

9.11.1   The Latin Square

The most important of these special cases is the Latin Square design, in which three factors are involved, each at s levels, where s may be 2, or 3, or 4, or .... The differential effect of levels i and j may be estimated for each factor, but no information can be obtained about interactions between the factors. The design is thus particularly appropriate to experiments in which the three factors may reasonably be assumed to be independent (i.e. with no interactions). An example would be an experiment to measure how tyre wear is affected by tyre pressure, position on the car, and speed of driving. In the first place we have to assume the absence of interactions between pressure, wheel location and speed (or any two of them). Secondly, since there are four wheels and hence four locations, the number of pressure levels and the number of test speeds must also be four. The combinations of pressures (P_1, P_2, P_3, P_4), locations (L_1, L_2, L_3, L_4) and speeds (S_1, S_2, S_3, S_4) may then be set out in a square 4×4 tableau such as the following:

	P_1	P_2	P_3	P_4
L_1	S_1	S_3	S_4	S_2
L_2	S_2	S_1	S_3	S_4
L_3	S_3	S_4	S_2	S_1
L_4	S_4	S_2	S_1	S_3

Here the four levels (P_1, P_2, P_3, P_4) of one of the factors are taken as defining four *columns*, the four levels (L_1, L_2, L_3, L_4) of another factor are taken as defining *rows*, and the levels S_1, S_2, S_3, S_4 of the remaining factor are allocated to the 4×4 grid defined by these rows and columns so as to form a Latin Square, that is in such a way that each of the four symbols S_1, S_2, S_3, S_4 occurs exactly once in each column, and each occurs exactly once in each row. (Latin Squares are usually described in terms of 'Latin' letters $A, B, C, D, \ldots$ Examples of 4×4 Latin Squares in this notation are:

A	B	C	D		A	D	C	B
B	C	D	A		D	B	A	C
C	D	A	B		B	C	D	A
D	A	B	C		C	A	B	D

Tables of Latin Squares of various sizes are available [see, e.g. Fisher and Yates (1957)—Bibliography G]. They are tabulated in standard forms: one of these may be selected at random, one of the family that may be generated from this selected by another randomization, and finally the allocation of factors to rows, columns and Latin letters performed randomly.)

A complete factorial design for three factors, each at four levels, would require $4^3 = 64$ treatment combinations in each replication. In the Latin Square only 16 treatment combinations are required, this economy being paid for of course by the sacrifice of information about the interactions.

If we denote the experimental measurement corresponding to the rth row, the cth column, and the tth level S_t of the 'Latin letter' factor by the symbol (r, c, t), $r, c, t = 1, 2, 3, 4$, the data for the square for locations L_r, pressures P_c, and speeds S_t shown above

$$
\begin{array}{llll}
(1,1,1) & (1,2,3) & (1,3,4) & (1,4,2) \\
(2,1,2) & (2,2,1) & (2,3,3) & (2,4,4) \\
(3,1,3) & (3,2,4) & (3,3,2) & (3,4,1) \\
(4,1,4) & (4,2,2) & (4,3,1) & (4,4,3)
\end{array}
$$

The absence of interactions between factors implies additivity of their effects. From either version of the tableaux it is easy to see that the sum of the yields in column c will estimate the effect of P_c averaged over all locations and speeds, $c = 1, 2, 3, 4$. Thus the differential effect of increasing the pressure from Level 1 to Level 2, averaged over location and speed, may be estimated from the difference between the Column 2 total and the Column 1 total. Similarly the difference between the r_1th and the r_2th row totals estimates the differential effect of wheel positions r_1 and r_2. Finally, if we take the sum of the four measurements $(1, 1, 1)$, $(2, 2, 1)$, $(4, 3, 1)$ $(3, 4, 1)$ corresponding to level 1 of the speed factor, and subtract it from the sum of the four measurements $(2, 1, 2)$, $(4, 2, 2)$, $(3, 3, 2)$, $(1, 4, 2)$ corresponding to S_2, we obtain an estimate of the differential effect of increasing the test speed of the car from Level 1 to Level 2.

A modification: *Dummy levels.* It may be mentioned that it is possible to relax the restriction that the number of levels must be the same for each factor. Suppose for example that in the tyre-wear measurements discussed above the test machinery is capable of running at only three speeds, namely S_1, S_2 and S_3. We may fit this into the Latin Square scheme, preserving the original design, by making S_4 equal to one of S_1, S_2, S_3; if, for example, it is the highest speed that is thought to require greatest attention, we take S_4 equal to S_3.

9.11.2. The Graeco–Latin Square

The $(s \times s)$ Latin Square design, which compares the effects of three factors each having s levels, may be generalized to allow *four* factors to be compared, again each at s levels, and again with the restriction that no inter-factor interactions can be estimated. The design is therefore properly applicable only in cases where the four factors act 'independently', i.e. additively. In combinatorial theory a Graeco–Latin Square is an array such as the following

$$A\alpha \quad B\gamma \quad C\beta$$
$$C\gamma \quad A\beta \quad B\alpha \qquad .$$
$$B\beta \quad C\alpha \quad A\gamma$$

Here the Latin letters, considered by themselves, form a Latin Square. So do the Greek letters. Further, the Greek letter α appears once with A, once with B, and once with C. The Greek letter β likewise appears once with A, once with B, and once with C. Similarly for γ.

In the application to experimental design, let us revert to the tyre-wear example considered earlier, but let us now add quality of road surface as a fourth factor. As before we label the four columns by the pressure levels P_i, the rows by the location indicators L_i, the speeds by the letters S_k (these now replacing the 'Latin letters'). The four levels Q_1, Q_2, Q_3, Q_4 of the new factor Q will now be regarded as 'Greek letters'. A suitable Graeco–Latin Square would then be:

	P_1	P_2	P_3	P_4
L_1	S_1Q_1	S_2Q_2	S_3Q_3	S_4Q_4
L_2	S_2Q_3	S_1Q_4	S_4Q_1	S_3Q_2
L_3	S_3Q_4	S_4Q_3	S_1Q_2	S_2Q_1
L_4	S_4Q_2	S_3Q_1	S_2Q_4	S_1Q_3

As before, the differential effect of increasing the pressure from the lowest to the second level, averaged now over all locations, speeds, and road qualities, is estimated from the difference between the Column-1 total and the Column-2 total. Similarly for location effects, etc. Finally, the total of those observations to which the letter Q_4 is attached, diminished by the corresponding total for Q_2 , provides an estimate of the differential effect of roads of quality 4 and quality 2.

As in the case of Latin Squares, Graeco–Latin Squares are tabulated at all sizes above 2×2, and excluding 6×6, that are likely to be required.

E.H.L.

9.12. FURTHER READING AND REFERENCES

The subject was invented by Fisher and Yates in the thirties, Fisher's book [Fisher (1951)] being first published in 1935 and Yates' classic monograph on factorial experiments in 1937. [Yates (1937)]. The celebrated 'Statistical Tables' by Fisher and Yates [Fisher and Yates (1957)—Bibliography G) first appeared in 1938. This invaluable volume is not merely a collection of 'standard' tables of statistical functions: it also contains tables of Latin Squares and Balanced Incomplete Blocks together with invaluable explanatory commentary.

A selection of the best available texts by other authors is appended.

Fisher, R. A. (1951). *The Design of Experiments* (1950). Sixth edition, Oliver and Boyd.

Finney, D. J. (1960). *An Introduction to the Theory of Experimental Design*, University of Chicago Press.

Cochran, W. G. and Cox, G. M. (1957. *Experimental Designs* (Accord, ed.), Wiley.

Cox, D. R. (1958). *Planning of Experiments*, Wiley.

Davies, O. L. (editor) (1954). *The Design and Analysis of Industrial Experiments*, Oliver and Boyd.

Finney, D. J. (1955). *Experimental Design and its Statistical Basis*, University of Chicago Press.

Kempthorne, O. (1952). *The Design and Analysis of Experiments*, Wiley.

Yates, F. (1937). *The Design and Analysis of Factorial Experiments*, Imperial Bureau of the Soil Science.

Vajda, S. *The Mathematics of Experimental Design: Incomplete Block Designs and Latin Squares*, Griffin.

CHAPTER 10

Least Squares and the Analysis of Statistical Experiments: Singular Models; Multiple Tests

10.1. SINGULAR MODELS

10.1.1. Introduction

All the methods discussed in Chapter 8 assume that the basic model $E(\mathbf{Y}) = \mathbf{A}\theta$ is of full rank [see I, § 5.6] i.e. the matrix $\mathbf{A}$ has rank p equal to the number of parameters to be estimated, or, equivalently, that $\mathbf{A}'\mathbf{A}$ is nonsingular [see I, § 6.4]. Many important models, particularly in the area of experimental design, are singular models, in which the matrix $\mathbf{A}$ has rank $s < p$. Thus $\mathbf{A}'\mathbf{A}$, which is $p \times p$, is singular since the rank of $\mathbf{A}'\mathbf{A}$ equals that of $\mathbf{A}$.

In several examples, we have considered the one-way layout, [Examples 8.2.6, 8.2.10, 8.2.14 and 8.3.8] with model

$$y_{ij} = \mu_i + e_{ij}; \qquad i = 1, \ldots, I; \qquad j = 1, \ldots, J_i. \qquad (10.1.1)$$

This parameterization is of full rank (here, $p = I$); the design matrix $\mathbf{A}$ was given in Example 8.2.6, and $\mathbf{A}'\mathbf{A} = \mathrm{diag}(J_1, J_2, \ldots, J_I)$ where J_i is the number of observations in group i. This is clearly a sensible parameterization: there are I groups of observations, one corresponding to each of I treatments, and the model allots a parameter to each treatment.

There is, however, an attractive alternative, which is to express μ_i as

$$\mu_i = a + b_i, \qquad i = 1, 2, \ldots, I;$$

that is, μ_i is expressed in terms of a fixed constant a, the same for all treatments, and a deviation from a, characterising the ith treatment. This parameterization has a certain symmetry but appears to involve $I + 1$ parameters where, clearly, we really need only I of them. We appear to have a *redundant* parameter. The fact is that the formula

$$\mu_i = a + b_i$$

459

as it stands is not meaningful since it is equivalent to

$$\mu_i = (a + \alpha) + (b_i - \alpha)$$

$$= a' + b_i', \quad \text{say}$$

whatever the value of α. It can be made meaningful, however, by imposing a linear restraint on the b_i. The simplest one, and one that is appropriate if each group contains the same number of observations is

$$\sum b_i = 0.$$

This could be used, if desired, to express one of the b_i in terms of the others and to substitute this value in the formula

$$\mu_i = a + b_i.$$

In practice we often prefer to retain the latter 'redundant' form, and to supplement it with the appropriate linear restraint such as $\sum b_i = 0$. (This procedure is discussed more fully in section 10.1.2.)

The model then becomes

$$y_{ij} = a + b_i + e_{ij}, \tag{10.1.2}$$

(with side condition imposing a linear restraint on the b_i.) Translating this into matrix notation, it becomes

$$y = A\theta + e,$$

or, explicitly,

$$
\begin{bmatrix} y_{11} \\ \vdots \\ y_{1J_1} \\ y_{21} \\ \vdots \\ y_{2J_2} \\ \vdots \\ y_{I1} \\ \vdots \\ y_{IJ_I} \end{bmatrix}
=
\begin{bmatrix}
1 & 1 & 0 & 0 & & 0 & 0 \\
\vdots & \vdots & \vdots & \vdots & \cdots & \vdots & \vdots \\
1 & 1 & 0 & 0 & & 0 & 0 \\
1 & 0 & 1 & 0 & & 0 & 0 \\
\vdots & \vdots & \vdots & \vdots & \cdots & \vdots & \vdots \\
1 & 0 & 1 & 0 & & 0 & 0 \\
& & & & \vdots & & \\
1 & 0 & 0 & 0 & & 0 & 1 \\
\vdots & \vdots & \vdots & \vdots & & \vdots & \vdots \\
1 & 0 & 0 & 0 & & 0 & 1
\end{bmatrix}
\begin{bmatrix} a \\ b_1 \\ b_2 \\ \vdots \\ b_I \end{bmatrix}
+
\begin{bmatrix} e_{11} \\ \vdots \\ e_{1J_1} \\ e_{21} \\ \vdots \\ e_{2J_2} \\ \vdots \\ e_{I1} \\ \vdots \\ e_{IJ_I} \end{bmatrix}, \tag{10.1.3}
$$

where $\theta = [a, b_1, b_2, \ldots, b_I]'$ has $(I+1)$ components, and the matrix A is of order $n \times (I+1)$, where $n = \sum J_i$ is the total number of observations. Since the first column of A is the sum of the remaining columns, rank $(A) < I+1$ and so we have a singular model ($p = I+1$ in this case).

Given a singular model $E(Y) = A\theta$, with rank $(A) = s < p$, we can always choose new parameters $\phi_1, \ldots, \phi_s$ to produce a full-rank model $E(Y) = B\phi$; the new parameters, which can be chosen in various ways, are independent linear functions of $\theta_1, \ldots, \theta_p$. For instance, starting with the singular formulation for the one-way classification, given above in (10.1.3), we can introduce new parameters

$$\mu_1 = a + b_1, \qquad \mu_2 = a + b_2, \ldots, \mu_I = a + b_I$$

and recast the model in the non-singular form given in (10.1.1). Another reparameterization is given in (10.1.5) below. The model can then be analysed by the standard theory given in Chapter 8.

However, as stated earlier, there are sometimes reasons for preferring a singular model to a full-rank formulation. In experimental design situations, the model is usually overparameterized because this is the most natural specification of the problem—the parameters used in such a model have simple interpretations in the context of the particular investigation. It is therefore of interest to apply least squares directly to the singular model.

10.1.2. Estimation. Estimable Functions

For the singular model $E(\mathbf{Y}) = \mathbf{A}\boldsymbol{\theta}$, with rank $(\mathbf{A}) = s < p \leq n$, if we proceed as before in section 8.1 and section 8.2 by minimizing $S(\mathbf{u}) = (\mathbf{y} - \mathbf{A}\mathbf{u})'(\mathbf{y} - \mathbf{A}\mathbf{u})$, we again find that the least squares estimate $\tilde{\boldsymbol{\theta}}$ of $\boldsymbol{\theta}$ satisfies the normal equations (8.2.2), viz.

$$\mathbf{A}'\mathbf{A}\tilde{\boldsymbol{\theta}} = \mathbf{A}'\mathbf{y}.$$

However, the solution cannot now be expressed in the form $\tilde{\boldsymbol{\theta}} = (\mathbf{A}'\mathbf{A})^{-1}\mathbf{A}'\mathbf{y}$ since $\mathbf{A}'\mathbf{A}$ has no inverse. In fact, the equations have no unique solution at all, but an infinite number of solutions.

At first this seems a discouraging result; a method which produces a profusion of different estimates of the same thing appears unlikely to be helpful. However, the proper interpretation is that the components of $\tilde{\boldsymbol{\theta}}$ are not appropriate quantities to estimate. This can easily be appreciated by considering the case of the one-way layout. The expected response for group i is $E(Y_i) = a + b_i$, and, as mentioned earlier, this is unchanged if we replace a by $(a + \alpha)$ and b_i by $(b_i - \alpha)$, for any constant, α. There is thus a basic indeterminacy in the model as specified (which leads to a corresponding indeterminacy in estimates of the parameters) arising out of the fact that more parameters are used (p) in the specification than is absolutely necessary (s); the singularity of $\mathbf{A}'\mathbf{A}$ is just an algebraic consequence of this feature of the model.

Nevertheless, there are some things which can be estimated, viz. certain linear functions of $\theta_1, \ldots, \theta_p$. These are known as *estimable functions*, and have the following properties

(a) Every element of $E(\mathbf{Y}) = \mathbf{A}\boldsymbol{\theta}$ is estimable.
(b) If $\boldsymbol{\lambda}'\boldsymbol{\theta} = \lambda_1 \theta_1 + \ldots + \lambda_p \theta_p$ (where $\boldsymbol{\lambda}$ is given) is an estimable function then it has a unique least squares estimate, viz. $\boldsymbol{\lambda}'\tilde{\boldsymbol{\theta}}$, which is the same whatever solution $\tilde{\boldsymbol{\theta}}$ of the normal equations is used; moreover, $\boldsymbol{\lambda}'\tilde{\boldsymbol{\Theta}}$ is the minimum variance unbiased linear estimator (MVULE) of $\boldsymbol{\lambda}'\boldsymbol{\theta}$.

(c) The maximum number of linearly independent estimable functions is s, so that any others can be obtained as linear combinations of these; any s linearly independent estimable functions can be used as new parameters in a full-rank reparameterization.

The one-way classification illustrates these points: $(a + b_i)$, $i = 1, \ldots, I$ are independent estimable functions, and linear combinations of these have MVULE's; we have already seen this in the nonsingular formulation in which $\mu_i = a + b_i$.

In addition, the above results imply that whichever solution, $\tilde{\boldsymbol{\theta}}$, of the normal equations we use, we get the same residuals $\tilde{\mathbf{e}} = (\mathbf{y} - \mathbf{A}\tilde{\boldsymbol{\theta}})$, and hence the same SSR, viz. $r = \tilde{\mathbf{e}}'\tilde{\mathbf{e}}$. Furthermore, as in the full-rank case, we can obtain

$$\mathbf{y}'\mathbf{y} = \tilde{\boldsymbol{\theta}}'\mathbf{A}'\mathbf{A}\tilde{\boldsymbol{\theta}} + r$$

since this result made use only of the normal equations themselves and not the explicit formula $\tilde{\boldsymbol{\theta}} = (\mathbf{A}'\mathbf{A})^{-1}\mathbf{A}'\mathbf{y}$. Hence, we also get the same sum of squares due to the model, viz. $\tilde{\boldsymbol{\theta}}\mathbf{A}'\mathbf{A}\tilde{\boldsymbol{\theta}}$, whichever $\tilde{\boldsymbol{\theta}}$ is used and this may also be expressed in the form $\text{SSM} = \tilde{\boldsymbol{\theta}}\mathbf{A}'\mathbf{y}$, as in the case when $r(\mathbf{A}) = p$. In fact, these values of SSR and SSM are precisely what we would get if the model were reparameterized in the full-rank form $E(\mathbf{Y}) = \mathbf{B}\boldsymbol{\phi}$, for which $\text{SSM} = \hat{\boldsymbol{\phi}}'\mathbf{B}'\mathbf{y}$ and $\text{SSR} = \mathbf{y}'\mathbf{y} - \hat{\boldsymbol{\phi}}'\mathbf{B}'\mathbf{y}$ [see § 8.3.4].

We can thus choose any particular solution of the normal equations we like, and this can be done by imposing conditions which will be satisfied by only one of the solutions. Specifically, we shall require $\boldsymbol{\theta}$ to satisfy independent linear restrictions, or *side conditions*. The number of such conditions required is the number of redundant parameters in the model, i.e. $(p - s)$. Explicitly, we have

$$g_{11}\theta_1 + g_{12}\theta_2 + \ldots + g_{1p}\theta_p = 0$$
$$g_{21}\theta_1 + g_{22}\theta_2 + \ldots + g_{2p}\theta_p = 0$$
$$\vdots$$
$$g_{p-s,1}\theta_1 + g_{p-s,2}\theta_2 + \ldots + g_{p-s,p}\theta_p = 0$$

where the coefficients g_{ij} are known and chosen so that the matrix $\mathbf{G} = \{g_{ij}\}$ has rank equal to $(p - s)$; in terms of $\mathbf{G}$, the conditions are simply $\mathbf{G}\boldsymbol{\theta} = \mathbf{0}$. It is clear that such restrictions will remove the indeterminacy in the model since the equations could be used to eliminate $(p - s)$ of the parameters from the model (by solving for them in terms of the rest), thereby producing a full-rank reparameterization.

We can now estimate $\boldsymbol{\theta}$ by choosing $\tilde{\boldsymbol{\theta}}$ to minimize

$$(\mathbf{y} - \mathbf{A}\mathbf{u})'(\mathbf{y} - \mathbf{A}\mathbf{u}) \quad \text{subject to} \quad \mathbf{G}\mathbf{u} = \mathbf{0}.$$

It turns out that, in addition to the normal equations, $\tilde{\boldsymbol{\theta}}$ must satisfy the natural requirement that $\mathbf{G}\tilde{\boldsymbol{\theta}} = \mathbf{0}$, i.e. $\tilde{\boldsymbol{\theta}}$ must satisfy the same conditions as the vector it is designed to estimate.

In summary then, we estimate $\boldsymbol{\theta}$ by the unique solution $\tilde{\boldsymbol{\theta}}$ of the normal equations

$$\mathbf{A}'\mathbf{A}\tilde{\boldsymbol{\theta}} = \mathbf{A}'\mathbf{y}$$

which satisfies the side conditions

$$\mathbf{G}\tilde{\boldsymbol{\theta}} = \mathbf{0}.$$

It can be shown that $\tilde{\boldsymbol{\theta}}$ is the MVULE of all $\boldsymbol{\theta}$ for which $\mathbf{G}\boldsymbol{\theta} = \mathbf{0}$.

We have already noted that the residual sum of squares may be calculated from

$$r = \mathbf{y}'\mathbf{y} - \tilde{\boldsymbol{\theta}}'\mathbf{A}'\mathbf{y}.$$

Under the usual assumptions for the distribution of the errors, the corresponding random variable, R, is distributed as $\sigma^2 \chi^2_{n-s}$ [see § 2.5.4(a)] since we can reparameterize to get a nonsingular model $E(\mathbf{Y}) = \mathbf{B}\boldsymbol{\phi}$, having rank s, which has the same residual. Hence

$$E(R) = (n-s)\sigma^2$$

and

$$\tilde{\sigma}^2 = r/(n-s).$$

EXAMPLE 10.1.1.   *The one-way classification model in 'redundant para-meter' form.*   We illustrate the above discussion with the one-way classification. In the full-rank formulation of (10.1.1) there are I parameters, while the singular version of (10.1.2) uses $(I+1)$ parameters. In this latter case, then, $p = I+1$ and $s = I$, which means that one linear relation between the parameters will remove the redundancy. The condition is usually taken to be

$$\sum_{i=1}^{I} J_i b_i = 0. \tag{10.1.4}$$

This form is chosen partly for algebraic convenience and partly because of its simple interpretation in contexts in which the model is used. For example, suppose that I anticorrosive treatments are compared by applying treatment i to J_i items in group i, $i = 1, \ldots, I$, with y_{ij} denoting the observation on the jth item of group i of an appropriate response variable. Now (10.1.4) implies that $E\{\sum_{i=1}^{I} \sum_{j=1}^{J_i} Y_{ij}\} = na$, so that in the model $E(Y_{ij}) = a + b_i$, the parameter a can be interpreted as the overall average for all the items treated, while b_i is the average variation from this value due to treatment i. The parameter b_i is then referred to as the 'effect of treatment i'; more accurately, it is the (average) effect of treatment i relative to the other treatments considered. The 'real' effect of treatment i presumably refers to the difference between the expected response when treatment i is applied and the expected response when no treatment is applied. If we want to estimate this, it is necessary to have a control group receiving no treatment. Supposing this to be group 1, the natural side condition is then $b_1 = 0$, so that a is now the expected response if no treatment is applied. The model is $y_{ij} = a + b_i + e_{ij}$; $b_1 = 0$, a full rank model with I independent parameters, in which b_i represents the 'real' effect of treatment i.

Returning to the model $y_{ij} = a + b_i + e_{ij}$ with side condition $\sum_{i=1}^{I} J_i b_i = 0$, we note that the side condition could be used to give a different reparameterization than $\mu_i = a + b_i$, for example, by writing $J_I b_I = -\sum_{i=1}^{I-1} J_i b_i$, and using this to eliminate b_I from the original model, in which case we would have

$$
\mathbf{A} = \atop{(n \times I)} \left[\begin{array}{cccc}
1 & 1 & 0 & 0 \\
\vdots & \vdots & \vdots & & \cdots & \vdots \\
1 & 1 & 0 & & & 0 \\
\hline
1 & 0 & 1 & & & 0 \\
\vdots & \vdots & \vdots & & \cdots & \vdots \\
1 & 0 & 1 & & & 0 \\
\hline
& & & \vdots & & \\
\hline
1 & 0 & 0 & & & 1 \\
\vdots & \vdots & \vdots & \vdots & \cdots & \vdots \\
1 & 0 & 0 & & & 1 \\
\hline
1 & -J_1/J_I & -J_2/J_I & & & -J_{I-1}/J_I \\
\vdots & \vdots & \vdots & & \cdots & \vdots \\
1 & -J_1/J_I & -J_2/J_I & & & -J_{I-1}/J_I
\end{array} \right]
\begin{array}{l}
\left.\rule{0pt}{18pt}\right\} J_1 \text{ rows} \\
\left.\rule{0pt}{18pt}\right\} J_2 \text{ rows} \\
\\
\left.\rule{0pt}{18pt}\right\} J_{I-1} \text{ rows} \\
\left.\rule{0pt}{18pt}\right\} J_I \text{ rows}
\end{array}
$$

$$
\boldsymbol{\theta} = \atop{(I \times 1)} \begin{bmatrix} a \\ b_1 \\ b_2 \\ \vdots \\ b_{I-1} \end{bmatrix} \tag{10.1.5}
$$

In this case $\mathbf{A}$ has rank I and we could apply the theory of Chapter 8; this is not done because symmetry is lost, and we no longer have a simple parameter representing the effect of the Ith treatment. Thus we take as our model

$$
y_{ij} = a + b_i + e_{ij}, \qquad \sum_{i=1}^{I} J_i b_i = 0.
$$

The design matrix $\mathbf{A}$ for this model is given in (10.1.3), whence

$$
\mathbf{A}'\mathbf{A} = \begin{bmatrix}
n & J_1 & J_2 & \cdots & J_I \\
J_1 & J_1 & 0 & \cdots & 0 \\
J_2 & 0 & J_2 & \cdots & 0 \\
\vdots & \vdots & \vdots & & \vdots \\
J_I & 0 & 0 & \cdots & J_I
\end{bmatrix}, \qquad
\mathbf{A}'\mathbf{y} = \begin{bmatrix}
ny.. \\
J_1 y_1. \\
J_2 y_2. \\
\vdots \\
J_I y_I.
\end{bmatrix}
$$

The vector of L.S. estimates $\tilde{\boldsymbol{\theta}} = (\tilde{a}, \tilde{b}_1, \ldots, \tilde{b}_I)'$ can now be found by solving

$$\mathbf{A}'\mathbf{A}\tilde{\boldsymbol{\theta}} = \mathbf{A}'\mathbf{y} \quad \text{and} \quad \sum_{i=1}^{I} J_i \tilde{b}_i = 0.$$

Explicitly, the normal equations are

$$n\tilde{a} + \sum J_i \tilde{b}_i = ny.., \qquad J_i \tilde{a} + J_i \tilde{b}_i = J_i y_{i.}, \qquad i = 1, \ldots, I.$$

Using the constraint we obtain

$$\tilde{a} = y.., \qquad \tilde{b}_i = y_{i.} - y.., \qquad i = 1, \ldots, I. \tag{10.1.6}$$

The associated estimators $\tilde{A} = Y..$, $\tilde{B}_i = Y_{i.} - Y..$ are MVULE of a and b_i, with variances $\sigma^2 n^{-1}$, $\sigma^2(J_i^{-1} - n^{-1})$ respectively; they are also uncorrelated. These results imply that the MVULE of $\mu_i = a + b_i$ is $\tilde{A} + \tilde{B}_i = Y_{i.}$, having variance $\sigma^2 J_i^{-1}$, which is what we obtained using the full-rank model parameterized by $\mu_1, \ldots, \mu_I$.

Further, the residual sum of squares is

$$r = \sum_{i=1}^{I} \sum_{j=1}^{J_i} (y_{ij} - \tilde{a} - \tilde{b}_i)^2 = \sum_{i=1}^{I} \sum_{j=1}^{J_i} (y_{ij} - y_{i.})^2, \tag{10.1.7}$$

again, the same as before. Using $r = \mathbf{y}'\mathbf{y} - \tilde{\boldsymbol{\theta}}'\mathbf{A}'\mathbf{y}$ we have the alternative expression

$$r = \sum \sum y_{ij}^2 - n\tilde{a}y.. - \sum \tilde{b}_i J_i y_{i.}; \tag{10.1.8}$$

the associated random variable R is distributed as $\sigma^2 \chi_{n-s}^2$, where $s = \text{rank } \mathbf{A} = I$; the error variance is estimated by $\tilde{\sigma}^2 = r/(n - I)$.

Making the usual assumptions concerning the errors, confidence intervals for the group means are readily obtained; thus, for μ_i, since $\tilde{\sigma}^{-1}(Y_{i.} - \mu_i)J_i^{1/2}$ is distributed as Student's t [see § 2.5.5] with $n - I$ d.f., a $100\,(1 - \alpha)\%$ central confidence interval for μ_i [see § 4.5] is

$$[Y_{i.} \pm \tilde{\sigma}J_i^{-1/2}t_{n-I}(1 - \alpha/2)].$$

10.1.3.   Tests of Hypotheses

Suppose we want to examine the general linear hypothesis H, specified by

$$\mathbf{H}\boldsymbol{\theta} = \mathbf{c},$$

where $\mathbf{H}$ is a given $m \times p$ matrix of rank h and $\mathbf{c}$ is a given $m \times 1$ vector. When such a hypothesis is tested in conjunction with a singular model subject to side conditions, we proceed as before, with the proviso that if any of the constraints specified by H are already included in the side conditions $\mathbf{G}\boldsymbol{\theta} = 0$, then the order of H is reduced by the number of independent conditions duplicated. If this is d, the order of H is thus $(h - d)$. We construct the likelihood-ratio test of H in the usual way [see § 5.5]:

(i) For the full model

$$y = A\theta + e \quad \text{subject to } G\theta = 0,$$

obtain r, the SSR, which is the minimum value of

$$S(u) = (y - Au)'(y - Au) \quad \text{subject to } Gu = 0.$$

Whether H is true or not, R is distributed as $\sigma^2 \chi^2_{n-s}$, where $s = \text{rank } A$.

(ii) For the model restricted by H, viz.

$$y = A\theta + e \quad \text{subject to } G\theta = 0 \text{ and } H\theta = c,$$

obtain r_H, the SSR under the hypothesis H, this being the minimum value of $S(u)$ subject to $Gu = 0$ and $Hu = c$. If H is true, the extra reduction $R_H - R$ is distributed independently of R as $\sigma^2 \chi^2_{h-d}$, where $(h - d)$ is the effective order of H (rank $H = h$, $d = $ number of restrictions common to H and the side conditions).

(iii) Calculate $[(r_H - r)/(h - d)]/[r/(n - s)]$; under H, this is an observation from the $F_{h-d,n-s}$ distribution [see § 2.5.6].

Once again, if H is true, this is just a comparison of two independent estimates of σ^2.

EXAMPLE 10.1.2.   '*Within and between*' *analysis*.   As an illustration using the one-way classification, consider the hypothesis that all treatments have the same (unknown) effect. In terms of the parameters of the singular model this is $H : b_1 = b_2 = \ldots = b_I = 0$. At first sight this hypothesis looks as if it has order I; in fact, the hypothesis of a common unknown effect has already been tested (Example 8.3.8) in the context of the full rank model, where we concluded the order was $(I - 1)$. This can be seen directly in the present case because the side condition $\sum J_i b_i = 0$ means that $b_1 = b_2 = \ldots = b_I = 0$ specifies only $(I - 1)$ independent constraints (explicitly, these could be $b_1 = b_2 = \ldots = b_{I-1} = 0$; since $\sum J_i b_i = 0$ then shows that b_I must also be zero).

To obtain r_H, we note that under H the model becomes $y_{ij} = a + e_{ij}$, which is the same as in the full rank case, where we found in Example 8.3.3 that

$$r_H = \sum_{i=1}^{I} \sum_{j=1}^{J_i} (y_{ij} - y_{..})^2 = y'y - ny_{..}^2.$$

Using (10.1.8), $r_H - r = \sum_i \bar{b}_i J_i y_{i\cdot} = \sum J_i y_{i\cdot}(y_{i\cdot} - y_{..}) = \sum J_i(y_{i\cdot} - y_{..})^2$. If H is true, $R_H - R$ is proportional to a χ^2 variable with degrees of freedom equal to the order of H, i.e. $I - 1$. The $L - R$ test statistic is thus

$$\frac{(r_H - r)/(I - 1)}{r/(n - I)} = \frac{\sum\limits_i J_i(y_{i\cdot} - y_{..})^2/(I - 1)}{\sum\limits_i \sum\limits_j (y_{ij} - y_{i\cdot})^2/(n - I)}$$

as found in Example 8.3.3.

In the usual tabular form, we have:

Source	Sum of squares	d.f.	ms
Reduction due to $\theta = (a, b_1, \ldots, b_I)'$	$r(\theta) = \tilde{\theta}'A'y$	I	
Reduction due to $\theta\|H$, $=$ reduction due to a	$r(\theta\|H) = r(a)$ $= ny_{..}^2$	1	
Extra reduction	$r(\theta) - r(\theta\|H) = r_H - r$ $= \sum J_i(y_{i.} - y_{..})^2$	$I-1$	$\dfrac{r_H - r}{I-1}$
Residual	$r = y'y - \tilde{\theta}'A'y$ $= \sum\sum_{i\ j}(y_{ij} - y_{i.})^2$	$n-I$	$\tilde{\sigma}^2 = \dfrac{r}{n-I}$
Total	$y'y$	n	

(10.1.9)

The parts of the table necessary for the test of H can be found quickly by the following device. Consider the identity

$$y_{ij} - y_{..} = (y_{ij} - y_{i.}) + (y_{i.} - y_{..}). \tag{10.1.10}$$

This expresses the variation of an observation from the overall sample mean as the sum of its variation from the mean of the group to which it belongs and the variation of that group mean from the overall mean. Squaring both sides of (10.1.10) and summing over i and j, we have

$$\sum_{i=1}^{I}\sum_{j=1}^{J_i}(y_{ij} - y_{..})^2 = \sum_i\sum_j(y_{ij} - y_{i.})^2 + \sum_i\sum_j(y_{i.} - y_{..})^2$$

$$+ 2\sum_i\sum_j(y_{ij} - y_{i.})(y_{i.} - y_{..}).$$

The product term is

$$2\sum_i\{(y_{i.} - y_{..})\sum_j(y_{ij} - y_{i.})\}$$

which is zero since $\sum_j(y_{ij} - y_{i.}) = 0$, $i = 1, \ldots, I.$
    Thus we have the identity

$$\sum_i\sum_j(y_{ij} - y_{..})^2 = \sum_i\sum_j(y_{ij} - y_{i.})^2 + \sum_i J_i(y_{i.} - y_{..})^2 \tag{10.1.11}$$

    The basic idea in this identity, and similar ones to be developed later, is that of dividing up a total sum of squared deviations of a variable from its sample mean into several distinct sums of squares, each corresponding to a different source of variation. In this particular identity, we note that

(a) $\sum_j (y_{ij} - y_{i\cdot})^2$ is a measure of the variability within group j, so that $\sum_i \sum_j (y_{ij} - y_{i\cdot})^2$ is a measure of overall variability within groups; it is known as the *within groups sum of squares* (WGSS), and is the residual in the above anova table.

(b) $\sum_i J_i(y_{i\cdot} - y_{\cdot\cdot})^2$ is a measure of variability between groups; accordingly, it is the *between groups sum of squares* (BGSS), and is the extra reduction term in the anova table.

In the absence of the least squares analysis, (10.1.10) can be used to suggest the test for equality of group means by arguing as follows. If the group means $\mu_i = a + b_i$ differed we would expect the BGSS to be larger than if they were the same; on the other hand, we would not expect the WGSS to be affected by this. Hence an intuitively reasonable procedure for examining whether the group means are the same is to consider the ratio BGSS/WGSS. Investigation of the distributions of these two statistics when the errors are independent Normals with zero means and common variance σ^2, shows that (as we already know from the L.S. approach)

(a) WGSS is a $\sigma^2 \chi^2_{n-I}$ variable [see 2.5.4(a)],
(b) BGSS is a $\sigma^2 \chi^2_{I-1}$ variable, independent of WGSS, if the group expectations are equal.

Hence a more convenient test function is

$$\frac{\text{BGSS}/(I-1)}{\text{WGSS}/(n-I)},$$

which will have the $F_{I-1,n-I}$ distribution [see § 2.5.6] when the expectations are equal.

From this point of view, it is natural to use a slightly different tabular form:

Source	Sum of squares	d.f.	ms	E(MS)
Between groups	$\sum_i J_i(y_{i\cdot} - y_{\cdot\cdot})^2$	$I-1$	$\dfrac{\text{BGSS}}{I-1}$	$\sigma^2 + \dfrac{\sum J_i b_i^2}{I-1}$
Within groups	$\sum_i \sum_j (y_{ij} - y_{i\cdot})^2$	$n-I$	$\dfrac{\text{WGSS}}{n-I}$	σ^2
Total (about mean)	$\sum_i \sum_j (y_{ij} - y_{\cdot\cdot})^2$	$n-1$		

$$(10.1.12)$$

The final column gives the expected values of the two mean squares, confirming that the expectation of BGSS is increased when the groups have unequal means, and suggesting a one-tailed test. When the group means are equal, the two mean squares are independent estimates of σ^2.

Although this singular model is not orthogonal (in fact, no singular model can be, since $\mathbf{A}'\mathbf{A}$ is not diagonal) it does possess the partial orthogonality which was referred to earlier in section 8.3.7. The SSM is partitioned in the table (10.1.9) as

$$r(\boldsymbol{\theta}) = \tilde{\boldsymbol{\theta}}'\mathbf{A}'\mathbf{y} = n y_{..}^2 + \sum J_i(y_{i.} - y_{..})^2,$$

the first term on the right being $r(a)$ and the second the sum of squares for testing the hypothesis that $b_1 = \ldots = b_I = 0$, since it is the extra reduction when the hypothesis is lifted. Now the L.S. estimate of a was the same in the full model as in the model restricted by the hypothesis, viz. $\tilde{a} = y_{...}$. If we consider a second hypothesis $a = 0$ (not usually of any interest), the L.S. procedure results in estimates of b_i ($i = 1, \ldots, I$) which are the same as in the full model, with $r(\mathbf{b}) = r(b_1, \ldots, b_I) = \sum J_i(y_{i.} - y_{..})^2$ being the reduction under the hypothesis and $r(a)$ the extra reduction when the restriction $a = 0$ is removed. Thus, the same partition $r(\boldsymbol{\theta}) = r(a) + r(\mathbf{b})$ of SSM is used for both hypotheses, the terms having reversed roles in the two cases. We can summarize the position as follows

Hypothesis H	$\tilde{a}$	$\tilde{b}_i$	$r(\boldsymbol{\theta}\|H)$	$r(\boldsymbol{\theta}) - r(\boldsymbol{\theta}\|H)$
$a = 0$		$y_{i.} - y_{..}$	$\sum_i J_i(y_{i.} - y_{..})^2$	$r(a) = n y_{..}^2$
$b_1 = \ldots = b_I = 0$	$y_{..}$		$n y_{..}^2$	$r(b) = \sum J_i(y_{i.} - y_{..})^2$

This amounts to saying that the parameters $a, b_1, b_2, \ldots, b_I$ fall into two orthogonal groups, viz. a and $(b_1, \ldots, b_I)$. In fact cov $(\tilde{A}, \tilde{B}_i) = 0$, $i = 1, \ldots, I$. The orthogonality of the two parameter groups is most easily recognized in the earlier full-rank version in which the side condition is used to eliminate b_I. The design matrix was given in (10.1.5), from which it can be seen that $\mathbf{A}'\mathbf{A}$ has the form

$$\begin{bmatrix} n & \vdots & 0 & \cdots & 0 \\ \hline 0 & \vdots & & & \\ \vdots & \vdots & & \mathbf{B} & \\ 0 & \vdots & & & \end{bmatrix}.$$

For other hypotheses, e.g. $b_1 = 0$, these special features of the anova would disappear as the estimates of $a, b_2, \ldots, b_I$ would not be the same in the full model and the model restricted by the hypothesis.

10.1.4. The Two-way Hierarchical Classification

In the one-way classification the data is divided into several groups by means of a group factor such as treatment received, and the structure is

Groups	1	2	...	I
Observations	×	×		×
	×	×		×
	×			×

where each cross represents an observation.

If a second factor is used to subdivide each group, we have a *two-way hierarchical* or *nested* classification, with the following structure

Groups		1				2		...	I	
Subgroups	1	2	3	4	1	2	3		1	2
Observations	×	×	×	×	×	×	×		×	×
	×	×		×	×	×			×	×
	×			×		×				×
	×									×

Here, group 1 has four subgroups, containing four, two, one and three observations, etc. As an illustration, the data might be measurements of air pollution in various cities of several countries, in which case the countries form the groups and the cities the subgroups.

We shall denote the kth observation in the jth subgroup of group i by y_{ijk}, and suppose that group i has J_i subgroups, the jth such comprising K_{ij} observations. Thus group 2 above has $J_2 = 3$ subgroups, the number of observations in the first of these being $K_{21} = 3$.

A full rank model for this situation is

$$y_{ijk} = \mu_{ij} + e_{ijk}$$

where the μ's are unknown parameters and the e's are independent observations from a distribution with zero mean; in other words, the observations in the jth subgroup of group i are a random sample from a distribution with mean μ_{ij}. The number of parameters, s, is thus the number of subgroups, i.e. $s = \sum_{i=1}^{I} J_i$.

This model is not very convenient. It does not reflect the structure of the data particularly well; it does not contain explicit parameters associated with the effects we usually want to investigate; and specification of the hypotheses considered below is cumbersome. Instead we shall consider the singular model

$$y_{ijk} = \mu + g_i + s_{ij} + e_{ijk}$$

where $i = 1, \ldots, I$; $j = 1, \ldots, J_i$; $k = 1, \ldots, K_{ij}$.

This new model has $p = 1 + I + \sum_{i=1}^{I} J_i$ parameters and is over-parameterized; if μ is replaced by $(\mu - \alpha)$, g_i by $(g_i - \alpha_i + \alpha)$ and s_{ij} by $(s_{ij} + \alpha_i)$ then μ_{ij} is unchanged; there are $(I + 1)$ *redundant* parameters.

We deal with this by imposing side conditions on the g's and s's, namely

$$\sum_{j=1}^{J_i} K_{ij} s_{ij} = 0, \quad \text{for } i = 1, \ldots, I$$

$$\sum_{i=1}^{I} K_{i+} g_i = 0,$$

where $K_{i+} = \sum_{j=1}^{J_i} K_{ij}$ is the number of observations in group i. The effect of these $(I + 1)$ side conditions is to produce unique L.S. estimates of μ, g_i and s_{ij}, and allow us to interpret the parameters as follows:

μ	is the expected value of the average of all the observations;
$\mu + g_i$	is the expected value of the average of the observations in group i; g_i therefore measures the 'effect of group i';
$\mu + g_i + s_{ij}$	is the expected value of the average of the observations in subgroup j of group i; s_{ij} therefore measures the 'effect of subgroup j of group i'.

The L.S. estimates of the parameters are found by minimizing $\sum_i \sum_j \sum_k \{y_{ijk} - \mu - g_i - s_{ij}\}^2$, subject to the side conditions noted above, the parameters being treated as algebraic variables for the purposes of this procedure. As might be anticipated, we find that

$\tilde{\mu} = y_{...}$,	the average of all the observations,
$\tilde{g}_i = y_{i..} - y_{...}$,	where $y_{i..}$ is the average for group i,
$\tilde{s}_{ij} = y_{ij.} - y_{i..}$,	where $y_{ij.}$ is the average for the jth subgroup of group i.

The residual sum of squares is

$$r = \sum_i \sum_j \sum_k (y_{ijk} - \tilde{\mu} - \tilde{g}_i - \tilde{s}_{ij})^2 = \sum_i \sum_j \sum_k (y_{ijk} - y_{ij.})^2$$

with degrees of freedom equal to the number of observations minus the number of independent parameters, i.e. to $\sum_i \sum_j K_{ij} - \sum_i J_i = n - s$, where n is the number of observations and s is the number of subgroups. The two main hypotheses of interest are

(i) $H_g : g_1 = g_2 = \ldots = g_I = 0$, i.e. there are no differences between groups as a whole,

(ii) $H_s :$ all $s_{ij} = 0$, i.e. for each group, there are no differences amongst its subgroups.

To test these hypotheses, we require to re-estimate the parameters under each hypothesis and find the corresponding residual sum of squares. The details of this procedure are straightforward and will be omitted; the main features are as follows. The parameters form three orthogonal groups, namely μ; $\mathbf{g} = \{g_i\}$, comprising I parameters, only $(I-1)$ of which are independent; $\mathbf{s} = \{s_{ij}\}$, consisting of s parameters, only $(s-I)$ of which are independent. Thus, the estimates of the g's under H_s and the estimates of the s's under H_g, are the same as in the full model, while the estimate of μ is $y...$ in all three cases. The amount by which $\mathbf{y'y}$ is reduced on fitting the g's is the same whether the s's are included in the model or not, and vice-versa. The sum of squares due to fitting the full model (SSM) can thus be expressed in the form

$$r(\mathbf{\theta}) = \mathbf{y'y} - r = r(\mu) + r(\mathbf{g}) + r(\mathbf{s})$$

where $r(\mathbf{g})$ is the reduction due to fitting $g_1, \ldots, g_I$, i.e. the increase in residual under H_g (so that $r_g = r + r(\mathbf{g})$), and $r(\mathbf{s})$ is the increase in residual under H_s. This anova may be displayed in the usual tabular form

Source	Sum of Squares	d.f.	MS
Due to mean	$r(\mu)$	1	
Due to groups	$r(\mathbf{g})$	$I-1$	$r(\mathbf{g})/(I-1)$
Due to subgroups	$r(\mathbf{s})$	$s-I$	$r(\mathbf{s})/(s-I)$
Residual	$r = \mathbf{y'y} - r(\mu) - r(\mathbf{g}) - r(\mathbf{s})$	$n-s$	$\tilde{\sigma}^2 = r/(n-s)$
Total	$\mathbf{y'y}$	n	

where $n = \sum_i \sum_j K_{ij}$ and $s = \sum_i J_i$. For each group the number of degrees of freedom is the number of independent parameters fitted. The tests are carried out in the usual way by comparing each mean square with $\tilde{\sigma}^2$, and referring the ratio to the appropriate F distribution.

Explicit expressions in terms of the observations for the various sums of squares emerge naturally as part of the least-squares analysis; they can also be found quickly by a similar device to that employed earlier for the one-way classification. We express the deviation of an observation from the overall average as the sum of three parts, namely

$$y_{ijk} - y... = (y_{ijk} - y_{ij.}) + (y_{ij.} - y_{i..}) + (y_{i..} - y...)$$

If we square both sides and sum over i, j, k we obtain

$$\sum_i \sum_j \sum_k (y_{ijk} - y...)^2 = \sum_i \sum_j \sum_k (y_{ijk} - y_{ij.})^2 + \sum_i \sum_j K_{ij}(y_{ij.} - y_{i..})^2$$

$$+ \sum_i K_{i+}(y_{i..} - y...)^2 \tag{10.1.13}$$

the cross-product terms being zero. The terms on the right-hand side of this equation are, respectively, r, $r(\mathbf{s})$ and $r(\mathbf{g})$. The left-hand side may be written as

$$\sum_i \sum_j \sum_k y_{ijk}^2 - n y_{...}^2 \; ;$$

as the reduction due to μ is $r(\mu) = n y_{...}^2$, a simple re-arrangement gives us the basic form

$$\mathbf{y}'\mathbf{y} = r(\mu) + r(\mathbf{g}) + r(\mathbf{s}) + r$$

It can be seen from (10.1.13) that the reduction, $r(\mathbf{g})$, due to fitting the group parameters, is just a measure of the variability between the groups; accordingly, $r(\mathbf{g}) = \sum_i K_{i+}(y_{i..} - y_{...})^2$ is called the *Between groups sum of squares* (BGSS). Similarly, $\sum_j K_{ij}(y_{ij.} - y_{i..})^2$ is a measure of variability between subgroups of group i; $r(\mathbf{s}) = \sum_i \sum_j K_{ij}(y_{ij.} - y_{i..})^2$ is the aggregate of these for all groups and is the *Between subgroups within groups sum of squares* (BSWGSS). The expectations of the BG and BSWG mean squares can be found by elementary methods, the results being

$$E(\text{BGMS}) = E[r(\mathbf{g})/(I-1)] = \sigma^2 + \sum_i K_{i+} g_i^2/(I-1),$$

$$E(\text{BSWGMS}) = E[r(\mathbf{s})/(s-I)] = \sigma^2 + \sum_i \sum_j K_{ij} s_{ij}^2/(s-I);$$

again, therefore, the tests against the residual are one-tailed.

10.1.5.   The Two-way Cross-classification

In the hierarchical classification described above, each group is associated with a different set of subgroups; thus, in the air pollution illustration, a (necessarily) different set of cities is chosen from each country. A different arrangement occurs when the same subgroups are involved in each group; for example, with the same group factor (countries) we might be interested in the annual mortality rates, over a period of several years, for a number of diseases common to all the countries considered. Such data forms a *cross-classification*:

		\multicolumn{4}{c}{Countries}			
		1	2	3	...
	A	× ×	× × × ×	×	
Diseases	B	×	× × ×	× × × ×	
	C	× ×	× × × ×	× ×	
	⋮				

Each cross represents the mortality rate for a different year; we may not have the same amount of data in each cell, i.e. for each combination of country and disease. We shall refer to a 'row' factor and a 'column' factor; in this example, these are 'diseases' and 'countries', respectively. If the diseases were not common to all the countries considered (as would be the case if tropical diseases and temperature countries were both included), the above arrangement would present a number of empty cells and the cross-classification would be *incomplete*; it would in fact, revert to a nested classification in which certain subgroups were common to several groups.

We denote the kth observation in the $i-j$th cell (row i column j) by y_{ijk}, where $i=1,\ldots,I$; $j=1,\ldots,J$; $k=1,\ldots,K_{ij}$. Thus, there are IJ cells, and cell (i,j) contains K_{ij} observations. A full rank model is

$$y_{ijk} = \mu_{ij} + e_{ijk},$$

which assigns one parameter to each cell, this being the theoretical cell mean. As in the nested case, we shall use instead a singular model which reflects the data structure and contains explicit parameters for the effects of interest. This model is

$$y_{ijk} = \mu + R_i + C_j + (RC)_{ij} + e_{ijk},$$

where the parameters $R_1,\ldots,R_I$ represent row effects, $C_1,\ldots,C_J$ represent column effects and $(RC)_{11},\ldots,(RC)_{IJ}$ represent *interactions* [see § 8.9.1] between rows and columns. These latter parameters are included to deal with the common situation in which row and column effects do not act additively. If we omit the interaction parameters the model is $y_{ijk} = \mu + R_i + C_j + e_{ijk}$. This is less general than the full rank model, as it uses only $(I+J)$ parameters, and implies that

$$E(Y_{ijk} - Y_{i'jk}) = R_i - R_{i'}, \qquad \text{for each } j,$$

$$E(Y_{ijk} - Y_{ij'k}) = C_j - C_{j'}, \qquad \text{for each } i,$$

For the example considered above, it is thus adequate only if the difference in expected mortality rates for any two diseases is the same for all countries, and the difference in mortality rates for any two countries is the same for each disease. (Of course, none of these models would be appropriate if the underlying rates were changing with time.)

To illustrate further the nature of interaction, we consider the case of two drugs, each tried in combination with two dietary regimes $(I = J = 2)$:

	Drugs	
	1	2
Diet 1	$y_{111},\ldots,y_{11K_{11}}$	$y_{121},\ldots,y_{12K_{12}}$
Diet 2	$y_{211},\ldots,y_{21K_{21}}$	$y_{221},\ldots,y_{22K_{22}}$

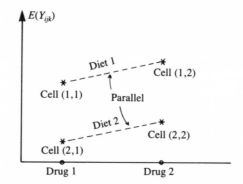

Figure 10.1.1: Effect of treatments when there is no interaction.

When no interaction is present, the situation is as indicated in Figure 10.1.1. In this case, the expected difference between diets is the same for both drugs, and the expected difference between drugs is the same for both diets; diet 1 is always better than diet 1, drug 2 is always better than drug 1. When interaction exists, however, this is no longer necessarily the case, as indicated in Figure 10.1.2. In the first case, there is not much difference between diets or between drugs, but for drug 1, diet 2 is best, while for drug 2, diet 1 is best, i.e. drugs and diets interact. In the second case, there is no overall difference between drugs, but diet 1 is always better than diet 2, and this is particularly so with drug 1, again showing interaction of drugs and diets. In short, interaction means that the difference between drugs depends on the diet and vice versa.

The response y_{ijk} is unchanged on replacing μ by $(\mu - \alpha)$, R_i by $(R_i - \alpha_i)$, C_j by $(C_j - \beta_j)$ and $(RC)_{ij}$ by $[(RC)_{ij} + \alpha + \alpha_i + \beta_j]$, thus showing that there are $(I + J + 1)$ redundant parameters in the model (alternatively, we can just subtract the number of parameters in the full rank model, i.e. IJ, from the corresponding figure $(IJ + I + J + 1)$ for the singular model). Hence $(I + J + 1)$

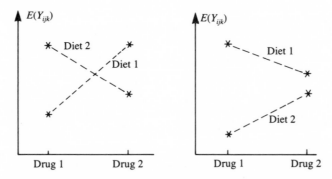

Figure 10.1.2: Effect of treatments when there is interaction.

linearly independent side conditions will produce unique L.S. estimates of the parameters. In general, explicit expressions for the L.S. estimates in terms of the observations cannot be given; their numerical values are obtained by solving the usual equations

$$A'A\tilde{\theta} = A'y,$$

$$G\tilde{\theta} = 0,$$

where A is the design matrix and $G\theta = 0$ the side conditions. However, in the case of *proportional frequencies*, explicit expressions can be given.

Proportional frequencies

The case of proportional frequencies is said to occur when

$$K_{ij}/K_{i+} \quad \text{is independent of } i,$$

and

$$K_{ij}/K_{+j} \quad \text{is independent of } j,$$

where $K_{i+} = \sum_j K_{ij}$ is the total number of observations in the cells of row i, and $K_{+j} = \sum_j K_{ij}$ is the total number of observations in the cells of column j. In other words, the cell numbers in a given column (row) are proportional to the row (column) totals. Equivalently, we have

$$K_{ij} = K_{i+}K_{+j}/n.$$

In particular, this holds when all cells have the *same* number of observations, i.e. $K_{ij} = K$, which is known as the *balanced* case [cf. § 9.7].

When this proportional frequencies condition holds, we choose the following $(I+J+1)$ independent side conditions:

$$\sum_i K_{i+}R_i = 0,$$

$$\sum_j K_{+j}C_j = 0,$$

$$\sum_i K_{ij}(RC)_{ij} = 0, \qquad j = 1, \ldots, J,$$

$$\sum_j K_{ij}(RC)_{ij} = 0, \qquad i = 1, \ldots, I-1,$$

The conditions on the interaction parameters imply that $\sum_j K_{ij}(RC)_{ij} = 0$ also in the case $i = I$. In the balanced case $(K_{ij} = K)$, these conditions are just

$$\sum_i R_i = \sum_j C_j = \sum_i (RC)_{ij} = \sum_j (RC)_{ij} = 0.$$

With the above side conditions, the parameters have simple and convenient interpretations; thus μ is the expectation of the average of all the observations, $\mu + R_i$ is the expectation of the average of the observations in row i (so that R_i is the 'effect of row i') and $\mu + C_j$ is the expectation of the average of the observations in column j (so that C_j is the 'effect of column j'). The mean of the observations in cell (i, j) has expectation $\mu + R_i + C_j + (RC)_{ij}$; when there is no interaction we see that this differs from μ by just the sum of the row and column effects, a situation referred to briefly as 'additivity'.

The least squares estimates of the parameters are obtained by minimizing (with respect to μ, R_i, C_j, $(RC)_{ij}$), the sum of squares

$$\sum_i \sum_j \sum_k \{y_{ijk} - \mu - R_i - C_j - (RC)_{ij}\}^2,$$

subject to the above side conditions. This gives

$$\tilde{\mu} = y_{\dots}$$
$$\tilde{R}_i = y_{i\cdot\cdot} - y_{\dots}$$
$$\tilde{C}_j = y_{\cdot j\cdot} - y_{\dots}$$
$$\widetilde{(RC)}_{ij} = y_{ij\cdot} - y_{i\cdot\cdot} - y_{\cdot j\cdot} + y_{\dots},$$

These results are not surprising, in view of the interpretations of the parameters given above. The residual sum of squares is thus

$$r = \sum_i \sum_j \sum_k \{y_{ijk} - y_{ij\cdot}\}^2.$$

The total number of observations is $n = \sum_i \sum_j K_{ij}$ and the number of independent parameters is IJ; hence r is proportional to a χ^2-variate with $n-IJ$ degrees of freedom and $\tilde{\sigma}^2 = r/(n-IJ)$. There are three hypotheses of particular interest:

$$H_R : R_1 = R_2 = \dots = R_I = 0 \quad \text{(no row effects)}$$
$$H_C : C_1 = C_2 = \dots = C_J = 0 \quad \text{(no column effects)}$$
$$H_I : (RC)_{ij} = 0 \text{ for } i = 1 \dots I, j = 1 \dots J \quad \text{(no interaction)}$$

The parameters form four orthogonal groups:

$$\mu, \quad \mathbf{R} = \{R_i\}, \quad \mathbf{C} = \{C_j\}, \quad (\mathbf{RC}) = \{(RC)_{ij}\}.$$

The SS due to fitting the model is thus of the form

$$r(\theta) = r(\mu) + r(\mathbf{R}) + r(\mathbf{C}) + r((\mathbf{RC})),$$

and each of these gives the reduction in $\mathbf{y'y}$ (the total SS) achieved by fitting the parameters in that group, irrespective of which of the remaining parameter-groups are included in the model. Thus, we can test the three hypotheses stated above by comparing the appropriate mean square term with $\tilde{\sigma}^2$. For each term, the number of independent parameters involved (taking account of the side conditions) gives the corresponding degrees of freedom. Thus, $r(\mu)$ has 1 d.f., $r(\mathbf{R})$ has $I-1$ d.f., $r(\mathbf{C})$ has $J-1$ d.f. and $r((\mathbf{RC}))$ has $IJ - (I+J-1) = (I-1)(J-1)$ d.f.

We can find explicit expressions for these various sums of squares in the partition of SSM, without going through the details of the least squares procedure, by using the following break-up

$$y_{ijk} - y... = (y_{ijk} - y_{ij\cdot}) + (y_{i\cdot\cdot} - y...)$$
$$+ (y_{\cdot j\cdot} - y...) + (y_{ij\cdot} - y_{i\cdot\cdot} - y_{\cdot j\cdot} + y...)$$

Squaring both sides and summing over i, j, k we have

$$\sum_i \sum_j \sum_k (y_{ijk} - y...)^2 = \sum_i \sum_j K_{ij}(y_{ijk} - y_{ij\cdot})^2 + \sum_i K_{i+}(y_{i\cdot\cdot} - y...)^2$$

$$+ \sum_j K_{+j}(y_{\cdot j\cdot} - y...)^2 + \sum_i \sum_j K_{ij}(y_{ij\cdot} - y_{i\cdot\cdot} - y_{\cdot j\cdot} + y...)^2 \qquad (10.1.14)$$

all the cross product terms being zero. As in the examples previously considered, the reduction due to μ, viz. $r(\mu) = ny^2...$, has been absorbed into the total SS about the mean on the LHS, while on the right we have r, $r(\mathbf{R})$, $r(\mathbf{C})$ and $r((\mathbf{RC}))$, respectively. The three latter sums of squares are known as the *Between rows* SS, the *Between columns* SS and the *Interaction* SS, respectively. The anova table is given in Table 10.1.1.

Source	Sum of squares	d.f.	$E(MS)$
Mean	$r(\mu) = ny^2..$	1	$\sigma^2 + n\mu^2$
Between rows	$r(\mathbf{R}) = \sum_i K_{i+}(y_{i\cdot\cdot} - y...)^2$	$I-1$	$\sigma^2 + \sum_i K_{i+}R_i^2/(I-1)$
Between columns	$r(\mathbf{C}) = \sum_j K_{+j}(y_{\cdot j\cdot} - y...)^2$	$J-1$	$\sigma^2 + \sum_j K_{+j}C_j^2/(J-1)$
Interaction	$r((\mathbf{RC}))$ $= \sum_i \sum_j K_{ij}(y_{ij\cdot} - y_{i\cdot\cdot} + y_{\cdot j\cdot} - y...)^2$	$(I-1)(J-1)$	$\sigma^2 + \sum_i \sum_j K_{ij}(RC)_{ij}^2/ (I-1)(J-1)$
Residual	$r = y'y - r(\mu) - r(\mathbf{R}) - r(\mathbf{C}) - r((\mathbf{RC}))$	$n-IJ$	σ^2
Total	$y'y$	n	

Table 10.1.1

The hypotheses are tested by comparing each mean square with $\tilde{\sigma}^2 = r/(n-IJ)$, the column of expected mean squares showing that, in each case, the upper tail of the F-distribution is used. When H_R (no row effects) is true,

$$\frac{r(\mathbf{R})/(I-1)}{\tilde{\sigma}^2} \quad \text{has the } F_{I-1,n-IJ} \text{ distribution.}$$

When H_C (no column effects) is true,

$$\frac{r(\mathbf{C})/(J-1)}{\tilde{\sigma}^2} \quad \text{has the } F_{J-1,n-IJ} \text{ distribution.}$$

When H_I (no interaction effects) is true,

$$\frac{r((\mathbf{RC}))/(I-1)(J-1)}{\tilde{\sigma}^2} \quad \text{has the } F_{(I-1)(J-1),n-IJ} \text{ distribution.}$$

The interpretation of these tests is along the following lines. We have already said that absence of interaction means that row and column effects act additively. Suppose, for illustrative purposes, that rows correspond to different varieties of wheat and columns correspond to different fertilizers. Then if H_I is not rejected but H_R is rejected, we conclude that there are real differences amongst varieties and the difference between two given varieties of wheat will be the same for all fertilizers, i.e. if variety A is better than variety B, it is better by the same amount for all fertilizers. On the other hand, if interaction is present and H_R is rejected, the difference between two varieties will depend on the fertilizer considered, and the rejection of H_R is because there are differences in the effects of varieties when these are averaged over fertilizers; overall (i.e. averaged over the fertilizers considered) variety A may be better than variety B, although variety B could be superior to A for some fertilizers. Clearly the interpretation of the anova is much simpler when there is no interaction. If H_I were rejected but H_R and H_C were not, the conclusion would be that there are differences between cell means, but varieties do not show differences when averaged over fertilizers, and vice versa.

A joint test of all three hypotheses (i.e. of a common cell mean, μ) is obtained using $w = r(\mathbf{R}) + r(\mathbf{C}) + r((\mathbf{RC}))$. When all three hypotheses are true, w/σ^2 is an observation from a χ^2 distribution with d.f. equal to $(I-1)+(J-1)+(I-1)(J-1) = IJ-1$. The test function is thus $w/(IJ-1)\tilde{\sigma}^2$, which is referred to the $F_{IJ-1,n-IJ}$ distribution.

In the particular case of equal cell frequencies (all $K_{ij} = K$), the anova table simplifies to

Source	Sum of Squares	d.f.	$E(MS)$
Mean	$r(\mu) = n\bar{y}^2_{...}$	1	$\sigma^2 + IJK\mu^2$
Between rows	$r(\mathbf{R}) = JK \sum_i (y_{i..} - y...)^2$	$I-1$	$\sigma^2 + JK \sum R_i^2/(I-1)$
Between columns	$r(\mathbf{C}) = IK \sum_j (y_{.j.} - y...)^2$	$J-1$	$\sigma^2 + IK \sum C_j^2/(J-1)$
Interaction	$r((\mathbf{RC})) = K \sum_i \sum_j (y_{ij.} - y_{i..} - y_{.j.} + y_{...})^2$	$(I-1)(J-1)$	$\sigma^2 + K \sum \sum (RC)^2_{ij}/$ $(I-1)(J-1)$
Residual	$r = K \sum_i \sum_j (y_{ijk} - y_{ij.})^2$ (by subtraction).	$IJ(K-1)$	σ^2
Total	$\mathbf{y'y}$	IJK	

If each cell contains only one observation $(K=1)$, none of the above tests can be carried out since the residual is now $r = \sum \sum (y_{ijk} - y_{ij.})^2$, which is identically zero. This occurs because the number of observations is now IJ, the same as the number of independent parameters; the data can therefore be fitted exactly, without residual variation. However, if there is no interaction (all $(RC)_{ij} = 0$), the model becomes

$$y_{ij} = \mu + R_i + C_j + e_{ij}, \quad i = 1, \ldots, I; \quad j = 1, \ldots, J,$$

in which there are $(I+J+1)$ parameters, subject to two side conditions (viz. $\sum R_i = \sum C_j = 0$), leaving $(I+J-1)$ independent parameters. The application of least squares to this model gives estimates

$$\tilde{\mu} = y.., \qquad \tilde{R}_i = y_{i\cdot} - y.., \qquad \tilde{C}_j = y._j - y..,$$

and the residual turns out to be the same as the interaction SS in the more general model. The anova is thus

Source	Sum of squares	d.f.	$E(MS)$
Mean	$r(\mu) = ny_{..}^2.$	1	$\sigma^2 + IJ\mu^2$
Between rows	$r(\mathbf{R}) = J \sum_i (y_{i\cdot} - y..)^2$	$I-1$	$\sigma^2 + J \sum R_i^2/(I-1)$
Between columns	$r(\mathbf{C}) = I \sum_j (y._j - y..)^2$	$J-1$	$\sigma^2 + I \sum C_j^2/(J-1)$
Residual	$r = \sum_i \sum_j (y_{ij} - y_{i\cdot} - y._j + y..)^2$	$(I-1)(J-1)$	σ^2
	(by subtraction)		
Total	$\mathbf{y}'\mathbf{y}$	IJ	

We can therefore test for row and column effects in the usual way by comparing mean squares and using the F-distribution. It must be emphasized that this procedure is only valid when an additive model is appropriate; when interactions are present, the interaction mean square is not a suitable estimate of σ^2 and it is necessary to ensure that $K > 1$ in order to test for row and column effects.

Returning to the general model, with unequal cell frequencies, when the proportional frequencies condition is not satisfied, the parameters do not form orthogonal groups. To test H_R, we need the extra reduction in $\mathbf{y}'\mathbf{y}$ when the row parameters are fitted in addition to the remaining parameters, and similarly for the other hypotheses. As a result of the non-orthogonality, these sums of squares will no longer add up to SSM. The full least squares procedure must now be used, as the procedure based on (10.1.14) no longer gives the required sums of squares because the cross product terms do not disappear in the absence of proportional frequencies. In these circumstances there is no particular reason to keep the side conditions used in the proportional frequencies case, and the (now) numerical solution of the normal equations is usually carried out subject to the simpler constraints

$$\sum_i R_i = \sum_j C_j = \sum_i (RC)_{ij} = \sum_j (RC)_{ij} = 0.$$

Details may be found in Kendall & Stuart (1966), Scheffé (1959).

10.1.6. Higher Order Classifications

The analysis rapidly gets more tedious as further classificatory factors are introduced. We can repeat the hierarchical set-up to involve a third factor within which our previous example is nested. The third factor might be 'continents', the model used being of the form

$$y_{ijkl} = \mu + g_i + s_{ij} + t_{ijk} + e_{ijkl}$$

where the main group parameters $\{g_i\}$ refer to continents, the sub-group parameters $\{s_{ij}\}$ to countries, and the sub-sub-group parameters $\{t_{ijk}\}$ to cities.

Similarly, we can consider a three-factor cross-classification in which each pair of factors gives a cross-classification as discussed above in section 10.1.5; this case is discussed in greater detail below. Both of these extensions are 'pure'; one is completely hierarchical and the other is completely crossed. A new feature appears with three or more factors: we can have mixtures of cross- and hierarchical classifications. As an illustration of this possibility, consider the following model which might be used to investigate strikes in international manufacturing companies:

$$y_{ijkl} = \mu + C_i + M_j + (CM)_{ij} + F_{ijk} + e_{ijkl}$$

where the factors are: (a) countries (C_i); (b) manufacturers (M_j); and (c) Factories (F_{ijk}). The response y_{ijkl} could be (man-days lost in strikes per year)/(man-days worked per year), over a period of l_{ijk} years for factory (i, j, k).

Supposing that each company has manufacturing plant in each country, 'countries' and 'manufacturers' form a two-way cross-classification, in which each cell, i.e. combination of country and company, is a main group in a two-way hierarchical classification in which factories are the subgroups. Details of the analysis of such mixed classifications, as well as of purely nested ones with three factors, are given in Scheffé (1959).

We now consider more fully the three-way cross-classification. Data relating to obesity might be cross-classified by drug treatments, dietary regime and personality type. The model is

$$y_{ijkl} = \mu + D_i + F_j + P_k + (DF)_{ij} + (DP)_{ik} + (FP)_{jk} + (DFP)_{ijk} + e_{ijkl}$$

where $i = 1, \ldots, I$; $j = 1, \ldots, J$; $k = 1, \ldots, K$; $l = 1, \ldots, L_{ijk}$. Here, the D_i's refer to drugs, the F_j's represent the various dietary regimes and the P_k's are associated with different personality types. Note that, in addition to the three sets of parameters needed to represent *first-order interactions* between each pair of factors, we also require parameters $(DFP)_{ijk}$ to deal with the *second-order interaction* between all three factors, without which the model is unable to represent the general case of different expectations in all IJK cells. More particularly, the absence of the second-order interaction implies that the first-order interaction between, say, drugs and diets, is the same for each personality type (i.e. for each k, drugs and diets interact in the same way, this being expressed only through the parameters $(DF)_{ij}$); and similarly for the other first-order interactions between the remaining factor-pairs. We shall consider only the *balanced* case, in which all cells have the same number of observations ($L_{ijk} = L$). The number of parameters in the model is

$$1 + I + J + K + IJ + IK + JK + IJK,$$

and the number of cells is IJK. The number of redundant parameters is thus

$$1 + I + J + K + IJ + IK + JK.$$

We use the following side conditions

$$\sum_i D_i = \sum_j F_j = \sum_k P_k = 0$$

$$\sum_i (DF)_{ij} = \sum_j (DF)_{ij} = 0$$

$$\sum_i (DP)_{ik} = \sum_k (DP)_{ik} = 0$$

$$\sum_j (FP)_{jk} = \sum_k (FP)_{jk} = 0$$

$$\sum_i (DFP)_{ijk} = \sum_j (DFP)_{ijk} = \sum_k (DFP)_{ijk} = 0.$$

The L.S. estimates of the parameters are obtained by minimizing

$$\sum_i \sum_j \sum_k \sum_l \{ y_{ijkl} - \mu - D_i - F_j - P_k - (DF)_{ij} - (DP)_{ik} - (FP)_{jk} - (DFP)_{ijk} \}^2$$

subject to the above side conditions, the parameters in this expression being treated as algebraic variables. The least-squares estimates are

$$\tilde{\mu} = y_{....}$$
$$\tilde{D}_i = y_{i...} - y_{....}$$
$$(\widetilde{DF})_{ij} = y_{ij..} - y_{i...} - y_{.j..} + y_{....}$$
$$(\widetilde{DFP})_{ijk} = y_{ijk.} - y_{ij..} - y_{i.k.} - y_{.jk.} + y_{i...} + y_{.j..} + y_{..k.} - y_{....}$$

with corresponding expressions for the remaining parameter-estimates. The residual sum of squares is then

$$r = \sum_i \sum_j \sum_k \sum_l (y_{ijk} - y_{ijk.})^2$$

The number of observations is $IJKL$ and the number of independent parameters is IJK; thus SSR has $IJK(L-1)$ degrees of freedom.

As in the two-way cross-classification, the parameters form orthogonal groups only when a 'proportional frequencies' condition is satisfied. This applies in the present case of equal cell-frequencies, and we have, in the usual notation

$$\text{SMM} = r(\mu) + r(\mathbf{D}) + r(\mathbf{F}) + r(\mathbf{P}) + r[(\mathbf{DF})] + r[(\mathbf{DP})] + r[(\mathbf{FP})] + r[(\mathbf{DFP})]$$

The various sums of squares in this expression can be found by re-minimization under appropriate hypotheses, or by using the device of expressing the deviation $(y_{ijkl} - y....)$ as a sum of various components. The anova table is found to be as in Table 10.1.2.

Source	Sum of squares	d.f.	$E(MS)$
Mean	$r(\mu) = ny^2....$	1	
Between drugs (D)	$r(\mathbf{D}) = JKL \sum_i \tilde{D}_i^2$	$I-1$	$\sigma^2 + \sigma_D^2$
	$= JKL \sum_i (y_{i...} - y....)^2$		
Between diets (F)	$r(\mathbf{F}) = IKL \sum_j \tilde{F}_j^2$	$J-1$	$\sigma^2 + \sigma_F^2$
	$= IKL \sum_j (y._{j\cdot} - y....)^2$		
Between personalities	$r(\mathbf{P}) = IJL \sum_k \tilde{P}_k^2$	$K-1$	$\sigma^2 + \sigma_P^2$
(P)	$= IJL \sum_k (y.._{k\cdot} - y....)^2$		
Interaction (DF)	$r[(\mathbf{DF})] = KL \sum_i \sum_j (\tilde{DF})_{ij}^2$	$(I-1)(J-1)$	$\sigma^2 + \sigma_{DF}^2$
Interaction (DP)	$r[(\mathbf{DP})] = JL \sum_i \sum_k (\tilde{DP})_{ik}^2$	$(I-1)(K-1)$	$\sigma^2 + \sigma_{DP}^2$
Interaction (FP)	$r[(\mathbf{FP})] = IL \sum_j \sum_k (\tilde{FP})_{jk}^2$	$(J-1)(K-1)$	$\sigma^2 + \sigma_{FP}^2$
Interaction (DFP)	$r[(\mathbf{DFP})] = L \sum_i \sum_j \sum_k (\tilde{DFP})_{ijk}^2$	$(I-1)(J-1)$ $(K-1)$	$\sigma^2 + \sigma_{DFP}^2$
Residual	$r = \mathbf{y'y} - SSM$	$IJK(L-1)$	σ^2
	$= \sum_i \sum_j \sum_k \sum_l (y_{ijkl} - y_{ijk\cdot})^2$		
Total	$\mathbf{y'y}$	$IJKL$	

Table 10.1.2

The degrees of freedom in each case (apart from the last two lines) is the number of independent parameters involved, i.e. the total number of parameters associated with each effect less the number of independent restrictions imposed.

The column of expected mean squares uses the following notation:

$$\sigma_D^2 = (I-1)^{-1} \sum_i D_i^2$$

$$\sigma_{DF}^2 = (I-1)^{-1}(J-1)^{-1} \sum_i \sum_j (DF)_{ij}^2$$

$$\sigma_{DFP}^2 = (I-1)^{-1}(J-1)^{-1}(K-1)^{-1} \sum_i \sum_j \sum_k (DFP)_{ijk}^2$$

with similar definitions of σ_F^2, σ_P^2, σ_{DP}^2, σ_{FP}^2 (N.B. these are not variances!)

Under the appropriate hypothesis, each sum of squares is distributed independently of the residual and is proportional to a chi-squared variate; F-tests are constructed in the usual way. For example, to test for differences between drugs, we use the fact that in the absence of such differences (i.e. when all D_i are zero), $r(\mathbf{D})/\tilde{\sigma}^2(I-1)$ has the $F_{(I-1),IJK(L-1)}$ distribution, where $\tilde{\sigma}^2 = r/IJK(L-1)$. Similarly, to test for 'no interaction between drugs and personalities', we refer the statistic $r[(\mathbf{DP})]/\tilde{\sigma}^2(I-1)(K-1)$ to the $F_{(I-1)(K-1),IJK(L-1)}$ distribution. Rejection of the hypothesis implies that some drugs are particularly effective (relative to the others under consideration) with certain personality types.

If $L=1$ the residual is identically zero, and all the tests are void. However, if we are willing to assume that the second-order interaction is zero, the remaining hypotheses can still be tested; for in this case $r[(\mathbf{DFP})]$ assumes the role of residual, and the tests are carried out accordingly. For example, to examine the possibility of a dietary effect, we use $(I-1)(K-1)r(\mathbf{F})/r[(\mathbf{DFP})]$, which has the $F_{(J-1),(I-1)(J-1)(K-1)}$ distribution when there are no differences between diets.

10.1.7. Analysis of Covariance

In comparing group effects in a one-way classification it is often possible to take into account supplementary information on related variables. For example, in the case of the I anti-corrosive treatments previously mentioned, other information which would be relevant to the comparison might include temperature, time of exposure, moisture content, etc.; differences in the values of these latter variables could be expected to be partly responsible for observed differences between treatments. The analysis of covariance (anocova) makes allowances for such effects. We shall give the analysis when observations on one such *concomitant variable* (or, *covariate*), say time of exposure, X, are available. The model is then

$$y_{ij} = a + b_i + \beta x_{ij} + e_{ij}, \tag{10.1.15}$$

where x_{ij} is the time of exposure of the jth item in group i; the side condition is unchanged, viz.

$$\sum_{i=1}^{I} J_i b_i = 0.$$

As far as the hypothesis of equal group effects is concerned (all $b_i = 0$), the L.S. analysis of this model is essentially a method of making allowances for the differing times of exposure so that the effects of these are eliminated in the comparison of treatments. The design matrix, $\mathbf{A}$, is obtained by adding a new column to the design matrix for the one-way classification without covariates, given in (10.1.3). The new column is

$$(x_{11}, \ldots, x_{1J_1}, x_{21}, \ldots, x_{2J_2}, \ldots, x_{I1}, \ldots, x_{IJ_I})'.$$

To obtain the L.S. estimates of the parameters, we solve the equations

$$\mathbf{A}'\mathbf{A}\tilde{\boldsymbol{\theta}} = \mathbf{A}'\mathbf{y}, \qquad \sum_{i=1}^{I} J_i \tilde{b}_i = 0,$$

where $\boldsymbol{\theta} = (a, b_1, \ldots, b_I, \beta)'$. In this case, we have

$$
\mathbf{A}'\mathbf{A} =
\begin{bmatrix}
n & J_1 & J_2 & \cdots & J_I & nx.. \\
J_1 & J_1 & 0 & \cdots & 0 & J_1x_1. \\
J_2 & 0 & J_2 & \cdots & 0 & J_2x_2. \\
\vdots & \vdots & \vdots & & \vdots & \vdots \\
J_I & 0 & 0 & \cdots & J_I & J_Ix_I. \\
nx.. & J_1x_1. & J_2x_2. & \cdots & J_Ix_I. & \sum_i\sum_j x_{ij}^2
\end{bmatrix}
\qquad
\mathbf{A}'\mathbf{y} =
\begin{bmatrix}
ny.. \\
J_1y_1. \\
J_2y_2. \\
\vdots \\
J_Iy_I. \\
\sum_i\sum_j x_{ij}y_{ij}
\end{bmatrix}
$$

whence

(i) $\tilde{a} + \tilde{\beta}x.. = y..,$

(ii) $\tilde{a} + \tilde{b}_i + \tilde{\beta}x_i. = y_i., \qquad i = 1, \ldots, I,$

(iii) $nx..\tilde{a} + \sum_{i=1}^{I} J_ix_i.\tilde{b}_i + \tilde{\beta}\sum_i\sum_j x_{ij}^2 = \sum_i\sum_j x_{ij}y_{ij}.$

We may re-write (i) as $\tilde{a} = y.. - \tilde{\beta}x..$; substituting this in (ii) gives $\tilde{b}_i = y_i. - y.. - \tilde{\beta}(x_i. - x..)$. These should be compared with the corresponding estimates when there is no covariate (10.1.6). Substitution in (iii) gives an equation for $\tilde{\beta}$ which can be solved to give

$$
\tilde{\beta} = \frac{\sum_i\sum_j (x_{ij} - x_i.)(y_{ij} - y_i.)}{\sum_i\sum_j (x_{ij} - x_i.)^2}.
$$

The residual sum of squares is thus

$$
r = \sum_i\sum_j (y_{ij} - \tilde{a} - \tilde{b}_i - \tilde{\beta}x_{ij})^2.
$$

The model employs $I+2$ parameters and one side condition; the number of independent parameters is thus $I+1$, and the residual is proportional to a χ^2-variate on $n-I-1$ degrees of freedom. The estimate of the error variance is therefore $\tilde{\sigma}^2 = r/(n-I-1)$.

It turns out that we require the quantities in the following table.

Source	SS(x^2)	SS(xy)	SS(y^2)
Mean	$nx_{..}^2$	$nx..y..$	$ny_{..}^2$
Groups	$G_{xx} = \sum_i n_i(x_i. - x..)^2$	$G_{xy} = \sum_i n_i(x_i. - x..)(y_i. - y..)$	$G_{yy} = \sum_i n_i(y_i. - y..)^2$
Residual	$R_{xx} = \sum_i\sum_j (x_{ij} - x_i.)^2$	$R_{xy} = \sum_i\sum_j (x_{ij} - x_i.)(y_{ij} - y_i.)$	$R_{yy} = \sum_i\sum_j (y_{ij} - y_i.)^2$
Total	$\sum_i\sum_j x_{ij}^2$	$\sum_i\sum_j x_{ij}y_{ij}$	$\sum_i\sum_j y_{ij}^2$

The final column is the usual anova which would be performed in the absence of the supplemantary observations on X; the other two columns give the corresponding calculations for x^2 and xy, whence the term *analysis of covariance*. Using this notation, we have $\tilde{\beta} = R_{xy}/R_{xx}$. Further, it can be shown that

$$r = R_{yy} - \tilde{\beta}R_{xy},$$

which may alternatively be written in the forms

$$r = R_{yy} - \tilde{\beta}^2 R_{xx} = R_{yy} - R_{xy}^2/R_{xx}.$$

Let us recall the identity established earlier for the group classification without concomitant observations, namely

$$\sum_i \sum_j (y_{ij} - y..)^2 = \sum_i \sum_j (y_{ij} - y_{i.})^2 + \sum_i n_i(y_{i.} - y..)^2.$$

Denoting the L.H.S. (which is the total sum of squares about the mean) by T_{yy}, and using the notation introduced in the table above, this reads

$$T_{yy} = R_{yy} + G_{yy}.$$

Denoting

$$\sum_i \sum_j (x_{ij} - x..)^2 \text{ and } \sum_i \sum_j (x_{ij} - x..)(y_{ij} - y..)$$

by T_{xx} and T_{xy}, respectively, we can establish (by the same technique as was used for the above identity) analogous identities

$$T_{xx} = R_{xx} + G_{xx}$$
$$T_{xy} = R_{xy} + G_{xy}.$$

These show (since $T_{xx} = \sum \sum x_{ij}^2 - nx..^2$ and $T_{xy} = \sum \sum x_{ij}y_{ij} - nx..y..$) that in each column the residual term may be found by subtracting the first two terms from the total.

To test the hypothesis, H, that all group effects are the same, i.e. $b_1 = b_2 = \ldots = b_I = 0$, we consider the following anova (Table 10.1.3).

Source	SS	d.f.
Reduction due to $a, \mathbf{b}, \beta$	$R(a, \mathbf{b}, \beta)$	$I+1$
Reduction due to a, β,	$R(a, \beta)$	2
$\quad$ = Reduction under $H: \mathbf{b} = \mathbf{0}$		
Extra reduction	$R(\mathbf{b}\|a, \beta) = r_H - r$	$I-1$
Residual	$r = \mathbf{y'y} - R(a, \mathbf{b}, \beta)$	$n-I-1$
Total	$\mathbf{y'y}$	n

Table 10.1.3: Analysis of covariance

Under H, we consider the reduced model

$$y_{ij} = a + \beta x_{ij} + e_{ij}.$$

Now this model postulates linear regression of Y on X, so from Example 8.2.4 the L.S. estimates of a and β under H (denoted by a^* and β^*) are given by

$$a^* = y.. - \beta^* x..$$

$$\beta^* = \frac{\sum_i \sum_j (x_{ij} - x..)(y_{ij} - y..)}{\sum_i \sum_j (x_{ij} - x..)^2},$$

and from Example 8.12.13, the residual is

$$r_H = \sum_i \sum_j (y_{ij} - y..)^2 - \beta^* \sum_i \sum_j (x_{ij} - x..)(y_{ij} - y..)$$

Thus

$$\beta^* = \frac{T_{xy}}{T_{xx}} = \frac{R_{xx} + G_{xx}}{R_{xy} + G_{xy}}$$

and

$$r_H = T_{yy} - T_{xy}^2 / T_{xx}.$$

The increase in residual due to H is therefore

$$r_H - r = T_{yy} - T_{xy}^2 / T_{xx} - (R_{yy} - R_{xy}^2 / R_{xx})$$

$$= G_{yy} - \frac{(R_{xy} + G_{xy})^2}{R_{xx} + G_{xx}} + \frac{R_{xy}^2}{R_{xx}}.$$

The order of H is $I - 1$, so that the test function is

$$F = \frac{(r_H - r)/(I-1)}{r/(n - I - 1)}.$$

If H is true, F has the $F_{(I-1),(n-I-1)}$ distribution.

We can summarize the main points in the following table for testing $H : b_1 = b_2 = \ldots = b_I = 0$:

Source	SS	d.f.	MS	F
Residual under H	$r_H = T_{yy} - T_{xy}^2 / T_{xx}$	$n-2$		
Residual	$r = R_{yy} - R_{xy}^2 / R_{xx}$	$n - I - 1$	$\tilde{\sigma}^2 = \dfrac{r}{n - I - 1}$	
Extra reduction without H	$r_H - r$	$I - 1$	$w = \dfrac{r_H - r}{I - 1}$	$\dfrac{w}{\tilde{\sigma}^2}$

Although interest naturally centres on the investigation of the group effects, there is also some interest in testing the hypothesis $H': \beta = 0$, since rejection of H' means that inclusion of the covariate in the model is worthwhile.

Under H', the model is just the usual model for the one-way classification without the covariate, so that $r_{H'} = R_{yy}$, whence the extra reduction is

$$r_{H'} - r = \tilde{\beta}^2 R_{xx} = R_{xy}^2 / R_{xx}.$$

Now, as H' is of order 1, the test of H' is based on $F' = \tilde{\beta}^2 R_{xx} / \tilde{\sigma}^2$, which has the $F_{1, n-I-1}$ distribution when $\beta = 0$.

Analogous methods are applied to models involving further clasificatory factors and/or concomitant variables. For instance, in a two-way cross-classification with two covariates W and X, and the same number of observations in each cell, we would use the model

$$y_{ijk} = \mu + R_i + C_j + (RC)_{ij} + \alpha w_{ijk} + \beta x_{ijk} + e_{ijk},$$

subject to the side conditions

$$\sum_i R_i = \sum_j C_j = \sum_i (RC)_{ij} = \sum_j (RC)_{ij} = 0.$$

Hypotheses are tested in the usual way, e.g. to test for 'no row effect' we would obtain the residual under this hypothesis and examine the increase in this over the residual for the full model. The basic computations required may be set out in the form of a table in which one column is the usual anova for the cross-classification without covariates, together with further columns giving corresponding calculations for w^2, x^2, wy, xy, wx.

10.2. MULTIPLE TESTS AND COMPARISONS

10.2.1. Introduction

We have seen in the preceding examples of the least squares analysis of data that, usually, several hypotheses are tested and confidence statements constructed. We can distinguish between two types of inferential questions, namely

(a) those specified in advance of the experiment being performed;
(b) those formulated in the light of the data.

Thus, in the case of data classified by group factors, the anova tests via the F distribution are used to detect the presence of various effects which the experiment is designed to investigate; appropriate hypotheses and tests are therefore specified in advance of the data being collected. For example, in a complete two-way cross classification comparing four drugs ($A, B, C, D,$ say) for each of three dietary regimes, we would certainly test the hypotheses associated with the anova given earlier, i.e.

(i) no interaction between drugs and diets;
(ii) no differences between drugs;
(iii) no differences between diets.

We could also ask for confidence intervals for the mean effect of each drug and diet. All these would constitute group (a). Additionally, we might also formulate other questions suggested by the results. Thus, if the F test indicated differences between the drugs (assume these are associated with rows) by rejecting $H_R: R_1 = R_2 = R_3 = R_4$, we would want to explore this further to find out which drugs (if any) had the same effects and which had different effects. For instance, suppose that the row totals show that drugs A, B and D gave rather similar results while drug C was somewhat better than these. We might ask: do A, B, D have the same effect? If so, is C better than all three? For the first question we could consider the hypothesis $H: R_1 = R_2 = R_4$, which would involve the usual procedure of re-estimation under H so as to determine r_H. Quite apart from the tedium of this (possibly repeated for other hypotheses in category (b)), the nominal significance level (say 5%) at which H is tested is not the true significance level and could be seriously misleading. Why is this? Essentially it is because H has been suggested by the results and the test is therefore conditional on the observed data. If H were always tested (in hypothetical repetitions of the experiment), then the significance level would be 5%; i.e. if H were true, on 5% of occasions in the long run H would be incorrectly rejected. In fact, we would only test H when (i) the F test had rejected H_R; and (ii) it seemed pertinent to test H; for instance, we would certainly not bother if A, B, D gave clearly very different results—yet this could happen if H were true and is explicitly allowed for in the 5% probability of incorrect rejection when H is true. The result of all this is that the actual rejection rate of H when true is reduced to an unknown figure below 5%, the value depending on the experimenter's subjective decision on when to test H. In the case of the second question, a confidence interval for $R_3 - (R_1 + R_2 + R_4)/3$ might be constructed. Here again the nominal confidence level is only appropriate if such an interval is constructed whatever the experimental results and not just in this particular case.

What is needed, therefore, is a procedure which will permit such tests and confidence statements as are deemed appropriate on analysing the data to be carried out with a given overall level of significance or confidence. Such a procedure is described below in 10.2.3.

10.2.2. Combination of Tests, and Overall Size

We consider first the case of several category (a) hypotheses $H_1, H_2, \ldots, H_k$, formulated for testing as part of the design, before the data is analysed. If H_i is tested at level α_i, then even though all the α_i are small, there may be an appreciable probability of at least one Type I error, unless k is small, i.e. if enough hypotheses are tested we are likely to obtain some 'significant results' even when no effects are present. Assuming that the test statistics of the hypotheses are independent, the probability that none of these hypotheses is rejected, when all are true, is

$$\prod_{i=1}^{k} (1-\alpha_i).$$

The probability that at least one is rejected when all are true is thus

$$1-\prod_{i=1}^{k} (1-\alpha_i)$$

If all $\alpha_i = \alpha$, this is simply

$$P_k(\alpha) = 1-(1-\alpha)^k. \qquad (10.2.1)$$

Provided $k\alpha$ is not too big (say $k\alpha \le \frac{1}{4}$), this is approximately equal to $k\alpha$, as the following table illustrates:

k	$P_k(0.05)$	$P_k(0.01)$
1	0·05	0·01
2	0·0975	0·0199
4	0·1855	0·0394
8	0·3366	0·0773
16	0·5599	0·1485

Clearly, when several hypotheses are tested, the value of $P_k(\alpha)$ is a valuable aid in the interpretation of the tests, and it may be preferable to fix the overall size $P_k(\alpha)$ at an appropriate level by suitable choice of α, instead of using the conventional levels for the individual tests. For example, if four tests are to be made and the overall size required is 0·05, we solve $0.05 = P_4(\alpha) = 1-(1-\alpha)^4$. The right-hand side is approximately 4α, so $\alpha \approx 0.05/4 = 0.0125$ (the exact value is 0·0127). By using this value of α for each test we ensure that the probability of incorrectly rejecting one or more of the four hypotheses tested (when all are true) is 0·05.

There is an immediate problem in applying this technique to the tests in an anova table, as these tests all use the same residual and are therefore not independent. Fortunately, however, it can be shown that (10.2.1) still holds as an approximation in this case, so that a test of overall size α' is obtained approximately if each of the tests in the anova table has size α, where $\alpha = 1-\{1-\alpha'\}^{1/k} \approx \alpha'/k.$

10.2.3.   Multiple Comparisons

Also in category (a) we may wish to compare various parameters by initially specifying several linear combinations of them (e.g. group means or differences between these) for which confidence statements $C_1, C_2, \ldots, C_m$ are required. The same kind of problem arises; for while the confidence attaching to each is appropriate when considering it in isolation from the others, we also want to know the overall confidence in all of them i.e. the probability that all the statements are simultaneously correct. Once again, dependencies between them may make it difficult to answer this question precisely, but we can at least say

$$P(\text{all } C_i \text{ are true}) \geq 1 - \sum P(C_i \text{ is untrue})$$

so if the same confidence coefficient $1 - \alpha = P(C_i \text{ is true})$ is used for each, we have $P(\text{all are true}) \geq 1 - m\alpha$. We therefore have a lower bound to the required probability and we can guarantee overall confidence $1 - \alpha'$ by choosing $\alpha = \alpha'/m$. This is known as the Bonferroni method of constructing simultaneous confidence intervals as the inequality stated above is an example of Bonferroni's inequality.

As an illustration of this approach, we consider the one-way layout with model

$$y_{ij} = \mu_i + e_{ij}, \qquad i = 1, 2, \ldots, I; \qquad j = 1, 2, \ldots, J_i.$$

If we construct $100(1 - \alpha)\%$ confidence intervals for each group mean, i.e. $C_i = \{y_{i\cdot} \pm \tilde{\sigma} J_i^{-1/2} t_{n-I}(1 - \alpha/2)\}$, where $\tilde{\sigma}^2 = r/(n - I)$, then how much confidence can we have in the assertion $(\mu_i \in C_i, i = 1, 2, \ldots, I)$? In this case we have $m = I$, so it will be at least $1 - I\alpha$, and we can take $\alpha = \alpha'/I$ to obtain overall confidence at least $1 - \alpha'$.

When the F-test rejects the hypothesis $H : \mu_1 = \mu_2 = \ldots = \mu_I$ of equal group effects, we would usually want to know which groups can be regarded as the same and which are different. A $100(1 - \alpha)\%$ confidence interval for the difference between two means, say $\mu_i - \mu_{i'}$, $(i \neq i')$, is (since $\text{var}(Y_{i\cdot} - Y_{i'\cdot}) = \sigma^2(J_i^{-1} + J_{i'}^{-1})$) given by

$$D_{i,i'} = \left\{ (y_{i\cdot} - y_{i'\cdot}) \pm \tilde{\sigma} \sqrt{\left(\frac{1}{J_i} + \frac{1}{J_{i'}} \right)} \, t_{n-I}(1 - \alpha/2) \right\}.$$

We regard the two means as different if $D_{i,i'}$ excludes zero. When all groups have the same size $(J_i = J)$, this just involves comparing each difference $y_{i\cdot} - y_{i'\cdot}$ with a Least Significant Difference, $L = \tilde{\sigma}(2/J)^{1/2} t_{n-I}(1 - \alpha/2)$, the group means being judged different when $|y_{i\cdot} - y_{i'\cdot}|$ exceeds L. If this is done for each pair (i, i') with $i \neq i'$ we have altogether $m = \frac{1}{2}I(I - 1)$ intervals, and we can guarantee overall confidence at least $1 - \alpha'$ by taking $\alpha = \alpha'/m$; unfortunately, when m is large, such intervals will be very wide and hence of limited use.

Furthermore the confidence coefficient $1-\alpha'$ is only correct when these intervals are constructed irrespective of the result of the F-test; if, as is reasonable, we only bother when the F-test rejects H, then the true confidence level is unknown.

EXAMPLE 10.2.1. *Simultaneous confidence intervals.* The following table gives measurements of the tensile strength of six alloys:

	1	2	3	4	5	6
	15·1	15·1	14·9	15·3	15·2	15·2
	15·0	15·3	15·4	14·9	14·8	15·1
	15·4	15·5	15·2	15·0	15·4	15·3
	15·0	15·6	15·5	14·9	15·0	15·1
		15·1		15·2	15·0	
				14·9		
J_i	4	5	4	6	5	4
$y_i.$	15·12	15·32	15·25	15·03	15·08	15·17

The anova table is:

Source	SS	d.f.	MS
Between groups	0·9143	5	0·1828
Within groups (residual)	0·2942	22	$0·0134 = \tilde{\sigma}^2$
Total (about mean)	1·2085	27	

The ratio of mean squares is 13·6; $F_{5,22}(0·99) = 3·99$ so there is strong evidence of differences in strength amongst the alloys. Using $\tilde{\sigma} = 0·1156$, the estimated standard errors of the means are, respectively 0·058, 0·052, 0·058, 0·047, 0·052, 0·058; whence, (using $t_{22}(0·975) = 2·074$), we find the following individual 95% confidence intervals for $\mu_1, \mu_2, \ldots, \mu_6$: $(15·12 \pm 0·12)$, $(15·32 \pm 0·11)$, $(15·25 \pm 0·12)$, $(15·03 \pm 0·10)$, $(15·08 \pm 0·11)$, $(15·17 \pm 0·12)$, respectively.

If we want six intervals with joint confidence at least 95%, we use $\alpha = 0·05/6 = 0·0083$, so we re-calculate using $t_{22}(0·9958) = 2·9$ (see Appendix T5). The required intervals (which are thus the same as separate intervals each with confidence coefficient 0·9917) are then: $(15·12 \pm 0·17)$, $(15·32 \pm 0·15)$, $(15·25 \pm 0·17)$, $(15·03 \pm 0·14)$, $(15·08 \pm 0·15)$, $(15·17 \pm 0·17)$.

If we consider intervals for differences between all pairs of means, then $m = \frac{1}{2} \times 6 \times 5 = 15$, so overall confidence $1 - \alpha' \geq 0·95$ requires $\alpha = 0·05/15 = 0·0033$ and we need $t_{22}(0·9983) = 3·3$. The resulting confidence intervals are fairly wide; for example in comparing the first two groups we find $D_{1,2} = (-0·20 \pm 0·25)$ while for groups 2 and 4 we have $D_{2,4} = (0·29 \pm 0·23)$.

Sometimes it is appropriate a priori to consider a fixed subset of these comparisons, thereby reducing m and resulting in narrower intervals. In this example, group 4 actually relates to observations on a standard alloy, with the others being measurements for five new alloys. If we restrict attention to comparisons between each new alloy and the standard, we have $m = 5$ and so $\alpha = 0·05/5 = 0·01$ for overall confidence 0·95. The intervals are

$$\mu_1 - \mu_4: 0·09 \pm 0·21$$
$$\mu_2 - \mu_4: 0·29 \pm 0·20$$
$$\mu_3 - \mu_4: 0·22 \pm 0·21$$
$$\mu_5 - \mu_4: 0·05 \pm 0·20$$
$$\mu_6 - \mu_4: 0·14 \pm 0·21.$$

The joint confidence in these intervals (i.e. in the five statements typified by $-0·12 \leq \mu_1 - \mu_4 \leq 0·30$) is 0·95.

As the interval for $\mu_2 - \mu_4$ shows, by restricting attention to these five comparisons, we obtain narrower intervals for them than when all pairs were considered.

For situations in which it is difficult to specify in advance which comparisons are of interest, it is very useful to have available procedures for obtaining intervals relevant to questions posed after the data has been examined. The *S-method* of Scheffé is an approach to this problem. [See Scheffé (1959).]

We introduce the term *contrast* among the parameters $\mu_1, \mu_2, \ldots, \mu_I$ for any linear function of the μ_i, say $\phi = \sum_{i=1}^{I} c_i \mu_i$, with the property that $\sum_{i=1}^{I} c_i = 0$, the c_i's being known constants.

For example, the difference $\mu_i - \mu_{i'}$ between means of any two groups is a contrast; the mean of one group minus the average of the means of the remaining groups, e.g. $\mu_1 - (\mu_2 + \mu_3 + \ldots + \mu_I)/(I - 1)$ is a contrast; and more generally the average of the means of several groups minus the average of the means of several other groups. It can be shown that $H: \mu_1 = \mu_2 = \ldots = \mu_I$ is equivalent to H': all possible contrasts are zero. Essentially, Scheffé's method provides intervals for all possible contrasts, with given overall confidence; thus any contrasts we choose (either in advance or after inspecting the data) are automatically included.

Now the least squares estimate of ϕ is $\hat{\phi} = \sum c_i \hat{\mu}_i = \sum c_i y_i.$; the corresponding estimator is $\tilde{\Phi} = \sum c_i Y_i.$, which has $E(\tilde{\Phi}) = \phi$ and $\text{var } \tilde{\Phi} = \sum c_i^2 \text{ var } Y_i. = \sum c_i^2 (\sigma^2/J_i)$. We estimate this variance by $\tilde{\text{var}} \, \tilde{\Phi} = \tilde{\sigma}^2 \sum c_i^2/J_i$, where $\tilde{\sigma}^2$ is the usual estimate of σ^2 based on r, viz. $\tilde{\sigma}^2 = r/(n - I)$.

Scheffé has shown that, as ϕ ranges over all possible contrasts, there is probability $1 - \alpha$ that all inequalities of the form

$$\tilde{\Phi} - S\sqrt{\frac{R\sum c_i^2/J_i}{n-I}} \le \phi \le \tilde{\Phi} + S\sqrt{\frac{R\sum c_i^2/J_i}{n-I}} \qquad (10.2.2)$$

are simultaneously satisfied, where S is given by $S^2 = (I-1)F_{I-1,n-I}(1-\alpha)$, n being the total number of observations $\sum J_i$. We note that the multiplier $(I-1)$ is the order of $H: \mu_1 = \mu_2 = \ldots = \mu_I$.

For any contrast ϕ, let H_ϕ denote the hypothesis $\phi = 0$. We reject H_ϕ at the simultaneous level α if the calculated interval for ϕ, viz. $\tilde{\phi} \pm S\sqrt{\mathrm{var}\, \tilde{\Phi}}$, does not include the value $\phi = 0$. It can be shown that the F-test rejects H if, and only if, at least one H_ϕ is rejected. So, when all contrasts are actually zero, there is probability α that at least one H_ϕ is rejected. We can apply these results to any contrasts we like so as to find out if they were involved in the rejection of H. We note that if attention is confined to a single contrast, ϕ, nominated in advance and not suggested by the results, then (10.2.2) gives the usual $100(1-\alpha)\%$ confidence interval for ϕ if $I-1$ is replaced by 1 in the definition of S^2. If we want to construct simultaneous intervals for all linear combinations of $\mu_1, \mu_2, \ldots, \mu_I$ (i.e. not just contrasts) we only have to replace $I-1$ by I in the definition of S^2.

EXAMPLE 10.2.2.   *Scheffé's S-method.*   Let us apply the S-method to the data used previously in Example 10.2.1. We have $S\tilde{\sigma} = (5\,F_{5,22}(0.95)\,0.0134)^{1/2} = 0.4222$.

Using 10.2.2 we obtain the following:

	Contrast	$\sum c_i^2/J_i$	Interval
1	$\mu_1 - \mu_4$	0.42	0.09 ± 0.27
2	$\mu_2 - \mu_4$	0.37	0.29 ± 0.26
3	$\mu_3 - \mu_4$	0.42	0.22 ± 0.27
4	$\mu_5 - \mu_4$	0.37	0.05 ± 0.26
5	$\mu_6 - \mu_4$	0.42	0.14 ± 0.27
6	$\mu_2 - \mu_6$	0.45	0.15 ± 0.28
7	$\mu_2 - \mu_5$	0.40	0.24 ± 0.27
8	$\frac{1}{2}(\mu_2 + \mu_3) - \mu_4$	0.3125	0.255 ± 0.236
9	$\frac{1}{3}(\mu_1 + \mu_5 + \mu_6) - \mu_4$	0.2444	0.123 ± 0.209
10	$\frac{1}{2}(\mu_2 + \mu_3) - \frac{1}{3}(\mu_1 + \mu_5 + \mu_6)$	0.1903	0.160 ± 0.180

Now groups $1, 5, 6$ relate to an alloy with different amounts of a common constituent (call this alloy Type 1) and groups 2 and 3 also have differing amounts of a common constituent (not the same as $1, 5, 6$); call this alloy

Type 2. We can summarize as follows. In comparing the five alloys with the standard (first five contrasts) there is evidence that alloy 2 is superior to the standard; with the others there is no clear superiority, but e.g. if alloy 1 is better than the standard, the difference is not likely to exceed $0 \cdot 36$; similarly for the other three. The sixth contrast shows that there is insufficient evidence to conclude that the best Type 2 alloy is superior to the best Type 1 alloy, though the difference could be as much as $0 \cdot 43$. The interval for contrast 8 shows that Type 2 alloys are better than the standard; interval 9 shows that there is not enough evidence to conclude that Type 1 alloys are generally better than the standard. The final contrast compares Type 1 alloys with Type 2 alloys; the difference in favour of Type 2 could be as much as $0 \cdot 34$, although the possibility of Type 1 being superior is not excluded—more evidence is needed. The over-all confidence in all these statements together (and any more we care to make by computing intervals for other contrasts) is 95%.

As far as the first five contrasts are concerned, shorter intervals were achieved by the Bonferroni method, but this was at the cost of drawing no conclusions about the others.

When all groups are the same size, another method (due to Tukey) based on the Studentized range [see § 2.5.9] can be used to provide intervals for all contrasts. The intervals it gives are shorter than those of the S-method for contrasts of the form $\mu_i - \mu'_i$, but are otherwise longer; details may be found in Scheffé (1959).

The S-method can be applied generally to any set of effects in a linear model to provide simultaneous confidence statements of the type given in (10.2.2); the general theory is given in Scheffé (1959). As a final illustration, consider a balanced two-way cross classification:

$$y_{ijk} = \mu + R_i + C_j + (RC)_{ij} + e_{ijk}; \qquad i = 1, 2, \ldots I; \qquad j = 1, 2, \ldots J;$$
$$k = 1, 2, \ldots K.$$

If H_R: all $R_i = 0$ is rejected we can apply the S-method to contrasts among the row means $\mu + R_i$, employing intervals of the form

$$\tilde{\phi} \pm S \sqrt{\tilde{\operatorname{vâr}} \, \tilde{\Phi}},$$

where

$$\phi = \sum c_i(\mu + R_i) = \sum c_i R_i,$$
$$\tilde{\phi} = \sum c_i(y_{i\cdot\cdot} - y_{\cdot\cdot\cdot}) = \sum c_i y_{i\cdot\cdot},$$

vâr $\tilde{\Phi}$ is the estimated variance of $\sum c_i Y_{i\cdot\cdot}$, based on $\tilde{\sigma}^2 = r / IJ(K - 1)$, and

$$S^2 = (I - 1) F_{I-1, IJ(K-1)}(1 - \alpha).$$

If H_c: all $C_j = 0$ is rejected we can similarly deal with contrasts among the column means $\mu + C_j$, this time using $S^2 = (J - 1) F_{J-1, IJ(K-1)}(1 - \alpha)$.

10.3.   THE ASSUMPTIONS ABOUT THE ERRORS

10.3.1.   The Basic Assumptions

We conclude with a brief further consideration of the assumptions underlying the analyses presented earlier; a detailed treatment of these topics is given in Kendall and Stuart (1966) and Scheffé (1959). To re-iterate, the assumptions made regarding the errors $\varepsilon_1, \varepsilon_2, \ldots, \varepsilon_n$ in the linear model $\mathbf{Y} = \mathbf{A}\boldsymbol{\theta} + \boldsymbol{\varepsilon}$ are as follows

(1) the errors have zero expectations;
(2) the errors are independent;
(3) the errors have the same variance (homoscedasticity)
(4) the errors are Normally distributed.

We remind the reader that this last assumption is needed for tests of hypotheses and confidence intervals but is not required for the results given in section 8.2 concerning the means and second order properties of the L.S. estimators.

The first assumption presents no problems; for if the errors have expectation $\mu \neq 0$, we can introduce this parameter as an additive constant in the model or, if one is already present, modify it. In the first case this means replacing $(\theta_1, \theta_2, \ldots, \theta_p)'$ by $(\mu, \theta_1, \theta_2, \ldots, \theta_p)'$ and augmenting $\mathbf{A}$ by a new first column consisting entirely of units; in the second case, supposing that θ_1 is already a component of $E(Y_i)$ for each i, we merely replace θ_1 by $\theta_1 + \mu$ and treat the latter as the parameter of interest.

The second assumption is very important; we usually have to satisfy ourselves that the physical set-up or method of sampling makes this reasonable. The analyses of variance presented earlier are purely algebraic identities and are thus free of assumptions 1–4; however, the use we have made of them in testing hypotheses rested on assumptions 3 and 4. The assumption of constant error variances is important; without it—when the errors are heteroscedastic— the methods may be seriously misleading as a result of distorted significance levels unless the design is balanced (i.e. has the same number of observations in each group or cell). If, as in section 8.2.6, the variances are known up to a constant factor, we can use the technique described to transform to a new model to which the methods apply. In practice, such precise information will often be lacking, but various variance-stabilizing transformations [see § 2.7.3] can be used to deal with particular types of heteroscedasticity commonly encountered with classified data; for instance, when the variance in a group is proportional to the mean of the group we transform the observations by $y' = y^{1/2}$, while when the standard deviation is proportional to the group mean we use $y' = \log y$. Further details may be found in Kendall & Stuart (1966), Scheffé (1959).

The final assumption, of Normal error distributions will only be approximately true in practice, but fortunately anova procedures are robust to non-Normality, i.e. considerable departures from exact Normality can be tolerated before standard anova tests are seriously affected.

10.3.2. Residual Analysis

The analysis of data is necessarily incomplete and provisional in nature until the above assumptions are shown to be reasonable, which in practice means an absence of evidence of clear departure from them. This is most easily checked by means of simple graphical techniques involving the residuals; a clear discussion of this topic is given in Draper & Smith (1966).

Assumptions 1–4 may be summarized as: $e_1, e_2, \ldots, e_n$, is a random sample from $N(0, \sigma^2)$. The residuals $\tilde{e}_1, \tilde{e}_2, \ldots, \tilde{e}_n$, as our best estimates of these quantities should therefore resemble a random sample from $N(0, \sigma^2)$, and we can plot the $\tilde{e}_i$ on Normal probability paper [see § 3.2.2(d)] to see if this is

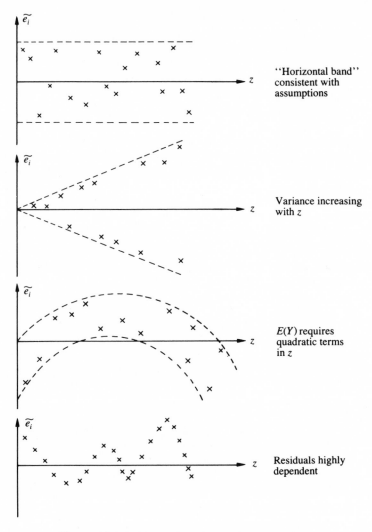

Figure 10.3.1: Examples of graphical procedures.

the case. Some departure is to be expected, as even if the assumptions are exactly right, the residuals will not be a random sample from $N(0, \sigma^2)$ owing to dependencies between them, which can be seen on writing the normal equations as

$$\sum_1^n \tilde{e}_i a_{ij} = 0, \qquad j = 1, 2, \ldots, p.$$

Thus, for any model with an additive constant θ_1, the residuals must sum to zero.
Other graphical procedures are as follows.

1. Plot $\tilde{e}_i$ against the estimated mean

$$\tilde{E}(Y_i) = \sum_{j=1}^p a_{ij}\tilde{\theta}_j;$$

2. Plot $\tilde{e}_i$ against i if the order of observations is meaningful. (Thus, if observations are taken successively in time, plot the residuals in chronological order.)
3. In the case of regression models, with regressors $X_1, X_2, \ldots, X_s$ (say), plot the residuals against each regressor in turn, i.e. $\tilde{e}_i$ against X_{ij}, for $j = 1, 2, \ldots, s$.

The diagram (Figure 10.3.1) illustrates some of the possibilities (z denotes the variable chosen for the plot). For classified observations, we can split the residuals up according to the group they relate to and then examine each subset by the above methods; we can also look for differences between such groups of residuals.

D.W.

10.4. FURTHER READING AND REFERENCES

There is a huge literature dealing with the subject matter of this chapter, under such titles as analysis of variance, analysis of statistical experiments, regression analysis, correlation analysis, the linear hypothesis, etc., etc. The reader will find useful information, and bibliographies in, for example, the following works cited in Bibliography C: Davies (1957), Graybill (1976), Hald (1957), and Volume 3 of Kendall and Stuart (1976).

Other excellent references are the following:

Draper, N. and Smith, H. (1966). *Applied Regression Analysis*, Wiley.
Edwards, A. L. (1976). *An Introduction to Linear Regression and Correlation*, Freeman.
Plackett, R. L. (1960). *Principles of Regression Analysis*, Oxford, Clarendon Press.
Scheffé, H. (1959) *The Analysis of Variance*, Wiley.
Seber, G. A. F. (1980). *The Linear Hypothesis: A General Theory* (Second edition), Griffen.

Appendix — Statistical Tables

APPENDIX T1

Cumulative Binomial Tables Probabilities [see II: § 5.2.2]

The tabulated quantity is

$$P\{R(n, \theta) \geq r\} = \sum_{s=r}^{n} \binom{n}{s} \theta^s (1-\theta)^{n-s}, \qquad r = 0, 1, \ldots, n,$$

where $R(n, \theta)$ denotes a random variable which has the Binomial (n, θ) distribution. Thus $P\{R(n, \theta) \geq r\}$ is the probability of r *or more* successes in n independent trials when the probability of a success in a particular trial is θ. The tables give values of P for $\theta = 0 \cdot 01(0 \cdot 01)0 \cdot 10(0 \cdot 05)0 \cdot 50$. For values of θ exceeding $0 \cdot 50$, use the relation

$$P\{R(n, \theta) \geq r\} = P\{R(n, 1-\theta) \leq n-r\}$$

$$= 1 - P\{R(n, 1-\theta) \geq n-r+1\}, \qquad r = 1, 2, \ldots, n.$$

For example,

$$P\{R(20, 0 \cdot 7) \geq 12\} = 1 - P\{R(20, 0 \cdot 3) \geq 9\}$$

$$= 1 - 0 \cdot 1133 = 0 \cdot 8867.$$

Related quantities that can be obtained from the table are

$$P\{R(n, \theta) \leq r\} = 1 - P\{R(n, \theta) \geq r+1\}, \qquad r = 0, 1, \ldots, n-1,$$

and

$$P\{R(n, \theta) = r\} = P\{R(n, \theta) \geq r\} - P\{R(n, \theta) \geq r+1\}, \qquad r = 0, 1, \ldots, n-1.$$

A1

Cumulative Binomial probabilities

(Reproduced by permission of Macmillan Publishers Ltd. from *Statistical Tables for Science, Engineering and Management* by J. Murdoch and J. A. Barnes.)

θ = probability of success in a single trial; n = number of trials. The table gives the probability of obtaining *r or more* successes in n independent trials, i.e.

$$\sum_{x=r}^{n} \binom{n}{x} \theta^x (1-\theta)^{n-x}$$

When there is no entry for a particular pair of values of *r* and θ, this indicates that the appropriate probability is less than 0·000 05. Similarly, except for the case $r = 0$, when the entry is exact, a tabulated value of 1·0000 represents a probability greater than 0·999 95.

θ		0·01	0·02	0·03	0·04	0·05	0·06	0·07	0·08	0·09
$n=2$	$r=0$	1·0000	1·0000	1·0000	1·0000	1·0000	1·0000	1·0000	1·0000	1·0000
	1	0·0199	0·0396	0·0591	0·0784	0·0975	0·1164	0·1351	0·1536	0·1719
	2	0·0001	0·0004	0·0009	0·0016	0·0025	0·0036	0·0049	0·0064	0·0081
$n=5$	$r=0$	1·0000	1·0000	1·0000	1·0000	1·0000	1·0000	1·0000	1·0000	1·0000
	1	0·0490	0·0961	0·1413	0·1846	0·2262	0·2661	0·3043	0·3409	0·3760
	2	0·0010	0·0038	0·0085	0·0148	0·0226	0·0319	0·0425	0·0544	0·0674
	3		0·0001	0·0003	0·0006	0·0012	0·0020	0·0031	0·0045	0·0063
	4						0·0001	0·0001	0·0002	0·0003
$n=10$	$r=0$	1·0000	1·0000	1·0000	1·0000	1·0000	1·0000	1·0000	1·0000	1·0000
	1	0·0956	0·1829	0·2626	0·3352	0·4013	0·4614	0·5160	0·5656	0·6106
	2	0·0043	0·0162	0·0345	0·0582	0·0861	0·1176	0·1517	0·1879	0·2254
	3	0·0001	0·0009	0·0028	0·0062	0·0115	0·0188	0·0283	0·0401	0·0540
	4			0·0001	0·0004	0·0010	0·0020	0·0036	0·0058	0·0088
	5					0·0001	0·0002	0·0003	0·0006	0·0010
	6									0·0001

n = 20

r									
0	1·0000	1·0000	1·0000	1·0000	1·0000	1·0000	1·0000	1·0000	1·0000
1	0·1821	0·3324	0·4562	0·5580	0·6415	0·7099	0·7658	0·8113	0·8484
2	0·0169	0·0599	0·1198	0·1897	0·2642	0·3395	0·4131	0·4831	0·5484
3	0·0010	0·0071	0·0210	0·0439	0·0755	0·1150	0·1610	0·2121	0·2666
4		0·0006	0·0027	0·0074	0·0159	0·0290	0·0471	0·0706	0·0993
5			0·0003	0·0010	0·0026	0·0056	0·0107	0·0183	0·0290
6				0·0001	0·0003	0·0009	0·0019	0·0038	0·0068
7						0·0001	0·0003	0·0006	0·0013
8								0·0001	0·0002

n = 50

r									
0	1·0000	1·0000	1·0000	1·0000	1·0000	1·0000	1·0000	1·0000	1·0000
1	0·3950	0·6358	0·7819	0·8701	0·9231	0·9547	0·9734	0·9845	0·9910
2	0·0894	0·2642	0·4447	0·5995	0·7206	0·8100	0·8735	0·9173	0·9468
3	0·0138	0·0784	0·1892	0·3233	0·4595	0·5838	0·6892	0·7740	0·8395
4	0·0016	0·0178	0·0628	0·1391	0·2396	0·3527	0·4673	0·5747	0·6697
5	0·0001	0·0032	0·0168	0·0490	0·1036	0·1794	0·2710	0·3710	0·4723
6		0·0005	0·0037	0·0144	0·0378	0·0776	0·1350	0·2081	0·2928
7		0·0001	0·0007	0·0036	0·0118	0·0289	0·0583	0·1019	0·1596
8			0·0001	0·0008	0·0032	0·0094	0·0220	0·0438	0·0768
9				0·0001	0·0008	0·0027	0·0073	0·0167	0·0328
10					0·0002	0·0007	0·0022	0·0056	0·0125
11						0·0002	0·0006	0·0017	0·0043
12							0·0001	0·0005	0·0013
13								0·0001	0·0004
14									0·0001

θ		0.10	0.15	0.20	0.25	0.30	0.35	0.40	0.45	0.50
$n=2$	$r=0$	1·0000	1·0000	1·0000	1·0000	1·0000	1·0000	1·0000	1·0000	1·0000
	1	0·1900	0·2775	0·3600	0·4375	0·5100	0·5775	0·6400	0·6975	0·7500
	2	0·0100	0·0225	0·0400	0·0625	0·0900	0·1225	0·1600	0·2025	0·2500
$n=5$	$r=0$	1·0000	1·0000	1·0000	1·0000	1·0000	1·0000	1·0000	1·0000	1·0000
	1	0·4095	0·5563	0·6723	0·7627	0·8319	0·8840	0·9222	0·9497	0·9688
	2	0·0815	0·1648	0·2627	0·3672	0·4718	0·5716	0·6630	0·7438	0·8125
	3	0·0086	0·0266	0·0579	0·1035	0·1631	0·2352	0·3174	0·4069	0·5000
	4	0·0005	0·0022	0·0067	0·0156	0·0308	0·0540	0·0870	0·1312	0·1875
	5		0·0001	0·0003	0·0010	0·0024	0·0053	0·0102	0·0185	0·0313
$n=10$	$r=0$	1·0000	1·0000	1·0000	1·0000	1·0000	1·0000	1·0000	1·0000	1·0000
	1	0·6513	0·8031	0·8926	0·9437	0·9718	0·9865	0·9940	0·9975	0·9990
	2	0·2639	0·4557	0·6242	0·7560	0·8507	0·9140	0·9536	0·9767	0·9893
	3	0·0702	0·1798	0·3222	0·4744	0·6172	0·7384	0·8327	0·9004	0·9453
	4	0·0128	0·0500	0·1209	0·2241	0·3504	0·4862	0·6177	0·7430	0·8281
	5	0·0016	0·0099	0·0328	0·0781	0·1503	0·2485	0·3669	0·4956	0·6230
	6	0·0001	0·0014	0·0064	0·0197	0·0473	0·0949	0·1662	0·2616	0·3770
	7		0·0001	0·0009	0·0035	0·0106	0·0260	0·0548	0·1020	0·1719
	8			0·0001	0·0004	0·0016	0·0048	0·0123	0·0274	0·0547
	9				0·0001	0·0005	0·0017	0·0045	0·0107
	10					0·0001	0·0003	0·0010
$n=20$	$r=0$	1·0000	1·0000	1·0000	1·0000	1·0000	1·0000	1·0000	1·0000	1·0000
	1	0·8784	0·9612	0·9885	0·9968	0·9992	0·9998	1·0000	1·0000	1·0000
	2	0·6083	0·8244	0·9308	0·9757	0·9924	0·9979	0·9995	0·9999	1·0000
	3	0·3231	0·5951	0·7939	0·9087	0·9645	0·9879	0·9964	0·9991	0·9998
	4	0·1330	0·3523	0·5886	0·7748	0·8929	0·9556	0·9840	0·9951	0·9987

5	0·9941	0·9811	0·9490	0·8818	0·7625	0·5852	0·3704	0·1702	0·0432
6	0·9793	0·9447	0·8744	0·7546	0·5836	0·3828	0·1958	0·0673	0·0113
7	0·9423	0·8701	0·7500	0·5834	0·3920	0·2142	0·0867	0·0219	0·0024
8	0·8684	0·7480	0·5841	0·3990	0·2277	0·1018	0·0321	0·0059	0·0004
9	0·7483	0·5857	0·4044	·02376	0·1133	0·0409	0·0100	0·0013	0·0001
10	0·5881	0·4086	0·2447	0·1218	0·0480	0·0139	0·0026	0·0002	
11	0·4119	0·2493	0·1275	0·0532	0·0171	0·0039	0·0006		
12	0·2517	0·1308	0·0565	0·0196	0·0051	0·0009	0·0001		
13	0·1316	0·0580	0·0210	0·0060	0·0013	0·0002			
14	0·0577	0·0214	0·0065	0·0015	0·0003				
15	0·0207	0·0064	0·0016	0·0003					
16	0·0059	0·0015	0·0003						
17	0·0013	0·0003							
18	0·0002								

$n = 50$

$r =$									
0	1·0000	1·0000	1·0000	1·0000	1·0000	1·0000	1·0000	1·0000	1·0000
1	1·0000	1·0000	1·0000	1·0000	1·0000	1·0000	1·0000	0·9997	0·9948
2	1·0000	1·0000	1·0000	1·0000	1·0000	1·0000	0·9998	0·9971	0·9662
3	1·0000	1·0000	1·0000	1·0000	1·0000	0·9999	0·9987	0·9858	0·8883
4	1·0000	1·0000	1·0000	1·0000	1·0000	0·9995	0·9943	0·9540	0·7497
5	1·0000	1·0000	1·0000	1·0000	0·9998	0·9979	0·9815	0·8879	0·5688
6	1·0000	1·0000	1·0000	0·9999	0·9993	0·9930	0·9520	0·7806	0·3839
7	1·0000	1·0000	1·0000	0·9998	0·9975	0·9806	0·8966	0·6387	0·2298
8	1·0000	1·0000	0·9999	0·9992	0·9927	0·9547	0·8096	0·4812	0·1221
9	1·0000	1·0000	0·9998	0·9975	0·9817	0·9084	0·6927	0·3319	0·0579
10	1·0000	0·9999	0·9992	0·9933	0·9598	0·8363	0·5563	0·2089	0·0245
11	1·0000	0·9998	0·9978	0·9840	0·9211	0·7378	0·4164	0·1199	0·0094
12	1·0000	0·9994	0·9943	0·9658	0·8610	0·6184	0·2893	0·0628	0·0032
13	0·9998	0·9982	0·9867	0·9339	0·7771	0·4890	0·1861	0·0301	0·0010
14	0·9995	0·9955	0·9720	0·8837	0·6721	0·3630	0·1106	0·0132	0·0003

θ	0·10	0·15	0·20	0·25	0·30	0·35	0·40	0·45	0·50
15	0·0001	0·0053	0·0607	0·2519	0·5532	0·8122	0·9460	0·9896	0·9987
16		0·0019	0·0308	0·1631	0·4308	0·7199	0·9045	0·9780	0·9967
17		0·0007	0·0144	0·0983	0·3161	0·6111	0·8439	0·9573	0·9923
18		0·0002	0·0063	0·0551	0·2178	0·4940	0·7631	0·9235	0·9836
19		0·0001	0·0025	0·0287	0·1406	0·3784	0·6644	0·8727	0·9675
20			0·0009	0·0139	0·0848	0·2736	0·5535	0·8026	0·9405
21			0·0003	0·0063	0·0478	0·1861	0·4390	0·7138	0·8987
22			0·0001	0·0026	0·0251	0·1187	0·3299	0·6100	0·8389
23				0·0010	0·0123	0·0710	0·2340	0·4981	0·7601
24				0·0004	0·0056	0·0396	0·1562	0·3866	0·6641
25				0·0001	0·0024	0·0207	0·0978	0·2840	0·5561
26					0·0009	0·0100	0·0573	0·1966	0·4439
27					0·0003	0·0045	0·0314	0·1279	0·3359
28					0·0001	0·0019	0·0160	0·0780	0·2399
29						0·0007	0·0076	0·0444	0·1611
30						0·0003	0·0034	0·0235	0·1013
31						0·0001	0·0014	0·0116	0·0595
32							0·0005	0·0053	0·0325
33							0·0002	0·0022	0·0164
34							0·0001	0·0009	0·0077
35								0·0003	0·0033
36								0·0001	0·0013
37									0·0005
38									0·0002

APPENDIX T2

Cumulative Poisson Probabilities [see II: § 5.4]

The tabulated quantity is

$$P\{S(\lambda) \geq r\} = \sum_{s=r}^{\infty} e^{-\lambda} \lambda^s / s!,$$

where $S(\lambda)$ denotes a random variable which has the Poisson (λ) distribution. Thus $P\{S(\lambda) \geq r\}$ is the probability of *r or more* occurrences. Related quantities are

$$P\{S(\lambda) \leq r\} = 1 - P\{S(\lambda) \geq r+1\}, \qquad r = 0, 1, \ldots$$

and

$$P\{S(\lambda) = r\} = P\{S(\lambda) \geq r\} - P\{S(\lambda) \geq r+1\}, \qquad r = 0, 1, \ldots.$$

Cumulative Poisson probabilities

(Reproduced by permission of Macmillan Publishers Ltd. from *Statistical Tables for Science, Engineering and Management* by J. Murdoch and J. A. Barnes.)

The table gives the probability that *r or more* random events are contained in an interval when the average number of such events per interval is λ, i.e.

$$\sum_{x=r}^{\infty} e^{-\lambda}\frac{\lambda^x}{x!}$$

Where there is no entry for a particular pair of values of *r* and λ, this indicates that the appropriate probability is less than 0·000 05. Similarly, except for the case *r* = 0 when the entry is exact, a tabulated value of 1·0000 represents a probability greater than 0·999 95.

λ	0·1	0·2	0·3	0·4	0·5	0·6	0·7	0·8	0·9	1·0
r = 0	1·0000	1·0000	1·0000	1·0000	1·0000	1·0000	1·0000	1·0000	1·0000	1·0000
1	0·0952	0·1813	0·2592	0·3297	0·3935	0·4512	0·5034	0·5507	0·5934	0·6321
2	0·0047	0·0175	0·0369	0·0616	0·0902	0·1219	0·1558	0·1912	0·2275	0·2642
3	0·0002	0·0011	0·0036	0·0079	0·0144	0·0231	0·0341	0·0474	0·0629	0·0803
4		0·0001	0·0003	0·0008	0·0018	0·0034	0·0058	0·0091	0·0135	0·0190
5				0·0001	0·0002	0·0004	0·0008	0·0014	0·0023	0·0037
6							0·0001	0·0002	0·0003	0·0006
7										0·0001

λ	1·1	1·2	1·3	1·4	1·5	1·6	1·7	1·8	1·9	2·0
r = 0	1·0000	1·0000	1·0000	1·0000	1·0000	1·0000	1·0000	1·0000	1·0000	1·0000
1	0·6671	0·6988	0·7275	0·7534	0·7769	0·7981	0·8173	0·8347	0·8504	0·8647
2	0·3010	0·3374	0·3732	0·4082	0·4422	0·4751	0·5068	0·5372	0·5663	0·5940
3	0·0996	0·1205	0·1429	0·1665	0·1912	0·2166	0·2428	0·2694	0·2963	0·3233
4	0·0257	0·0338	0·0431	0·0537	0·0656	0·0788	0·0932	0·1087	0·1253	0·1429
5	0·0054	0·0077	0·0107	0·0143	0·0186	0·0237	0·0296	0·0364	0·0441	0·0527
6	0·0010	0·0015	0·0022	0·0032	0·0045	0·0060	0·0080	0·0104	0·0132	0·0166
7	0·0001	0·0003	0·0004	0·0006	0·0009	0·0013	0·0019	0·0026	0·0034	0·0045
8			0·0001	0·0001	0·0002	0·0003	0·0004	0·0006	0·0008	0·0011
9							0·0001	0·0001	0·0002	0·0002

λ	2·1	2·2	2·3	2·4	2·5	2·6	2·7	2·8	2·9	3·0
r = 0	1·0000	1·0000	1·0000	1·0000	1·0000	1·0000	1·0000	1·0000	1·0000	1·0000
1	0·8775	0·8892	0·8997	0·9093	0·9179	0·9257	0·9328	0·9392	0·9450	0·9502
2	0·6204	0·6454	0·6691	0·6916	0·7127	0·7326	0·7513	0·7689	0·7854	0·8009
3	0·3504	0·3773	0·4040	0·4303	0·4562	0·4816	0·5064	0·5305	0·5540	0·5768
4	0·1614	0·1806	0·2007	0·2213	0·2424	0·2640	0·2859	0·3081	0·3304	0·3528
5	0·0621	0·0725	0·0838	0·0959	0·1088	0·1226	0·1371	0·1523	0·1682	0·1847
6	0·0204	0·0249	0·0300	0·0357	0·0420	0·0490	0·0567	0·0651	0·0742	0·0839
7	0·0059	0·0075	0·0094	0·0116	0·0142	0·0172	0·0206	0·0244	0·0287	0·0335
8	0·0015	0·0020	0·0026	0·0033	0·0042	0·0053	0·0066	0·0081	0·0099	0·0119
9	0·0003	0·0005	0·0006	0·0009	0·0011	0·0015	0·0019	0·0024	0·0031	0·0038
10	0·0001	0·0001	0·0001	0·0002	0·0003	0·0004	0·0005	0·0007	0·0009	0·0011
11					0·0001	0·0001	0·0001	0·0002	0·0002	0·0003
12									0·0001	0·0001

λ	3·1	3·2	3·3	3·4	3·5	3·6	3·7	3·8	3·9	4·0
r = 0	1·0000	1·0000	1·0000	1·0000	1·0000	1·0000	1·0000	1·0000	1·0000	1·0000
1	0·9550	0·9592	0·9631	0·9666	0·9698	0·9727	0·9753	0·9776	0·9798	0·9817
2	0·8153	0·8288	0·8414	0·8532	0·8641	0·8743	0·8838	0·8926	0·9008	0·9084
3	0·5988	0·6201	0·6406	0·6603	0·6792	0·6973	0·7146	0·7311	0·7469	0·7619
4	0·3752	0·3975	0·4197	0·4416	0·4634	0·4848	0·5058	0·5265	0·5468	0·5665
5	0·2018	0·2194	0·2374	0·2558	0·2746	0·2936	0·3128	0·3322	0·3516	0·3712
6	0·0943	0·1054	0·1171	0·1295	0·1424	0·1559	0·1699	0·1844	0·1994	0·2149
7	0·0388	0·0446	0·0510	0·0579	0·0653	0·0733	0·0818	0·0909	0·1005	0·1107
8	0·0142	0·0168	0·0198	0·0231	0·0267	0·0308	0·0352	0·0401	0·0454	0·0511
9	0·0047	0·0057	0·0069	0·0083	0·0099	0·0117	0·0137	0·0160	0·0185	0·0214
10	0·0014	0·0018	0·0022	0·0027	0·0033	0·0040	0·0048	0·0058	0·0069	0·0081
11	0·0004	0·0005	0·0006	0·0008	0·0010	0·0013	0·0016	0·0019	0·0023	0·0028
12	0·0001	0·0001	0·0002	0·0002	0·0003	0·0004	0·0005	0·0006	0·0007	0·0009
13				0·0001	0·0001	0·0001	0·0001	0·0002	0·0002	0·0003
14									0·0001	0·0001

λ	4·1	4·2	4·3	4·4	4·5	4·6	4·7	4·8	4·9	5·0
r=0	1·0000	1·0000	1·0000	1·0000	1·0000	1·0000	1·0000	1·0000	1·0000	1·0000
1	0·9834	0·9850	0·9864	0·9877	0·9889	0·9899	0·9909	0·9918	0·9926	0·9933
2	0·9155	0·9220	0·9281	0·9337	0·9389	0·9437	0·9482	0·9523	0·9561	0·9596
3	0·7762	0·7898	0·8026	0·8149	0·8264	0·8374	0·8477	0·8575	0·8667	0·8753
4	0·5858	0·6046	0·6228	0·6408	0·6577	0·6743	0·6903	0·7058	0·7207	0·7350
5	0·3907	0·4102	0·4296	04488	0·4679	0·4868	0·5054	0·5237	0·5418	0·5595
6	0·2307	0·2469	0·2633	0·2801	0·2971	0·3142	0·3316	0·3490	0·3665	0·3840
7	0·1214	0·1325	0·1442	0·1564	0·1689	0·1820	0·1954	0·2092	0·2233	0·2378
8	0·0573	0·0639	0·0710	0·0786	0·0866	0·0951	0·1040	0·1133	0·1231	0·1334
9	0·0245	0·0279	0·0317	0·0358	0·0403	0·0451	0·0503	0·0558	0·0618	0·0681
10	0·0095	0·0111	0·0129	0·0149	0·0171	0·0195	0·0222	0·0251	0·0283	0·0318
11	0·0034	0·0041	0·0048	0·0057	0·0067	0·0078	0·0090	0·0104	0·0120	0·0137
12	0·0011	0·0014	0·0017	0·0020	0·0024	0·0029	0·0034	0·0040	0·0047	0·0055
13	0·0003	0·0004	0·0005	0·0007	0·0008	0·0010	0·0012	0·0014	0·0017	0·0020
14	0·0001	0·0001	0·0002	0·0002	0·0003	0·0003	0·0004	0·0005	0·0006	0·0007
15				0·0001	0·0001	0·0001	0·0001	0·0001	0·0002	0·0002
16									0·0001	0·0001

λ	5·2	5·4	5·6	5·8	6·0	6·2	6·4	6·6	6·8	7·0
r=0	1·0000	1·0000	1·0000	1·0000	1·0000	1·0000	1·0000	1·0000	1·0000	1·0000
1	0·9945	0·9955	0·9963	0·9970	0·9975	0·9980	0·9983	0·9986	0·9989	0·9991
2	0·9658	0·9711	0·9756	0·9794	0·9826	0·9854	0·9877	0·9897	0·9913	0·9927
3	0·8912	0·9052	0·9176	0·9285	0·9380	0·9464	0·9537	0·9600	0·9656	0·9704
4	0·7619	0·7867	0·8094	0·8300	0·8488	0·8658	0·8811	0·8948	0·9072	0·9182
5	0·5939	0·6267	0·6579	0·6873	0·7149	0·7408	0·7649	0·7873	0·8080	0·8270
6	0·4191	0·4539	0·4881	0·5217	0·5543	0·5859	0·6163	0·6453	0·6730	0·6993
7	0·2676	0·2983	0·3297	0·3616	0·3937	0·4258	0·4577	0·4892	0·5201	0·5503
8	0·1551	0·1783	0·2030	0·2290	0·2560	0·2840	0·3127	0·3419	0·3715	0·4013
9	0·0819	0·0974	0·1143	0·1328	0·1528	0·1741	0·1967	0·2204	0·2452	0·2709
10	0·0397	0·0488	0·0591	0·0708	0·0839	0·0984	0·1142	0·1314	0·1498	0·1695
11	0·0177	0·0225	0·0282	0·0349	0·0426	0·0514	0·0614	0·0726	0·0849	0·0985
12	0·0073	0·0096	0·0125	0·0160	0·0201	0·0250	0·0307	0·0373	0·0448	0·0534
13	0·0028	0·0038	0·0051	0·0068	0·0088	0·0113	0·0143	0·0179	0·0221	0·0270
14	0·0010	0·0014	0·0020	0·0027	0·0036	0·0048	0·0063	0·0080	0·0102	0·0128
15	0·0003	0·0005	0·0007	0·0010	0·0014	0·0019	0·0026	0·0034	0·0044	0·0057
16	0·0001	0·0002	0·0002	0·0004	0·0005	0·0007	0·0010	0·0014	0·0018	0·0024
17		0·0001	0·0001	0·0001	0·0002	0·0003	0·0004	0·0005	0·0007	0·0010
18					0·0001	0·0001	0·0001	0·0002	0·0003	0·0004
19								0·0001	0·0001	0·0001

λ	7·2	7·4	7·6	7·8	8·0	8·2	8·4	8·6	8·8	9·0
$r=0$	1·0000	1·0000	1·0000	1·0000	1·0000	1·0000	1·0000	1·0000	1·0000	1·0000
1	0·9993	0·9994	0·9995	0·9996	0·9997	0·9997	0·9998	0·9998	0·9998	0·9999
2	0·9939	0·9949	0·9957	0·9964	0·9970	0·9975	0·9979	0·9982	0·9985	0·9988
3	0·9745	0·9781	0·9812	0·9839	0·9862	0·9882	0·9900	0·9914	0·9927	0·9938
4	0·9281	0·9368	0·9446	0·9515	0·9576	0·9630	0·9677	0·9719	0·9756	0·9788
5	0·8445	0·8605	0·8751	0·8883	0·9004	0·9113	0·9211	0·9299	0·9379	0·9450
6	0·7241	0·7474	0·7693	0·7897	0·8088	0·8264	0·8427	0·8578	0·8716	0·8843
7	0·5796	0·6080	0·6354	0·6616	0·6866	0·7104	0·7330	0·7543	0·7744	0·7932
8	0·4311	0·4607	0·4900	0·5188	0·5470	0·5746	0·6013	0·6272	0·6522	0·6761
9	0·2973	0·3243	0·3518	0·3796	0·4075	0·4353	0·4631	0·4906	0·5177	0·5443
10	0·1904	0·2123	0·2351	0·2589	0·2834	0·3085	0·3341	0·3600	0·3863	0·4126
11	0·1133	0·1293	0·1465	0·1648	0·1841	0·2045	0·2257	0·2478	0·2706	0·2940
12	0·0629	0·0735	0·0852	0·0980	0·1119	0·1269	0·1429	0·1600	0·1780	0·1970
13	0·0327	0·0391	0·0464	0·0546	0·0638	0·0739	0·0850	0·0971	0·1102	0·1242
14	0·0159	0·0195	0·0238	0·0286	0·0342	0·0405	0·0476	0·0555	0·0642	0·0739
15	0·0073	0·0092	0·0114	0·0141	0·0173	0·0209	0·0251	0·0299	0·0353	0·0415
16	0·0031	0·0041	0·0052	0·0066	0·0082	0·0102	0·0125	0·0152	0·0184	0·0220
17	0·0013	0·0017	0·0022	0·0029	0·0037	0·0047	0·0059	0·0074	0·0091	0·0111
18	0·0005	0·0007	0·0009	0·0012	0·0016	0·0021	0·0027	0·0034	0·0043	0·0053
19	0·0002	0·0003	0·0004	0·0005	0·0006	0·0009	0·0011	0·0015	0·0019	0·0024
20	0·0001	0·0001	0·0001	0·0002	0·0003	0·0003	00005	0·0006	0·0008	0·0011
21				0·0001	0·0001	0·0001	0·0002	0·0002	0·0003	0·0004
22							0·0001	0·0001	0·0001	0·0002
23										0·0001

λ	9·2	9·4	9·6	9·8	10·0	11·0	12·0	13·0	14·0	15·0
r = 0	1·0000	1·0000	1·0000	1·0000	1·0000	1·0000	1·0000	1·0000	1·0000	1·0000
1	0·9999	0·9999	0·9999	0·9999	1·0000	1·0000	1·0000	1·0000	1·0000	1·0000
2	0·9990	0·9991	0·9993	0·9994	0·9995	0·9998	0·9999	1·0000	1·0000	1·0000
3	0·9947	0·9955	0·9962	0·9967	0·9972	0·9988	0·9995	0·9998	0·9999	1·0000
4	0·9816	0·9840	0·9862	0·9880	0·9897	0·9951	0·9977	0·9990	0·9995	0·9998
5	0·9514	0·9571	0·9622	0·9667	0·9707	0·9849	0·9924	0·9963	0·9982	0·9991
6	0·8959	0·9065	0·9162	0·9250	0·9329	0·9625	0·9797	0·9893	0·9945	0·9972
7	0·8108	0·8273	0·8426	0·8567	0·8699	0·9214	0·9542	0·9741	0·9858	0·9924
8	0·6990	0·7208	0·7416	0·7612	0·7798	0·8568	0·9105	0·9460	0·9684	0·9820
9	0·5704	0·5958	0·6204	0·6442	0·6672	0·7680	0·8450	0·9002	0·9379	0·9626
10	0·4389	0·4651	0·4911	0·5168	0·5421	0·6595	0·7576	0·8342	0·8906	0·9301
11	0·3180	0·3424	0·3671	0·3920	0·4170	0·5401	0·6528	0·7483	0·8243	0·8815
12	0·2168	0·2374	0·2588	0·2807	0·3032	0·4207	0·5384	0·6468	0·7400	0·8152
13	0·1393	0·1552	0·1721	0·1899	0·2084	0·3113	0·4240	0·5369	0·6415	0·7324
14	0·0844	0·0958	0·1081	0·1214	0·1355	0·2187	0·3185	0·4270	0·5356	0·6368
15	0·0483	0·0559	0·0643	0·0735	0·0835	0·1460	0·2280	0·3249	0·4296	0·5343
16	0·0262	0·0309	0·0362	0·0421	0·0487	0·0926	0·1556	0·2364	0·3306	0·4319
17	0·0135	0·0162	0·0194	0·0230	0·0270	0·0559	0·1013	0·1645	0·2441	0·3359
18	0·0066	0·0081	0·0098	0·0119	0·0143	0·0322	0·0630	0·1095	0·1728	0·2511
19	0·0031	0·0038	0·0048	0·0059	0·0072	0·0177	0·0374	0·0698	0·1174	0·1805
20	0·0014	0·0017	0·0022	0·0028	0·0035	0·0093	0·0213	0·0427	0·0765	0·1248
21	0·0006	0·0008	0·0010	0·0012	0·0016	0·0047	0·0116	0·0250	0·0479	0·0830
22	0·0002	0·0003	0·0004	0·0005	0·0007	0·0023	0·0061	0·0141	0·0288	0·0531
23	0·0001	0·0001	0·0002	0·0002	0·0003	0·0010	0·0030	0·0076	0·0167	0·0327
24			0·0001	0·0001	0·0001	0·0005	0·0015	0·0040	0·0093	0·0195
25						0·0002	0·0007	0·0020	0·0050	0·0112
26						0·0001	0·0003	0·0010	0·0026	0·0062
27							0·0001	0·0005	0·0013	0·0033
28							0·0001	0·0002	0·0006	0·0017
29								0·0001	0·0003	0·0009
30									0·0001	0·0004
31									0·0001	0·0002
32										0·0001

APPENDIX T3

Cumulative Standard Normal Probabilities [see § 2.5 and II: § 11.4]

The tabulated quantity is

$$P(U \geq u) = \int_u^\infty (2\pi)^{-1/2} e^{-(1/2)y^2} \, dy$$

$$= 1 - \Phi(u),$$

where $\Phi(u)$ is the c.d.f. of the standard Normal variable U.
For example,

$$1 - \Phi(2 \cdot 32) = 0 \cdot 010170.$$

Entries in bold type take the same decimal prefix as entries in the following row. For example,

$$1 - \Phi(2 \cdot 36) = 0 \cdot 0091375.$$

The table gives values of $1 - \Phi(u)$ for $u \geq 0$. For negative values of u, use the relation

$$\Phi(u) = 1 - \Phi(-u).$$

For example,

$$\Phi(-2 \cdot 36) = 1 - \Phi(2 \cdot 36) = 0 \cdot 0091375.$$

If X is Normally distributed with expected value μ and standard deviation σ (variance σ^2),

$$P(X \leq x) = \Phi\left(\frac{x - \mu}{\sigma}\right), \qquad P(X \geq x) = 1 - \Phi\left(\frac{x - \mu}{\sigma}\right).$$

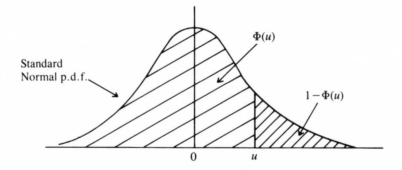

The Normal probability integral $1 - \Phi(x)$

(Reproduced by permission of Longman Group Ltd. from *Statistical Tables for Biological, Agricultural and Medical Research* by R. A. Fisher and F. Yates, 1974.)

x		0	1	2	3	4	5	6	7	8	9
0·0	0·0	50000	49601	49202	48803	48405	48006	47608	47210	46812	46414
0·1		46017	45620	45224	44828	44433	44038	43644	43251	42858	42465
0·2		42074	41683	41294	40905	40517	40129	39743	39358	38974	38591
0·3		38209	37828	37448	37070	36693	36317	35942	35569	35197	34827
0·4		34458	34090	33724	33360	32997	32636	32276	31918	31561	31207
0·5		30854	30503	30153	29806	29460	29116	28774	28434	28096	27760
0·6		27425	27093	26763	26435	26109	25785	25463	25143	24825	24510
0·7		24196	23885	23576	23270	22965	22663	22363	22065	21770	21476
0·8		21186	20897	20611	20327	20045	19766	19489	19215	18943	18673
0·9		18406	18141	17879	17619	17361	17106	16853	16602	16354	16109
1·0	0·0	15866	15625	15386	15151	14917	14686	14457	14231	14007	13786
1·1		13567	13350	13136	12924	12714	12507	12302	12100	11900	11702
1·2		11507	11314	11123	10935	10749	10565	10383	10204	10027	**98525**
1·3		96800	95098	93418	91759	90123	88508	86915	85343	83793	82264
1·4		80757	79270	77804	76359	74934	73529	72145	70781	69437	68112
1·5		66807	65522	64255	63008	61780	60571	59380	58208	57053	55917
1·6		54799	53699	52616	51551	50503	49471	48457	47460	46479	45514
1·7		44565	43633	42716	41815	40930	40059	39204	38364	37538	36727
1·8		35930	35148	34380	33625	32884	32157	31443	30742	30054	29379
1·9		28717	28067	27429	26803	26190	25588	24998	24419	23852	23295
2·0	0·0²	22750	22216	21692	21178	20675	20182	19699	19226	18763	18309
2·1		17864	17429	17003	16586	16177	15778	15386	15003	14629	14262
2·2		13903	13553	13209	12874	12545	12224	11911	11604	11304	11011
2·3		10724	10444	10170	**99031**	**96419**	**93867**	**91375**	**88940**	**86563**	**84242**
2·4		81975	79763	77603	75494	73436	71428	69469	67557	65691	63872

x		0	1	2	3	4	5	6	7	8	9
2·5		62097	60366	58677	57031	55426	53861	52336	50849	49400	47988
2·6		46612	45271	43965	42692	41453	40246	39070	37926	36811	35726
2·7		34670	33642	32641	31667	30720	29798	28901	28028	27179	26354
2·8		25551	24771	24012	23274	22557	21860	21182	20524	19884	19262
2·9		18658	18071	17502	16948	16411	15889	15382	14890	14412	13949
3·0	$0 \cdot 0^3$	13499	13062	12639	12228	11829	11442	11067	10703	10350	10008
3·1		96760	93544	90426	87403	84474	81635	78885	76219	73638	71136
3·2		68714	66367	64095	61895	59765	57703	55706	53774	51904	50094
3·3		48342	46648	45009	43423	41889	40406	38971	37584	36243	34946
3·4		33693	32481	31311	30179	29086	28029	27009	26023	25071	24151
3·5	$0 \cdot 0^4$	23263	22405	21577	20778	20006	19262	18543	17849	17180	16534
3·6		15911	15310	14730	14171	13632	13112	12611	12128	11662	11213
3·7		10780	10363	99611	95740	92010	88417	84957	81624	78414	75324
3·8		72348	69483	66726	64072	61517	59059	56694	54418	52228	50122
3·9		48096	46148	44274	42473	40741	39076	37475	35936	34458	33037
4·0	$0 \cdot 0^5$	31671	30359	29099	27888	26726	25609	24536	23507	22518	21569
4·1		20658	19783	18944	18138	17365	16624	15912	15230	14575	13948
4·2		13346	12769	12215	11685	11176	10689	10221	97736	93447	89337
4·3		85399	81627	78015	74555	71241	68069	65031	62123	59340	56675
4·4		54125	51685	49350	47117	44979	42935	40980	39110	37322	35612
4·5	$0 \cdot 0^6$	33977	32414	30920	29492	28127	26823	25577	24386	23249	22162
4·6		21125	20133	19187	18283	17420	16597	15810	15060	14344	13660
4·7		13008	12386	11792	11226	10686	10171	96796	92113	87648	83391
4·8		79333	75465	71779	68267	64920	61731	58693	55799	53043	50418
4·9		47918	45538	43272	41115	39061	37107	35247	33476	31792	30190

APPENDIX T4

Percentage Points of the Standard Normal Distribution

The table gives the 100α percentage points u_α of the standard Normal distribution, that is values u_α such that

$$P(U \geq u_\alpha) = \alpha.$$

In the notation of T3,

$$1 - \Phi(u_\alpha) = \alpha,$$

or

$$u_\alpha = \Phi^{-1}(1 - \alpha).$$

A central interval of content $1 - \alpha$ requires upper and lower tails of size $\frac{1}{2}\alpha$, and the interval is $(-u_{\alpha/2}, u_{\alpha/2})$

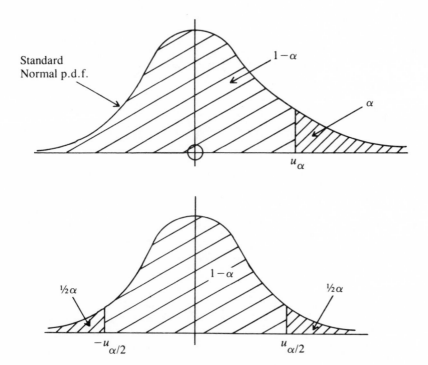

Percentage points of the Normal distribution

(Reproduced by permission of Macmillan Publishers Ltd. from Statistical Tables for Science, Engineering and Management by J. Murdoch and J. A. Barnes.)

α	u_α	α	u_α	α	u_α	α	u_α	α	u_α	α	u_α
0·50	0·0000	0·050	1·6449	0·030	1·8808	0·020	2·0537	0·010	2·3263	0·050	1·6449
0·45	0·1257	0·048	1·6646	0·029	1·8957	0·019	2·0749	0·009	2·3656	0·010	2·3263
0·40	0·2533	0·046	1·6849	0·028	1·9110	0·018	2·0969	0·008	2·4089	0·001	3·0902
0·35	0·3853	0·044	1·7060	0·027	1·9268	0·017	2·1201	0·007	2·4573	0·0001	3·7190
0·30	0·5244	0·042	1·7279	0·026	1·9431	0·016	2·1444	0·006	2·5121	0·00001	4·2649
0·25	0·6745	0·040	1·7507	0·025	1·9600	0·015	2·1701	0·005	2·5758	0·025	1·9600
0·20	0·8416	0·038	1·7744	0·024	1·9774	0·014	2·1973	0·004	2·6521	0·005	2·5758
0·15	1·0364	0·036	1·7991	0·023	1·9954	0·013	2·2262	0·003	2·7478	0·0005	3·2905
0·10	1·2816	0·034	1·8250	0·022	2·0141	0·012	2·2571	0·002	2·8782	0·00005	3·8906
0·05	1·6449	0·032	1·8522	0·021	2·0335	0·011	2·2904	0·001	3·0902	0·000005	4·4172

APPENDIX T5

Cumulative Probabilities of Student's Distribution (the '*t*-distribution')
[see § 2.5.5]

The tabulated quantity is

$$p_t(\nu) = P\{T(\nu) \le t\}$$

where the random variable $T(\nu)$ has Student's distribution on ν degrees of freedom.

The probability in the upper tail beyond t is $1 - p_t(\nu)$.

The table gives values of $p_t(\nu)$ for $\nu = 1(1)24, 30, 40, 60, 120, \infty$. For each of the tabulated values of ν, $p_t(\nu)$ is given for non-negative values of t. For negative values use

$$p_{-t}(\nu) = 1 - p_t(\nu).$$

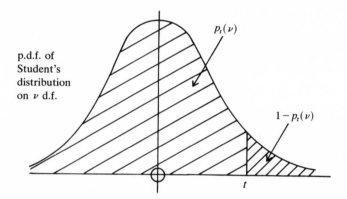

Probability integral, $p_t(\nu)$, of the t-distribution

t \ ν	1	2	3	4	5	6	7	8	9	10
0·0	0·50000	0·50000	0·50000	0·50000	0·50000	0·50000	0·50000	0·50000	0·50000	0·50000
0·1	0·53173	0·53527	0·53667	0·53742	0·53788	0·53820	0·53843	0·53860	0·53873	0·53884
0·2	0·56283	0·57002	0·57286	0·57438	0·57532	0·57596	0·57642	0·57676	0·57704	0·57726
0·3	0·59277	0·60376	0·60812	0·61044	0·61188	0·61285	0·61356	0·61409	0·61450	0·61484
0·4	0·62112	0·63608	0·64203	0·64520	0·64716	0·64850	0·64946	0·65019	0·65076	0·65122
0·5	0·64758	0·66667	0·67428	0·67834	0·68085	0·68256	0·68380	0·68473	0·68546	0·68605
0·6	0·67202	0·69529	0·70460	0·70958	0·71267	0·71477	0·71629	0·71745	0·71835	0·71907
0·7	0·69440	0·72181	0·73284	0·73875	0·74243	0·74493	0·74674	0·74811	0·74919	0·75006
0·8	0·71478	0·74618	0·75890	0·76574	0·76999	0·77289	0·77500	0·77659	0·77784	0·77885
0·9	0·73326	0·76845	0·78277	0·79050	0·79531	0·79860	0·80099	0·80280	0·80422	0·80536
1·0	0·75000	0·78868	0·80450	0·81305	0·81839	0·82204	0·82469	0·82670	0·82828	0·82955
1·1	0·76515	0·80698	0·82416	0·83346	0·83927	0·84325	0·84614	0·84834	0·85006	0·85145
1·2	0·77886	0·82349	0·84187	0·85182	0·85805	0·86232	0·86541	0·86777	0·86961	0·87110
1·3	0·79129	0·83838	0·85777	0·86827	0·87485	0·87935	0·88262	0·88510	0·88705	0·88862
1·4	0·80257	0·85177	0·87200	0·88295	0·88980	0·89448	0·89788	0·90046	0·90249	0·90412
1·5	0·81283	0·86380	0·88471	0·89600	0·90305	0·90786	0·91135	0·91400	0·91608	0·91775
1·6	0·82219	0·87464	0·89605	0·90758	0·91475	0·91964	0·92318	0·92587	0·92797	0·92966
1·7	0·83075	0·88439	0·90615	0·91782	0·92506	0·92998	0·93354	0·93622	0·93833	0·94002
1·8	0·83859	0·89317	0·91516	0·92688	0·93412	0·93902	0·94256	0·94522	0·94731	0·94897
1·9	0·84579	0·90109	0·92318	0·93488	0·94207	0·94691	0·95040	0·95302	0·95506	0·95669
2·0	0·85242	0·90825	0·93034	0·94194	0·94903	0·95379	0·95719	0·95974	0·96172	0·96331
2·1	0·85854	0·91473	0·93672	0·94817	0·95512	0·95976	0·96306	0·96553	0·96744	0·96896
2·2	0·86420	0·92060	0·94241	0·95367	0·96045	0·96495	0·96813	0·97050	0·97233	0·97378
2·3	0·86945	0·92593	0·94751	0·95853	0·96511	0·96945	0·97250	0·97476	0·97650	0·97787
2·4	0·87433	0·93077	0·95206	0·96282	0·96919	0·97335	0·97627	0·97841	0·98005	0·98134
2·5	0·87888	0·93519	0·95615	0·96662	0·97275	0·97674	0·97950	0·98153	0·98307	0·98428
2·6	0·88313	0·93923	0·95981	0·96998	0·97587	0·97967	0·98229	0·98419	0·98563	0·98675
2·7	0·88709	0·94292	0·96311	0·97295	0·97861	0·98221	0·98468	0·98646	0·98780	0·98884

2·9	0·89430	0·94941	0·96875	0·97794	0·98310	0·98633	0·98851	0·99005	0·99120	0·99208
3·0	0·89758	0·95227	0·97116	0·98003	0·98495	0·98800	0·99003	0·99146	0·99252	0·99333
3·1	0·90067	0·95490	0·97335	0·98189	0·98657	0·98944	0·99134	0·99267	0·99364	0·99437
3·2	0·90359	0·95733	0·97533	0·98355	0·98800	0·99070	0·99247	0·99369	0·99459	0·99525
3·3	0·90634	0·95958	0·97713	0·98503	0·98926	0·99180	0·99344	0·99457	0·99539	0·99599
3·4	0·90895	0·96166	0·97877	0·98636	0·99037	0·99275	0·99428	0·99532	0·99606	0·99661
3·5	0·91141	0·96358	0·98026	0·98755	0·99136	0·99359	0·99500	0·99596	0·99664	0·99714
3·6	0·91376	0·96538	0·98162	0·98862	0·99223	0·99432	0·99563	0·99651	0·99713	0·99758
3·7	0·91598	0·96705	0·98286	0·98958	0·99300	0·99496	0·99617	0·99698	0·99754	0·99795
3·8	0·91809	0·96860	0·98400	0·99045	0·99369	0·99552	0·99664	0·99738	0·99789	0·99826
3·9	0·92010	0·97005	0·98504	0·99123	0·99430	0·99601	0·99705	0·99773	0·99819	0·99852
4·0	0·92202	0·97141	0·98600	0·99193	0·99484	0·99644	0·99741	0·99803	0·99845	0·99874
4·2	0·92560	0·97386	0·98768	0·99315	0·99575	0·99716	0·99798	0·99850	0·99885	0·99909
4·4	0·92887	0·97602	0·98912	0·99415	0·99649	0·99772	0·99842	0·99886	0·99914	0·99933
4·6	0·93186	0·97792	0·99034	0·99498	0·99708	0·99815	0·99876	0·99912	0·99936	0·99951
4·8	0·93462	0·97962	0·99140	0·99568	0·99756	0·99850	0·99902	0·99932	0·99951	0·99964
5·0	0·93717	0·98113	0·99230	0·99625	0·99795	0·99877	0·99922	0·99947	0·99963	0·99973
5·2	0·93952	0·98248	0·99309	0·99674	0·99827	0·99899	0·99937	0·99959	0·99972	0·99980
5·4	0·94171	0·98369	0·99378	0·99715	0·99853	0·99917	0·99950	0·99968	0·99978	0·99985
5·6	0·94375	0·98478	0·99437	0·99750	0·99875	0·99931	0·99959	0·99975	0·99983	0·99989
5·8	0·94565	0·98577	0·99490	0·99780	0·99893	0·99942	0·99967	0·99980	0·99987	0·99991
6·0	0·94743	0·98666	0·99536	0·99806	0·99908	0·99952	0·99973	0·99984	0·99990	0·99993
6·2	0·94910	0·98748	0·99577	0·99828	0·99920	0·99959	0·99978	0·99987	0·99992	0·99995
6·4	0·95066	0·98822	0·99614	0·99847	0·99931	0·99966	0·99982	0·99990	0·99994	0·99996
6·6	0·95214	0·98890	0·99646	0·99863	0·99940	0·99971	0·99985	0·99992	0·99995	0·99997
6·8	0·95352	0·98953	0·99675	0·99878	0·99948	0·99975	0·99987	0·99993	0·99996	0·99998
7·0	0·95483	0·99010	0·99701	0·99890	0·99954	0·99979	0·99990	0·99994	0·99997	0·99998
7·2	0·95607	0·99063	0·99724	0·99901	0·99960	0·99982	0·99991	0·99995	0·99997	0·99999
7·4	0·95724	0·99111	0·99745	0·99911	0·99964	0·99984	0·99993	0·99996	0·99998	0·99999
7·6	0·95836	0·99156	0·99764	0·99920	0·99969	0·99986	0·99994	0·99997	0·99998	0·99999
7·8	0·95941	0·99198	0·99781	0·99927	0·99972	0·99988	0·99995	0·99997	0·99999	0·99999
8·0	0·96042	0·99237	0·99796	0·99934	0·99975	0·99990	0·99996	0·99998	0·99999	0·99999

Probability integral of the t-distribution (continued)

t \ ν	11	12	13	14	15	16	17	18	19	20
0·0	0·50000	0·50000	0·50000	0·50000	0·50000	0·50000	0·50000	0·50000	0·50000	0·50000
0·1	0·53893	0·53900	0·53907	0·53912	0·53917	0·53921	0·53924	0·53928	0·53930	0·53933
0·2	0·57744	0·57759	0·57771	0·57782	0·57792	0·57800	0·57807	0·57814	0·57820	0·57825
0·3	0·61511	0·61534	0·61554	0·61571	0·61585	0·61598	0·61609	0·61619	0·61628	0·61636
0·4	0·65159	0·65191	0·65217	0·65240	0·65260	0·65278	0·65293	0·65307	0·65319	0·65330
0·5	0·68654	0·68694	0·68728	0·68758	0·68783	0·68806	0·68826	0·68843	0·68859	0·68873
0·6	0·71967	0·72017	0·72059	0·72095	0·72127	0·72155	0·72179	0·72201	0·72220	0·72238
0·7	0·75077	0·75136	0·75187	0·75230	0·75268	0·75301	0·75330	0·75356	0·75380	0·75400
0·8	0·77968	0·78037	0·78096	0·78146	0·78190	0·78229	0·78263	0·78293	0·78320	0·78344
0·9	0·80630	0·80709	0·80776	0·80833	0·80883	0·80927	0·80965	0·81000	0·81031	0·81058
1·0	0·83060	0·83148	0·83222	0·83286	0·83341	0·83390	0·83433	0·83472	0·83506	0·83537
1·1	0·85259	0·85355	0·85436	0·85506	0·85566	0·85620	0·85667	0·85709	0·85746	0·85780
1·2	0·87233	0·87335	0·87422	0·87497	0·87562	0·87620	0·87670	0·87715	0·87756	0·87792
1·3	0·88991	0·89099	0·89191	0·89270	0·89339	0·89399	0·89452	0·89500	0·89542	0·89581
1·4	0·90546	0·90658	0·90754	0·90836	0·90907	0·90970	0·91025	0·91074	0·91118	0·91158
1·5	0·91912	0·92027	0·92125	0·92209	0·92282	0·92346	0·92402	0·92452	0·92498	0·92538
1·6	0·93105	0·93221	0·93320	0·93404	0·93478	0·93542	0·93599	0·93650	0·93695	0·93736
1·7	0·94140	0·94256	0·94354	0·94439	0·94512	0·94576	0·94632	0·94683	0·94728	0·94768
1·8	0·95034	0·95148	0·95245	0·95328	0·95400	0·95463	0·95518	0·95568	0·95612	0·95652
1·9	0·95802	0·95914	0·96008	0·96089	0·96158	0·96220	0·96273	0·96321	0·96364	0·96403
2·0	0·96460	0·96567	0·96658	0·96736	0·96803	0·96861	0·96913	0·96959	0·97000	0·97037
2·1	0·97020	0·97123	0·97209	0·97283	0·97347	0·97403	0·97452	0·97495	0·97534	0·97569
2·2	0·97496	0·97593	0·97675	0·97745	0·97805	0·97858	0·97904	0·97945	0·97981	0·98014
2·3	0·97898	0·97990	0·98067	0·98132	0·98189	0·98238	0·98281	0·98319	0·98352	0·98383
2·4	0·98238	0·98324	0·98396	0·98457	0·98509	0·98554	0·98594	0·98629	0·98660	0·98688
2·5	0·98525	0·98604	0·98671	0·98727	0·98775	0·98816	0·98853	0·98885	0·98913	0·98938
2·6	0·98765	0·98839	0·98900	0·98951	0·98995	0·99033	0·99066	0·99095	0·99121	0·99144

x										
2·7	0·98967	0·99035	0·99090	0·99137	0·99177	0·99211	0·99241	0·99267	0·99290	0·99311
2·8	0·99136	0·99198	0·99249	0·99291	0·99327	0·99358	0·99385	0·99408	0·99429	0·99447
2·9	0·99278	0·99334	0·99380	0·99418	0·99450	0·99478	0·99502	0·99523	0·99541	0·99557
3·0	0·99396	0·99447	0·99488	0·99522	0·99551	0·99576	0·99597	0·99616	0·99632	0·99646
3·1	0·99495	0·99541	0·99578	0·99608	0·99634	0·99656	0·99675	0·99691	0·99705	0·99718
3·2	0·99577	0·99618	0·99652	0·99679	0·99702	0·99721	0·99738	0·99752	0·99764	0·99775
3·3	0·99646	0·99683	0·99713	0·99737	0·99757	0·99774	0·99789	0·99801	0·99812	0·99821
3·4	0·99703	0·99737	0·99763	0·99784	0·99802	0·99817	0·99830	0·99840	0·99850	0·99858
3·5	0·99751	0·99781	0·99804	0·99823	0·99839	0·99852	0·99863	0·99872	0·99880	0·99887
3·6	0·99791	0·99818	0·99838	0·99855	0·99869	0·99880	0·99890	0·99898	0·99905	0·99911
3·7	0·99825	0·99848	0·99867	0·99881	0·99893	0·99903	0·99911	0·99918	0·99924	0·99929
3·8	0·99853	0·99874	0·99890	0·99902	0·99913	0·99921	0·99928	0·99934	0·99939	0·99944
3·9	0·99876	0·99895	0·99909	0·99920	0·99929	0·99936	0·99942	0·99948	0·99952	0·99956
4·0	0·99896	0·99912	0·99924	0·99934	0·99942	0·99948	0·99954	0·99958	0·99962	0·99965
4·2	0·99926	0·99938	0·99948	0·99955	0·99961	0·99966	0·99970	0·99973	0·99976	0·99978
4·4	0·99947	0·99957	0·99964	0·99970	0·99974	0·99978	0·99980	0·99983	0·99985	0·99986
4·6	0·99962	0·99969	0·99975	0·99979	0·99983	0·99985	0·99987	0·99989	0·99990	0·99991
4·8	0·99972	0·99978	0·99983	0·99986	0·99988	0·99990	0·99992	0·99993	0·99994	0·99995
5·0	0·99980	0·99985	0·99988	0·99990	0·99992	0·99993	0·99995	0·99995	0·99996	0·99997
5·2	0·99985	0·99989	0·99992	0·99993	0·99995	0·99996	0·99996	0·99997	0·99997	0·99998
5·4	0·99989	0·99992	0·99994	0·99995	0·99996	0·99997	0·99998	0·99998	0·99998	0·99999
5·6	0·99992	0·99994	0·99996	0·99997	0·99997	0·99998	0·99998	0·99999	0·99999	0·99999
5·8	0·99994	0·99996	0·99997	0·99998	0·99998	0·99999	0·99999	0·99999	0·99999	0·99999
6·0	0·99995	0·99997	0·99998	0·99998	0·99999	0·99999	0·99999	0·99999		
6·2	0·99997	0·99998	0·99998	0·99999	0·99999	0·99999				
6·4	0·99997	0·99998	0·99999	0·99999	0·99999					
6·6	0·99998	0·99999	0·99999	0·99999						
6·8	0·99998	0·99999	0·99999							
7·0	0·99999									

Probability integral of the t-distribution (continued)

t	ν 20	21	22	23	24	30	40	60	120	∞
0·00	0·50000	0·50000	0·50000	0·50000	0·50000	0·50000	0·50000	0·50000	0·50000	0·50000
0·05	0·51969	0·51970	0·51971	0·51972	0·51973	0·51977	0·51981	0·51986	0·51990	0·51994
0·10	0·53933	0·53935	0·53938	0·53939	0·53941	0·53950	0·53958	0·53966	0·53974	0·53983
0·15	0·55887	0·55890	0·55893	0·55896	0·55899	0·55912	0·55924	0·55937	0·55949	0·55962
0·20	0·57825	0·57830	0·57834	0·57838	0·57842	0·57858	0·57875	0·57892	0·57909	0·57926
0·25	0·59743	0·59749	0·59755	0·59760	0·59764	0·59785	0·59807	0·59828	0·59849	0·59871
0·30	0·61636	0·61644	0·61650	0·61656	0·61662	0·61688	0·61713	0·61739	0·61765	0·61791
0·35	0·63500	0·63509	0·63517	0·63524	0·63530	0·63561	0·63591	0·63622	0·63652	0·63683
0·40	0·65330	0·65340	0·65349	0·65358	0·65365	0·65400	0·65436	0·65471	0·65507	0·65542
0·45	0·67122	0·67134	0·67144	0·67154	0·67163	0·67203	0·67243	0·67283	0·67324	0·67364
0·50	0·68873	0·68886	0·68898	0·68909	0·68919	0·68964	0·69009	0·69055	0·69100	0·69146
0·55	0·70579	0·70594	0·70607	0·70619	0·70630	0·70680	0·70731	0·70782	0·70833	0·70884
0·60	0·72238	0·72254	0·72268	0·72281	0·72294	0·72349	0·72405	0·72462	0·72518	0·72575
0·65	0·73846	0·73863	0·73879	0·73893	0·73907	0·73968	0·74030	0·74091	0·74153	0·74215
0·70	0·75400	0·75419	0·75437	0·75453	0·75467	0·75534	0·75601	0·75668	0·75736	0·75804
0·75	0·76901	0·76921	0·76940	0·76957	0·76973	0·77045	0·77118	0·77191	0·77264	0·77337
0·80	0·78344	0·78367	0·78387	0·78405	0·78422	0·78500	0·78578	0·78657	0·78735	0·78814
0·85	0·79731	0·79754	0·79776	0·79796	0·79814	0·79897	0·79981	0·80065	0·80149	0·80234
0·90	0·81058	0·81084	0·81107	0·81128	0·81147	0·81236	0·81325	0·81414	0·81504	0·81594
0·95	0·82327	0·82354	0·82378	0·82401	0·82421	0·82515	0·82609	0·82704	0·82799	0·82894
1·00	0·83537	0·83565	0·83591	0·83614	0·83636	0·83735	0·83834	0·83934	0·84034	0·84134
1·05	0·84688	0·84717	0·84744	0·84769	0·84791	0·84895	0·84999	0·85104	0·85209	0·85314
1·10	0·85780	0·85811	0·85839	0·85864	0·85888	0·85996	0·86105	0·86214	0·86323	0·86433
1·15	0·86814	0·86846	0·86875	0·86902	0·86926	0·87039	0·87151	0·87265	0·87378	0·87493
1·20	0·87792	0·87825	0·87855	0·87882	0·87907	0·88023	0·88140	0·88257	0·88375	0·88493

1·25	0·89435	0·89313	0·89192	0·89072	0·88952	0·88832	0·88807	0·88778	0·88747	0·88714
1·30	0·90320	0·90195	0·90071	0·89948	0·89825	0·89703	0·89676	0·89647	0·89616	0·89581
1·35	0·91149	0·91022	0·90896	0·90770	0·90644	0·90519	0·90492	0·90463	0·90431	0·90395
1·40	0·91924	0·91795	0·91667	0·91539	0·91411	0·91285	0·91257	0·91227	0·91194	0·91158
1·45	0·92647	0·92517	0·92387	0·92257	0·92128	0·92000	0·91972	0·91942	0·91908	0·91872
1·50	0·93319	0·93188	0·93057	0·92927	0·92797	0·92667	0·92639	0·92608	0·92575	0·92538
1·55	0·93943	0·93811	0·93680	0·93549	0·93419	0·93289	0·93260	0·93230	0·93196	0·93159
1·60	0·94520	0·94389	0·94257	0·94127	0·93996	0·93866	0·93838	0·93807	0·93773	0·93736
1·65	0·95053	0·94922	0·94792	0·94661	0·94531	0·94401	0·94373	0·94342	0·94309	0·94272
1·70	0·95543	0·95414	0·95284	0·95155	0·95026	0·94897	0·94869	0·94839	0·94805	0·94768
1·75	0·95994	0·95866	0·95738	0·95611	0·95483	0·95355	0·95327	0·95297	0·95264	0·95228
1·80	0·96407	0·96281	0·96156	0·96030	0·95904	0·95778	0·95750	0·95720	0·95688	0·95652
1·85	0·96784	0·96661	0·96538	0·96414	0·96291	0·96167	0·96140	0·96110	0·96078	0·96043
1·90	0·97128	0·97008	0·96888	0·96767	0·96646	0·96524	0·96498	0·96469	0·96437	0·96403
1·95	0·97441	0·97325	0·97207	0·97089	0·96971	0·96852	0·96827	0·96798	0·96767	0·96733
2·0	0·97725	0·97612	0·97498	0·97384	0·97269	0·97153	0·97128	0·97100	0·97070	0·97037
2·1	0·98214	0·98109	0·98003	0·97896	0·97788	0·97679	0·97655	0·97629	0·97601	0·97569
2·2	0·98610	0·98514	0·98416	0·98318	0·98218	0·98116	0·98094	0·98070	0·98043	0·98014
2·3	0·98928	0·98841	0·98753	0·98663	0·98571	0·98478	0·98457	0·98435	0·98410	0·98383
2·4	0·99180	0·99103	0·99024	0·98943	0·98860	0·98774	0·98756	0·98735	0·98712	0·98688
2·5	0·99379	0·99312	0·99241	0·99169	0·99094	0·99017	0·99000	0·98982	0·98961	0·98938
2·6	0·99534	0·99475	0·99414	0·99350	0·99284	0·99215	0·99200	0·99183	0·99164	0·99144
2·7	0·99653	0·99603	0·99550	0·99494	0·99436	0·99375	0·99361	0·99346	0·99329	0·99311
2·8	0·99744	0·99702	0·99657	0·99608	0·99557	0·99504	0·99492	0·99478	0·99463	0·99447
2·9	0·99813	0·99778	0·99740	0·99698	0·99654	0·99607	0·99596	0·99585	0·99572	0·99557
3·0	0·99865	0·99836	0·99804	0·99768	0·99730	0·99690	0·99681	0·99670	0·99659	0·99646
3·1	0·99903	0·99879	0·99853	0·99823	0·99791	0·99756	0·99748	0·99739	0·99729	0·99718
3·2	0·99931	0·99912	0·99890	0·99865	0·99838	0·99808	0·99801	0·99793	0·99785	0·99775
3·3	0·99952	0·99936	0·99918	0·99898	0·99875	0·99849	0·99844	0·99837	0·99829	0·99821
3·4	0·99966	0·99954	0·99940	0·99923	0·99904	0·99882	0·99877	0·99871	0·99865	0·99858

t \ ν	20	21	22	23	24	30	40	60	120	∞
3·5	0·99887	0·99893	0·99899	0·99904	0·99908	0·99926	0·99942	0·99956	0·99967	0·99977
3·6	0·99911	0·99916	0·99920	0·99925	0·99928	0·99943	0·99957	0·99968	0·99977	0·99984
3·7	0·99929	0·99933	0·99937	0·99941	0·99944	0·99957	0·99967	0·99976	0·99984	0·99989
3·8	0·99944	0·99948	0·99951	0·99954	0·99956	0·99967	0·99976	0·99983	0·99989	0·99993
3·9	0·99956	0·99959	0·99961	0·99964	0·99966	0·99975	0·99982	0·99988	0·99992	0·99995
4·0	0·99965	0·99967	0·99970	0·99972	0·99974	0·99981	0·99987	0·99991	0·99995	0·99997
5·0	0·99997	0·99997	0·99998	0·99998	0·99998	0·99999	0·99999			

Upper percentage points of t

$1 - p_t(\nu)$	$\nu = 1$	2	3	4	5	6	7	8	9	10
10^{-3}	318·3	22·33	10·21	7·17	5·89	5·21	4·79	4·50	4·30	4·14
10^{-4}	3183	70·7	22·20	13·03	9·68	8·02	7·06	6·44	6·01	5·69
10^{-5}	31831	224	47·91	23·33	15·54	12·03	10·11	8·90	8·10	7·53
5×10^{-6}	63652	316	60·40	27·82	17·89	13·55	11·22	9·79	8·83	8·15

APPENDIX T6

Percentage Points of the χ^2-distribution [see § 2.5.4(a)]

The table gives the 100α percentage points $\chi^2(\alpha; \nu)$ of the χ^2-distribution on ν degrees of freedom, that is, values $\chi^2(\alpha; \nu)$ such that

$$P\{X_\nu^2 \geq \chi^2(\alpha; \nu)\} = \alpha,$$

where the random variable X_ν^2 has the chi-squared distribution on ν degrees of freedom.

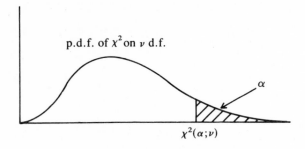

p.d.f. of χ^2 on ν d.f.

α

$\chi^2(\alpha; \nu)$

Percentage points of the χ^2 distribution

(Reproduced by permission of Longman Group Ltd. from *Statistical Tables for Biological, Agricultural and Medical Research* by R. A. Fisher and F. Yates, 1974.)

ν	0·99	0·98	0·95	0·90	0·80	0·70	0·50	0·30	0·20	0·10	0·05	0·02	0·01	0·001
							Probability							
1	0·0³157	0·0³628	0·00393	0·0158	0·0642	0·148	0·455	1·074	1·642	2·706	3·841	5·412	6·635	10·827
2	0·0201	0·404	0·103	0·211	0·446	0·713	1·386	2·408	3·219	4·605	5·991	7·824	9·210	13·815
3	0·115	0·185	0·352	0·584	1·005	1·424	2·366	3·665	4·642	6·251	7·815	9·837	11·345	16·266
4	0·297	0·429	0·711	1·064	1·649	2·195	3·357	4·878	5·989	7·779	9·488	11·688	13·277	18·467
5	0·544	0·752	1·145	1·610	2·343	3·000	4·351	6·064	7·289	9·236	11·070	13·388	15·086	20·515
6	0·872	1·134	1·635	2·204	3·070	3·828	5·348	7·231	8·558	10·645	12·592	15·033	16·812	22·457
7	1·239	1·564	2·167	2·833	3·822	4·671	6·346	8·383	9·803	12·017	14·067	16·622	18·475	24·322
8	1·646	2·032	2·733	3·490	4·594	5·527	7·344	9·524	11·030	13·362	15·507	18·168	20·090	26·125
9	2·088	2·532	3·325	4·168	5·380	6·393	8·343	10·656	12·242	14·684	16·919	19·679	21·666	27·877
10	2·558	3·059	3·940	4·865	6·179	7·267	9·342	11·781	13·442	15·987	18·307	21·161	23·209	29·588
11	3·053	3·609	4·575	5·578	6·989	8·148	10·341	12·899	14·631	17·275	19·675	22·618	24·725	31·264
12	3·571	4·178	5·226	6·304	7·807	9·034	11·340	14·011	15·812	18·549	21·026	24·054	26·217	32·909
13	4·107	4·765	5·892	7·042	8·634	9·926	12·340	15·119	16·985	19·812	22·362	25·472	27·688	34·528
14	4·660	5·368	6·571	7·790	9·467	10·821	13·339	16·222	18·151	21·064	23·685	26·873	29·141	36·123
15	5·229	5·985	7·261	8·547	10·307	11·721	14·339	17·322	19·311	22·307	24·996	28·259	30·578	37·697
16	5·812	6·614	7·962	9·312	11·152	12·624	15·338	18·418	20·465	23·542	26·296	29·633	32·000	39·252
17	6·408	7·255	8·672	10·085	12·002	13·531	16·338	19·511	21·615	24·769	27·587	30·995	33·409	40·790
18	7·015	7·906	9·390	10·865	12·857	14·440	17·338	20·601	22·760	25·989	28·869	32·346	34·805	42·312
19	7·633	8·567	10·117	11·651	13·716	15·352	18·338	21·689	23·900	27·204	30·144	33·687	36·191	43·820
20	8·260	9·237	10·851	12·443	14·578	16·266	19·337	22·775	25·038	28·412	31·410	35·020	37·566	45·315
21	8·897	9·915	11·591	13·240	15·445	17·182	20·337	23·858	26·171	29·615	32·671	36·343	38·932	46·797
22	9·542	10·600	12·338	14·041	16·314	18·101	21·337	24·939	27·301	30·813	33·924	37·659	40·289	48·268
23	10·196	11·293	13·091	14·848	17·187	19·021	22·337	26·018	28·429	32·007	35·172	38·968	41·638	49·728
24	10·856	11·992	13·848	15·659	18·062	19·943	23·337	27·096	29·553	33·196	36·415	40·270	42·980	51·179
25	11·524	12·697	14·611	16·473	18·940	20·867	24·337	28·172	30·675	34·382	37·652	41·566	44·314	52·620

ν														
26	12·198	13·409	15·379	17·292	19·820	21·792	25·336	29·246	31·795	35·563	38·885	42·856	45·642	54·052
27	12·879	14·125	16·151	18·114	20·703	22·719	26·336	30·319	32·912	36·741	40·113	44·140	46·963	55·476
28	13·565	14·847	16·928	18·939	21·588	23·647	27·336	31·391	34·027	37·916	41·337	45·419	48·278	56·893
29	14·256	15·574	17·708	19·768	22·475	24·577	28·336	32·461	35·139	39·087	42·557	46·693	49·588	58·302
30	14·953	16·306	18·493	20·599	23·364	25·508	29·336	33·530	36·250	40·256	43·773	47·962	50·892	59·703
32	16·362	17·783	20·072	22·271	25·148	27·373	31·336	35·665	38·466	42·585	46·194	50·487	53·486	62·487
34	17·789	19·275	21·664	23·952	26·938	29·242	33·336	37·795	40·676	44·903	48·602	52·995	56·061	65·247
36	19·233	20·783	23·269	25·643	28·735	31·115	35·336	39·922	42·879	47·212	50·999	55·489	58·619	67·985
38	20·691	22·304	24·884	27·343	30·537	32·992	37·335	42·045	45·076	49·513	53·384	57·969	61·162	70·703
40	22·164	23·838	26·509	29·051	32·345	34·872	39·335	44·165	47·269	51·805	55·759	60·436	63·691	73·402
42	23·650	25·383	28·144	30·765	34·157	36·755	41·335	46·282	49·456	54·090	58·124	62·892	66·206	76·084
44	25·148	26·939	29·787	32·487	35·974	38·641	43·335	48·396	51·639	56·369	60·481	65·337	68·710	78·750
46	26·657	28·504	31·439	34·215	37·795	40·529	45·335	50·507	53·818	58·641	62·830	67·771	71·201	81·400
48	28·177	30·080	33·098	35·949	39·621	42·420	47·335	52·616	55·993	60·907	65·171	70·197	73·683	84·037
50	29·707	31·664	34·764	37·689	41·449	44·313	49·335	54·723	58·164	63·167	67·505	72·613	76·154	86·661
52	31·246	33·256	36·437	39·433	43·281	46·209	51·335	56·827	60·332	65·422	69·832	75·021	78·616	89·272
54	32·793	34·856	38·116	41·183	45·117	48·106	53·335	58·930	62·496	67·673	72·153	77·422	81·069	91·872
56	34·350	36·464	39·801	42·937	46·955	50·005	55·335	61·031	64·658	69·919	74·468	79·815	83·513	94·461
58	35·913	38·078	41·492	44·696	48·797	51·906	57·335	63·129	66·816	72·160	76·778	82·201	85·950	97·039
60	37·485	39·699	43·188	46·459	50·641	53·809	59·335	65·227	68·972	74·397	79·082	84·580	88·379	99·607
62	39·063	41·327	44·889	48·226	52·487	55·714	61·335	67·322	71·125	76·630	81·381	86·953	90·802	102·166
64	40·649	42·960	46·595	49·996	54·336	57·620	63·335	69·416	73·276	78·860	83·675	89·320	93·217	104·716
66	42·240	44·599	48·305	51·770	56·188	59·527	65·335	71·508	75·424	81·085	85·965	91·681	95·626	107·258
68	43·838	46·244	50·020	53·548	58·042	61·436	67·335	73·600	77·571	83·308	88·250	94·037	98·028	109·791
70	45·442	47·893	51·739	55·329	59·898	63·346	69·334	75·689	79·715	85·527	90·531	96·388	100·425	112·317

For odd values of ν between 30 and 70 the mean of the tabular values for ν − 1 and ν + 1 may be taken. For larger values of ν, the expression $\sqrt{2\chi^2} - \sqrt{2\nu - 1}$ may be used as a normal deviate with unit variance.

Percentage points of the F distribution [see § 2.5.6]

The tables give the 100α percentage point $x_\alpha(m, n)$ of the $F_{m,n}$ distribution with $\alpha = 0.05$, 0.01 and 0.001, that is the upper 5%, 1% and 0.1% points of the distribution. The tables give the values of $x_\alpha(m, n)$ exceeding unity. For values less than unity use the result

$$x_\alpha(m, n) = 1/x_{1-\alpha}(n, m).$$

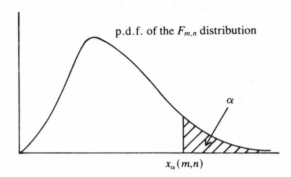

p.d.f. of the $F_{m,n}$ distribution

α

$x_\alpha(m,n)$

$\nu_2 \backslash \nu_1$	1	2	3	4	5	6	7	8	9	10	12	15	20	24	30	40	60	120	∞
1	161·4	199·5	215·7	224·6	230·2	234·0	236·8	238·9	240·5	241·9	243·9	245·9	248·0	249·1	250·1	251·1	252·2	253·3	254·3
2	18·51	19·00	19·16	19·25	19·30	19·33	19·35	19·37	19·38	19·40	19·41	19·43	19·45	19·45	19·46	19·47	19·48	19·49	19·50
3	10·13	9·55	9·28	9·12	9·01	8·94	8·89	8·85	8·81	8·79	8·74	8·70	8·66	8·64	8·62	8·59	8·57	8·55	8·53
4	7·71	6·94	6·59	6·39	6·26	6·16	6·09	6·04	6·00	5·96	5·91	5·86	5·80	5·77	5·75	5·72	5·69	5·66	5·63
5	6·61	5·79	5·41	5·19	5·05	4·95	4·88	4·82	4·77	4·74	4·68	4·62	4·56	4·53	4·50	4·46	4·43	4·40	4·36
6	5·99	5·14	4·76	4·53	4·39	4·28	4·21	4·15	4·10	4·06	4·00	3·94	3·87	3·84	3·81	3·77	3·74	3·70	3·67
7	5·59	4·74	4·35	4·12	3·97	3·87	3·79	3·73	3·68	3·64	3·57	3·51	3·44	3·41	3·38	3·34	3·30	3·27	3·23
8	5·32	4·46	4·07	3·84	3·69	3·58	3·50	3·44	3·39	3·35	3·28	3·22	3·15	3·12	3·08	3·04	3·01	2·97	2·93
9	5·12	4·26	3·86	3·63	3·48	3·37	3·29	3·23	3·18	3·14	3·07	3·01	2·94	2·90	2·86	2·83	2·79	2·75	2·71
10	4·96	4·10	3·71	3·48	3·33	3·22	3·14	3·07	3·02	2·98	2·91	2·85	2·77	2·74	2·70	2·66	2·62	2·58	2·54
11	4·84	3·98	3·59	3·36	3·20	3·09	3·01	2·95	2·90	2·85	2·79	2·72	2·65	2·61	2·57	2·53	2·49	2·45	2·40
12	4·75	3·89	3·49	3·26	3·11	3·00	2·91	2·85	2·80	2·75	2·69	2·62	2·54	2·51	2·47	2·43	2·38	2·34	2·30
13	4·67	3·81	3·41	3·18	3·03	2·92	2·83	2·77	2·71	2·67	2·60	2·53	2·46	2·42	2·38	2·34	2·30	2·25	2·21
14	4·60	3·74	3·34	3·11	2·96	2·85	2·76	2·70	2·65	2·60	2·53	2·46	2·39	2·35	2·31	2·27	2·22	2·18	2·13
15	4·54	3·68	3·29	3·06	2·90	2·79	2·71	2·64	2·59	2·54	2·48	2·40	2·33	2·29	2·25	2·20	2·16	2·11	2·07
16	4·49	3·63	3·24	3·01	2·85	2·74	2·66	2·59	2·54	2·49	2·42	2·35	2·28	2·24	2·19	2·15	2·11	2·06	2·01
17	4·45	3·59	3·20	2·96	2·81	2·70	2·61	2·55	2·49	2·45	2·38	2·31	2·23	2·19	2·15	2·10	2·06	2·01	1·96
18	4·41	3·55	3·16	2·93	2·77	2·66	2·58	2·51	2·46	2·41	2·34	2·27	2·19	2·15	2·11	2·06	2·02	1·97	1·92
19	4·38	3·52	3·13	2·90	2·74	2·63	2·54	2·48	2·42	2·38	2·31	2·23	2·16	2·11	2·07	2·03	1·98	1·93	1·88
20	4·35	3·49	3·10	2·87	2·71	2·60	2·51	2·45	2·39	2·35	2·28	2·20	2·12	2·08	2·04	1·99	1·95	1·90	1·84
21	4·32	3·47	3·07	2·84	2·68	2·57	2·49	2·42	2·37	2·32	2·25	2·18	2·10	2·05	2·01	1·96	1·92	1·87	1·81
22	4·30	3·44	3·05	2·82	2·66	2·55	2·46	2·40	2·34	2·30	2·23	2·15	2·07	2·03	1·98	1·94	1·89	1·84	1·78
23	4·28	3·42	3·03	2·80	2·64	2·53	2·44	2·37	2·32	2·27	2·20	2·13	2·05	2·01	1·96	1·91	1·86	1·81	1·76
24	4·26	3·40	3·01	2·78	2·62	2·51	2·42	2·36	2·30	2·25	2·18	2·11	2·03	1·98	1·94	1·89	1·84	1·79	1·73
25	4·24	3·39	2·99	2·76	2·60	2·49	2·40	2·34	2·28	2·24	2·16	2·09	2·01	1·96	1·92	1·87	1·82	1·77	1·71
26	4·23	3·37	2·98	2·74	2·59	2·47	2·39	2·32	2·27	2·22	2·15	2·07	1·99	1·95	1·90	1·85	1·80	1·75	1·69
27	4·21	3·35	2·96	2·73	2·57	2·46	2·37	2·31	2·25	2·20	2·13	2·06	1·97	1·93	1·88	1·84	1·79	1·73	1·67
28	4·20	3·34	2·95	2·71	2·56	2·45	2·36	2·29	2·24	2·19	2·12	2·04	1·96	1·91	1·87	1·82	1·77	1·71	1·65
29	4·18	3·33	2·93	2·70	2·55	2·43	2·35	2·28	2·22	2·18	2·10	2·03	1·94	1·90	1·85	1·81	1·75	1·70	1·64
30	4·17	3·32	2·92	2·69	2·53	2·42	2·33	2·27	2·21	2·16	2·09	2·01	1·93	1·89	1·84	1·79	1·74	1·68	1·62
40	4·08	3·23	2·84	2·61	2·45	2·34	2·25	2·18	2·12	2·08	2·00	1·92	1·84	1·79	1·74	1·69	1·64	1·58	1·51
60	4·00	3·15	2·76	2·53	2·37	2·25	2·17	2·10	2·04	1·99	1·92	1·84	1·75	1·70	1·65	1·59	1·53	1·47	1·39
120	3·92	3·07	2·68	2·45	2·29	2·17	2·09	2·02	1·96	1·91	1·83	1·75	1·66	1·61	1·55	1·50	1·43	1·35	1·25
∞	3·84	3·00	2·60	2·37	2·21	2·10	2·01	1·94	1·88	1·83	1·75	1·67	1·57	1·52	1·46	1·39	1·32	1·22	1·00

F distribution: Upper 1% points

$\nu_2 \backslash \nu_1$	1	2	3	4	5	6	7	8	9	10	12	15	20	24	30	40	60	120	∞
1	4052	4999·5	5403	5625	5764	5859	5928	5981	6022	6056	6106	6157	6209	6235	6261	6287	6313	6339	6366
2	98·50	99·00	99·17	99·25	99·30	99·33	99·36	99·37	99·39	99·40	99·42	99·43	99·45	99·46	99·47	99·47	99·48	99·49	99·50
3	34·12	30·82	29·46	28·71	28·24	27·91	27·67	27·49	27·35	27·23	27·05	26·87	26·69	26·60	26·50	26·41	26·32	26·22	26·13
4	21·20	18·00	16·69	15·98	15·52	15·21	14·98	14·80	14·66	14·55	14·37	14·20	14·02	13·93	13·84	13·75	13·65	13·56	13·46
5	16·26	13·27	12·06	11·39	10·97	10·67	10·46	10·29	10·16	10·05	9·89	9·72	9·55	9·47	9·38	9·29	9·20	9·11	9·02
6	13·75	10·92	9·78	9·15	8·75	8·47	8·26	8·10	7·98	7·87	7·72	7·56	7·40	7·31	7·23	7·14	7·06	6·97	6·88
7	12·25	9·55	8·45	7·85	7·46	7·19	6·99	6·84	6·72	6·62	6·47	6·31	6·16	6·07	5·99	5·91	5·82	5·74	5·65
8	11·26	8·65	7·59	7·01	6·63	6·37	6·18	6·03	5·91	5·81	5·67	5·52	5·36	5·28	5·20	5·12	5·03	4·95	4·86
9	10·56	8·02	6·99	6·42	6·06	5·80	5·61	5·47	5·35	5·26	5·11	4·96	4·81	4·73	4·65	4·57	4·48	4·40	4·31
10	10·04	7·56	6·55	5·99	5·64	5·39	5·20	5·06	4·94	4·85	4·71	4·56	4·41	4·33	4·25	4·17	4·08	4·00	3·91
11	9·65	7·21	6·22	5·67	5·32	5·07	4·89	4·74	4·63	4·54	4·40	4·25	4·10	4·02	3·94	3·86	3·78	3·69	3·60
12	9·33	6·93	5·95	5·41	5·06	4·82	4·64	4·50	4·39	4·30	4·16	4·01	3·86	3·78	3·70	3·62	3·54	3·45	3·36
13	9·07	6·70	5·74	5·21	4·86	4·62	4·44	4·30	4·19	4·10	3·96	3·82	3·66	3·59	3·51	3·43	3·34	3·25	3·17
14	8·86	6·51	5·56	5·04	4·69	4·46	4·28	4·14	4·03	3·94	3·80	3·66	3·51	3·43	3·35	3·27	3·18	3·09	3·00
15	8·68	6·36	5·42	4·89	4·56	4·32	4·14	4·00	3·89	3·80	3·67	3·52	3·37	3·29	3·21	3·13	3·05	2·96	2·87
16	8·53	6·23	5·29	4·77	4·44	4·20	4·03	3·89	3·78	3·69	3·55	3·41	3·26	3·18	3·10	3·02	2·93	2·84	2·75
17	8·40	6·11	5·18	4·67	4·34	4·10	3·93	3·79	3·68	3·59	3·46	3·31	3·16	3·08	3·00	2·92	2·83	2·75	2·65
18	8·29	6·01	5·09	4·58	4·25	4·01	3·84	3·71	3·60	3·51	3·37	3·23	3·08	3·00	2·92	2·84	2·75	2·66	2·57
19	8·18	5·93	5·01	4·50	4·17	3·94	3·77	3·63	3·52	3·43	3·30	3·15	3·00	2·92	2·84	2·76	2·67	2·58	2·49
20	8·10	5·85	4·94	4·43	4·10	3·87	3·70	3·56	3·46	3·37	3·23	3·09	2·94	2·86	2·78	2·69	2·61	2·52	2·42
21	8·02	5·78	4·87	4·37	4·04	3·81	3·64	3·51	3·40	3·31	3·17	3·03	2·88	2·80	2·72	2·64	2·55	2·46	2·36
22	7·95	5·72	4·82	4·31	3·99	3·76	3·59	3·45	3·35	3·26	3·12	2·98	2·83	2·75	2·67	2·58	2·50	2·40	2·31
23	7·88	5·66	4·76	4·26	3·94	3·71	3·54	3·41	3·30	3·21	3·07	2·93	2·78	2·70	2·62	2·54	2·45	2·35	2·26
24	7·82	5·61	4·72	4·22	3·90	3·67	3·50	3·36	3·26	3·17	3·03	2·89	2·74	2·66	2·58	2·49	2·40	2·31	2·21
25	7·77	5·57	4·68	4·18	3·85	3·63	3·46	3·32	3·22	3·13	2·99	2·85	2·70	2·62	2·54	2·45	2·36	2·27	2·17
26	7·72	5·53	4·64	4·14	3·82	3·59	3·42	3·29	3·18	3·09	2·96	2·81	2·66	2·58	2·50	2·42	2·33	2·23	2·13
27	7·68	5·49	4·60	4·11	3·78	3·56	3·39	3·26	3·15	3·06	2·93	2·78	2·63	2·55	2·47	2·38	2·29	2·20	2·10
28	7·64	5·45	4·57	4·07	3·75	3·53	3·36	3·23	3·12	3·03	2·90	2·75	2·60	2·52	2·44	2·35	2·26	2·17	2·06
29	7·60	5·42	4·54	4·04	3·73	3·50	3·33	3·20	3·09	3·00	2·87	2·73	2·57	2·49	2·41	2·33	2·23	2·14	2·03
30	7·56	5·39	4·51	4·02	3·70	3·47	3·30	3·17	3·07	2·98	2·84	2·70	2·55	2·47	2·39	2·30	2·21	2·11	2·01
40	7·31	5·18	4·31	3·83	3·51	3·29	3·12	2·99	2·89	2·80	2·66	2·52	2·37	2·29	2·20	2·11	2·02	1·92	1·80
60	7·08	4·98	4·13	3·65	3·34	3·12	2·95	2·82	2·72	2·63	2·50	2·35	2·20	2·12	2·03	1·94	1·84	1·73	1·60
120	6·85	4·79	3·95	3·48	3·17	2·96	2·79	2·66	2·56	2·47	2·34	2·19	2·03	1·95	1·86	1·76	1·66	1·53	1·38
∞	6·63	4·61	3·78	3·32	3·02	2·80	2·64	2·51	2·41	2·32	2·18	2·04	1·88	1·79	1·70	1·59	1·47	1·32	1·00

ν_2 \ ν_1	1	2	3	4	5	6	7	8	9	10	12	15	20	24	30	40	60	120	∞
1	4053*	5000*	5404*	5625*	5764*	5859*	5929*	5981*	6023*	6056*	6107*	6158*	6209*	6235*	6261*	6287*	6313*	6340*	6366*
2	998·5	999·0	999·2	999·2	999·3	999·3	999·4	999·4	999·4	999·4	999·4	999·4	999·4	999·5	999·5	999·5	999·5	999·5	999·5
3	167·0	148·5	141·1	137·1	134·6	132·8	131·6	130·6	129·9	129·2	128·3	127·4	126·4	125·9	125·4	125·0	124·5	124·0	123·5
4	74·14	61·25	56·18	53·44	51·71	50·53	49·66	49·00	48·47	48·05	47·41	46·76	46·10	45·77	45·43	45·09	44·75	44·40	44·05
5	47·18	37·12	33·20	31·09	29·75	28·84	28·16	27·64	27·24	26·92	26·42	25·91	25·39	25·14	24·87	24·60	24·33	24·06	23·79
6	35·51	27·00	23·70	21·92	20·81	20·03	19·46	19·03	18·69	18·41	17·99	17·56	17·12	16·89	16·67	16·44	16·21	15·99	15·75
7	29·25	21·69	18·77	17·19	16·21	15·52	15·02	14·63	14·33	14·08	13·71	13·32	12·93	12·73	12·53	12·33	12·12	11·91	11·70
8	25·42	18·49	15·83	14·39	13·49	12·86	12·40	12·04	11·77	11·54	11·19	10·84	10·48	10·30	10·11	9·92	9·73	9·53	9·33
9	22·86	16·39	13·90	12·56	11·71	11·13	10·70	10·37	10·11	9·89	9·57	9·24	8·90	8·72	8·55	8·37	8·19	8·00	7·81
10	21·04	14·91	12·55	11·28	10·48	9·92	9·52	9·20	8·96	8·75	8·45	8·13	7·80	7·64	7·47	7·30	7·12	6·94	6·76
11	19·69	13·81	11·56	10·35	9·58	9·05	8·66	8·35	8·12	7·92	7·63	7·32	7·01	6·85	6·68	6·52	6·35	6·17	6·00
12	18·64	12·97	10·80	9·63	8·89	8·38	8·00	7·71	7·48	7·29	7·00	6·71	6·40	6·25	6·09	5·93	5·76	5·59	5·42
13	17·81	12·31	10·21	9·07	8·35	7·86	7·49	7·21	6·98	6·80	6·52	6·23	5·93	5·78	5·63	5·47	5·30	5·14	4·97
14	17·14	11·78	9·73	8·62	7·92	7·43	7·08	6·80	6·58	6·40	6·13	5·85	5·56	5·41	5·25	5·10	4·94	4·77	4·60
15	16·59	11·34	9·34	8·25	7·57	7·09	6·74	6·47	6·26	6·08	5·81	5·54	5·25	5·10	4·95	4·80	4·64	4·47	4·31
16	16·12	10·97	9·00	7·94	7·27	6·81	6·46	6·19	5·98	5·81	5·55	5·27	4·99	4·85	4·70	4·54	4·39	4·23	4·06
17	15·72	10·66	8·73	7·68	7·02	6·56	6·22	5·96	5·75	5·58	5·32	5·05	4·78	4·63	4·48	4·33	4·18	4·02	3·85
18	15·38	10·39	8·49	7·46	6·81	6·35	6·02	5·76	5·56	5·39	5·13	4·87	4·59	4·45	4·30	4·15	4·00	3·84	3·67
19	15·08	10·16	8·28	7·26	6·62	6·18	5·85	5·59	5·39	5·22	4·97	4·70	4·43	4·29	4·14	3·99	3·84	3·68	3·51
20	14·82	9·95	8·10	7·10	6·46	6·02	5·69	5·44	5·24	5·08	4·82	4·56	4·29	4·15	4·00	3·86	3·70	3·54	3·38
21	14·59	9·77	7·94	6·95	6·32	5·88	5·56	5·31	5·11	4·95	4·70	4·44	4·17	4·03	3·88	3·74	3·58	3·42	3·26
22	14·38	9·61	7·80	6·81	6·19	5·76	5·44	5·19	4·99	4·83	4·58	4·33	4·06	3·92	3·78	3·63	3·48	3·32	3·15
23	14·19	9·47	7·67	6·69	6·08	5·65	5·33	5·09	4·89	4·73	4·48	4·23	3·96	3·82	3·68	3·53	3·38	3·22	3·05
24	14·03	9·34	7·55	6·59	5·98	5·55	5·23	4·99	4·80	4·64	4·39	4·14	3·87	3·74	3·59	3·45	3·29	3·14	2·97
25	13·88	9·22	7·45	6·49	5·88	5·46	5·15	4·91	4·71	4·56	4·31	4·06	3·79	3·66	3·52	3·37	3·22	3·06	2·89
26	13·74	9·12	7·36	6·41	5·80	5·38	5·07	4·83	4·64	4·48	4·24	3·99	3·72	3·59	3·44	3·30	3·15	2·99	2·82
27	13·61	9·02	7·27	6·33	5·73	5·31	5·00	4·76	4·57	4·41	4·17	3·92	3·66	3·52	3·38	3·23	3·08	2·92	2·75
28	13·50	8·93	7·19	6·25	5·66	5·24	4·93	4·69	4·50	4·35	4·11	3·86	3·60	3·46	3·32	3·18	3·02	2·86	2·69
29	13·39	8·85	7·12	6·19	5·59	5·18	4·87	4·64	4·45	4·29	4·05	3·80	3·54	3·41	3·27	3·12	2·97	2·81	2·64
30	13·29	8·77	7·05	6·12	5·53	5·12	4·82	4·58	4·39	4·24	4·00	3·75	3·49	3·36	3·22	3·07	2·92	2·76	2·59
40	12·61	8·25	6·60	5·70	5·13	4·73	4·44	4·21	4·02	3·87	3·64	3·40	3·15	3·01	2·87	2·73	2·57	2·41	2·23
60	11·97	7·76	6·17	5·31	4·76	4·37	4·09	3·87	3·69	3·54	3·31	3·08	2·83	2·69	2·55	2·41	2·25	2·08	1·89
120	11·38	7·32	5·79	4·95	4·42	4·04	3·77	3·55	3·38	3·24	3·02	2·78	2·53	2·40	2·26	2·11	1·95	1·76	1·54
∞	10·83	6·91	5·42	4·62	4·10	3·74	3·47	3·27	3·10	2·96	2·74	2·51	2·27	2·13	1·99	1·84	1·66	1·45	1·00

* Multiply these entries by 100.

This 0·1% table is based on the following sources: Colcord & Deming (1935); Fisher & Yates (1953, Table V) used with the permission of the authors and of Messrs Oliver and Boyd; Norton (1952).

APPENDIX T8

Random Numbers [see II: §§ 5.1, 11.1]

The table gives 5000 'Random Digits' which are independent realizations of a random variable N which has the discrete uniform distribution on $(0, 1, \ldots, 9)$; thus

$$P(N = n) = \tfrac{1}{10}, \qquad n = 0, 1, \ldots, 9.$$

The table may be read in any direction, starting at any point. A d-decimal random number may be obtained by placing a decimal point before any d consecutive digits. For example, starting at the ninth digit in the seventh row and taking $d = 5$ gives the random number 0·31572. This may be taken as a rounded realization of a continuous uniform $(0, 1)$ random variable Z, for which the p.d.f. at z is

$$f(z) = \begin{cases} 1, & 0 \le z \le 1 \\ 0, & \text{otherwise.} \end{cases}$$

Random numbers

(Reproduced by permission of Longman Group Ltd. from *Statistical Tables for Biological, Agricultural and Medical Research* by R. A. Fisher and F. Yates, 1974.)

03 47 43 73 86	36 96 47 36 61	46 98 63 71 62	33 26 16 80 45	60 11 14 10 95
97 74 24 67 62	42 81 14 57 20	42 53 32 37 32	27 07 36 07 51	24 51 79 89 73
16 76 62 27 66	56 50 26 71 07	32 90 79 78 53	13 55 38 58 59	88 97 54 14 10
12 56 85 99 26	96 96 68 27 31	05 03 72 93 15	57 12 10 14 21	88 26 49 81 76
55 59 56 35 64	38 54 82 46 22	31 62 43 09 90	06 18 44 32 53	23 83 01 30 30
16 22 77 94 39	49 54 43 54 82	17 37 93 23 78	87 35 20 96 43	84 26 34 91 64
84 42 17 53 31	57 24 55 06 88	77 04 74 47 67	21 76 33 50 25	83 92 12 06 76
63 01 63 78 59	16 95 55 67 19	98 10 50 71 75	12 86 73 58 07	44 39 52 38 79
33 21 12 34 29	78 64 56 07 82	52 42 07 44 38	15 51 00 13 42	99 66 02 79 54
57 60 86 32 44	09 47 27 96 54	49 17 46 09 62	90 52 84 77 27	08 02 73 43 28
18 18 07 92 46	44 17 16 58 09	79 83 86 19 62	06 76 50 03 10	55 23 64 05 05
26 62 38 97 75	84 16 07 44 99	83 11 46 32 24	20 14 85 88 45	10 93 72 88 71
23 42 40 64 74	82 97 77 77 81	07 45 32 14 08	32 98 94 07 72	93 85 79 10 75
32 36 28 19 95	50 92 26 11 97	00 56 76 31 38	80 22 02 53 53	86 60 42 04 53
37 85 94 35 12	83 39 50 08 30	42 34 07 96 88	54 42 06 87 98	35 85 29 48 39
70 29 17 12 13	40 33 20 38 26	13 89 51 03 74	17 76 37 13 04	07 74 21 19 30
56 62 18 37 35	96 83 50 87 75	97 12 25 93 47	70 33 24 03 54	97 77 46 44 80
99 49 57 22 77	88 42 95 45 72	16 64 36 16 00	04 43 18 66 79	94 77 24 21 90
16 08 15 04 72	33 27 14 34 09	45 59 34 68 49	12 72 07 34 45	99 27 72 95 14
31 16 93 32 43	50 27 89 87 19	20 15 37 00 49	52 85 66 60 44	38 68 88 11 80
68 34 30 13 70	55 74 30 77 40	44 22 78 84 26	04 33 46 09 52	68 07 97 06 57
74 57 25 65 76	59 29 97 68 60	71 91 38 67 54	13 58 18 24 76	15 54 55 95 52
27 42 37 86 53	48 55 90 65 72	96 57 69 36 10	96 46 92 42 45	97 60 49 04 91
00 39 68 29 61	66 37 32 20 30	77 84 57 03 29	10 45 65 04 26	11 04 96 67 24
29 94 98 94 24	68 49 69 10 82	53 75 91 93 30	34 25 20 57 27	40 48 73 51 92

Random numbers (cont.)

16 90 82 66 59	83 62 64 11 12	67 19 00 71 74	60 47 21 29 68	02 02 37 03 31
11 27 94 75 06	06 09 19 74 66	02 94 37 34 02	76 70 90 30 86	38 45 94 30 38
35 24 10 16 20	33 32 51 26 38	79 78 45 04 91	16 92 53 56 16	02 75 50 95 98
38 23 16 86 38	42 38 97 01 50	87 75 66 81 41	40 01 74 91 62	48 51 84 08 32
31 96 25 91 47	96 44 33 49 13	34 86 82 53 91	00 52 43 48 85	27 55 26 89 62
66 67 40 67 14	64 05 71 95 86	11 05 65 09 68	76 83 20 37 90	57 16 00 11 66
14 90 84 45 11	75 73 88 05 90	52 27 41 14 86	22 98 12 22 08	07 52 74 95 80
68 05 51 18 00	33 96 02 75 19	07 60 62 93 55	59 33 82 43 90	49 37 38 44 59
20 46 78 73 90	97 51 40 14 02	04 02 33 31 08	39 54 16 49 36	47 95 93 13 30
64 19 58 97 79	15 06 15 93 20	01 90 10 75 06	40 78 78 89 62	02 67 74 17 33
05 26 93 70 60	22 35 85 15 13	92 03 51 59 77	59 56 78 06 83	52 91 05 70 74
07 97 10 88 23	09 98 42 99 64	61 71 62 99 15	06 51 29 16 93	58 05 77 09 51
68 71 86 85 85	54 87 66 47 54	73 32 08 11 12	44 95 92 63 16	29 56 24 29 48
26 99 61 65 53	58 37 78 80 70	42 10 50 67 42	32 17 55 85 74	94 44 67 16 94
14 65 52 68 75	87 59 36 22 41	26 78 63 06 55	13 08 27 01 50	15 29 39 39 43
17 53 77 58 71	71 41 61 50 72	12 41 94 96 26	44 95 27 36 99	02 96 74 30 83
90 26 59 21 19	23 52 23 33 12	96 93 02 18 39	07 02 18 36 07	25 99 32 70 23
41 23 52 55 99	31 04 49 69 96	10 47 48 45 88	13 41 43 89 20	97 17 14 49 17
60 20 50 81 69	31 99 73 68 68	35 81 33 03 76	24 30 12 48 60	18 99 10 72 34
91 25 38 05 90	94 58 28 41 36	45 37 59 03 09	90 35 57 29 12	82 62 54 65 60
34 50 57 74 37	98 80 33 00 91	09 77 93 19 82	74 94 80 04 04	45 07 31 66 49
85 22 04 39 43	73 81 53 94 79	33 62 46 86 28	08 31 54 46 31	53 94 13 38 47
09 79 13 77 48	73 82 97 22 21	05 03 27 24 83	72 89 44 05 60	35 80 39 94 88
88 75 80 18 14	22 95 75 42 49	39 32 82 22 49	02 48 07 70 37	16 04 61 67 87
90 96 23 70 00	39 00 03 06 90	55 85 78 38 36	94 37 30 69 32	90 89 00 76 33

APPENDIX T9

Random Standard Normal Deviates

The table gives 500 independent realizations, rounded to three decimals, of the Standard Normal random variable U for which the p.d.f. at u is

$$\varphi(u) = \frac{1}{\sqrt{(2\pi)}} e^{-(1/2)u^2}, \qquad -\infty < u < \infty.$$

A realization u of the Standard Normal U may be transformed to a realization x of a Normal (μ, σ) variable by setting $x = \mu + \sigma u$.

Random standard Normal deviates

(Reproduced by permission of Macmillan Publishers Ltd. from *Statistical Tables for Science, Engineering and Management* by J. Murdoch and J. A. Barnes.)

	0	1	2	3	4	5	6	7	8	9
00	−0·179	−0·399	−0·235	−0·098	−0·465	+1·563	−1·085	+0·860	+0·388	+0·710
01	+0·421	+1·454	+0·904	+0·437	−2·120	+1·085	−0·277	−2·170	+0·018	−0·722
02	+0·210	−0·556	+0·465	−1·812	−2·748	−0·345	−0·251	+0·622	−1·015	+0·762
03	−1·598	+0·919	−0·266	−0·999	+0·308	−0·592	+0·817	−0·454	+1·598	+0·240
04	+1·717	+1·514	−0·012	−0·852	+0·118	+0·399	−0·123	+0·432	−0·470	+0·776
05	−0·308	+0·867	−0·372	+0·697	−1·787	+0·568	−0·002	−0·133	+0·545	−0·824
06	−0·421	+0·516	−0·038	+1·200	+0·063	−0·377	−1·007	−0·334	+1·299	+0·038
07	−0·776	+0·874	−1·265	−0·580	+0·377	−0·697	−2·226	−1·299	−0·796	−0·628
08	+0·640	−0·522	+0·023	−0·393	−1·142	−2·457	−1·580	+1·160	+0·008	+0·487
09	−0·319	+0·889	+1·180	−0·404	+1·322	+0·410	+1·468	+0·235	−0·810	−1·131
10	+0·610	−0·383	+1·812	+0·729	+0·204	−0·225	+0·169	−0·729	−0·432	+0·634
11	−0·174	−0·154	+0·098	+0·393	−3·090	+1·762	+1·530	+0·028	+0·950	−0·935
12	+2·576	−0·684	−1·200	+0·002	+0·261	−0·415	+0·598	−0·769	−0·169	−1·498
13	−1·103	+1·398	−0·653	+1·739	+0·476	+0·510	+0·782	−0·634	+0·562	−0·053
14	+1·635	+0·448	−1·530	−0·043	+2·290	−0·063	−1·695	+0·199	+1·211	−1·360
15	−0·068	−0·860	−0·194	−1·616	+0·334	+0·189	+0·927	−1·454	+0·958	+0·404
16	−1·960	+1·076	−0·671	−0·103	+1·041	+2·226	+1·838	−0·510	−1·322	+2·366
17	+0·443	−0·912	+0·251	−0·574	+1·131	−0·204	−0·324	−0·487	−1·287	+0·522
18	+1·360	+0·533	+1·094	+0·671	+0·852	−2·576	−0·539	−0·568	+0·225	−0·545
19	+0·810	+0·319	−1·514	+0·556	+1·112	−0·210	+0·292	+0·749	+0·882	−0·033
20	+0·616	+1·347	−1·866	−0·755	+0·329	+0·148	−0·058	−0·199	+0·048	+1·546
21	−0·598	−2·366	−0·831	+0·454	−0·118	−1·762	+0·493	+1·103	+0·361	+0·113
22	+0·426	+1·580	−1·112	+0·550	−1·254	−0·033	+0·143	−1·141	+0·366	−0·073
23	+0·831	−0·516	−1·717	−0·340	+1·655	+0·194	−0·388	−0·942	−1·243	−0·292
24	−0·640	−0·128	+1·276	−1·838	−0·410	+0·646	+2·075	−0·159	+1·695	+0·527

Random standard Normal deviates (continued)

	0	1	2	3	4	5	6	7	8	9
25	-0·927	+0·838	-1·546	+0·246	-0·742	-0·143	+2·457	+0·043	-1·058	-0·867
26	+1·232	+2·170	+0·088	-0·803	+0·574	+0·058	+0·282	+0·356	+0·350	-1·927
27	+0·935	+0·665	+2·034	-1·995	+0·703	-0·083	-1·468	+0·078	-0·966	-0·303
28	-1·739	-0·622	-1·563	+0·313	+0·220	-0·586	+0·272	+0·789	-1·335	+1·440
29	+0·990	-1·483	+0·154	-1·372	-1·896	+1·385	-1·041	+0·974	+0·482	-1·211
30	-0·189	-0·240	+0·133	-2·290	-0·616	-0·437	+0·459	-0·499	-0·845	+0·383
31	+1·866	-1·398	+0·068	+0·053	-2·034	+1·426	+1·254	+1·067	+0·592	+0·174
32	-0·018	+0·628	+0·230	+0·659	-0·298	+1·927	-0·282	+0·769	-0·690	+1·675
33	-0·646	-0·350	+0·324	-1·675	+1·190	-1·076	+1·287	-1·426	+0·345	-0·215
34	-1·150	-0·220	-0·533	+0·912	-0·710	-0·904	-0·817	-1·160	-0·919	-0·659
35	+0·103	-0·361	+1·024	-0·6-0·482	-0·562	+0·277	-1·440	-0·366	-0·256	+2·120
37	-0·093	-1·190	+0·580	-1·276	+0·653	-0·048	+0·742	-1·170	+1·960	+1·787
38	-0·261	-0·194	+0·303	+0·340	+1·498	-1·232	-0·078	-0·443	+1·141	+1·995
39	-0·230	-0·550	+0·266	-1·655	+0·999	-1·067	+1·058	+0·796	+0·415	
40	-0·148	+0·504	-0·028	+0·083	+0·824	-1·024	+1·412	-0·164	+1·150	-0·272
41	+1·122	+0·896	-0·789	+0·215	-0·426	-1·049	-0·974	+0·586	+1·311	-0·736
42	+0·499	-1·032	+0·159	+0·123	+2·748	-0·749	-0·665	-1·221	-1·180	+1·049
43	+0·678	-0·782	+0·470	+0·256	+0·298	-0·990	+0·287	+0·942	+0·128	+1·372
44	-1·347	+3·090	-0·896	+0·138	-0·838	+0·690	+1·007	+0·184	+0·164	+0·179
45	-1·094	-0·610	-0·287	+0·755	-0·459	-1·635	-0·108	-0·246	+1·032	-0·527
46	-0·088	-0·889	+0·803	-1·311	-0·703	+1·170	-0·113	+0·108	-0·874	+0·372
47	+0·093	-0·476	+1·265	-0·448	+1·015	-0·313	-0·958	+0·716	+1·483	+0·722
48	-0·950	-0·008	+0·012	+0·073	-0·762	-0·493	+1·896	+0·982	+1·616	+1·221
49	-0·329	-0·138	-0·504	-0·678	+1·335	-2·075	-1·385	-0·023	-0·356	-0·982

APPENDIX T10

Confidence Limits for Binomial Parameter [see § 4.7]

Charts providing confidence limits for the parameter θ in a Binomial (n, θ) distribution, given a sample fraction r/n.

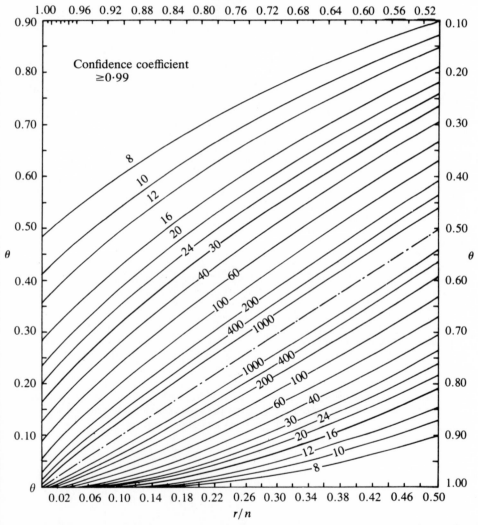

The numbers printed along the curves indicate the sample size, n. If for a given value of r/n, θ' and θ'' are the ordinates read from the appropriate lower and upper curves, then (θ', θ'') is a confidence interval for θ, with confidence interval $\geq 99\%$.

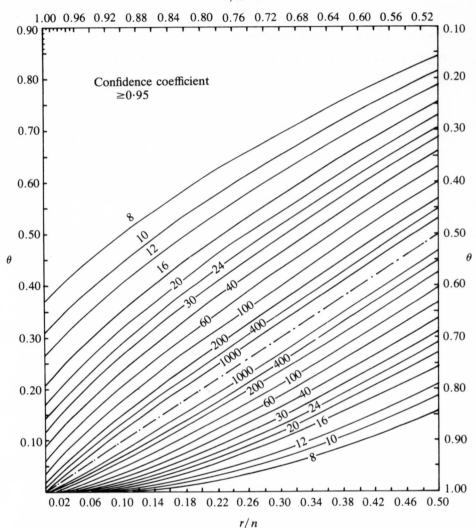

The numbers printed along the curves indicate the sample size, *n*. If for a given value of *r/n*, θ' and θ'' are the ordinates read from the appropriate lower and upper curves, then (θ', θ'') is a confidence interval for θ, with confidence level $\geq 95\%$.

APPENDIX T11

Confidence Limits for the Expectation of a Poisson Variable [see § 4.7]

Confidence limits for the parameter λ in a Poisson (λ) distribution given c occurrences of the Poisson event. The table gives lower and upper confidence limits for λ, at confidence level $1 - 2\alpha$.

Confidence limits for the expectation of a Poisson variable

(Reproduced by permission of Biometrika Trustees from *Biometrika Tables for Statisticians*, Vol. 1, 3rd edition, 1966.)

$1-2\alpha$	0·998		0·99		0·98		0·95		0·90		$1-2\alpha$
α	0·001		0·005		0·01		0·025		0·05		α
c	Lower	Upper	Lower	Upper	Lower	Upper	Lower	Upper	Lower	Upper	c
0	0·00000	6·91	0·00000	5·30	0·0000	4·61	0·0000	3·69	0·0000	3·00	0
1	0·00100	9·23	0·00501	7·43	0·0101	6·64	0·0253	5·57	0·0513	4·74	1
2	0·0454	11·23	0·103	9·27	0·149	8·41	0·242	7·22	0·355	6·30	2
3	0·191	13·06	0·338	10·98	0·436	10·05	0·619	8·77	0·818	7·75	3
4	0·429	14·79	0·672	12·59	0·823	11·60	1·09	10·24	1·37	9·15	4
5	0·739	16·45	1·08	14·15	1·28	13·11	1·62	11·67	1·97	10·51	5
6	1·11	18·06	1·54	15·66	1·79	14·57	2·20	13·06	2·61	11·84	6
7	1·52	19·63	2·04	17·13	2·33	16·00	2·81	14·42	3·29	13·15	7
8	1·97	21·16	2·57	18·58	2·91	17·40	3·45	15·76	3·98	14·43	8
9	2·45	22·66	3·13	20·00	3·51	18·78	4·12	17·08	4·70	15·71	9
10	2·96	24·13	3·72	21·40	4·13	20·14	4·80	18·39	5·43	16·96	10
11	3·49	25·59	4·32	22·78	4·77	21·49	5·49	19·68	6·17	18·21	11
12	4·04	27·03	4·94	24·14	5·43	22·82	6·20	20·96	6·92	19·44	12
13	4·61	28·45	5·58	25·50	6·10	24·14	6·92	22·23	7·69	20·67	13
14	5·20	29·85	6·23	26·84	6·78	25·45	7·65	23·49	8·46	21·89	14
15	5·79	31·24	6·89	28·16	7·48	26·74	8·40	24·74	9·25	23·10	15
16	6·41	32·62	7·57	29·48	8·18	28·03	9·15	25·98	10·04	24·30	16
17	7·03	33·99	8·25	30·79	8·89	29·31	9·90	27·22	10·83	25·50	17
18	7·66	35·35	8·94	32·09	9·62	30·58	10·67	28·45	11·63	26·69	18
19	8·31	36·70	9·64	33·38	10·35	31·85	11·44	29·67	12·44	27·88	19

20	8·96	38·04	10·35	34·67	11·08	33·10	12·22	30·89	13·25	29·06	20
21	9·62	39·38	11·07	35·95	11·82	34·36	13·00	32·10	14·07	30·24	21
22	10·29	40·70	11·79	37·22	12·57	35·60	13·79	33·31	14·89	31·42	22
23	10·96	42·02	12·52	38·48	13·33	36·84	14·58	34·51	15·72	32·59	23
24	11·65	43·33	13·25	39·74	14·09	38·08	15·38	35·71	16·55	33·75	24
25	12·34	44·64	14·00	41·00	14·85	39·31	16·18	36·90	17·38	34·92	25
26	13·03	45·94	14·74	42·25	15·62	40·53	16·98	38·10	18·22	36·08	26
27	13·73	47·23	15·49	43·50	16·40	41·76	17·79	39·28	19·06	37·23	27
28	14·44	48·52	16·24	44·74	17·17	42·98	18·61	40·47	19·90	38·39	28
29	15·15	49·80	17·00	45·98	17·96	44·19	19·42	41·65	20·75	39·54	29
30	15·87	51·08	17·77	47·21	18·74	45·40	20·24	42·83	21·59	40·69	30
35	19·52	57·42	21·64	53·32	22·72	51·41	24·38	48·68	25·87	46·40	35
40	23·26	63·66	25·59	59·36	26·77	57·35	28·58	54·47	30·20	52·07	40
45	27·08	69·83	29·60	65·34	30·88	63·23	32·82	60·21	34·56	57·69	45
50	30·96	75·94	33·66	71·27	35·03	69·07	37·11	65·92	38·96	63·29	50

Index

(Part A = pages 1–498; Part B = pages 499–942)